LATITUDE & LONGITUDE
BY THE NOON SIGHT

ALSO BY HEWITT SCHLERETH:

Commonsense Celestial Navigation
Commonsense Sailboat Buying
Commonsense Coastal Navigation

The Cruising Navigator—No. 1

LATITUDE & LONGITUDE BY THE NOON SIGHT

by Hewitt Schlereth

Drawings by Umberto Bertoli

SEVEN SEAS PRESS, INC.
NEWPORT, RI   02840

TO ERATOSTHENES

Who discovered the noon sight and used it to measure the world—
in the third century, B.C.

PUBLISHED BY Seven Seas Press, Inc.
               Newport, Rhode Island 02840

Copyright © 1982 by Hewitt Schlereth

1 3 5 7 9 HL/JB 0 8 6 4 2

Library of Congress Cataloging in Publication Data

Schlereth, Hewitt.
    Latitude & longitude by the noon sight.

    (The Cruising navigator; no. 1)
      1. Latitude. 2. Longitude. 3. Navigation.
4. Geographical positions. I. Title. II. Title:
Latitude and longitude by the noon sight.
III. Series.
VK565.S37   1982   623.89   82-10632
ISBN 0-915160-51-X

Designed by Irving Perkins Associates

Printed in the United States of America

Contents

Part V COMPLETE EXAMPLE: LONGITUDE & LATITUDE BY NOON SIGHT

Part VI TABLES

Part VII WORK FORMS

LIST OF ILLUSTRATIONS

PART II LATITUDE BY NOON SIGHT

PART III HIGH ALTITUDE SUN SIGHTS

PART IV LONGITUDE BY SUN SIGHT

PART V COMPLETE EXAMPLE: LONGITUDE & LATITUDE BY NOON SIGHT

PART VI TABLES

PART VII   WORK FORMS

ABBREVIATIONS USED IN THIS BOOK:

DR   Dead Reckoning Position

DIP   Difference between True and Sea Horizon

Dec   Declination

GHA   Greenwich Hour Angle

GMT   Greenwich Mean Time

GP   Geographical Position

Hs   Sextant Reading (uncorrected)

ha   Apparent Altitude (Hs corrected for Index Error and DIP)

Ho   Observed Altitude (ha corrected for Refraction and Semi-Diameter)

HE   Height of Eye

IC   Index Correction

IE   Index Error

R   Refraction

SD   Semi-Diameter

z   Zenith Distance

Preface

THE IDEA for this book came about as the result of a number of discussions I had in a period of several years with people who were living on their boats and cruising about pretty much as their fancy took them, generally following the sun—north in the summer and south in the winter.

For the most part, I met these folks aboard their boats at the municipal dock in the town of St. George, Bermuda, a very civilized island and a natural way station for ocean voyagers.

Anyway, in chats with these voyagers about life on the ocean in general and celestial navigation in particular, three themes kept recurring: (1) most of these blue water cruisers navigated exclusively by the sun, (2) for a great number, finding the latitude at noon and comparing it to the DR was the *sole* method of offshore navigation and (3) almost everyone complained about the annual chore of chasing down a *Nautical Almanac* in remote areas of the world—indeed, it is not always easy to get hold of the *Nautical Almanac* (the source of the basic information on the sun's position needed for celestial navigation) even in the United States and in Great Britain, where they are published.

When I mentioned the long-term almanacs in *The American Practical Navigator* and HO 249, most of these people had not heard about them, and those that had said that the amount of interpolation needed vitiated their usefulness in day-to day

navigating—just simply too time-consuming and fraught with arithmetical traps.

To make a long story short, it occurred to me that there was a practical way to make a long-term almanac that would reduce the interpolation to one step. This, in turn, led to the concept of this book, because the unique advantage of the noon sight—the reason for its simplicity—is that it does not require the so-called "sight reduction" process (and allied tables) of full-fledged celestial navigation—sight reduction being the big hurdle for most people trying to learn celestial navigation. Thus, it is possible to get—within the confines of one smallish volume—everything (except sextant) required for determination of latitude, and everything (except sextant and chronometer) for the determination of longitude.

Of course, no system is all virtues; the noon sight has its disadvantages, the principal one being that you only get one time a day (noon) to use it, and it may happen that clouds fill the sky at the critical moment. Also, it takes time to get the sight, 20 to 40 minutes, as against only 2 to 5 minutes required for the standard, or "line-of-position," methods.

It is my experience, though, that the first shortcoming is compensated for by the fact that in moderate to long yacht voyages (600 or more nautical miles), the time of passage ensures clear weather often enough for sufficient noon sights. Often, even a watery sun seen through thick haze or clouds becomes a sharp, useable disk when seen through a sextant's telescope. Further, a simple table lets you determine latitude from a sight taken as much as one half hour before or after noon. As to the second shortcoming—the time involved in taking the sights—this is largely offset by the fact that the work you have to do *after* taking the sights is *considerably* reduced. There is no sight reduction and no plotting in the usual sense, because the sights yield longitude and/or latitude directly.

Finally, in addition to the fact that it is the simplest practical

technique of celestial navigation, *the noon sight is also the logical foundation for acquiring the rest of the art.* The *fundamental* theory is all contained in the mechanics and geometry of the sun's position at noon. As I noted in the beginning, however, lots of people have voyaged and are voyaging with the noon sight alone. With this method you can become one of them.

In learning and teaching over the years I have found that the best way to study a technical subject is to go quickly through the *entire* text at least once. Subconsciously, your mind picks up and retains a lot of information. Then, as you begin a more concentrated effort, much of this early information surfaces to facilitate your comprehension. The structure of this book is made to assist this method of study: once you've read the text, the illustrations alone teach the subject; once you've read the illustrations, the key ⚷ concepts are all you need to remember in actually going through the steps to find your latitude and longitude by the noon sight.

Part I

BASICS

Latitude & Longitude

AS I mentioned in the introduction, one of the advantages of navigation by noon sight is that it gives a position directly, without the intermediate plotting steps required by the line-of-position methods. The reason is that at noon the sun is uniquely aligned with the basic position-locating grid of the earth —latitude and longitude. Although most of us have a good general notion of latitude and longitude, I discovered while teaching courses in celestial navigation that it is a good idea to review these basic concepts, because celestial requires a specific, rather than general, way of understanding latitude and longitude.

What is generally understood is that the earth is divided vertically by "lines of longitude" (running north and south) that begin at Greenwich, England, and proceed around the globe eastward and westward to a longitude line (in the Pacific) located 180° from Greenwich, this 180° line also being called the International Date Line. If a drawing is made showing the longitude lines at some handy interval, say every 30°, the earth looks a lot like a skinned orange (Figure 1), with these "meridians of longitude" (as these lines are called) meeting at the poles.

This grid of longitude lines is one half of the system for locating places on the surface of the earth. The 90°W line, for instance, runs close to Pelly Bay in Northern Canada; Thunder

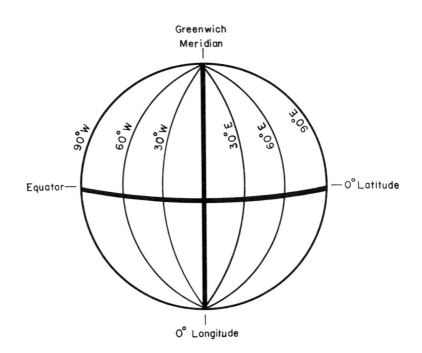

Figure 1. Meridians of longitude.

Bay in Southern Canada; St. Louis, Memphis and New Orleans in the United States; Campeche, Mexico, and the Galapagos Islands off Equador. The 90°E line runs through Western Mongolia, near Lhasa in Tibet, Dacca in Bangladesh and across the Bay of Bengal.

The other part of the grid that covers the globe is the latitude system, beginning at the equator (0° latitude) and moving both northward and southward to the two poles (90° north latitude and 90° south latitude, respectively). Combining these two sets of lines gives us a method of designating any point on the surface of the earth. Thus, although Tokyo, San Francisco, Norfolk, Virginia, Gibraltar and Athens, Greece, are all near 36°N latitude, they are easily distinguished by their widely differing longitudes (Figure 2).

4   Latitude & Longitude by the Noon Sight

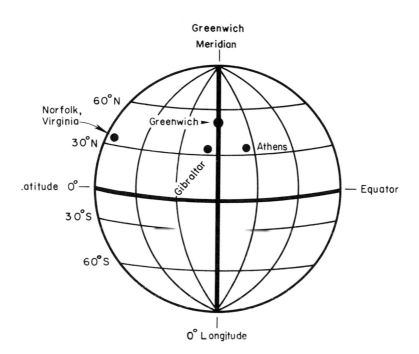

Figure 2. Parallels of latitude.

The foregoing seems to be pretty general knowledge, espe
cially among those who have an interest in navigation. What
does not seem to be as well understood is the underlying geometry of the latitude-longitude grid system. Why is it, for instance, that we talk about *degrees* (a unit of angular
measurement) when we are designating *distances* on the surface of a *sphere?*

*The reason is that latitude and longitude are the surface man-
ifestations of an internal system based on angles.* Look at Figure
3. It shows the earth in cross section, as if sliced in half vertically
along a longitude line. It is easy to see that there is a 90° angle
between the lines from the center to the North Pole, and to the
equator. Figure 4 shows 40°N (the latitude of New York City
and Lisbon, Portugal, approximately), and 40°S (roughly the

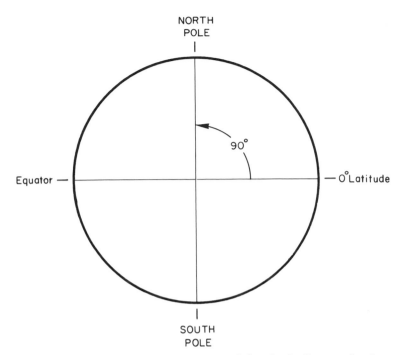

NORTH
POLE

Equator —

90°

— 0° Latitude

SOUTH
POLE

Figure 3. The earth in cross section, as if sliced in half vertically along a longitude line. It is easy to see that there is 90° between the lines from the center to the North Pole and the center to the equator. Thus, we conventionally say that the North Pole is 90° north latitude and that the equator is 0° latitude.

latitude of Wellington, New Zealand). The complete latitude *line* is made up of *all* the imaginary points on the surface of the globe which are 40° of arc above the equator. Now, since the internal angles are dividing up a circle and since circles are considered to have 360° in a complete circuit, it is obvious that there is a one-to-one correspondence between the *angle* at the center and the *arc* on the surface. We know that the circumference of the earth is about 22,000 nautical miles. Using Figure 4 and arithmetic, therefore, it is clear that the *length* in nautical miles of the 40° arc is 40/360th's of 22,000, or 2,444. Hence,

⚷ ARC = ANGLE

Arc (distance on surface of earth) and internal angle are equal and both are measured in degrees and minutes.

in the latitude scheme, there is an equivalence between the plane angle as measured at the center of the earth and the length of the arc (created by that angle) on the surface.

Figure 4. This view of the earth, still sliced through along a meridian, shows two points on the earth's surface, 40° north latitude and 40° south latitude. The interior lines are there to show how the latitude angles are measured. This drawing also shows the equivalence between the internal angle and the length of the arc on the surface of the earth.

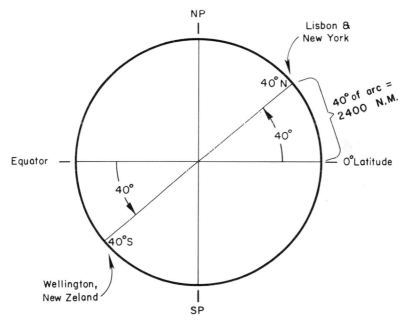

"MILE A MINUTE"

By international convention, 1° of latitude equals 60 nautical miles. So, 1 minute of latitude equals 1 nautical mile, a "mile a minute," so to speak.

As a matter of fact, 1° of arc on the surface of the earth has been defined (by international agreement) to equal 60 nautical miles. Because 1° is—again by convention—further divided into 60' (minutes), the result is that one nautical mile equals 1' of latitude.

The longitude part of the positional grid proceeds from a similar division of the earth by internal angles, but there is an important difference. Look at Figure 5, which shows the earth as seen from above and sliced parallel to the equator at 40°N latitude. The longitude system divides up the circle of latitude just like the latitude system divided in the circle of longitude, but this circle at 40°N is *not* equal to the (equator) circumference of the earth; and the higher the latitude, the greater the difference. The only latitude circle that is equal to the earth's circumference is the equator latitude. We can still talk about degrees of longitude east or west of Greenwich (0° longitude), but there is no easy equivalence between 1' of longitude and 1 nautical mile *except at the equator*. It is important to keep this distinction in mind when using charts and plotting positions on them so that you don't inadvertently mix up the two scales or use the same divider setting to plot both latitude and longitude.

THE DISTANCE SCALE:

The latitude scale is the distance scale, since 1' of latitude equals 1 nautical mile.

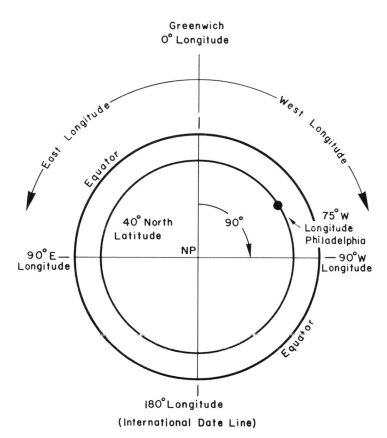

Figure 5. View of the earth from *above* the North Pole.

On the following pages are sections of three charts which show the difference between latitude and longitude scales. Take your dividers and check this scale difference for yourself.

COUNTING WEST, COUNTING EAST

West longitudes increase numerically right to left on charts.

East longitudes increase left to right on charts.

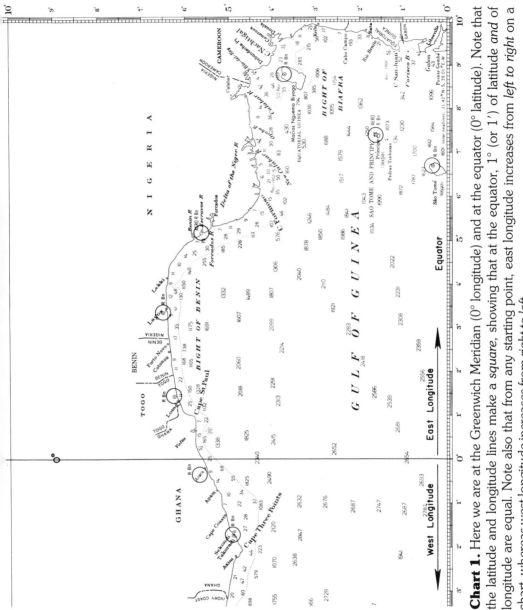

Chart 1. Here we are at the Greenwich Meridian (0° longitude) and at the equator (0° latitude). Note that the latitude and longitude lines make a *square*, showing that at the equator, 1° (or 1') of latitude *and of* longitude are equal. Note also that from any starting point, east longitude increases from *left to right* on a chart, whereas west longitude increases from *right to left*.

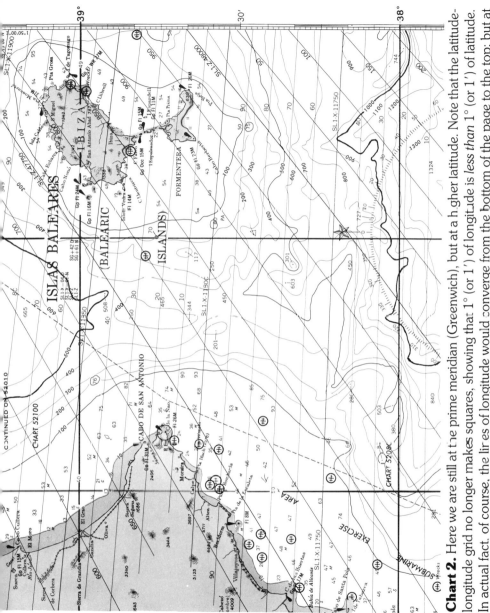

Chart 2. Here we are still at the prime meridian (Greenwich), but at a higher latitude. Note that the latitude-longitude grid no longer makes squares, showing that 1° (or 1') of longitude is *less than* 1° (or 1') of latitude. In actual fact, of course, the lines of longitude would converge from the bottom of the page to the top; but at this scale, that convergence is undetectable.

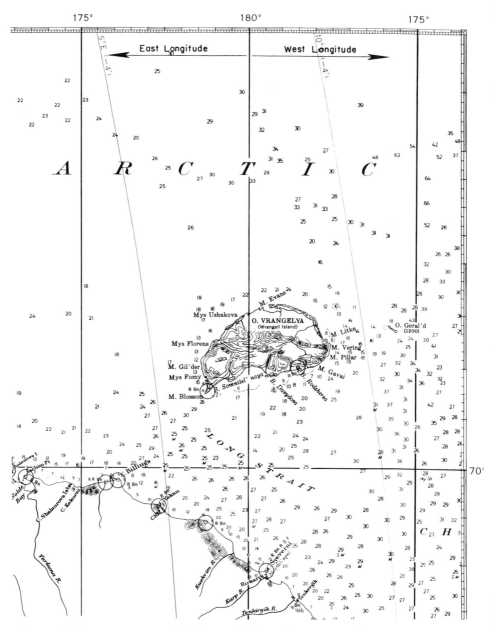

Chart 3. Here we are at 180° longitude (exactly opposite the prime meridian at Greenwich) and at a very high latitude in the Arctic Ocean. The difference between the latitude-longitude grid is dramatic. Note again that longitude increases *left to right* in east longitudes and *right to left* in west longitudes.

After you have checked out the sample charts, I think you'll be interested in the short quiz that follows. This is the first of seven which wrap up each chapter in Part I of this book. Folks I've taught in the past have found such quizzes informative and entertaining; so, if only for comic relief, give them a go before plunging on.

QUIZ

1. At approximately what latitude are the following places:
    a. The California-Oregon border
    b. Great Salt Lake, Utah
    c. Des Moines, Iowa
    d. Cleveland, Ohio
    e. New York, New York
    f. Madrid, Spain
    g. Naples, Italy
    h. Istanbul, Turkey
    i. Caspian Sea, Russia
    j. Peking, China
    k. Aomori, Japan
2. In what direction from these places is the equator?
3. Approximately how far is it in nautical miles from these places to the equator?
4. At approximately what longitude are the following places:
    a. Thüle, Greenland
    b. Bangor, Maine
    c. Puerto Rico
    d. Caracas, Venezuela
    e. La Paz, Bolivia
    f. Cape Horn
    g. Palmer Island, Antarctica

5. What is the latitude and longitude of your home town? How many nautical miles from it to the equator?

QUIZ ANSWERS

1. About 40°N. A globe is very useful as an aid in this sort of exercise. Most Americans are quite surprised to realize how far north most of Europe is vis-a-vis the United States. We tend to think of places like Spain and Italy as being south, yet they are in the same band of latitudes as the United States Midwest.
2. South.
3. Since 1° of latitude equals 60 nautical miles, these places are 40 × 60, or 2400, nautical miles from the equator.
4. 70°W.
5. Take your latitude in degrees and minutes and remember: 60 NM = 1°; 1 NM = 1′.

Declination & Greenwich Hour Angle

JUST AS there is a system for designating places on the surface of the earth, there is also a system for designating the location of the sun relative to the surface of the earth. This is the Declination-Greenwich Hour Angle grid which, in west longitudes, has a one-to-one equivalence with the latitude-longitude grid. In east longitudes, equivalence is established by one step of arithmetic.

For reasons which will become clear a little later, when working up a noon sight the thing we need to know is the point on the surface of the earth that is directly underneath the sun at the time of the sight—i.e. we would like to know the latitude and longitude of this spot. Now, obviously, a table could be constructed that would give this information in terms of latitude and longitude, but this would not be strictly logical, since latitude-longitude is a fixed grid and, relative to the earth, the sun is

DECLINATION = LATITUDE

Declination is the latitude of the point directly under the center of the sun.

constantly moving. Besides, calculating the position of the sun is the job of astronomers and they have their conventions to observe, so we are stuck with having to understand their scheme of notation—Declination and Greenwich Hour Angle. Let's tackle declination first.

Very simply, the declination of the sun is the latitude of the point on earth that lies directly beneath the sun at any given moment. The equivalence between latitude and declination is one-to-one, *whatever the longitude*. Declination is named north or south, just like latitude. The difference is that declination changes slowly during the course of the year. Figures 6, 7 and 8 show the three major declinations of the sun that traditionally demarcate the seasons—summer solstice (maximum north declination), winter solstice (maximum south declination), and equinox (declination zero).

Graphically, you can visualize that on any given day the sun rolls along directly over the latitude that is equal to its declination. This is not strictly true, of course, otherwise there could be no solstices or equinoxes, but this picture of it does help to visualize the situation. In any case, the maximum rate of change in declination is about 1′ of arc per hour. In Figures 9 and 10 are three tables listing the declinations of the sun at every hour. The first (Figure 9) just happens to be for the day in which the sun crosses the equator—and designation changes from south to north. At this time of the year the change in declination is also most rapid—about 1′ per hour.

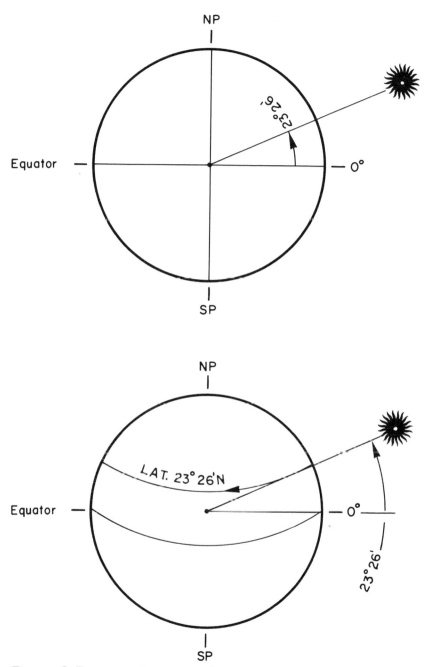

Figure 6. Diagrams of summer solstice. Around June 21st, the sun reaches its maximum northerly declination of 23°26′N. This is one solstice; and on this day, the sun's path is directly over latitude 23°26′N—the so called Tropic of Cancer.

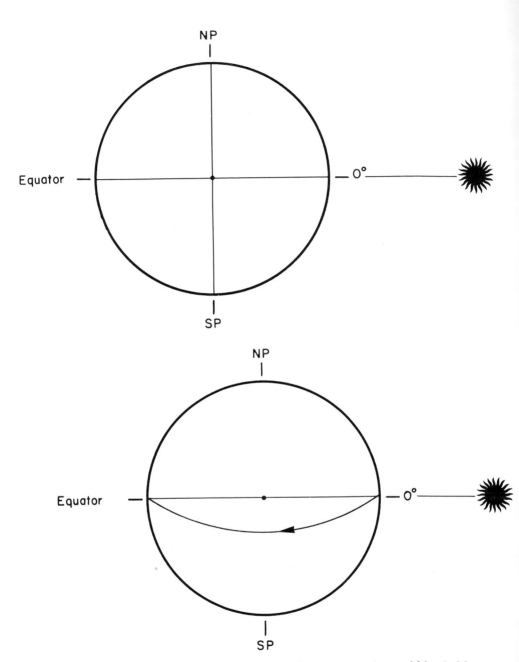

Figure 7. Diagrams of spring and fall equinoxes. Around March 20 and then again around September 23, the sun is directly over the equator (declination 0°). These are the days of the spring and fall equinoxes, respectively. On these days, the sun's path is effectively directly over the equator.

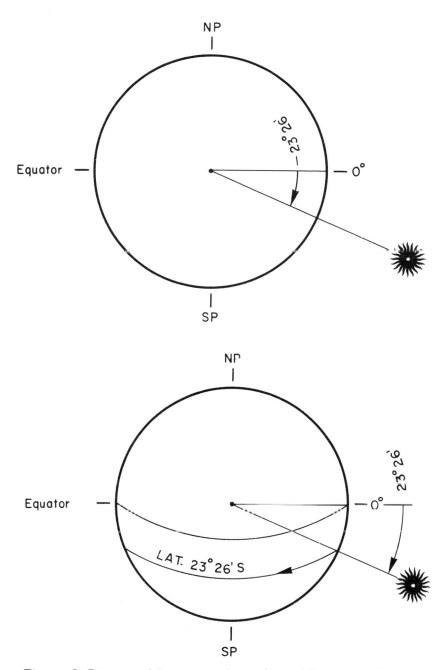

Figure 8. Diagram of the winter solstice. Around December 22 the sun reaches its maximum southerly declination of 23°26′S. This is the other annual solstice; and on this day the sun's path is directly over latitude 23°26′S—the so called Tropic of Capricorn.

Dec.

d	h	°	′
20	00	S 0	17.5
	01		16.5
	02		15.5
	03	··	14.5
	04		13.5
	05		12.6
	06	S 0	11.6
	07		10.6
	08		09.6
S	09	··	08.6
U	10		07.6
N	11		06.6
D	12	S 0	05.6
A	13		04.6
Y	14		03.7
	15	··	02.7
	16		01.7
	17	S 0	00.7
	18	N 0	00.3
	19		01.3
	20		02.3
	21	··	03.3
	22		04.2
	23		05.2

Figure 9. The spring equinox. This table lists the change in the sun's declination during the day of March 20, 1977. The time (GMT) is given in the left-hand column and is the time at Greenwich, England. As you can see, at midnight (00 hour), the sun is over a latitude just a little south of the equator, 17′.5 (read as seventeen and five-tenths *minutes* south latitude). One hour later (01 or 1 a.m. in the morning) it has moved 1′ of latitude north. This progress steadily continues until the sun actually passes over the equator between 5 and 6 in the afternoon (between 17 and 18 hours) and, by 11 that evening (23 hours), it is over latitude 5′.2 N (this convention of putting the minute sign (′) in the middle of a number with a decimal point in it is done simply to save space in this text).

Dec.

d	h	°	'
21	00	N23	26.3
	01		26.3
	02		26.3
	03	..	26.3
	04		26.3
	05		26.3
	06	N23	26.3
	07		26.3
	08		26.4
	09	..	26.4
	10		26.4
	11		26.4
	12	N23	26.4
	13		26.4
	14		26.4
	15	..	26.4
	16		26.4
	17		26.3
	18	N23	26.3
	19		26.3
	20		26.3
	21	..	26.3
	22		26.3
	23		26.3

(TUESDAY)

Dec.

d	h	°	'
21	00	S23	26.1
	01		26.2
	02		26.2
	03	..	26.2
	04		26.2
	05		26.2
	06	S23	26.2
	07		26.3
	08		26.3
	09	..	26.3
	10		26.3
	11		26.3
	12	S23	26.3
	13		26.3
	14		26.3
	15	..	26.3
	16		26.3
	17		26.3
	18	S23	26.4
	19		26.4
	20		26.4
	21	..	26.4
	22		26.4
	23		26.4
22	00	S23	26.4
	01		26.4
	02		26.4
	03	..	26.4
	04		26.4
	05		26.4
	06	S23	26.3
	07		26.3
	08		26.3
	09	..	26.3
	10		26.3
	11		26.3
	12	S23	26.3
	13		26.3
	14		26.3
	15	..	26.3
	16		26.3
	17		26.2
	18	S23	26.2
	19		26.2
	20		26.2
	21	..	26.2
	22		26.2
	23		26.1

(WEDNESDAY / THURSDAY)

Figure 10. Here we have two tables that show the sun at its solstices—literally, the days on which movement of the declination seems to halt for a while before changing direction.

The table on the left is for June 21, 1977, and begins and ends with the sun at N 23°26ʹ.3. Between 8 in the morning and 4 in the afternoon the sun is exactly over latitude 23°26ʹ.4 N. As you can see, the rate of change of the declination here is a lot less than the rate at the equinoxes (one-tenth of a minute per hour, versus one minute per hour).

The table on the right covers the two days in December, 1977, during which the sun reached its maximum southerly declination between 6 in the evening on the 21st and 5 in the morning on the 22nd.

DECLINATION MARKS THE SEASONS:

Sun's declination continually changes, ranging between 23°26'N and 23°26'S annually. Four moments have been used since pre-history to demarcate the seasons:

— Day of declination 0°, sun moving north = beginning of spring.
— Day declination reaches 23°26' north = beginning of summer.
— Day of declination 0°, sun moving south = beginning of fall.
— Day declination reaches 23°26' south = beginning of winter.

For folks in the southern hemisphere, terminology is just the reverse.

Slow motion is not the characteristic of the other part of this positional system—the sun's change in longitude amounts to 15° per hour and this is the reason that the coordinate is called an Hour Angle. We've already seen the equivalence between

GHA = LONGITUDE = TIME

The sun crosses 15° of longitude in one hour, 1° of longitude in four minutes, 15' of longitude in one minute, 1' of longitude in four seconds:

Arc (angle)	Time
15°	1 Hour
1°	4 Minutes
15'	1 Minute
1'	4 Seconds

distance on the surface of the earth and an internally-measured angle, so the fact that we can talk about an angle in terms of the amount of time it takes the sun to cover or move through it should not be a great shock. This second coordinate is called Greenwich Hour Angle because it, just like longitude, is measured from Greenwich, England—Greenwich Hour Angle zero being the same as longitude zero. The difference between Greenwich Hour Angle (GHA) and longitude is that GHA is measured *westwards only,* through a full 360°, not west and east to 180°. This is logical because, to an observer on earth, this is the way the sun—and the spot directly under it—appears to move. Figure 11 shows several hour angles and the correspondence is evident—15° of Greenwich Hour Angle equals

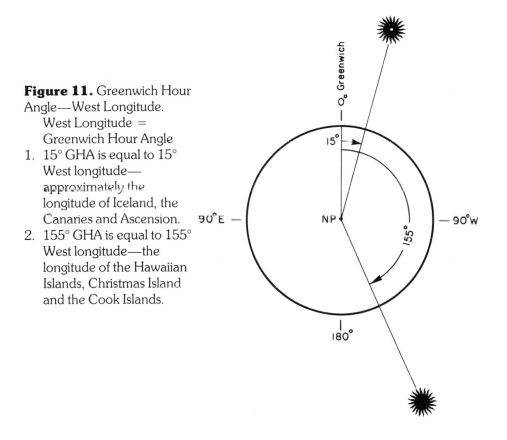

Figure 11. Greenwich Hour Angle—West Longitude.
    West Longitude = Greenwich Hour Angle
1. 15° GHA is equal to 15° West longitude— approximately the longitude of Iceland, the Canaries and Ascension.
2. 155° GHA is equal to 155° West longitude—the longitude of the Hawaiian Islands, Christmas Island and the Cook Islands.

15° of longitude. It's easy to see that, in west longitudes, GHA and longitude are *numerically* equal. In east longitudes, Figure 12 shows that GHA and longitude are related this way: 360° minus GHA equals longitude east. Figure 13 is a table showing the GHA of the sun for a period of three days so that you can readily see the complete cycle of GHA's.

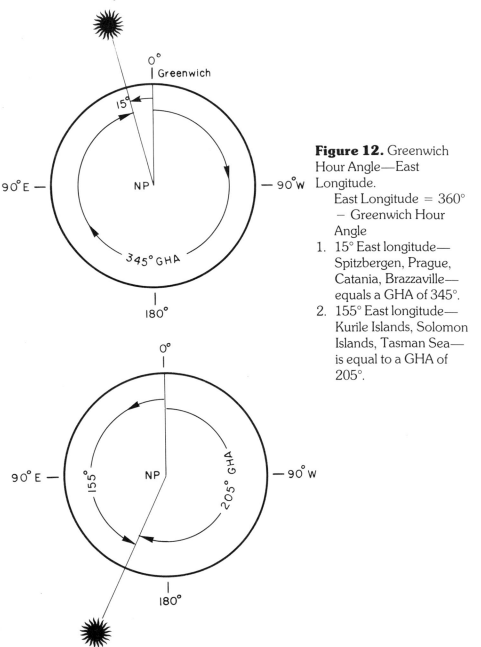

Figure 12. Greenwich Hour Angle—East Longitude.

> East Longitude = 360° − Greenwich Hour Angle
>
> 1. 15° East longitude— Spitzbergen, Prague, Catania, Brazzaville— equals a GHA of 345°.
> 2. 155° East longitude— Kurile Islands, Solomon Islands, Tasman Sea— is equal to a GHA of 205°.

G.M.T.	G.H.A.	
d h	°	′
4 00	180	12.3
01	195	12.5
02	210	12.7
03	225	12.9
04	240	13.1
05	255	13.3
06	270	13.6
07	285	13.8
08	300	14.0
S 09	315	14.2
U 10	330	14.4
N 11	345	14.6
D 12	0	14.8
A 13	15	15.0
Y 14	30	15.2
15	45	15.4
16	60	15.6
17	75	15.8
18	90	16.0
19	105	16.2
20	120	16.4
21	135	16.7
22	150	16.9
23	165	17.1

G.M.T.	G.H.A.	
d h	°	′
5 00	180	17.3
01	195	17.5
02	210	17.7
03	225	17.9
04	240	18.1
05	255	18.3
06	270	18.5
07	285	18.7
08	300	18.9
M 09	315	19.1
O 10	330	19.4
N 11	345	19.6
D 12	0	19.8
A 13	15	20.0
Y 14	30	20.2
15	45	20.4
16	60	20.6
17	75	20.8
18	90	21.0
19	105	21.2
20	120	21.4
21	135	21.7
22	150	21.9
23	165	22.1

G.M.T.	G.H.A.	
d h	°	′
6 00	180	22.3
01	195	22.5
02	210	22.7
03	225	22.9
04	240	23.1
05	255	23.3
06	270	23.5
07	285	23.8
08	300	24.0
T 09	315	24.2
U 10	330	24.4
E 11	345	24.6
S 12	0	24.8
D 13	15	25.0
A 14	30	25.2
Y 15	45	25.4
16	60	25.7
17	75	25.9
18	90	26.1
19	105	26.3
20	120	26.5
21	135	26.7
22	150	26.9
23	165	27.1

Figure 13. Greenwich Hour Angle of the sun for three days.

GHA & LONGITUDE

West longitudes: GHA = Longitude

East longitudes: 360 – GHA = Longitude

At the beginning of the day (midnight in Greenwich), the sun is near longitude 180°—i.e., it is on the opposite side of the earth from Greenwich. At 6:00 a.m. (near dawn in England) it is at 90° East, at 1200 (noon) its hour angle is around 360°— i.e., it is on the same longitude as Greenwich and would be due south, at 6:00 p.m. (near sunset in Greenwich) it is at longitude 90° West (its noon in New Orleans) and six hours later (midnight), the sun is back at 180° longitude—over the Pacific Ocean.

Figure 14 is a table which combines both GHA and declination. This listing obviously gives us the ability to determine the latitude and longitude of the spot on earth that is directly under the sun at any hour and, by interpolation, at any moment at all.

Up to now, I have been talking about the position of the "spot" on the earth directly under the sun because it is knowing the location of this spot that enables you to find latitude and longitude from a noon sight. We now know the coordinate system that designates such a "spot," and we have had a glimpse at the tables that enable us to know its latitude and

THE GP IS . . .

Point directly under the center of the sun is called the geographical position (GP).

d h	G.H.A. ° '	Dec. ° '
16 00	180 01.5	N10 00.1
01	195 01.6	01.0
02	210 01.8	01.9
03	225 01.9	·· 02.8
04	240 02.1	03.7
05	255 02.2	04.5
06	270 02.4	N10 05.4
07	285 02.5	06.3
S 08	300 02.7	07.2
A 09	315 02.8	·· 08.1
T 10	330 03.0	09.0
U 11	345 03.1	09.9
R 12	0 03.2	N10 10.8
D 13	15 03.4	11.6
A 14	30 03.5	12.5
Y 15	45 03.7	·· 13.4
16	60 03.8	14.3
17	75 04.0	15.2
18	90 04.1	N10 16.1
19	105 04.3	16.9
20	120 04.4	17.8
21	135 04.6	·· 18.7
22	150 04.7	19.6
23	165 04.9	20.5

d h	G.H.A. ° '	Dec. ° '
17 00	180 05.0	N10 21.4
01	195 05.1	22.2
02	210 05.3	23.1
03	225 05.4	·· 24.0
04	240 05.6	24.9
05	255 05.7	25.8
06	270 05.9	N10 26.6
07	285 06.0	27.5
08	300 06.2	28.4
S 09	315 06.3	·· 29.3
U 10	330 06.4	30.2
N 11	345 06.6	31.0
D 12	0 06.7	N10 31.9
A 13	15 06.9	32.8
Y 14	30 07.0	33.7
15	45 07.2	·· 34.6
16	60 07.3	35.4
17	75 07.4	36.3
18	90 07.6	N10 37.2
19	105 07.7	38.1
20	120 07.9	38.9
21	135 08.0	·· 39.8
22	150 08.2	40.7
23	165 08.3	41.6

d h	G.H.A. ° '	Dec. ° '
18 00	180 08.4	N10 42.4
01	195 08.6	43.3
02	210 08.7	44.2
03	225 08.9	·· 45.1
04	240 09.0	45.9
05	255 09.1	46.8
06	270 09.3	N10 47.7
07	285 09.4	48.6
08	300 09.6	49.4
M 09	315 09.7	·· 50.3
O 10	330 09.8	51.2
N 11	345 10.0	52.0
D 12	0 10.1	N10 52.9
A 13	15 10.3	53.8
Y 14	30 10.4	54.7
15	45 10.5	·· 55.5
16	60 10.7	56.4
17	75 10.8	57.3
18	90 10.9	N10 58.1
19	105 11.1	59.0
20	120 11.2	10 59.9
21	135 11.4	11 00.7
22	150 11.5	01.6
23	165 11.6	02.5

Figure 14. Combined table—GHA & declination.

⚷ NOT A HERESY:

Kepler to the contrary, for us navigators the sun goes around the earth, not vice versa.

longitude. I have avoided telling you the technical name of the spot because, up until now, it would not have conveyed much. Now we need to name it. It is generally called the geographical position and is abbreviated GP. In some more technical works it is called the *sub-solar* point. In this book, we'll use GP. It's the point on the earth directly under the sun at any moment, and the tables we've studied give, effectively, its latitude and longitude.

QUIZ

1. Assume you live in Key West, Florida (Latitude 24°33'N; Longitude 81°49'W). On the day of the summer solstice, how close to you does the spot directly underneath the sun (GP) pass? Answer in degrees and minutes; also in nautical miles. At its closest approach, in what direction (true) does the spot lie? What is its GHA?
2. Same question for Honolulu, Hawaii (Latitude 21°18'N; Longitude 157°52'W).
3. Suppose you live in Brisbane, Australia (Latitude 27°28'S; Longitude 153°02'E). On the day of the summer solstice, how close does the spot directly underneath the sun (GP) pass? At its closest approach, in what direction (true) does the spot lie? What is its GHA?

QUIZ ANSWERS

1. On the day of the solstice (around June 22), the sun is directly over latitude 23°26'N. Therefore, at its closest approach, the spot

directly underneath the sun (GP) would be 1°07′ away (24°33′ minus 23°26′), or 67 nautical miles. True bearing to the spot would be south. At the time of its closest approach, the spot would be on your meridian and, therefore, its GHA would be 81°49′, the same as your longitude. Once again, a globe is very helpful in visualizing these relationships.

2. On the day of the solstice, the sun passes north of Honolulu; the declination is greater than the latitude. Distance to the GP is 23°26′ minus 21°18′, or 2°08′; in nautical miles, it's 128. Since Honolulu's longitude is west, GHA = longitude: 157°52′.

3. On the day of the solstice (around December 22), the sun is directly over latitude 23°26′S. Therefore, it passes North of Brisbane at a distance of 27°28′ minus 23°26′, or 4°02′; 242 nautical miles. Since Brisbane is at longitude 153°02′ East, GHA = 360° − 153°02′, or 206°58′.

3

Time

IT STRIKES most people as odd, but the fact is that there are *two kinds of time*—what is called *apparent* time and what is called *mean* time. Briefly, apparent time is time based on the rotation of the *real* earth, and one unit of such time is determined by the interval between two passages of the sun (or a star) past a designated longitude (usually Greenwich). Mean time is based on the rotation of an *imaginary* earth which turns, in an unreal circular orbit, rather than in its actual elliptical orbit, and whose unit is determined by the average (mean) of 24 man-made atomic clocks like the one at the Bureau of Standards in Fort Collins, Colorado. The relation between apparent and mean time is given by the "equation of time."

Obviously, this subject can get very vexing, but, fortunately for navigators, there is only one time—Greenwich Mean Time (GMT). This is the time given in the left column of the tables shown previously and is also called Universal Coordinated Time (UTC) and Zulu (Z).

Anyway, just as longitude begins at Greenwich and hour angle begins at Greenwich, time begins at Greenwich. You have to start somewhere and, for good, historical reasons, modern navigation begins at Greenwich, so it is only logical that

THE ONLY TIME . . .

The only time a navigator needs to know is Greenwich Mean Time (GMT).

Other names for GMT:
• Coordinated Universal Time (UTC)
• Zulu Time (Z)
• Greenwich Civil Time (GCT)

these three vital aspects of navigation should originate there, too.

In this day of highly accurate, cheap quartz-crystal watches, most practicing navigators have one (or more) watches set to the Greenwich time and tucked away in a secure, dry spot(s). Continuous time information is available via shortwave radio from station WWV on 2.5, 5, 10, 15, 30 and 25 MH_z and from CHU on 3.3 and 6.7 MH_z. WWV also broadcasts very useful storm information at 8, 9 and 10 minutes after every hour. This last is an aside, but I mention it because it is not generally known. On one occasion WWV had information about a storm which was evidently unknown to my usual sources, and so saved me from setting out into a near hurricane.

To get back to the subject at hand, there is one thing you need to know to use Greenwich Mean Time (GMT), and that is that it is kept in the 24-hour clock system. In this system the day begins at midnight and is designated hour 0000. One in the

CONTINUOUS SHORT-WAVE TIME ON:

WWV frequencies: 2.5, 5, 10, 15, 25 MH_z

KEEPING GMT ABOARD THE VESSEL

If the arbitrariness of time conventions annoy you, get an inexpensive digital watch that keeps time in 24-hour system, and displays hour-minute-second and date. Set it to GMT and tuck it away in a safe corner.

morning is 0100 (pronounced "oh one hundred"). One thirty is 0130. Two is 0200, etc., right on through 23-59-59 to 2400 or 0000 again. Thus, the 24 hours of a day look like this:

Midnight	0000	1300	1 PM
	0100	1400	
	0200	1500	
	0300	1600	
	0400	1700	
	0500	1800	
	0600	1900	
	0700	2000	
	0800	2100	
	0900	2200	
	1000	2300	11 PM
	1100		
Noon	1200		

This system has the advantages of immediately indicating AM or PM and, thereby, also simplifying calculations of elapsed times. In fact, the "24-hour clock" is the system in general use throughout much of the world other than the United States.

In the event that you do not have access to a short-wave radio but do have an accurate watch (one that gains or loses only a few tenths of a second a day) and a source of precise time for your time zone (often the phone company has this service), you can always figure out GMT.

Since the sun moves through 15° of longitude in one hour,

THE 24-HOUR CLOCK:

In using this system, 9:45 a.m. is usually written 09-45 and, for navigation, the seconds are tacked on. Thus, 09-45-35 means 45 minutes and 35 seconds past nine in the morning.

time zones have been established around the world every 15° and time within any zone is based on one or the other of the boundary lines of longitude which are generally called "meridians" in this context. For instance, time in New York City is based on longitude 75°W (75th meridian), which means it is five hours different from Greenwich, because, traveling at 15° per hour, it takes the sun five hours to move through the 75° of longitude. Since on any given day the sun has already passed Greenwich by the time it reaches 75°W (or any West longitude), it is later in Greenwich than New York City, so five hours has to be added to whatever the time is in New York City to get GMT. Daylight time effectively moves the reference meridian one zone (15°) East (toward the sun) and, therefore, in New York City, Daylight Savings Time would be four hours from Greenwich.

In east longitudes, the reverse of the above is true. The day begins when the sun is 180° away from Greenwich, so by the time it reaches Greenwich, it has already gone by all places with east longitudes and it is earlier in Greenwich. You must subtract from local time to get Greenwich. Daylight time (also called summer time) would increase the difference between Greenwich and places with east longitudes, since moving the time meridian east would increase the distance in time and longitude, whereas in west longitudes, moving the meridian east decreases the distance in time and longitude.

On the following pages is a list for those readers who need to find GMT from their Local Mean Time:

The times given ⎱ *added* to G.M.T. to give Standard Time.
below should be ⎰ *subtracted* from Standard Time to give G.M.T.

	h	m		h	m
Admiralty Islands	10		Egypt (United Arab Republic) ...	02	
Afghanistan	04	30	Equatorial Guinea, Republic of ...	01	
Albania*	01		Estonia	03	
Amirante Islands	04		Ethiopia	03	
Andaman Islands	05	30			
Angola	01		Fernando Póo‡	01	
Annobon Island‡	01		Fiji	12	
Arabian Emirates, Federation of ...	04		Finland	02	
Australia			France‡*	01	
Australian Capital Territory* ...	10		French Territory of the Afars and Issas	03	
New South Wales[1]*	10		Friendly Islands	13	
Northern Territory	09	30			
Queensland	10		Gabon	01	
South Australia*	09	30	Germany	01	
Tasmania*	10		Gibraltar‡	01	
Victoria*	10		Gilbert Islands	12	
Western Australia	08		Greece*	02	
Austria	01		Guam	10	
Balearic Islands‡*	01		Holland (The Netherlands)	01	
Bangladesh	06		Hong Kong* ... .:.	08	
Belgium	01		Hungary	01	
Benin (Dahomey)	01				
Botswana, Republic of	02		India	05	30
Brunei	08		Indonesia, Republic of		
Bulgaria	02		Bali, Bangka, Billiton, Java,		
Burma	06	30	Lombok, Madura, Sumatra ...	07	
Burundi	02		Borneo, Celebes, Flores, Sumba,		
			Sumbawa, Timor	08	
			Aru, Kei, Moluccas, Tanimbar,		
Cambodia (Khmer Republic)	07		Irian Jaya	09	
Cameroun Republic	01		Iran	03	30
Caroline Islands, east of long. E. 160°	12		Iraq	03	
west of long. E. 160°	10		Irish Republic†	01	
Truk, Ponape ...	11		Israel	02	
Central African Republic	01		Italy*	01	
Chad	01				
Chagos Archipelago	05		Japan	09	
Chatham Islands‡	12	45	Jordan*	02	
China[2]	08				
Christmas Island, Indian Ocean ...	07		Kamchatka Peninsula	12	
Cocos Keeling Islands	06	30	Kenya	03	
Comoro Islands	03		Khmer Republic (Cambodia)	07	
Congo Republic	01		Korea	09	
Corsica‡	01		Kuril Islands	11	
Crete	02		Kuwait	03	
Cyprus, North*	02				
South	02		Laccadive Islands	05	30
Czechoslovakia	01		Ladrone Islands	10	
			Laos	07	
Denmark	01		Latvia	03	

  * Summer time may be kept in these countries.
  † Winter time is kept in this country.
  ‡ The legal time may differ from that given here.
  [1] Except Broken Hill Area which keeps 09ʰ 30ᵐ.
  [2] All the coast, but some areas may keep summer time.

Figure 15. Finding GMT from Local Time.

	h	m
Lebanon*	02	
Lesotho	02	
Libya‡	02	
Liechtenstein	01	
Lithuania	03	
Lord Howe Island	10	30
Luxembourg	01	
Macao*	08	
Malagasy Republic	03	
Malawi	02	
Malaysia		
Malaya	07	30
Sabah, Sarawak	08	
Maldive Republic	05	
Malta*	01	
Manchuria	09	
Mariana Islands	10	
Marshall Islands[1]	12	
Mauritius	04	
Monaco‡	01	
Mozambique	02	
Muscat and Oman, Sultanate of	04	
Namibia (South West Africa)	02	
Nauru	11	30
Netherlands, The	01	
New Caledonia	11	
New Hebrides	11	
New Zealand*	12	
Nicobar Islands	05	30
Niger	01	
Nigeria, Republic of	01	
Norfolk Island	11	30
Norway	01	
Novaya Zemlya	05	
Ocean Island	11	30
Okinawa	09	
Pakistan	05	
Papua New Guinea	10	
Pescadores Islands	08	
Philippine Republic	08	
Poland	01	
Portugal	01	
Réunion	04	
Rhodesia, Republic of	02	
Romania	02	
Rwanda	02	
Ryukyu Islands	09	
Sakhalin	11	
Santa Cruz Islands	11	
Sardinia	01	

	h	m
Saudi Arabia	03	
Schouten Islands	09	
Seychelles	04	
Sicily*	01	
Singapore	07	30
Socotra	03	
Solomon Islands	11	
Somali Republic	03	
South Africa, Republic of	02	
Southern Yemen	03	
South West Africa (Namibia)	02	
Spain‡*	01	
Spitsbergen (Svalbard)	01	
Sri Lanka	05	30
Sudan, Republic of	02	
Swaziland	02	
Sweden	01	
Switzerland	01	
Syria* (Syrian Arab Republic)	02	
Taiwan	08	
Tanzania	03	
Thailand	07	
Timor	08	
Tonga Islands	13	
Truk	11	
Tunisia	01	
Turkey*	02	
Tuvalu Islands	12	
Uganda	03	
Union of Soviet Socialist Republics[2]		
west of long. E. 40°	03	
long. E. 40° to E. 52° 30'	04	
long. E. 52° 30' to E. 67° 30'	05	
long. E. 67° 30' to E. 82° 30'	06	
long. E. 82° 30' to E. 97° 30'	07	
long. E. 97° 30' to E. 112° 30'	08	
long. E. 112° 30' to E. 127° 30'	09	
long. E. 127° 30' to E. 142° 30'	10	
long. E. 142° 30' to E. 157° 30'	11	
long. E. 157° 30' to E. 172° 30'	12	
east of long. E. 172° 30'	13	
Vietnam, Northern	07	
Southern‡	07	
Wrangell Island	13	
Yugoslavia	01	
Zaire		
Kinshasa, Mbandaka	01	
Orientale, Kivu, Katanga, Kasai	02	
Zambia, Republic of	02	

* Summer time may be kept in these countries.
‡ The legal time may differ from that given here.
[1] Except the islands of Kwajalein and Eniwetok which keep a time 24ʰ slow on that of the rest of the islands.
[2] The boundaries between the zones are irregular; the longitudes given are approximate only.

PLACES NORMALLY KEEPING G.M.T.

Algeria	Gambia	Ifni	Mali	St. Helena	Tangier
Ascension Island	Ghana	Ireland, Northern[1]	Mauritania	São Tomé	Togo Republic
Canary Islands‡*	Great Britain[1]	Ivory Coast	Morocco*	Senegal	Tristan da Cunha
Channel Islands[1]	Guinea Republic	Liberia	Principe	Sierra Leone	Upper Volta
Faeroes, The	Iceland	Madeira	Rio de Oro‡	Spanish Sahara	

* Summer time may be kept in these countries.
‡ The legal time may differ from that given here.
[1] Summer time, one hour in advance of G.M.T., is kept from March 20d 02h to October 23d 02h G.M.T.

PLACES SLOW ON G.M.T. (WEST OF GREENWICH)

The times given } subtracted from G.M.T. to give Standard Time.
below should be } added to Standard Time to give G.M.T.

	h	m		h	m
Argentina	03		Chile*	04	
Austral Islands[1]	10		Christmas Island, Pacific Ocean	10	
Azores	01		Colombia	05	
			Cook Islands, except Niue	10	30
Bahamas*	05		Costa Rica	06	
Barbados	04		Cuba*	05	
Belize	06		Curaçao Island	04	
Bermuda*	04				
Bolivia	04		Dominican Republic‡	04	
Brazil, eastern[2]	03		Dutch Guiana (Surinam)	03	30
Territory of Acre	05				
western	04		Easter Island (I. de Pascua)	07	
British Antarctic Territory[3]	03		Ecuador	05	
Canada					
Alberta*	07		Falkland Islands[4]	04	
British Columbia*	08		Fanning Island	10	
Labrador*	04		Fernando de Noronha Island	02	
Manitoba*	06		French Guiana‡	03	
New Brunswick*	04				
Newfoundland*	03	30	Galápagos Islands	05	
Northwest Territories*			Greenland, Scoresby Sound	02	
east of long. W. 68°	04		Angmagssalik and west coast	03	
long. W. 68° to W. 85°	05		Thule area	04	
long. W. 85° to W. 102°	06		Grenada	04	
west of long. W. 102°	07		Guadeloupe	04	
Nova Scotia*	04		Guatemala	06	
Ontario*, east of long. W. 90°	05		Guiana, Dutch	03	30
west of long. W. 90°	06		French‡	03	
Prince Edward Island*	04		Guyana, Republic of	03	45
Quebec*, east of long. W. 63°	04				
west of long. W. 63°	05		Haiti	05	
Saskatchewan*			Honduras	06	
east of long. W. 106°	06		Honduras, British (Belize)	06	
west of long. W. 106°	07				
Yukon, east of long. W. 138°	08				
west of long. W. 138°	09		Jamaica*	05	
Cape Verde Islands*	02		Jan Mayen Island	01	
Cayman Islands	05				

* Summer time may be kept in these countries.
‡ The legal time may differ from that given here.
[1] This is the legal standard time, but local mean time is generally used.
[2] Including all the coast and Brasilia.
[3] Except South Georgia which keeps 02h.
[4] Port Stanley keeps summer time September to March.

	h	m
Johnston Island	10	
Juan Fernandez Islands	04	
Leeward Islands	04	
Low Archipelago	10	
Marquesas Islands[1]	09	30
Martinique	04	
Mexico[2]	06	
Midway Islands	11	
Miquelon	03	
Nicaragua‡*	06	
Niue Island	11	
Panama Canal Zone	05	
Panama, Republic of	05	
Paraguay*	04	
Peru	05	
Portuguese Guinea	01	
Puerto Rico	04	
Rarotonga	10	30
St. Pierre and Miquelon	03	
Salvador, El	06	
Samoa	11	
Society Islands[1]	10	
South Georgia	02	
Surinam (Dutch Guiana)	03	30
Tobago	04	
Trindade Island, South Atlantic	02	
Trinidad	04	
Tuamotu Archipelago[1]	10	
Tubuai Islands[1]	10	
Turks and Caicos Islands	05	
United States of America		
Alabama[3]	06	
Alaska[3], east of long. W. 137°	08	
long. W. 137° to W. 141°	09	
long. W. 141° to W. 161°	10	
long. W. 161° to W. 172° 30′	11	
Aleutian Islands	11	
Arizona	07	
Arkansas[3]	06	
California[3]	08	
Colorado[3]	07	
Connecticut[3]	05	
Delaware[3]	05	
District of Columbia[3]	05	

	h	m
United States of America (continued)		
Florida[3,4]	05	
Georgia[3]	05	
Hawaii	10	
Idaho[3,4]	07	
Illinois[3]	06	
Indiana[4]	05	
Iowa[3]	06	
Kansas[3,4]	06	
Kentucky[3,4]	05	
Louisiana[3]	06	
Maine[3]	05	
Maryland[3]	05	
Massachusetts[3]	05	
Michigan[3,4]	05	
Minnesota[3]	06	
Mississippi[3]	06	
Missouri[3]	06	
Montana[3]	07	
Nebraska[3,4]	06	
Nevada[3]	08	
New Hampshire[3]	05	
New Jersey[3]	05	
New Mexico[3]	07	
New York[3]	05	
North Carolina[3]	05	
North Dakota[3,4]	06	
Ohio[3]	05	
Oklahoma[3]	06	
Oregon[3,4]	08	
Pennsylvania[3]	05	
Rhode Island[3]	05	
South Carolina[3]	05	
South Dakota[3], eastern part	06	
western part	07	
Tennessee[3,4]	06	
Texas[3]	06	
Utah[3,4]	07	
Vermont[3]	05	
Virginia[3]	05	
Washington, D.C.[3]	05	
Washington[3]	08	
West Virginia[3]	05	
Wisconsin[3]	06	
Wyoming[3]	07	
Uruguay	03	
Venezuela	04	
Virgin Islands	04	
Windward Islands	04	

* Summer time may be kept in these countries.
‡ The legal time may differ from that given here.
[1] This is the legal standard time, but local mean time is generally used.
[2] Except the states of Sonora, Sinaloa, Nayarit, and the Southern District of Lower California which keep 07$^\mathrm{h}$, and the Northern District of Lower California which keeps 08$^\mathrm{h}$.
[3] Emergency daylight-saving time, one hour fast on the given time, is kept in these states effective from 1974 January 6 at 02$^\mathrm{h}$ 00$^\mathrm{m}$ local clock time. Summer (daylight-saving) time, one hour fast on the time given, is kept in these states from the last Sunday in April to the last Sunday in October, changing at 02$^\mathrm{h}$ 00$^\mathrm{m}$ local clock time.
[4] This applies to the greater portion of the state.

QUIZ

1. The longitude of your home town is 90°W. How long does it take for the sun to get from the longitude of Greenwich, England, to you?
2. The longitude of your home town is 91°W. How long does it take the sun to get from Greenwich to you?
3. The longitude of your home town is 91°15'W. How long does it take the sun to get from Greenwich to you?
4. The longitude of your home town is 91°16'W. How long does it take the sun to get from Greenwich to you?
5. Your home town is at 90° East longitude. How long does it take the sun to travel from Greenwich to you? From you to Greenwich?
6. Your digital quartz watch is set to Eastern Standard Time (EST = GMT − 5 hours) and is accurate. You take the evening flight for London and wake up just as the wheels hit the runway at Heathrow Airport. You've missed the stewardess' announcement of the local time. You look at your watch: 03-10-11. What is the Greenwich Mean Time?

QUIZ ANSWERS

1. 6 hours. The sun travels across 15° of longitude in one hour.
2. 6 hours, 4 minutes. The sun crosses 1° of longitude in four minutes.
3. 6 hours, 5 minutes. It takes one minute of time for the sun to cover 15' of longitude; or, saying it another way, for the sun's GHA to increase 15' of arc.
4. 6 hours, 5 minutes, 4 seconds (written 6-05-04). It takes the sun four seconds to cover 1' of longitude, or for its GHA to increase 1'.
5. 18 hours. 90° East longitude equals a GHA of 270°. At 15°/hour, the sun needs 18 hours to cover 270°. 6 hours. Thus proving that the sun goes around the earth (Kepler, be still!) in 24 hours.

6. 08-10-11. There is a certain amount of "eye-wash" in the statement of this question. You haven't changed your watch in flight, so the difference between your watch and Greenwich is the same as it was in New York. To find GMT in New York, you would add five hours to your watch time. You do the same now. Your watch doesn't know you've flown to England. Remember, the *only* time a navigator needs to know is GMT.

4

Noon

FOR MOST of us, noon is when our watch says twelve during daytime, and that's good enough for normal purposes. In celestial navigation, however, noon has a much stricter and narrower definition.

First of all, noon is related strictly to *you,* the observer of the sun, and it comes at the moment when you and the sun are on the same line of longitude. That means that for an observer somewhere to the west of you it is not yet noon, and somewhere to the east of you it has already been noon. It sounds a little absurd, but, strictly speaking, noon occurs at a different moment for people just a few feet to the east or west of you.

WHEN IS NOON?

Noon is the moment the sun crosses *your* meridian (rarely agrees with 1200 on watch or chronometer). Thus, it is also called *meridian passage.*

Using the concepts discussed in the previous chapters, you can see that since there is a simple relationship between GHA and longitude, if you can determine the Greenwich Mean Time of the moment at which the sun is exactly over your longitude, you can look up in a table the sun's GHA and, thus, find your longitude.

The determination of the moment at which the sun is on your longitude is made possible by the fact that at that moment the sun reaches its highest point in the sky. With a sextant, you can find this moment, and we'll get to the specifics of doing it later. Figures 16 and 17 show and explain the instants of noon at several GMT's, and explain the procedure in broad outline.

Now, consider for a moment that you have the sun on your meridian (longitude). Visualize the situation shown in Figure 18 —a day when you are in the northern hemisphere and the sun's declination is 0°.

Imagine the world sliced through vertically along your meridian of longitude. Figure 19 shows the situation. The sun is directly over the equator and a ray of light from it is represented by the line through the center of the earth and the GP. Your location is also on a line through the surface and the center of the earth. Because the sun is so very far from the earth, sunlight falls in parallel rays. Therefore, the ray of light that you see as you look at the sun is parallel to the ray from the sun through the GP and angle A equals angle B. Remember latitude? Well, angle B is your latitude. Figures 20, 21 and 22 show what happens on days when the declination is not zero and it is obvious if you have a means of finding angle A and a table of the sun's declinations, you can find your latitude. Interestingly, you find A, called *zenith distance,* by the way, with the sextant, and we'll take that up next.

13	**00**	180	02.3	N23	11.4
	01	195	02.2		11.6
	02	210	02.1		11.7
	03	225	01.9	··	11.8
	04	240	01.8		12.0
	05	255	01.7		12.1
	06	270	01.5	N23	12.3
	07	285	01.4		12.4
	08	300	01.3		12.6
M	09	315	01.1	··	12.7
O	10	330	01.0		12.8
N	11	345	00.9		13.0
D	12	0	00.8	N23	13.1
A	13	15	00.6		13.3
Y	14	30	00.5		13.4
	15	45	00.4	··	13.5
	16	60	00.2		13.7
	17	75	00.1		13.8
	18	90	00.0	N23	13.9
	19	104	59.8		14.1
	20	119	59.7		14.2
	21	134	59.6	··	14.3
	22	149	59.4		14.5
	23	164	59.3		14.6

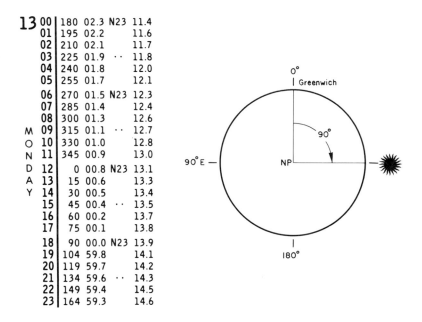

Figure 16. Noon in west longitudes: GHA = longitude.

The table shows that at 1800 GMT (6 p.m. in Greenwich) on May 13, 1977, the sun's GHA was 90° exactly (90°00′.0). This means that the sun was directly over 90° W longitude at that moment and it was noon for everyone on that longitude line. For everyone on that longitude at that moment the sun would be at its highest point in the sky —i.e. everyone shooting the sun with a sextant would measure the greatest altitude of the day.

Reversing the situation, if you noted that the sun reached its highest altitude for you when your (accurate) watch said it was 1800 hours GMT, you could look in the table and find that since it's GHA was 90°00′.0, your longitude must be 90°00′.0 W.

Likewise, the table shows the GMT of noon at longitudes 104°59′.8 W, 119°59′.7 W, etc.

25	00	180 01.9 S23 24.2
	01	195 01.6 24.2
	02	210 01.3 24.1
	03	225 01.0 ·· 24.0
	04	240 00.7 24.0
	05	255 00.4 23.9
	06	270 00.1 S23 23.8
	07	284 59.8 23.8
	08	299 59.5 23.7
S	09	314 59.2 ·· 23.7
U	10	329 58.8 23.6
N	11	344 58.5 23.5
D	12	359 58.2 S23 23.4
A	13	14 57.9 23.4
Y	14	29 57.6 23.3
	15	44 57.3 ·· 23.2
	16	59 57.0 23.2
	17	74 56.7 23.1
	18	89 56.4 S23 23.0
	19	104 56.1 22.9
	20	119 55.8 22.9
	21	134 55.4 ·· 22.8
	22	149 55.1 22.7
	23	164 54.8 22.6

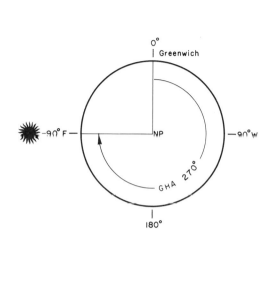

Figure 17. Noon in east longitudes: 360° − GHA = longitude.

The table shows that at 0600 GMT (6 a.m. in Greenwich) on Christmas Day, 1977, the sun's GHA was 270°00′.1. This means that the sun was directly over longitude 89°59′.9 E (effectively 90° E) and it was noon for everyone on that meridian of longitude. For everyone on that meridian, the sun would be at its highest point in the sky for the day—i.e. people with a sextant would measure the greatest altitude of the day.

Reversing the situation, if you noted that the sun reached its highest altitude for you when your chronometer said it was 0600 GMT, you could look in the table and find that since the sun's GHA at that time was 270°00′.1, your longitude must be 360°00′.0 minus 270°00′.1, or 89°59′.9 E.

Likewise, the table shows the GMT of noon at longitudes 75°00′.2 E, 45°00′.8 E, etc.

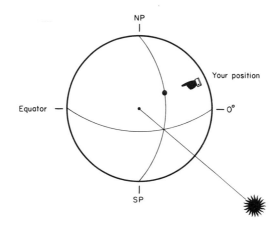

Figure 18. A typical example of noon: sun at an equinox and observer in the northern hemisphere.

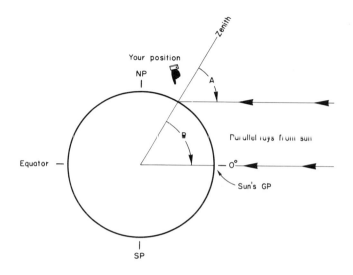

Figure 19. Latitude when the sun is over the equator.

Sun directly over equator. If you like, imagine yourself standing next to a flag pole, then the flag pole is part of the line running from the center of the earth through your GP, out into space toward a point vertically over your head—the point called your *zenith*.

At declination 0°—sun over the equator—your latitude equals the angle (A) between your zenith and the sun—the zenith distance. This is due to the fact that the sun's rays (for all practical purposes) are parallel and, therefore, A = B, your latitude.

To visualize the situation for south latitudes, turn the drawing upside down.

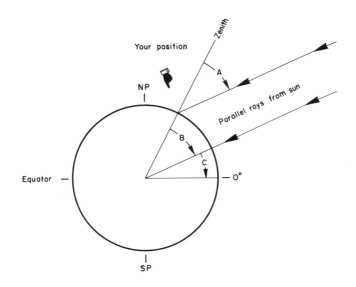

Figure 20. Latitude when sun is between you and equator.

Here the sun has moved toward you from the equator—i.e. it's declination is no longer zero. Now, your latitude—your distance in degrees and minutes of arc from the equator—is equal to angle B plus angle C. C is the declination of the sun and A and B are equal. Therefore, latitude = zenith distance (A) + declination (C).

For an equivalent situation in south latitudes with south declination, turn the drawing upside down.

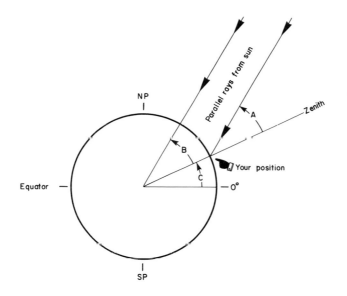

Figure 21. Latitude when *you* are between the sun and the equator.

Here you are between the sun and the equator. Your latitude is C. The sun's declination is B plus C. Therefore, your latitude is declination minus zenith distance (A). Expressed in algebra, it goes like this:

$$C = (B + C) - B$$
$$B = A$$
$$(B + C) = \text{declination}$$
$$C = \text{declination} - B$$
$$C = \text{declination} - A$$

For the equivalent situation in south latitudes, turn this diagram upside down.

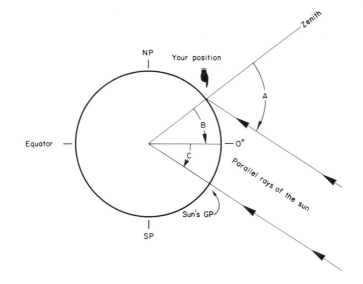

Figure 22. Latitude when sun is on the *opposite* side of the equator from you.

Here the sun is on the other side of the equator from you, as it would be in the winter. Your latitude is equal to angle B. C is the declination of the sun and A (zenith distance) is equal to B plus C. Thus, latitude = zenith distance minus declination. Algebraically:

$$B = (B + C) - C$$
$$(B + C) = A$$
$$C = \text{declination}$$
$$B = A - \text{declination}$$

For the southern hemisphere equivalent, turn the drawing upside down.

(NOTE: This is the *strictly correct* way of visualizing the geometry of the noon sight. If you find it a bit puzzling, relax. When you actually start taking noon sights at sea, there's another way to visualize this. It's not technically correct, but many students find it easier, and it produces the same result—the correct latitude. See Chapter 7.)

QUIZ

1. You live at longitude 90°W. Assuming that the sun was on the Greenwich meridian on a Monday at a GMT of 12-00-00, what will the GMT be when the sun crosses your meridian? That is to say, what is the GMT of your noon?
2. As above, but you are at 91°W.
3. As above, but you are at 91°15'W.
4. As above, but you are at 91°16'W.
5. As above, but you are at 180° longitude.
6. As above, but you are at 90°E.

QUIZ ANSWERS

1. 18-00-00. It takes the sun six hours to cross 90° of longitude.
2. 18-04-00. Monday.
3. 18-05-00. Monday.
4. 18-05-04. Monday.
5. 24-00-00. (Midnight).
6. 06-00-00. The next day, Tuesday.

The Sextant

THE MARINE sextant is a precision instrument for measuring angles. In celestial navigation, it is used to measure the angle between the horizon, your eye and the sun. Graphically, the horizon is represented (in Figure 23) by the line tangent to the surface of the earth at your position. Note that this line is at right angles to the line between your location and the center of the earth. Figure 24 shows the angles between the horizon and the sun. *(Remember, the sun's rays are parallel).* Now, applying the geometric concepts of the previous chapter, we find the following to be true:

$$\text{Zenith Distance} = A = B = 90° - S$$

Therefore, the sextant allows us to find the zenith distance— the angle we need to combine with declination to find latitude. All that is necessary now is to learn to use a sextant.

WHAT THE SEXTANT DOES

In effect, your sextant measures the *distance* along a line of longitude between the GP and you. This distance is called the Zenith Distance and, since it is the length of an arc, is usually expressed in degrees (°) and minutes (').

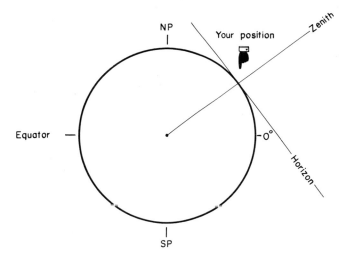

Figure 23. Horizon & zenith.

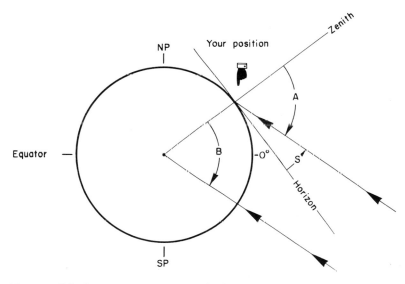

Figure 24. Sextant measures angle S.

In any case, I found that the least frustrating way of learning to use a sextant is on land. This cuts down on the number of things you have to do to adapt to the process of celestial navigation on a boat. By practicing on land, until the procedures and techniques of taking a sight become second nature. When you move aboard a boat, you only have to accommodate to one new factor—the motion of the boat.

When I first began teaching celestial navigation, I thought that learning to handle a sextant was the difficult part, but I soon discovered that most people get the hang of it rather quickly and, in many cases, are better than the teacher after a very short period of time.

What I suggest you do is get a sextant that seems suitable for your purpose. (I have found, by the way, that the plastic ones are perfectly practical.) Having done this, sit down some evening in a comfortable chair and read the pamphlet that comes with the sextant and become familiar with the names of its various parts. To help you orient yourself if you have not yet purchased a sextant, Figure 25 is a labeled illustration.

The index arm has a quick-release trigger at the bottom for making the arm move freely and also a "fine-tune knob" that moves the arm very slowly and almost imperceptibly.

Along the curved bottom part of the sextant is the main scale; each line represents 1°. One entire turn of the "fine-tune" knob (micrometer drum) is also 1°. This drum is divided into 60 parts, so each little line is one minute (1′).

As a beginning exercise, position the arm somewhere in the middle of the bottom arc, using the trigger, and try to determine the sextant reading (See Figure 26). Chances are you will not land right on an exact degree, but will fall in between two degree marks. The micrometer drum will tell you *exactly* where you are between the degree marks. Thus, in the picture shown in Figure 26, the sextant reads an angle of 29°42′. Turn the drum

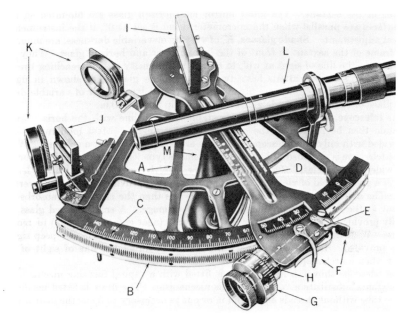

Figure 25

A : Frame
B : Grooves in which drum screw turns to move index arm (I) slowly
C : Arc
D : Index arm
E : Axis of the drum worm-screw
F : Quick-release trigger
G : Micrometer drum ("fine tune" knob)
H : Vernier scale (for reading tenths of a minute of arc)
 I : Index mirror
J : Horizon mirror
K : Shades
L : Telescope
M : Handle

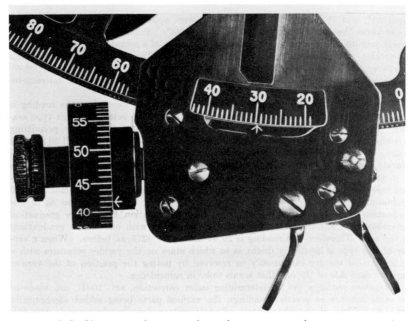

Figure 26. Close-up of main scale and micrometer drum on a typical marine sextant. Arrow on main scale shows sextant has measured an angle between 29° and 30°. Micrometer drum shows a reading of 42′. Thus, the sextant has measured an angle of 29°42′. Note: On many sextants, instead of the arrow on the micrometer drum you will find a zero (0) *and* additional graduations called a vernier scale. This scale enables you to read your sextant to 1/10 of a minute, a precison rarely needed in small-boat navigation.

through a full revolution, and you will see that when the drum reads 0, the index arm is on an exact, whole degree.

Now, sit down in a chair where you can face a window across the room. Set the index arm and the micrometer drum both to zero. Fold the shades away from the mirrors. Hold the sextant in your right hand by the handle and look through the telescope at the windowsill. Focus the scope if necessary. You will see the right-hand side of the sill reflected in the mirror in front of the scope and the left-hand side of the sill directly through the

unsilvered part of the mirror (horizon mirror) in front of the telescope. With your left hand, turn the micrometer drum this way and that and observe what happens.

Now, turn your watch around so that it is on the *inside* of your left wrist (or put it there if you wear it on your right wrist). Put the telescope back to your eye and turn the micrometer drum so that the sextant reading *increases*. Continue to do this until you see the top of the window reflected in the silvered side of the horizon mirror and in line with the sill. When you have lined them up exactly, look at your watch. Note the time. Read the sextant.

You have just measured the angle between the sill, your eye and the top of the window, and you have taken the time at which you measured this angle. In carrying out this exercise, you have just gone through *all* the actions of taking an actual celestial observation at sea. The reason for having the watch on the inside of your wrist is that it simplifies taking the time of the sight—just roll your palm outward a bit and you can see the face of your watch.

The next logical exercise with your sextant is to take it out in your backyard (or to a local park or beach) and actually observe the sun. Some afternoon or morning when the sun is visible from your yard, go out with your sextant, put your watch on the inside of your left wrist and pick out some horizontal line directly below the sun—a "practice" horizon. This can be the top edge of a fence, a neighbor's roof or a convenient windowsill. If you live in an apartment and have access to the roof, use the roof line of another apartment. For this exercise, it is not necessary that your reference line be at eye level; it can be considerably above or below.

The next step is to pick up the sun's image in the sextant. There are two ways of doing this. If you are good (or even fair) at judging the number of degrees in an angle, estimate the angle between the sun and your reference line (practice horizon) and

set the sextant to that angle. Then put the darkest shade down between the index mirror and the horizon mirror, hold up the sextant to your eye, and move it slowly from side to side and up and down until you pick up the sun. Through the telescope it will appear as a bright disk, the color of whatever shade you selected.

If you are not practiced at estimating angles, try this procedure: Make a fist with one hand and hold it out in front of you. Align the bottom edge of your fist with your reference line and estimate how many "fists" there are between there and the sun. A fist held at arm's length fills about 8°, so, if there is space for about five fists between your reference line and the sun, the angle is *about* 40°. Whatever you get, set this angle on your sextant and try, as above, to pick up the image of the sun.

If you are unable to get the sun with either method, then do this: Set your sextant on zero. Put down the darkest horizon-mirror shade, in addition to the darkest index-mirror shade, and aim the scope directly at the sun. With your left hand, release the index arm and move it forward as you move the sextant down. Once you get the image close to your reference line, flip away the horizon shade.

Now, use the micrometer drum to set the image of the sun on the reference line so that the lower edge of the sun's disk just touches the line (Figure 27). Look at your watch and jot down the time and sextant reading. I use a pocket-size pad for this and keep a pencil behind my left ear (this procedure works at sea, too).

Now, look through your telescope again. You will notice that in the time it took you to jot down the time and sextant reading, the sun has moved and is no longer resting on your reference line. With the micrometer drum, put the disc of the sun *back in position* so that the lower edge (called the "lower limb" by astronomers) is exactly touching. Again, note the time and reading.

Figure 27. What you see at the moment you take your sight. The sun sits exactly on the horizon. This is called the "moment of tangency."

If you compare the two readings, you will see that the angle of the sun has changed by only a few minutes of arc, and you should have a pretty good idea of how accurately your sextant is able to measure angles. As a further exercise, you might try lining everything up correctly and then determining how small a change in the sun's altitude you can detect. The results will probably amaze you. You'll find that you can detect a very small amount of change.

There is one more thing that you should be doing when you're practicing this way with your sextant. When you have the disk balanced on the reference line, rock the sextant slightly, clockwise and counter-clockwise, around the axis of the telescope. You will see the image of the sun rise off the reference line as though it were a ball at the bottom of a pendulum. You want to be sure that when the disk is on the line *it is also at the deepest part* of the pendulum's swing (Figure 28). This ensures that at the moment you take the time and make the sight, the sun is in a vertical line, and you are measuring the proper angle and not a slightly larger angle.

A further small, but important, factor you must consider is called "index error," and this is not your doing, but a built-in shortcoming of the sextant. To check for index error, set the sextant at 0°0′ and look through the telescope at the horizon, or at a horizontal line at least two miles away. You should see a

DETERMINING INDEX ERROR:

Micrometer drum in area 1'–10' means sextant reading that many minutes *too much,* so correction is *subtracted*—e.g. drum reads 4' when horizon line continuous means index correction is − 4' *(minus).*

If drum reads in area 59'–50' means *opposite*—i.e. say drum shows 56'. This means sextant reading 4' *too little,* so correction is + 4' *(plus).*

straight line in both sides of the horizon mirror. If you do, there is no index error. If the line is broken, turn the micrometer drum until you see a continuous straight line and read the drum. Assume that the drum shows 5'. This means that the sextant is reading an angle of 5' when it should be reading zero, and you will have to subtract 5' from the sextant reading. Such an error is said to be "on the arc" because it is to the left of the 0° mark.

If the drum had read 55', this would mean that the sextant showed an angle of − 5' (5' to the right of the 0° mark) and 5' would have to be *added* to the sextant reading after taking a sight.

One of the most difficult things to determine when you are learning celestial navigation on your own is how much progress

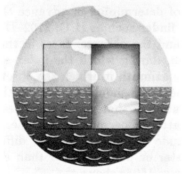

Figure 28. "Swinging the Arc." This is the action you take with your sextant (rotating the sextant, mentioned in the text) to determine the moment of tangency.

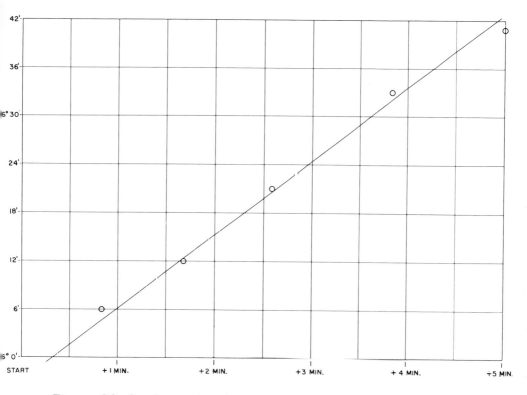

Figure 29. Graphing sights. As you practice, you will probably find that you can take sights at the rate of about one per minute. Using graph paper and a convenient scale—five minutes of time usually works—plot the sights at their respective times and use a ruler to draw in a line-of-best-fit by eye. You can then see if your sights are consistent. For practice on land your goal should be to have the sights within 1′ of the line. Close to noon the path of the sun becomes much more curved, so use this technique before 10 a.m. or after 2 p.m. local time. This technique is a graphical method of accomplishing what statisticians call a "linear regression."

you are making. After all, you probably can't stand alongside an experienced navigator and compare your sights with his; nor are there any fixed standards, as there are in some sports, in which you can compare yourself against a known performance.

When I was learning celestial navigation, I plotted my back-yard sights on graph paper. I would put a series of sights down and draw a straight line through them as nearly as I could. Later, when I actually started navigating from a boat at sea, I continued this practice and recommend it highly. You can tell how your sights are falling and you will get a "feel" for how consistent you are. After a while, you will find that your sights on land will fall very nearly on a straight line, and then even at sea in rough conditions.

QUIZ

1. After a practice session, you check for index error and find that when the horizon makes a continuous line in the mirror the micrometer drum says 6'. What is the index error (IE)? What is the index correction (IC)?
2. Same as above, except the micrometer drum says 54'.

QUIZ ANSWERS

1. IE = +6'. The drum is telling you that when the sextant should be reading 0°, it is showing an angle of 6'. Thus, the index correction (IC) will be −6'. This case is known as an index error "on the arc"—i.e. to the *left* of the 0° mark on the minor scale.
2. IE = −6'. Here the drum is telling you that the sextant is showing an angle of −6' instead of 0°. The correction (IC) is +6' and the error is called "off the arc"—i.e. to the *right* of the 0° mark on the main scale.

6

The DR

IT IS probably an indication of its supreme importance that the most useful factor in navigation has the least prepossessing name; in fact, no name at all: just the initials—DR. These are generally supposed to stand for "dead reckoning" position, a phrase further supposed to be a corruption of an earlier phrase, "deduced reckoning." Scholars of navigational history, however, cannot agree on even this much and I take it as proof positive that the concept is very ancient and that antiquity further proves its utility. The DR has not survived to this day for no reason at all, and the reason is this: navigation consists of two parts—the DR and the things a navigator does to check up on the DR. Contrary to the generally held impression, using bearings, ranges, celestial sights, RDF, Loran, radar, etc., is not navigating—it is one-half of navigating and probably not the most important half. Keeping your DR is the most important part of navigation.

Well, what is the DR? Very simply, *it is the position resulting from your course and distance from your last confirmed position.* And a confirmed position is one in which the DR and the celestial sights, radar bearings, Loran lines or whatever, reconcile themselves to the satisfaction of the navigator.

Put another way, the DR position is where your boat would

be if the course had been steered precisely and the distance run were known accurately. Of course, we know that both these factors (course and distance) are subject to error, but not as much as you might suppose, and not nearly enough to invalidate the concept. For instance, in the early efforts to chart the direction and velocity of the Gulf Stream, these two factors were taken to account for the difference between the DR and the celestial fixes, because a large ship can be steered very precisely and her speed is very accurately reflected by propeller RPM.

Even in a small boat, courses tend to be steered pretty well. Yes, boats do wander to one side or the other of the desired course, but the *average* is close to the planned compass bearing. Distances, too, tend to come out fairly well. The longer you're on a given boat, the more used to her you become, and the more occasions you have to *compare the DR with your noon fixes,* the better you will get at estimating the distance you've run in any period of time.

For the purposes of this book, the DR has two uses. First, it is your bench-mark against which to compare the positions given by the daily noon fixes to see if the difference between them—which there will always be—is reasonable and consistent with past experience. Second, a knowledge of your approximate longitude is very useful in predicting the time of noon. That knowledge will save needless sitting around on deck during the hottest part of the day.

To plot the DR position from your last confirmed position, you lay out your course and the distance run along that course.

"KEEP THE FAITH"

Keeping the DR is the most important part of navigation!

In the absence of celestial or other fixes, this is your position—don't think of it as a point, but as an area. When you get a noon fix, or simply a latitude, compare with the DR, and if it makes sense to you, begin a new DR track from the position of that fix. If it *doesn't* make sense, carefully rework the sight(s) or carry on with the DR until your next opportunity to take a noon sight. We'll have more to say about this as we now take up the three specific noon-sight techniques.

QUIZ

1. Below is a part of the passage record (log book) of a boat bound south from Morehead City, North Carolina. Noon sights were taken for latitude and longitude beginning at 1130 Tuesday. What was the DR (latitude & longitude) at 1200?

	Time	Course	Speed	Remarks
Monday	0700	205°	4	Depart Sea Buoy 34°32′N 76°40′W
	0900	205°	3	
	1100	205°	2	
	1300	205°	2	Cloudy—no noon sight
	1500	205°	1	
	1700	205°	2	Drizzle
	1900	205°	3	
	2100	205°	3	
	2300	205°	4	Ship crossed well ahead
Tuesday	0100	205°	5	
	0300	205°	5	Clearing
	0500	205°	6	Frying Pan Shoals Light abeam —stbd.
	0700	205°	4	
	0900	205°	4	Northbound ship hull down to port
	1100	205°	3	
	1300	205°	3	Noon Sight. 33°05′N 77°29′W

True courses are used in this example because if you don't have a chart, you may want to construct a blank one (plotting sheet). Ordinarily on small boats, courses are written down in magnetic because you are usually steering with a magnetic compass.

The approximate ratio between latitude and longitude at this latitude (35°, roughly) is 4:3, so a plotting sheet can be made by ruling out a grid of 4″ × 3″ rectangles. This makes 1″ equal 15 nautical miles.

QUIZ ANSWERS

1. The total of the speed column is 54. Since speeds were entered every two hours, the total distance logged was 108 NM. Because 1200 fell between two entries during which the speed was 3 knots, the distance run from 0700 Monday to 1200 Tuesday was approximately 105 NM.

    After constructing a plotting sheet and marking off 105 NM (remember the distance scale is the space between *latitude* lines) along a course of 205° True from the starting point, I come up with a noon DR of 33°01′ N, 77° 40′W.

    Admittedly, this is a bit crude, but it's the idea that counts. For a navigator, keeping the DR is keeping the faith.

    Working on a passage chart (scale 1:3,500,000), I get Latitude 33°07′N and Longitude 77°30′W. It can also be argued that an entry speed should not be made at the outset, since that tends to throw off the total; but since the speed of a sailboat varies a lot over the course of a passage, great precision is not possible in any case. If you happen to be on a boat that has an electronic "log" (odometer), you'll more than likely find that it doesn't record the distance run any closer than 10%, either. I just write down what the speedometer says at the time of making the notation and save the "fudging" for later. In small boat navigation you're always dealing with *useful approximations*. That's why it is so important to check one thing against another. If you've never heard it, the term "Naviguesser" is not a bad one.

7

The Almanac

THE TABLE referred to earlier, which gives the GP of the sun for each hour of Greenwich Mean Time, is called an almanac; and the most widely used is published by the U.S. Naval Observatory and is called the *Nautical Almanac*.

It is a curious fact that the sun's GP (geographical position, in case you've forgotten) recurs nearly exactly every four years. *So, if you have an almanac that is exactly four years out of date, you can use it to navigate by the noon sight.* Of course, nothing is exact, so the dated almanac only closely approximates the actual GP of the sun; but the approximation is close enough for practical navigation and, by application of a correction factor for each four-year interval of time, four almanacs will serve you for about *thirty years!*

Tables 5, 6, 7 and 8 at the back of this book are almanacs of the sun (tables of the GP of the sun for each hour of GMT) for the years 1976, 1977, 1978 and 1979 respectively. Since the sun will be over the same spot on earth at the same date every four years, this means that the 1976 table is good (once a small correction is made) for the years 1980, 1984, 1988, etc. The

THE NAUTICAL ALMANAC, BRIEFLY:

Nautical Almanac is a tabulation of the declination and Greenwich Hour Angle of the sun for *every* hour of *every* day of the year. Declinations and GHA's for times *between* the hours are found by interpolation.

1977 almanac will provide data of the sun's geographic position for the years 1981, 1985, 1989, etc. Similarly, the almanacs (Tables 7 and 8) for the years 1978 and 1979 will provide the GP's of the sun at four-year intervals, following their respective base years, once the appropriate correction is applied. The correction is given in Table 9 at three-day intervals corresponding to the way the almanac is laid out in groups of three days and is applied *once* for each four years that have elapsed since the year of the basic almanac. Thus, if you wanted to look up the Greenwich Hour Angle and declination of the sun at 2100 hours GMT on February 25, 1983, you would turn to Table 8 (base year 1979 plus 4 = 1983), look at 2100 on February 25 and find:

$$GHA = 131° 43.'2 \qquad Dec. = S 9° 04.'0$$

MARK UP THE ALMANAC

Mark the almanac entry you need with a pencil. Don't worry about your pencil marks. They improve your accuracy, and they won't get in your way because it's highly unlikely that you'll ever go back to the same line again.

CRAZY LITTLE SYMBOL

Although it looks like an exclamation point (!) throughout this book, *it's not*. It comes from the Nautical Almanac, in which GHA and Dec. are given to the nearest *tenth* of a minute ('). The tenths require a decimal point (.) and, *to save space,* the minute sign (') is put *over* the decimal point (.). Result: !

Turning to Table 9 you would find that in four years the GHA would have increased a little more than three tenths of a minute of arc (+.36') and the declination would have decreased six tenths of a minute of arc (−.60').

Applying these changes to the GHA and declination four years earlier would give:

Base year 1979 GHA	=	131°	43!2
Change in four years	=	+	.36
GHA 25 Feb. 1983	=	131°	43!56
	=	131°	43!6
Base year 1979 Dec.	=	S 9°	04!0
Change in four years	=	−	.60
Declination 25 Feb. 1981	=	S 9°	03!4

QUIZ

1. Find GHA and declination for GMT 13-00-00, April 28, 1982?
2. Same as above, but for 13-30-00.
3. Same as above, but for 13-36-00.
4. Same as above, but for 13-36-17.

QUIZ ANSWERS

1. From Table 7 (base year, 1978):

GHA at 13-00-00 =	15°37.'9	Dec =	N 14°08.'2
4-Year correction	+ .28		+ .34
(Table 9)			
Answer	15°38.'18		N 14°08.'54
Answer (rounded)	15°38.'2		N 14°08.'5
Answer (rounded again)	15°38'		N 14°09'

To me, the column to the right of the decimal point represents an opportunity for an arithmetic blunder, so I prefer to round to the *nearest minute,* as soon as possible. Remember also that minutes of arc total to 60 (not 100) to become a degree.

2. From Table 3:

30 minutes of time = 7°30' of GHA
Since GHA is always increasing with time, *add* 7°30' to answer(s) from question 1:
GHA at 13-00-00 = 15°38'
                30 m = +7°30'

Answer: GHA at 13-30-00 = 22°68' = 23°08'

The intermediate value for the declination is easy to find by eye. From Table 7:

Declination at 1300 – N 14°08.'2
Declination at 1400 = N 14°09.'0

Change in 1 hour = .'8 *increase*
Change in ½ hour = .'4 *increase*
So, *add* .'4 to answer(s) from question:
Question 1: 14°09' + .'4 = 14°09.'4

Answer: Declination at 13-30-00 = 14°09.'4

Answer (rounded): Declination at 13-30-00 = 14°09'

The declination of the sun never changes at a rate of more than 1' per hour. So, for practical navigation, if you forget this interpolation, it won't matter much.

3.   15°38'
                          + 9°00' (Table 3)
                    GHA = 24°38'   Declination = 14°09.'4

Declination is the same as before. Fussing about a few tenths of a
minute (') change is not worth the effort.
4. From Table 3:

17 seconds = 4.'25
 GHA at 13-00-00 h = 15°38'
          36-00 m = 9°00'
               17 s = 4.'25

GHA at 13-36-17 = 24°42'.25
GHA at 13-36-17 = 24°42'
Declination = N 14°09'

Part II

LATITUDE BY NOON SIGHT

8

Following the Sun Up

AS MENTIONED earlier, at noon the sun reaches its highest altitude above the horizon. The most direct way to determine this altitude is to start taking sights at some time *before* the sun crosses your longitude and "follow" or monitor the increasing altitude until it reaches its maximum. Using this maximum altitude, the distance between you and the sun's GP (the zenith distance) is determined, the latitude of the sun's GP (declination) is applied, and the result is your latitude.

Obviously, the thing you must avoid is "missing" noon by starting too late—i.e. at a time when the sun has already passed your longitude. One way to avoid this is to watch the compass and start taking sights when the sun approaches a bearing of *true* north or *true* south.

Another way is to determine the approximate GMT of noon. To do this for *west* longitudes, determine from the almanac the time when the GHA of the sun will be equal to your approximate longitude. In *east* longitude, noon occurs when 360° minus GHA is equal to your longitude. The sun moves westward 1° every four minutes, so get on deck at least 15 minutes before you estimate that it will be noon.

Then, since this process of "following" the sun takes time, make yourself as comfortable as possible. Sit on the cabin top

if you can and take sights at intervals until you see that the altitude is not changing much between sights. Carefully rock the sextant until the sun "sits" correctly on the horizon (check Figure 27 again). Continue to do this, and, as the sun creeps upward, you will notice an interval of time during which you will not be able to discern any change in altitude. Take your hand from the micrometer drum and continue to swing the arc until the sun's disc dips into the horizon. That means the sun has reached its maximum altitude (noon, by definition) and has started descending. Your sextant now reads the maximum altitude of the sun for that day.

Assume for the moment that the maximum altitude is 60°, and that it is December 11, 1979, between 2000 and 2100 hours GMT. Your DR has you at a position just a little north of 7° North latitude: 7°03′N.

Subtracting 60° from 90° gives a zenith distance of 30° between you and the latitude of the sun's GP. Going to Table 8 in the back of the book, you'll find that the declination (latitude of the GP) of the sun at 2000 hours GMT is south 22°59′9 and, at 2100 hours, is south 23°00′1. Simple inspection shows that the change in this one hour interval is a mere *two-tenths* of a minute, so that, for practical purposes, the declination at your noon is S 23°.

To figure your latitude from this data, look at Figure 30, which shows the arc (zenith distance) on the surface of the earth between you and the sun's GP.

The length of this arc is 30°, and the equator is *between* you and the sun. The declination of the sun and, therefore, of the arc between the equator and the sun, is 23°. Therefore, your latitude is 7°N.

A quick way of doing this problem without graphics is to see what you have to do with the numbers for zenith distance and declination to come out close to the latitude of your DR (as I said, the DR is *very* useful). In this case, it's obvious that the

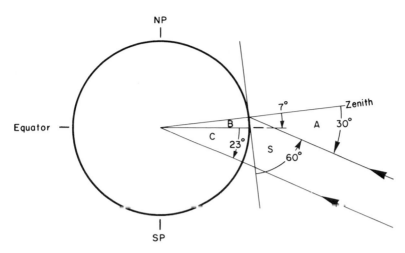

Figure 30. Example of finding latitude at noon:
- Sextant measures angle S between horizon and sun (60° in this example).
- Zenith distance (A) is, therefore, 30°.
- Zenith distance is equal to the sum of angles B (your latitude) and C (the declination of the sun).
- Since you can find the declination of the sun for any time of any day in the *Nautical Almanac,* you know the value of angle C (23° in this example).
- Because you also know (having indirectly measured it by subtracting your sextant angle from 90°) the totals of angles C and B, you can then find the value of angle B by subtracting declination (C) from zenith distance (A).
- Therefore: Latitude B = A − C.

declination (23°) has to be subtracted from the zenith distance (30°) to get 7°. No other combination makes sense.

I mentioned earlier there is another way to visualize the geometry of the noon sight. Here it is. Although not as technically correct as the other, it works. Imagine a right triangle in which one side is the curve of the earth (Figure 31). In this triangle, the sum of the angles is equal to 180°, and the arc representing the curve of the earth (zenith distance) is equal to

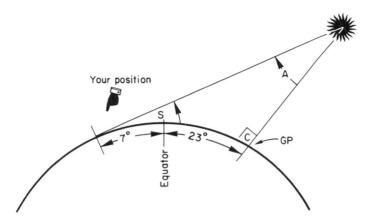

Figure 31. This drawing and Figures 32, 33, 34 and 35 show another way of visualizing the noon sight. This drawing shows the same situation as Figure 30 and is a specific example of the general situation shown in Figure 22.

Angle A. Since the sun is vertically over the GP, Angle C is 90°, and arc SC (representing the earth) is equal to 90°-S (the angle, or altitude, measured by the sextant). Note that Figure 32 portrays the same navigational situation as Figure 30, but in 32 we are using degrees of arc on the surface of the earth—the ultimately practical view—rather than the internal angles shown in Figure 30. Figures 32 through 35 show the four possible relationships of the sun and you at noon.

Up to this point, our discussion of the use of the sextant has considered only *one* correction to the initial reading taken off the instrument—index error—which is caused by the sextant itself. There are, however, several other factors which affect the angle observed by the sextant, and which must be eliminated so that the angle you use to compute latitude is *the* angle required by the theory we have been discussing. The theory is pure geometry, but the initial angle (called the sextant altitude) you measure with the sextant differs from real geometry for these reasons:

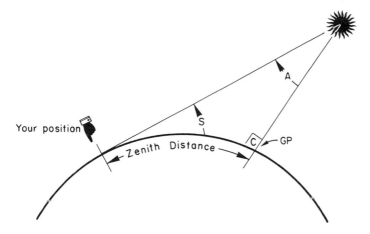

Figure 32. This is *another way* of representing the general picture of the noon sight. It is not technically correct, but produces the same result and many of my students have found it easier to visualize *this way*.

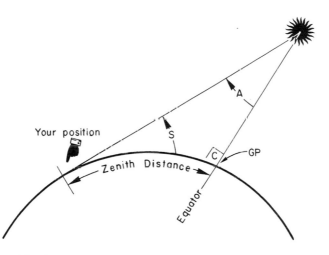

Figure 33. This is *another way* of representing the situation shown in Figure 19. Only when the sun is over the equator (i.e. declination = 0°) does your zenith distance equal your latitude.

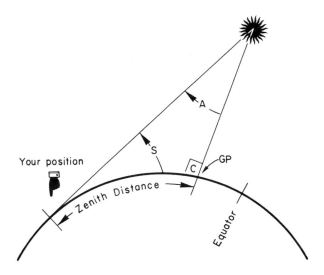

Figure 34. Another way of representing the situation which Figure 20 depicts in a geometrically correct manner.

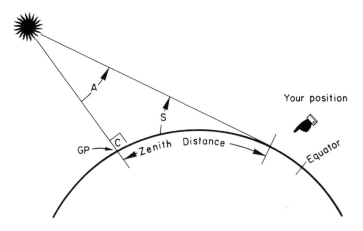

Figure 35. Another way of showing the situation which Figure 21 sets out in a more technically correct way.

First, in using the sea horizon as reference, your eye is not looking along an exact horizontal; and, so, your sextant is not measuring an angle from a geometrically true horizontal line. Because your eye is usually some distance *above* the surface of the water, the angle you measure from the horizon to the sun is *slightly greater* than the angle that would be measured from the true horizontal plane (Figure 36). A small number of minutes (′) must, therefore, always be subtracted from the raw sextant reading to correct it to the "*true* geometric horizon." The amount of this correction varies with the height of your eye above the surface of the sea, and is given in the table shown in Figure 37. This adjustment to obtain the *actual* angle measured from the "true geometric horizon" is known as "correction for "Dip" (of the horizon).

Figure 36. DIP. The angle formed between the sea horizon (seen through your sextant) and the sun is *greater* than the angle you need to get the correct number (angle of sun above *true* horizon). This is a small angle, usually only a few minutes of arc—see drawing above. It is called "Dip" and is always subtracted from your initial sextant reading.

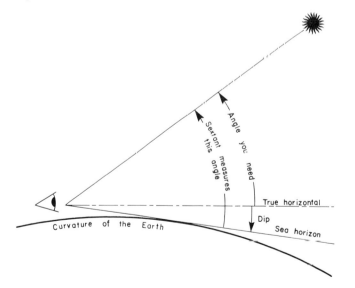

Ht. of Eye m	Corrⁿ	Ht. of Eye ft.	Ht. of Eye m	Corrⁿ	Ht. of Eye ft.
2·4		8·0	9·5		31·5
	−2·8			−5·5	
2·6		8·6	9·9		32·7
	−2·9			−5·6	
2·8		9·2	10·3		33·9
	−3·0			−5·7	
3·0		9·8	10·6		35·1
	−3·1			−5·8	
3·2		10·5	11·0		36·3
	−3·2			−5·9	
3·4		11·2	11·4		37·6
	−3·3			−6·0	
3·6		11·9	11·8		38·9
	−3·4			−6·1	
3·8		12·6	12·2		40·1
	−3·5			−6·2	
4·0		13·3	12·6		41·5
	−3·6			−6·3	
4·3		14·1	13·0		42·8
	−3·7			−6·4	
4·5		14·9	13·4		44·2
	−3·8			−6·5	
4·7		15·7	13·8		45·5
	−3·9			−6·6	
5·0		16·5	14·2		46·9
	−4·0			−6·7	
5·2		17·4	14·7		48·4
	−4·1			−6·8	
5·5		18·3	15·1		49·8
	−4·2			−6·9	
5·8		19·1	15·5		51·3
	−4·3			−7·0	
6·1		20·1	16·0		52·8
	−4·4			−7·1	
6·3		21·0	16·5		54·3
	−4·5			−7·2	
6·6		22·0	16·9		55·8
	−4·6			−7·3	
6·9		22·9	17·4		57·4
	−4·7			−7·4	
7·2		23·9	17·9		58·9
	−4·8			−7·5	
7·5		24·9	18·4		60·5
	−4·9			−7·6	
7·9		26·0	18·8		62·1
	−5·0			−7·7	
8·2		27·1	19·3		63·8
	−5·1			−7·8	
8·5		28·1	19·8		65·4
	−5·2			−7·9	
8·8		29·2	20·4		67·1
	−5·3			−8·0	
9·2		30·4	20·9		68·8
	−5·4			−8·1	
			21·4		70·5

Figure 37. Dip table. This table is extracted from the full table given in the *Nautical Almanac*. It covers heights of eye (above the water) from 8 feet to 70.5 feet. Height of eye in meters (m.) is shown in the left column. Height of eye in feet (ft.) is shown in the right column. Dip correction (*always* subtracted from sextant reading) is shown in the middle column. Note that the dip correction value always lies *between* two heights in either left or right column. As an example, for a height of eye from 10.5 ft. *through* 11.2 ft., the dip correction is 3.2 (This type of table is called a "critical" table).

DIP: A RULE OF THUMB:

For most cruising boats (height of eye from 6 feet to 12 feet) DIP is − 3'. (This correction is *always minus*.)

Second, the geometry of the noon sight ignores the fact that rays of light are bent, or *refracted,* by their passage through the atmosphere. This phenomenon makes the sun appear higher in the sky than it is (Figure 38), and so, another subtraction is required to bring what the sextant measures into line with the true geometry. The amount that a ray of light is bent by the air varies from zero when the sun is directly overhead (sextant angle or altitude of 90°), to a maximum of 34.5 at sunrise or

Figure 38. Refraction (bending) of the sun's rays as they pass through the atmosphere. S is the angle measured by your sextant, but the angle you get is actually greater than the true angle, S minus R, which is the sextant angle *minus* the amount of refraction of the sun's rays.

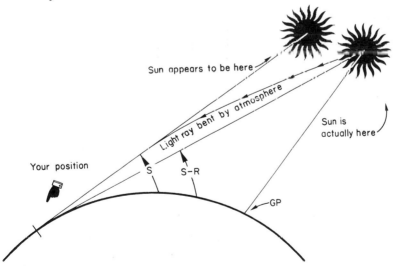

Sun appears to be here

Light ray bent by atmosphere

Sun is actually here

Your position

S

S−R

GP

sunset (altitude of 0°). The refraction table, which gives corrections for various sextant altitudes is shown in Figure 39.

The third correction needed arises from the fact that you have measured the angle between the horizon and the *lower* edge of the sun, instead of the angle between the horizon and the *center* of the sun, which is the actual geometric angle you're after.

Figure 39. Refraction table. Like the dip table, this is also a "critical table." The correction found in the right-hand column is *always subtracted* from the sun's angle as measured by your sextant.

App. Alt.	Corr^n	App. Alt.	Corr^n
9 56		19 58	
	−5.3		−2.6
10 08		20 42	
	−5.2		−2.5
10 20		21 28	
	−5.1		−2.4
10 33		22 19	
	−5.0		−2.3
10 46		23 13	
	−4.9		2.2
11 00		24 11	
	−4.8		−2.1
11 14		25 14	
	−4.7		−2.0
11 29		26 22	
	−4.6		1.9
11 45		27 36	1.8
	−4.5	28 56	
12 01			−1.7
	−4.4	30 24	−1.6
12 18		32 00	
	−4.3		1.5
12 35		33 45	1.4
	−4.2	35 40	
12 54			1.3
	−4.1	37 48	1.2
13 13		40 08	
	−4.0		1.1
13 33		42 44	
	−3.9		−1.0
13 54		45 36	
	−3.8		0.9
14 16		48 47	
	−3.7		−0.8
14 40		52 18	
	−3.6		−0.7
15 04		56 11	
	−3.5		−0.6
15 30		60 28	
	−3.4		−0.5
15 57		65 08	
	−3.3		−0.4
16 26		70 11	
	−3.2		−0.3
16 56		75 34	
	3.1		−0.2
17 28		81 13	
	3.0		−0.1
18 02		87 03	0.0
	−2.9	90 00	
18 38			
	−2.8		
19 17			
	−2.7		

App. Alt. = Apparent altitude = Sextant altitude corrected for index error and dip.

REFRACTION REMINDER:

Refraction, like dip, is *always subtracted* from the raw sextant reading (Hs).

Since the sun has an apparent diameter of 32', your sextant reading is 16' too small, so 16' should be *added* to the sextant reading you get. (See Figure 40). If you have difficulty visualizing why you need this correction, imagine that you were going to measure to the *middle* of the sun. To do that, you would have to bring the sun down farther by turning the micrometer drum more; and you would, therefore, increase the angle measured. Since it is much easier visually—and, therefore, more accurate—to bring the *edge* of the sun down to the horizon, it is preferable to make the correction afterwards, instead of trying to measure to the middle of the sun. This correction is called the semi-diameter (one half the diameter) correction.

Figure 40. Correction for semi-diameter of the sun. You want to measure the angle between the horizon and the *center* of the sun. To do this you would have to *estimate* the center of the sun. Instead of *guessing* where the sun's center is, you use a point you can actually see—the bottom edge of its disk. This is the most accurate way to measure, but leaves you with an angle that is *greater* than the true angle by *half* of the sun's diameter, so you *subtract* that amount, or 16 minutes.

Actual   Required

Looked at from the standpoint of finding the true zenith distance between you and the GP, Figure 41 shows that in measuring to the lower edge (limb) of the sun, you are measuring the distance to the point directly under the *back edge* of the sun. What you want is the distance to the GP as given in the almanac, which is the point on the surface of the earth directly beneath the *center* of the sun.

Now, it occasionally happens, due to clouds, that you have to bring the *upper* edge (limb) of the sun's disc down to the horizon. In that case, you have measured *too great an altitude* (which would result in a zenith distance too small by 16′), so 16′ has to be *subtracted* from an altitude of a sight taken of the upper limb of the sun.

This stuff may seem a little bewildering at this point, but it is easy enough to keep under control by establishing a routine form. Let's say you have taken a noon sight and found the

Figure 41. Semi-diameter correction as seen from the standpoint of zenith distance rather than sextant altitude.

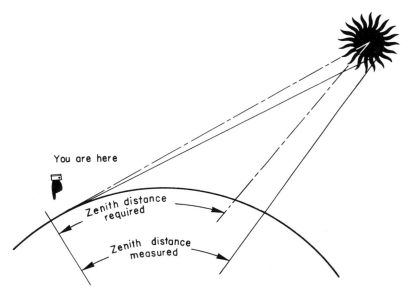

maximum altitude to be 67°36'. You estimate that your eye was about 8 feet above the water. Your Index Error was 3' on the arc. Here's a suggested form for these corrections:

Sextant Reading	:	67°36'
Index Correction $\begin{array}{l}+\text{ off}\\ -\text{ on}\end{array}$	:	− 3'
Difference (for eye height of 8') always −	:	− 2.7'
		67°30.'3
Refraction (always −)	:	− 0.'4
Semi-diameter $\begin{array}{l}\circ\;+\\ \circ\;-\end{array}$	:	+ 16.'0
True (geometric) altitude		67°45.'9

It is conventional to make the corrections in this order because the Table for the Refraction Correction is based on what is called the *apparent altitude.* This is defined as the sextant altitude (raw sextant reading), *net* of the corrections for index error and dip.

Since celestial navigation is an old art heavy with traditions, there are some further conventional terms (jargon) and symbols which I am going to set down here and which you can use or ignore *at absolutely no peril to yourself or your navigation.* Using the same data as before, then, here are the conventional terms and symbols (in parentheses).

Sextant Altitude (Hs)	:	67°36'
IC $\begin{array}{l}+\text{ off}\\ -\text{ on}\end{array}$	:	− 3'
Dip	:	− 2.7'
Apparent Altitude (ha)	:	67°30.'3
Refraction (R)	:	− 0.'4
Semi-diameter (SD)	:	+ 16.'0
Observed Altitude (Ho)	:	67°45.'9

The observed altitude is what you use to figure out the zenith distance. In this instance:

Zenith	: 89°60'	(90° written so that minutes can be subtracted)
Observed Altitude :	<u>67°45'9</u>	
Zenith Distance (z) :	22°14'1	

LATITUDE CHECKLIST

1. Sextant Reading (Hs)
   ± Index Correction
   − Dip
   _____
   Apparent Altitude (ha)
   + Semi-Diam: $\underline{o}$
   − Semi-Diam: $\overline{o}$
   − Refraction
   _____
   Observed (True) Altitude (Ho)
2. 89°60'
   − Ho
   _____
   Zenith Distance
3. Zenith Distance
   Declination
   _____
   Your Latitude

9

"But Clouds Got in the Way . . ." What to Do if You Miss Noon

SINCE CELESTIAL NAVIGATION is an ancient and widely practiced art, it's axiomatic that if you encounter a difficulty or are faced with repetitious calculations, somebody else in the past did, too, and, more than likely, devised a table to deal with such difficulties. A good case in point is the problem of clouds blocking the sun during the critical period when it reaches its maximum altitude.

A further axiom of navigation is that if a table exists, it will be found in *The American Practical Navigator,* a two-volume, 2000-page publication of the United States Defense Mapping Agency. The book is commonly referred to as "Bowditch," after its originator, the Salem shipmaster Nathaniel Bowditch. Bowditch Tables 29 and 30, which follow, are used for dealing with clouds covering the sun as you are trying to take a noon sight for latitude. It is a mathematical method of calculating from an altitude taken within one-half hour of noon, the maximum altitude that the sun *would* have reached—i.e., the altitude you would have measured if the clouds had not gotten in the way.

Obviously, it can also be used on a clear day if, for some reason or another, you miss the peak and find the sun's altitude already decreasing by the time you begin taking sights.

To find the missed maximum altitude, the first thing you need to know is the rate of change of the sun's altitude. This is given in Bowditch Table 29, which is divided into two sections, each covering a range of declinations of 0° to 24°, and range of latitudes from 0° to 60°. Which section you should use depends upon whether your position and the GP of the sun are on the *same* side of the equator, or on *opposite* sides. If you and the sun are both in north latitudes or both in south latitudes, then latitude and declination are said to have the *"same name."* But, if your latitude is north and the sun's declination is south or vice versa, latitude and declination are said to be *"contrary."* (*Same* and *Contrary* are noted in bold-face type above the columns.)

Let's say you are taking a noon sight on a day when the sun's declination is 23°N and your DR has you close to latitude 37°N. A cloud comes along and you miss the sun's highest altitude. You look at your watch and your best estimate, at the time you get your next sextant shot of the sun, and confirm that it is descending, is that the time is about 10 minutes *after* the sun passed your meridian.

Look at Table 29. At the point where the column for 23° (under the heading "Declination *Same* Name As Latitude"), and the row for latitude 37° intersect, you will see the number 6″0 (six seconds of arc).

Now, turn to Table 30 and look down the column at the left margin until you find 6″0. Then, look to the right under the column headed 10^m00^s (10 minutes 00 seconds) and you will find 10′0 (10 minutes of *arc*). Add this amount to the altitude you observed (Ho) ten minutes after noon, when the sun emerged from the clouds.

Using the same numbers for latitude (37°) and declination (23°), but taking an example in which they are of *contrary*

name, you find a rate of change (Table 29, Contrary Sections) of 1″.7 and a correction of 2′.8. To reach this correction, it is necessary to interpolate. Table 30 gives a value of 1′.7 for a rate of change of 1″.0, and 3′.3 for a rate of change of 2″.0. Our rate of change lies between the two.

Thus:	1″.0	=	1′.7
	2″.0	=	3′.3
	Difference	=	1′.6
	7/10 of Difference	=	1′.12
	Plus value for		
	rate of 1″.0	=	1′.7
	Correction	=	2′.8

Obviously, the correction is *always* added to your observed altitude, since if you miss noon, your observed altitude must be less than the altitude at noon.

Even that damned cloud did not stop you from getting an accurate Noon Position!

TABLE 29

Lati-tude	Declination same name as latitude, upper transit: add correction to observed altitude											
	0°	1°	2°	3°	4°	5°	6°	7°	8°	9°	10°	11°
°	"	"	"	"	"	"	"	"	"	"	"	"
0					28.1	22.4	18.7	16.0	14.0	12.4	11.1	10.1
1						28.0	22.4	18.6	16.0	13.9	12.4	11.1
2							28.0	22.3	18.6	15.9	13.9	12.3
3								27.9	22.3	18.5	15.8	13.8
4	28.1								27.8	22.2	18.5	15.8
5	22.4	28.0								27.7	22.1	18.4
6	18.7	22.4	28.0								27.6	22.0
7	16.0	18.6	22.3	27.9								27.4
8	14.0	16.0	18.6	22.3	27.8							
9	12.4	13.9	15.9	18.5	22.2	27.7						
10	11.1	12.4	13.9	15.8	18.5	22.1	27.6					
11	10.1	11.1	12.3	13.8	15.8	18.4	22.0	27.4				
12	9.2	10.1	11.1	12.3	13.8	15.7	18.3	21.9	27.3			
13	8.5	9.2	10.0	11.0	12.2	13.7	15.6	18.2	21.7	27.1		
14	7.9	8.5	9.2	10.0	10.9	12.1	13.6	15.5	18.0	21.6	26.9	
15	7.3	7.8	8.4	9.1	9.9	10.9	12.1	13.5	15.4	17.9	21.4	26.7
16	6.8	7.3	7.8	8.4	9.1	9.8	10.8	12.0	13.4	15.3	17.8	21.3
17	6.4	6.8	7.2	7.8	8.3	9.0	9.8	10.7	11.9	13.3	15.2	17.6
18	6.0	6.4	6.8	7.2	7.7	8.3	8.9	9.7	10.6	11.8	13.2	15.0
19	5.7	6.0	6.3	6.7	7.2	7.6	8.2	8.9	9.6	10.6	11.7	13.1
20	5.4	5.7	6.0	6.3	6.7	7.1	7.6	8.1	8.8	9.5	10.5	11.6
21	5.1	5.4	5.6	5.9	6.3	6.6	7.0	7.5	8.1	8.7	9.5	10.4
22	4.9	5.1	5.3	5.6	5.9	6.2	6.6	7.0	7.5	8.0	8.6	9.4
23	4.6	4.8	5.0	5.3	5.5	5.8	6.1	6.5	6.9	7.4	7.9	8.5
24	4.4	4.6	4.8	5.0	5.2	5.5	5.8	6.1	6.4	6.8	7.3	7.8
25	4.2	4.4	4.6	4.7	5.0	5.2	5.4	5.7	6.0	6.4	6.8	7.2
26	4.0	4.2	4.3	4.5	4.7	4.9	5.1	5.4	5.7	6.0	6.3	6.7
27	3.9	4.0	4.1	4.3	4.5	4.7	4.9	5.1	5.3	5.6	5.9	6.2
28	3.7	3.8	4.0	4.1	4.3	4.4	4.6	4.8	5.0	5.3	5.5	5.8
29	3.5	3.7	3.8	3.9	4.1	4.2	4.4	4.6	4.7	5.0	5.2	5.5
30	3.4	3.5	3.6	3.7	3.9	4.0	4.2	4.3	4.5	4.7	4.9	5.1
31	3.3	3.4	3.5	3.6	3.7	3.8	4.0	4.1	4.4	4.4	4.6	4.8
32	3.1	3.2	3.3	3.4	3.5	3.7	3.8	3.9	4.1	4.2	4.4	4.6
33	3.0	3.1	3.2	3.3	3.4	3.5	3.6	3.7	3.9	4.0	4.2	4.3
34	2.9	3.0	3.1	3.2	3.2	3.3	3.4	3.6	3.7	3.8	3.9	4.1
35	2.8	2.9	3.0	3.0	3.1	3.2	3.3	3.4	3.5	3.6	3.7	3.9
36	2.7	2.8	2.8	2.9	3.0	3.1	3.2	3.3	3.4	3.5	3.6	3.7
37	2.6	2.7	2.7	2.8	2.9	2.9	3.0	3.1	3.2	3.3	3.4	3.5
38	2.5	2.6	2.6	2.7	2.8	2.8	2.9	3.0	3.0	3.2	3.2	3.3
39	2.4	2.5	2.5	2.6	2.7	2.7	2.8	2.9	2.9	3.0	3.1	3.2
40	2.3	2.4	2.4	2.5	2.6	2.6	2.7	2.7	2.8	2.9	3.0	3.0
41	2.3	2.3	2.4	2.4	2.5	2.5	2.6	2.6	2.7	2.8	2.8	2.9
42	2.2	2.2	2.3	2.3	2.4	2.4	2.5	2.5	2.6	2.6	2.7	2.8
43	2.1	2.1	2.2	2.2	2.3	2.3	2.4	2.4	2.5	2.5	2.6	2.7
44	2.0	2.1	2.1	2.1	2.2	2.2	2.3	2.3	2.4	2.4	2.5	2.5
45	2.0	2.0	2.0	2.1	2.1	2.2	2.2	2.2	2.3	2.3	2.4	2.4
46	1.9	1.9	2.0	2.0	2.0	2.1	2.1	2.2	2.2	2.2	2.3	2.3
47	1.8	1.9	1.9	1.9	2.0	2.0	2.0	2.1	2.1	2.1	2.2	2.2
48	1.8	1.8	1.8	1.9	1.9	1.9	1.9	2.0	2.0	2.0	2.1	2.1
49	1.7	1.7	1.8	1.8	1.8	1.8	1.9	1.9	1.9	2.0	2.0	2.1
50	1.6	1.7	1.7	1.7	1.8	1.8	1.8	1.8	1.9	1.9	1.9	2.0
51	1.6	1.6	1.6	1.7	1.7	1.7	1.7	1.8	1.8	1.8	1.9	1.9
52	1.5	1.6	1.6	1.6	1.6	1.6	1.6	1.7	1.7	1.7	1.8	1.8
53	1.5	1.5	1.5	1.5	1.6	1.6	1.6	1.6	1.7	1.7	1.7	1.7
54	1.4	1.4	1.5	1.5	1.5	1.5	1.5	1.6	1.6	1.6	1.6	1.7
55	1.4	1.4	1.4	1.4	1.5	1.5	1.5	1.5	1.5	1.6	1.6	1.6
56	1.3	1.3	1.4	1.4	1.4	1.4	1.4	1.4	1.5	1.5	1.5	1.5
57	1.3	1.3	1.3	1.3	1.3	1.4	1.4	1.4	1.4	1.4	1.4	1.5
58	1.2	1.2	1.3	1.3	1.3	1.3	1.3	1.3	1.3	1.4	1.4	1.4
59	1.2	1.2	1.2	1.2	1.2	1.3	1.3	1.3	1.3	1.3	1.3	1.3
60	1.1	1.1	1.2	1.2	1.2	1.2	1.2	1.2	1.2	1.2	1.2	1.3

TABLE 29

Latitude	12°	13°	14°	15°	16°	17°	18°	19°	20°	21°	22°	23°	24°
°	"	"	"	"	"	"	"	"	"	"	"	"	"
0	9.2	8.5	7.9	7.3	6.8	6.4	6.0	5.7	5.4	5.1	4.9	4.6	4.4
1	10.1	9.2	8.5	7.8	7.3	6.8	6.4	6.0	5.7	5.4	5.1	4.8	4.6
2	11.1	10.0	9.2	8.4	7.8	7.2	6.8	6.3	6.0	5.6	5.3	5.0	4.8
3	12.3	11.0	10.0	9.1	8.4	7.8	7.2	6.7	6.3	5.9	5.6	5.3	5.0
4	13.8	12.2	10.9	9.9	9.1	8.3	7.7	7.2	6.7	6.3	5.9	5.5	5.2
5	15.7	13.7	12.1	10.9	9.8	9.0	8.3	7.6	7.1	6.6	6.2	5.8	5.5
6	18.3	15.6	13.6	12.1	10.8	9.8	8.9	8.2	7.6	7.0	6.6	6.1	5.8
7	21.9	18.2	15.5	13.5	12.0	10.7	9.7	8.9	8.1	7.5	7.0	6.5	6.1
8	27.3	21.7	18.0	15.4	13.4	11.9	10.6	9.6	8.8	8.1	7.5	6.9	6.4
9		27.1	21.6	17.9	15.3	13.3	11.8	10.6	9.5	8.7	8.0	7.4	6.8
10			26.9	21.4	17.8	15.2	13.2	11.7	10.5	9.5	8.6	7.9	7.3
11				26.7	21.3	17.6	15.0	13.1	11.6	10.4	9.4	8.5	7.8
12					26.5	21.1	17.5	14.9	13.0	11.5	10.3	9.3	8.4
13						26.2	20.9	17.3	14.8	12.8	11.3	10.1	9.2
14							26.0	20.7	17.1	14.6	12.7	11.2	10.0
15								25.7	20.4	16.9	14.4	12.5	11.1
16	26.5								25.4	20.2	16.7	14.3	12.4
17	21.1	26.2								25.1	20.0	16.5	14.1
18	17.5	20.9	26.0								24.8	19.7	16.3
19	14.9	17.3	20.7	25.7								24.5	19.5
20	13.0	14.8	17.1	20.4	25.4								24.2
21	11.5	12.8	14.6	16.9	20.2	25.1							
22	10.3	11.3	12.7	14.4	16.7	20.0	24.8						
23	9.3	10.1	11.2	12.5	14.3	16.5	19.7	24.5					
24	8.4	9.2	10.0	11.1	12.4	14.1	16.3	19.5	24.2				
25	7.7	8.3	9.0	9.9	10.9	12.2	13.9	16.1	19.2	23.8			
26	7.1	7.6	8.2	8.9	9.8	10.8	12.1	13.7	15.9	18.9	23.5		
27	6.6	7.0	7.5	8.1	8.8	9.6	10.6	11.9	13.5	15.6	18.6	23.1	
28	6.2	6.5	7.0	7.4	8.0	8.7	9.5	10.5	11.7	13.3	15.4	18.3	22.7
29	5.7	6.1	6.4	6.9	7.3	7.9	8.6	9.4	10.3	11.5	13.1	15.1	18.0
30	5.4	5.7	6.0	6.4	6.8	7.2	7.8	8.4	9.2	10.1	11.3	12.8	14.9
31	5.1	5.3	5.6	5.9	6.3	6.7	7.1	7.7	8.3	9.0	10.0	11.1	12.6
32	4.8	5.0	5.2	5.5	5.8	6.2	6.5	7.0	7.5	8.1	8.9	9.8	10.9
33	4.5	4.7	4.9	5.1	5.4	5.7	6.1	6.4	6.9	7.4	8.0	8.7	9.6
34	4.3	4.4	4.6	4.8	5.1	5.3	5.6	5.9	6.3	6.8	7.3	7.8	8.6
35	4.0	4.2	4.4	4.5	4.7	5.0	5.2	5.5	5.8	6.2	6.6	7.1	7.7
36	3.8	4.0	4.1	4.3	4.5	4.7	4.9	5.1	5.4	5.7	6.1	6.5	7.0
37	3.6	3.8	3.9	4.0	4.2	4.4	4.6	4.8	5.0	5.3	5.6	6.0	6.4
38	3.4	3.6	3.7	3.8	4.0	4.1	4.3	4.5	4.7	4.9	5.2	5.5	5.8
39	3.3	3.4	3.5	3.6	3.8	3.9	4.0	4.2	4.4	4.6	4.8	5.1	5.4
40	3.1	3.2	3.3	3.4	3.6	3.7	3.8	4.0	4.1	4.3	4.5	4.7	5.0
41	3.0	3.1	3.2	3.3	3.4	3.5	3.6	3.7	3.9	4.0	4.2	4.4	4.6
42	2.9	2.9	3.0	3.1	3.2	3.3	3.4	3.5	3.7	3.8	4.0	4.1	4.3
43	2.7	2.8	2.9	3.0	3.0	3.1	3.2	3.3	3.5	3.6	3.7	3.9	4.0
44	2.6	2.7	2.7	2.8	2.9	3.0	3.1	3.2	3.3	3.4	3.5	3.6	3.8
45	2.5	2.6	2.6	2.7	2.8	2.8	2.9	3.0	3.1	3.2	3.3	3.4	3.5
46	2.4	2.4	2.5	2.6	2.6	2.7	2.8	2.8	2.9	3.0	3.1	3.2	3.3
47	2.3	2.3	2.4	2.4	2.5	2.6	2.6	2.7	2.8	2.9	2.9	3.0	3.1
48	2.2	2.2	2.3	2.3	2.4	2.4	2.5	2.6	2.6	2.7	2.8	2.9	3.0
49	2.1	2.1	2.2	2.2	2.3	2.3	2.4	2.4	2.5	2.6	2.6	2.7	2.8
50	2.0	2.0	2.1	2.1	2.2	2.2	2.3	2.3	2.4	2.4	2.5	2.6	2.6
51	1.9	2.0	2.0	2.0	2.1	2.1	2.2	2.2	2.3	2.3	2.4	2.4	2.5
52	1.8	1.9	1.9	1.9	2.0	2.0	2.1	2.1	2.1	2.2	2.2	2.3	2.4
53	1.8	1.8	1.8	1.9	1.9	1.9	2.0	2.0	2.0	2.1	2.1	2.2	2.2
54	1.7	1.7	1.7	1.8	1.8	1.8	1.9	1.9	1.9	2.0	2.0	2.1	2.1
55	1.6	1.6	1.7	1.7	1.7	1.8	1.8	1.8	1.9	1.9	1.9	2.0	2.0
56	1.5	1.6	1.6	1.6	1.6	1.7	1.7	1.7	1.8	1.8	1.8	1.9	1.9
57	1.5	1.5	1.5	1.5	1.6	1.6	1.6	1.6	1.7	1.7	1.7	1.8	1.8
58	1.4	1.4	1.5	1.5	1.5	1.5	1.5	1.6	1.6	1.6	1.6	1.7	1.7
59	1.4	1.4	1.4	1.4	1.4	1.5	1.5	1.5	1.5	1.5	1.6	1.6	1.6
60	1.3	1.3	1.3	1.3	1.4	1.4	1.4	1.4	1.4	1.5	1.5	1.5	1.5
Latitude	12°	13°	14°	15°	16°	17°	18°	19°	20°	21°	22°	23°	24°

TABLE 29

Lati- tude	Declination **contrary** name to latitude, **upper** transit: **add** correction to observed altitude											
	0°	1°	2°	3°	4°	5°	6°	7°	8°	9°	10°	11°
°	ʺ	ʺ	ʺ	ʺ	ʺ	ʺ	ʺ	ʺ	ʺ	ʺ	ʺ	ʺ
0					28. 1	22. 4	18. 7	16. 0	14. 0	12. 4	11. 1	10. 1
1				28. 1	22. 4	18. 7	16. 0	14. 0	12. 4	11. 2	10. 1	9. 3
2			28. 1	22. 4	18. 7	16. 0	14. 0	12. 5	11. 2	10. 2	9. 3	8. 6
3		28. 1	22. 4	18. 7	16. 0	14. 0	12. 5	11. 2	10. 2	9. 3	8. 6	8. 0
4	28. 1	22. 4	18. 7	16. 0	14. 0	12. 5	11. 2	10. 2	9. 3	8. 6	8. 0	7. 4
5	22. 4	18. 7	16. 0	14. 0	12. 5	11. 2	10. 2	9. 3	8. 6	8. 0	7. 4	7. 0
6	18. 7	16. 0	14. 0	12. 5	11. 2	10. 2	9. 3	8. 6	8. 0	7. 5	7. 0	6. 6
7	16. 0	14. 0	12. 4	11. 2	10. 2	9. 3	8. 6	8. 0	7. 5	7. 0	6. 6	6. 2
8	14. 0	12. 4	11. 2	10. 2	9. 3	8. 6	8. 0	7. 5	7. 0	6. 6	6. 2	5. 9
9	12. 4	11. 2	10. 2	9. 3	8. 6	8. 0	7. 5	7. 0	6. 6	6. 2	5. 9	5. 6
10	11. 1	10. 1	9. 3	8. 6	8. 0	7. 4	7. 0	6. 6	6. 2	5. 9	5. 6	5. 3
11	10. 1	9. 3	8. 6	8. 0	7. 4	7. 0	6. 6	6. 2	5. 9	5. 6	5. 3	5. 1
12	9. 2	8. 5	7. 9	7. 4	7. 0	6. 5	6. 2	5. 9	5. 6	5. 3	5. 0	4. 8
13	8. 5	7. 9	7. 4	6. 9	6. 5	6. 2	5. 8	5. 6	5. 3	5. 0	4. 8	4. 6
14	7. 9	7. 4	6. 9	6. 5	6. 2	5. 8	5. 5	5. 3	5. 0	4. 8	4. 6	4. 4
15	7. 3	6. 9	6. 5	6. 1	5. 8	5. 5	5. 3	5. 0	4. 8	4. 6	4. 4	4. 2
16	6. 8	6. 5	6. 1	5. 8	5. 5	5. 2	5. 0	4. 8	4. 6	4. 4	4. 2	4. 1
17	6. 4	6. 1	5. 8	5. 5	5. 2	5. 0	4. 8	4. 6	4. 4	4. 2	4. 1	3. 9
18	6. 0	5. 7	5. 5	5. 2	5. 0	4. 8	4. 6	4. 4	4. 2	4. 1	3. 9	3. 8
19	5. 7	5. 4	5. 2	4. 9	4. 7	4. 5	4. 4	4. 2	4. 0	3. 9	3. 8	3. 6
20	5. 4	5. 1	4. 9	4. 7	4. 5	4. 3	4. 2	4. 0	3. 9	3. 8	3. 6	3. 5
21	5. 1	4. 9	4. 7	4. 5	4. 3	4. 2	4. 0	3. 9	3. 7	3. 6	3. 5	3. 4
22	4. 9	4. 7	4. 5	4. 3	4. 1	4. 0	3. 9	3. 7	3. 6	3. 5	3. 4	3. 3
23	4. 6	4. 4	4. 3	4. 1	4. 0	3. 8	3. 7	3. 6	3. 5	3. 4	3. 3	3. 2
24	4. 4	4. 2	4. 1	3. 9	3. 8	3. 7	3. 6	3. 5	3. 4	3. 3	3. 2	3. 1
25	4. 2	4. 1	3. 9	3. 8	3. 7	3. 5	3. 4	3. 3	3. 2	3. 1	3. 1	3. 0
26	4. 0	3. 9	3. 8	3. 6	3. 5	3. 4	3. 3	3. 2	3. 1	3. 0	3. 0	2. 9
27	3. 9	3. 7	3. 6	3. 5	3. 4	3. 3	3. 2	3. 1	3. 0	2. 9	2. 9	2. 8
28	3. 7	3. 6	3. 5	3. 4	3. 3	3. 2	3. 1	3. 0	2. 9	2. 8	2. 8	2. 7
29	3. 5	3. 4	3. 3	3. 2	3. 1	3. 1	3. 0	2. 9	2. 8	2. 8	2. 7	2. 6
30	3. 4	3. 3	3. 2	3. 1	3. 0	3. 0	2. 9	2. 8	2. 7	2. 7	2. 6	2. 5
31	3. 3	3. 2	3. 1	3. 0	2. 9	2. 9	2. 8	2. 7	2. 6	2. 6	2. 5	2. 5
32	3. 2	3. 1	3. 0	2. 9	2. 8	2. 8	2. 7	2. 6	2. 6	2. 5	2. 5	2. 4
33	3. 0	2. 9	2. 9	2. 8	2. 7	2. 7	2. 6	2. 5	2. 5	2. 4	2. 4	2. 3
34	2. 9	2. 8	2. 8	2. 7	2. 6	2. 6	2. 5	2. 5	2. 4	2. 4	2. 3	2. 3
35	2. 8	2. 7	2. 7	2. 6	2. 5	2. 5	2. 4	2. 4	2. 3	2. 3	2. 2	2. 2
36	2. 7	2. 6	2. 6	2. 5	2. 5	2. 4	2. 4	2. 3	2. 3	2. 2	2. 2	2. 1
37	2. 6	2. 5	2. 5	2. 4	2. 4	2. 3	2. 3	2. 2	2. 2	2. 2	2. 1	2. 1
38	2. 5	2. 5	2. 4	2. 4	2. 3	2. 3	2. 2	2. 2	2. 1	2. 1	2. 1	2. 0
39	2. 4	2. 4	2. 3	2. 3	2. 2	2. 2	2. 1	2. 1	2. 1	2. 0	2. 0	2. 0
40	2. 3	2. 3	2. 2	2. 2	2. 2	2. 1	2. 1	2. 0	2. 0	2. 0	1. 9	1. 9
41	2. 3	2. 2	2. 2	2. 1	2. 1	2. 1	2. 0	2. 0	1. 9	1. 9	1. 9	1. 8
42	2. 2	2. 1	2. 1	2. 1	2. 0	2. 0	2. 0	1. 9	1. 9	1. 9	1. 8	1. 8
43	2. 1	2. 1	2. 0	2. 0	2. 0	1. 9	1. 9	1. 9	1. 8	1. 8	1. 8	1. 7
44	2. 0	2. 0	2. 0	1. 9	1. 9	1. 9	1. 8	1. 8	1. 8	1. 7	1. 7	1. 7
45	2. 0	1. 9	1. 9	1. 9	1. 8	1. 8	1. 8	1. 7	1. 7	1. 7	1. 7	1. 6
46	1. 9	1. 9	1. 8	1. 8	1. 8	1. 7	1. 7	1. 7	1. 7	1. 6	1. 6	1. 6
47	1. 8	1. 8	1. 8	1. 7	1. 7	1. 7	1. 7	1. 6	1. 6	1. 6	1. 6	1. 6
48	1. 8	1. 7	1. 7	1. 7	1. 7	1. 6	1. 6	1. 6	1. 6	1. 6	1. 5	1. 5
49	1. 7	1. 7	1. 7	1. 6	1. 6	1. 6	1. 6	1. 5	1. 5	1. 5	1. 5	1. 5
50	1. 6	1. 6	1. 6	1. 6	1. 6	1. 5	1. 5	1. 5	1. 5	1. 5	1. 4	1. 4
51	1. 6	1. 6	1. 6	1. 5	1. 5	1. 5	1. 5	1. 5	1. 4	1. 4	1. 4	1. 4
52	1. 5	1. 5	1. 5	1. 5	1. 5	1. 4	1. 4	1. 4	1. 4	1. 4	1. 4	1. 3
53	1. 5	1. 5	1. 4	1. 4	1. 4	1. 4	1. 4	1. 4	1. 3	1. 3	1. 3	1. 3
54	1. 4	1. 4	1. 4	1. 4	1. 4	1. 3	1. 3	1. 3	1. 3	1. 3	1. 3	1. 3
55	1. 4	1. 4	1. 3	1. 3	1. 3	1. 3	1. 3	1. 3	1. 3	1. 2	1. 2	1. 2
56	1. 3	1. 3	1. 3	1. 3	1. 3	1. 3	1. 2	1. 2	1. 2	1. 2	1. 2	1. 2
57	1. 3	1. 3	1. 3	1. 2	1. 2	1. 2	1. 2	1. 2	1. 2	1. 2	1. 1	1. 1
58	1. 2	1. 2	1. 2	1. 2	1. 2	1. 2	1. 2	1. 1	1. 1	1. 1	1. 1	1. 1
59	1. 2	1. 2	1. 2	1. 2	1. 1	1. 1	1. 1	1. 1	1. 1	1. 1	1. 1	1. 1
60	1. 1	1. 1	1. 1	1. 1	1. 1	1. 1	1. 1	1. 1	1. 0	1. 0	1. 0	1. 0
Lati- tude	0°	1°	2°	3°	4°	5°	6°	7°	8°	9°	10°	11°

Declination **contrary** name to latitude, **upper** transit: **add** correction to observed altitude

TABLE 29

Latitude	12°	13°	14°	15°	16°	17°	18°	19°	20°	21°	22°	23°	24°
°	''	''	''	''	''	''	''	''	''	''	''	''	''
0	9.2	8.5	7.9	7.3	6.8	6.4	6.0	5.7	5.4	5.1	4.9	4.6	4.4
1	8.5	7.9	7.4	6.9	6.5	6.1	5.7	5.4	5.1	4.9	4.7	4.4	4.2
2	7.9	7.4	6.9	6.5	6.1	5.8	5.5	5.2	4.9	4.7	4.5	4.3	4.1
3	7.4	6.9	6.5	6.1	5.8	5.5	5.2	4.9	4.7	4.5	4.3	4.1	3.9
4	7.0	6.5	6.2	5.8	5.5	5.2	5.0	4.7	4.5	4.3	4.1	4.0	3.8
5	6.5	6.2	5.8	5.5	5.2	5.0	4.8	4.5	4.3	4.2	4.0	3.8	3.7
6	6.2	5.8	5.5	5.3	5.0	4.8	4.6	4.4	4.2	4.0	3.9	3.7	3.6
7	5.9	5.6	5.3	5.0	4.8	4.6	4.4	4.2	4.0	3.9	3.7	3.6	3.5
8	5.6	5.3	5.0	4.8	4.6	4.4	4.2	4.0	3.9	3.7	3.6	3.5	3.4
9	5.3	5.0	4.8	4.6	4.4	4.2	4.1	3.9	3.8	3.6	3.5	3.4	3.3
10	5.0	4.8	4.6	4.4	4.2	4.1	3.9	3.8	3.6	3.5	3.4	3.3	3.2
11	4.8	4.6	4.4	4.2	4.1	3.9	3.8	3.6	3.5	3.4	3.3	3.2	3.1
12	4.6	4.4	4.3	4.1	3.9	3.8	3.7	3.5	3.4	3.3	3.2	3.1	3.0
13	4.4	4.3	4.1	3.9	3.8	3.7	3.5	3.4	3.3	3.2	3.1	3.0	2.9
14	4.2	4.1	3.9	3.8	3.7	3.5	3.4	3.3	3.2	3.1	3.0	2.9	2.8
15	4.1	3.9	3.8	3.7	3.5	3.4	3.3	3.2	3.1	3.0	2.9	2.8	2.8
16	3.9	3.8	3.7	3.5	3.4	3.3	3.2	3.1	3.0	2.9	2.8	2.8	2.7
17	3.8	3.7	3.5	3.4	3.3	3.2	3.1	3.0	2.9	2.8	2.8	2.7	2.6
18	3.7	3.5	3.4	3.3	3.2	3.1	3.0	2.9	2.9	2.8	2.7	2.6	2.5
19	3.5	3.4	3.3	3.2	3.1	3.0	2.9	2.9	2.8	2.7	2.6	2.6	2.5
20	3.4	3.3	3.2	3.1	3.0	2.9	2.9	2.8	2.7	2.6	2.6	2.5	2.4
21	3.3	3.2	3.1	3.0	2.9	2.8	2.8	2.7	2.6	2.6	2.5	2.4	2.4
22	3.2	3.1	3.0	2.9	2.8	2.8	2.7	2.6	2.6	2.5	2.4	2.4	2.3
23	3.1	3.0	2.9	2.8	2.8	2.7	2.6	2.6	2.5	2.4	2.4	2.3	2.3
24	3.0	2.9	2.8	2.8	2.7	2.6	2.5	2.5	2.4	2.4	2.3	2.3	2.2
25	2.9	2.8	2.7	2.7	2.6	2.5	2.5	2.4	2.4	2.3	2.3	2.2	2.2
26	2.8	2.7	2.7	2.6	2.5	2.5	2.4	2.4	2.3	2.3	2.2	2.1	2.1
27	2.7	2.7	2.6	2.5	2.5	2.4	2.4	2.3	2.2	2.2	2.1	2.1	2.1
28	2.6	2.6	2.5	2.5	2.4	2.3	2.3	2.2	2.2	2.1	2.1	2.1	2.0
29	2.6	2.5	2.4	2.4	2.3	2.3	2.2	2.2	2.1	2.1	2.0	2.0	2.0
30	2.5	2.4	2.4	2.3	2.3	2.2	2.2	2.1	2.1	2.0	2.0	2.0	1.9
31	2.4	2.4	2.3	2.3	2.2	2.2	2.1	2.1	2.0	2.0	2.0	1.9	1.9
32	2.3	2.3	2.2	2.2	2.2	2.1	2.1	2.0	2.0	1.9	1.9	1.9	1.8
33	2.3	2.2	2.2	2.1	2.1	2.1	2.0	2.0	1.9	1.9	1.9	1.8	1.8
34	2.2	2.2	2.1	2.1	2.0	2.0	2.0	1.9	1.9	1.9	1.8	1.8	1.8
35	2.2	2.1	2.1	2.0	2.0	2.0	1.9	1.9	1.8	1.8	1.8	1.7	1.7
36	2.1	2.1	2.0	2.0	1.9	1.9	1.9	1.8	1.8	1.8	1.7	1.7	1.7
37	2.0	2.0	2.0	1.9	1.9	1.9	1.8	1.8	1.8	1.7	1.7	1.7	1.6
38	2.0	1.9	1.9	1.9	1.8	1.8	1.8	1.8	1.7	1.7	1.7	1.6	1.6
39	1.9	1.9	1.9	1.8	1.8	1.8	1.7	1.7	1.7	1.6	1.6	1.6	1.6
40	1.9	1.8	1.8	1.8	1.7	1.7	1.7	1.7	1.6	1.6	1.6	1.6	1.5
41	1.8	1.8	1.8	1.7	1.7	1.7	1.6	1.6	1.6	1.6	1.5	1.5	1.5
42	1.8	1.7	1.7	1.7	1.7	1.6	1.6	1.6	1.6	1.5	1.5	1.5	1.5
43	1.7	1.7	1.7	1.6	1.6	1.6	1.6	1.5	1.5	1.5	1.5	1.4	1.4
44	1.7	1.6	1.6	1.6	1.6	1.5	1.5	1.5	1.5	1.5	1.4	1.4	1.4
45	1.6	1.6	1.6	1.5	1.5	1.5	1.5	1.5	1.4	1.4	1.4	1.4	1.4
46	1.6	1.6	1.5	1.5	1.5	1.5	1.4	1.4	1.4	1.4	1.4	1.3	1.3
47	1.5	1.5	1.5	1.5	1.4	1.4	1.4	1.4	1.4	1.3	1.3	1.3	1.3
48	1.5	1.5	1.4	1.4	1.4	1.4	1.4	1.4	1.3	1.3	1.3	1.3	1.3
49	1.4	1.4	1.4	1.4	1.4	1.3	1.3	1.3	1.3	1.3	1.3	1.2	1.2
50	1.4	1.4	1.4	1.3	1.3	1.3	1.3	1.3	1.3	1.3	1.2	1.2	1.2
51	1.4	1.3	1.3	1.3	1.3	1.3	1.3	1.2	1.2	1.2	1.2	1.2	1.2
52	1.3	1.3	1.3	1.3	1.3	1.3	1.2	1.2	1.2	1.2	1.2	1.1	1.1
53	1.3	1.3	1.3	1.2	1.2	1.2	1.2	1.2	1.2	1.2	1.1	1.1	1.1
54	1.2	1.2	1.2	1.2	1.2	1.2	1.2	1.1	1.1	1.1	1.1	1.1	1.1
55	1.2	1.2	1.2	1.2	1.1	1.1	1.1	1.1	1.1	1.1	1.1	1.1	1.1
56	1.2	1.1	1.1	1.1	1.1	1.1	1.1	1.1	1.1	1.1	1.0	1.0	1.0
57	1.1	1.1	1.1	1.1	1.1	1.1	1.1	1.0	1.0	1.0	1.0	1.0	1.0
58	1.1	1.1	1.1	1.1	1.0	1.0	1.0	1.0	1.0	1.0	1.0	1.0	1.0
59	1.1	1.0	1.0	1.0	1.0	1.0	1.0	1.0	1.0	1.0	1.0	0.9	0.9
60	1.0	1.0	1.0	1.0	1.0	1.0	1.0	0.9	0.9	0.9	0.9	0.9	0.9
Latitude	12°	13°	14°	15°	16°	17°	18°	19°	20°	21°	22°	23°	24°

TABLE 30

Change of Altitude in Given Time from Meridian Transit

a (table 29)	5'	10'	15'	20'	25'	30'	35'	40'	45'	50'	55'	1°00'	1°05'	1°10'
	0ᵐ 20ˢ	0ᵐ 40ˢ	1ᵐ 00ˢ	1ᵐ 20ˢ	1ᵐ 40ˢ	2ᵐ 00ˢ	2ᵐ 20ˢ	2ᵐ 40ˢ	3ᵐ 00ˢ	3ᵐ 20ˢ	3ᵐ 40ˢ	4ᵐ 00ˢ	4ᵐ 20ˢ	4ᵐ 40ˢ
"	'	'	'	'	'	'	'	'	'	'	'	'	'	'
0. 1	0. 0	0. 0	0. 0	0. 0	0. 0	0. 0	0. 0	0. 0	0. 0	0. 0	0. 0	0. 0	0. 0	0. 0
0. 2	0. 0	0. 0	0. 0	0. 0	0. 0	0. 0	0. 0	0. 0	0. 0	0. 0	0. 0	0. 1	0. 1	0. 1
0. 3	0. 0	0. 0	0. 0	0. 0	0. 0	0. 0	0. 0	0. 0	0. 0	0. 1	0. 1	0. 1	0. 1	0. 1
0. 4	0. 0	0. 0	0. 0	0. 0	0. 0	0. 0	0. 0	0. 0	0. 1	0. 1	0. 1	0. 1	0. 1	0. 1
0. 5	0. 0	0. 0	0. 0	0. 0	0. 0	0. 0	0. 0	0. 1	0. 1	0. 1	0. 1	0. 1	0. 2	0. 2
0. 6	0. 0	0. 0	0. 0	0. 0	0. 0	0. 0	0. 1	0. 1	0. 1	0. 1	0. 1	0. 2	0. 2	0. 2
0. 7	0. 0	0. 0	0. 0	0. 0	0. 0	0. 0	0. 1	0. 1	0. 1	0. 1	0. 2	0. 2	0. 2	0. 3
0. 8	0. 0	0. 0	0. 0	0. 0	0. 0	0. 1	0. 1	0. 1	0. 1	0. 1	0. 2	0. 2	0. 3	0. 3
0. 9	0. 0	0. 0	0. 0	0. 0	0. 0	0. 1	0. 1	0. 1	0. 1	0. 2	0. 2	0. 2	0. 3	0. 3
1. 0	0. 0	0. 0	0. 0	0. 0	0. 0	0. 1	0. 1	0. 1	0. 2	0. 2	0. 2	0. 3	0. 3	0. 4
2. 0	0. 0	0. 0	0. 0	0. 1	0. 1	0. 1	0. 2	0. 2	0. 3	0. 4	0. 4	0. 5	0. 6	0. 7
3. 0	0. 0	0. 0	0. 0	0. 1	0. 1	0. 2	0. 3	0. 4	0. 4	0. 6	0. 7	0. 8	0. 9	1. 1
4. 0	0. 0	0. 0	0. 1	0. 1	0. 2	0. 3	0. 4	0. 5	0. 6	0. 7	0. 9	1. 1	1. 3	1. 5
5. 0	0. 0	0. 0	0. 1	0. 1	0. 2	0. 3	0. 5	0. 6	0. 8	0. 9	1. 1	1. 3	1. 6	1. 8
6. 0	0. 0	0. 0	0. 1	0. 2	0. 3	0. 4	0. 5	0. 7	0. 9	1. 1	1. 3	1. 6	1. 9	2. 2
7. 0	0. 0	0. 1	0. 1	0. 2	0. 3	0. 5	0. 6	0. 8	1. 0	1. 3	1. 6	1. 9	2. 2	2. 5
8. 0	0. 0	0. 1	0. 1	0. 2	0. 4	0. 5	0. 7	0. 9	1. 2	1. 5	1. 8	2. 1	2. 5	2. 9
9. 0	0. 0	0. 1	0. 2	0. 3	0. 4	0. 6	0. 8	1. 1	1. 4	1. 7	2. 0	2. 4	2. 8	3. 3
10. 0	0. 0	0. 1	0. 2	0. 3	0. 5	0. 7	0. 9	1. 2	1. 5	1. 9	2. 2	2. 7	3. 1	3. 6
11. 0	0. 0	0. 1	0. 2	0. 3	0. 5	0. 7	1. 0	1. 3	1. 6	2. 0	2. 5	2. 9	3. 4	4. 0
12. 0	0. 0	0. 1	0. 2	0. 4	0. 6	0. 8	1. 1	1. 4	1. 8	2. 2	2. 7	3. 2	3. 8	4. 4
13. 0	0. 0	0. 1	0. 2	0. 4	0. 6	0. 9	1. 2	1. 5	2. 0	2. 4	2. 9	3. 5	4. 1	4. 7
14. 0	0. 0	0. 1	0. 2	0. 4	0. 6	0. 9	1. 3	1. 7	2. 1	2. 6	3. 1	3. 7	4. 4	5. 1
15. 0	0. 0	0. 1	0. 2	0. 4	0. 7	1. 0	1. 4	1. 8	2. 2	2. 8	3. 4	4. 0	4. 7	5. 4
16. 0	0. 0	0. 1	0. 3	0. 5	0. 7	1. 1	1. 5	1. 9	2. 4	3. 0	3. 6	4. 3	5. 0	5. 8
17. 0	0. 0	0. 1	0. 3	0. 5	0. 8	1. 1	1. 5	2. 0	2. 6	3. 1	3. 8	4. 5	5. 3	6. 2
18. 0	0. 0	0. 1	0. 3	0. 5	0. 8	1. 2	1. 6	2. 1	2. 7	3. 3	4. 0	4. 8	5. 6	6. 5
19. 0	0. 0	0. 1	0. 3	0. 6	0. 9	1. 3	1. 7	2. 3	2. 8	3. 5	4. 3	5. 1	5. 9	6. 9
20. 0	0. 0	0. 1	0. 3	0. 6	0. 9	1. 3	1. 8	2. 4	3. 0	3. 7	4. 5	5. 3	6. 3	7. 3
21. 0	0. 0	0. 2	0. 4	0. 6	1. 0	1. 4	1. 9	2. 5	3. 2	3. 9	4. 7	5. 6	6. 6	7. 6
22. 0	0. 0	0. 2	0. 4	0. 7	1. 0	1. 5	2. 0	2. 6	3. 3	4. 1	4. 9	5. 9	6. 9	8. 0
23. 0	0. 0	0. 2	0. 4	0. 7	1. 1	1. 5	2. 1	2. 7	3. 4	4. 3	5. 2	6. 1	7. 2	8. 3
24. 0	0. 0	0. 2	0. 4	0. 7	1. 1	1. 6	2. 2	2. 8	3. 6	4. 4	5. 4	6. 4	7. 5	8. 7
25. 0	0. 0	0. 2	0. 4	0. 7	1. 2	1. 7	2. 3	3. 0	3. 8	4. 6	5. 6	6. 7	7. 8	9. 1
26. 0	0. 0	0. 2	0. 4	0. 8	1. 2	1. 7	2. 4	3. 1	3. 9	4. 8	5. 8	6. 9	8. 1	9. 4
27. 0	0. 0	0. 2	0. 4	0. 8	1. 2	1. 8	2. 4	3. 2	4. 0	5. 0	6. 0	7. 2	8. 4	9. 8
28. 0	0. 1	0. 2	0. 5	0. 8	1. 3	1. 9	2. 5	3. 3	4. 2	5. 2	6. 3	7. 5	8. 8	10. 2

TABLE 30

Change of Altitude in Given Time from Meridian Transit

a (table 29)	t, meridian angle													
	1°15′	1°20′	1°25′	1°30′	1°35′	1°40′	1°45′	1°50′	1°55′	2°00′	2°05′	2°10′	2°15′	2°20′
	5ᵐ00ˢ	5ᵐ20ˢ	5ᵐ40ˢ	6ᵐ00ˢ	6ᵐ20ˢ	6ᵐ40ˢ	7ᵐ00ˢ	7ᵐ20ˢ	7ᵐ40ˢ	8ᵐ00ˢ	8ᵐ20ˢ	8ᵐ40ˢ	9ᵐ00ˢ	9ᵐ20ˢ
0.1	0.0	0.0	0.1	0.1	0.1	0.1	0.1	0.1	0.1	0.1	0.1	0.1	0.1	0.1
0.2	0.1	0.1	0.1	0.1	0.1	0.1	0.2	0.2	0.2	0.2	0.2	0.3	0.3	0.3
0.3	0.1	0.1	0.2	0.2	0.2	0.2	0.2	0.3	0.3	0.3	0.3	0.4	0.4	0.4
0.4	0.2	0.2	0.2	0.2	0.3	0.3	0.3	0.4	0.4	0.4	0.5	0.5	0.5	0.6
0.5	0.2	0.2	0.3	0.3	0.3	0.4	0.4	0.5	0.5	0.5	0.6	0.6	0.7	0.7
0.6	0.2	0.3	0.3	0.4	0.4	0.4	0.5	0.5	0.6	0.6	0.7	0.8	0.8	0.9
0.7	0.3	0.3	0.4	0.4	0.5	0.5	0.6	0.6	0.7	0.7	0.8	0.9	0.9	1.0
0.8	0.3	0.4	0.4	0.5	0.5	0.6	0.7	0.7	0.8	0.9	0.9	1.0	1.1	1.2
0.9	0.4	0.4	0.5	0.5	0.6	0.7	0.7	0.8	0.9	1.0	1.0	1.1	1.2	1.3
1.0	0.4	0.5	0.5	0.6	0.7	0.7	0.8	0.9	1.0	1.1	1.2	1.3	1.4	1.5
2.0	0.8	0.9	1.1	1.2	1.3	1.5	1.6	1.8	2.0	2.1	2.3	2.5	2.7	2.9
3.0	1.2	1.4	1.6	1.8	2.0	2.2	2.4	2.7	2.9	3.2	3.5	3.8	4.0	4.4
4.0	1.7	1.9	2.1	2.4	2.7	3.0	3.3	3.6	3.9	4.3	4.6	5.0	5.4	5.8
5.0	2.1	2.4	2.7	3.0	3.3	3.7	4.1	4.5	4.9	5.3	5.8	6.3	6.8	7.3
6.0	2.5	2.8	3.2	3.6	4.0	4.4	4.9	5.4	5.9	6.4	6.9	7.5	8.1	8.7
7.0	2.9	3.3	3.7	4.2	4.7	5.2	5.7	6.3	6.9	7.5	8.1	8.8	9.4	10.2
8.0	3.3	3.8	4.3	4.8	5.3	5.9	6.5	7.2	7.8	8.5	9.3	10.0	10.8	11.6
9.0	3.8	4.3	4.8	5.4	6.0	6.7	7.4	8.1	8.8	9.6	10.4	11.3	12.2	13.1
10.0	4.2	4.7	5.4	6.0	6.7	7.4	8.2	9.0	9.8	10.7	11.6	12.5	13.5	14.5
11.0	4.6	5.2	5.9	6.6	7.4	8.1	9.0	9.9	10.8	11.7	12.7	13.8	14.8	16.0
12.0	5.0	5.7	6.4	7.2	8.0	8.9	9.8	10.8	11.8	12.8	13.9	15.0	16.2	17.4
13.0	5.4	6.2	7.0	7.8	8.7	9.6	10.6	11.7	12.7	13.9	15.0	16.3	17.6	18.9
14.0	5.8	6.6	7.5	8.4	9.4	10.4	11.4	12.5	13.7	14.9	16.2	17.5	18.9	20.3
15.0	6.2	7.1	8.0	9.0	10.0	11.1	12.2	13.4	14.7	16.0	17.4	18.8	20.2	21.8
16.0	6.7	7.6	8.6	9.6	10.7	11.9	13.1	14.3	15.7	17.1	18.5	20.0	21.6	23.2
17.0	7.1	8.1	9.1	10.2	11.4	12.6	13.9	15.2	16.7	18.1	19.7	21.3	23.0	24.7
18.0	7.5	8.5	9.6	10.8	12.0	13.3	14.7	16.1	17.6	19.2	20.8	22.5	24.3	26.1
19.0	7.9	9.0	10.2	11.4	12.7	14.1	15.5	17.0	18.6	20.3	22.0	23.8		
20.0	8.3	9.5	10.7	12.0	13.4	14.8	16.3	17.9	19.6	21.3	23.1			
21.0	8.8	10.0	11.2	12.6	14.0	15.6	17.2	18.8	20.6					
22.0	9.2	10.4	11.8	13.2	14.7	16.3	18.0	19.7	21.6					
23.0	9.6	10.9	12.3	13.8	15.4	17.0	18.8	20.6	21.5					
24.0	10.0	11.4	12.8	14.4	16.0	17.8	19.6	21.5						
25.0	10.4	11.9	13.4	15.0	16.7	18.5	20.4							
26.0	10.8	12.3	13.9	15.6	17.4	19.3								
27.0	11.2	12.8	14.4	16.2	18.0	20.0								

TABLE 30

Change of Altitude in Given Time from Meridian Transit

a (table 29)	2°25'	2°30'	2°35'	2°40'	2°45'	2°50'	2°55'	3°00'	3°05'	3°10'	3°15'	3°20'	3°25'	3°30'
	9m40s	10m00s	10m20s	10m40s	11m00s	11m20s	11m40s	12m00s	12m20s	12m40s	13m00s	13m20s	13m40s	14m00s
0.1	0.2	0.2	0.2	0.2	0.2	0.2	0.2	0.2	0.3	0.3	0.3	0.3	0.3	0.3
0.2	0.3	0.3	0.4	0.4	0.4	0.4	0.5	0.5	0.5	0.5	0.6	0.6	0.6	0.7
0.3	0.5	0.5	0.5	0.6	0.6	0.6	0.7	0.7	0.8	0.8	0.8	0.9	0.9	1.0
0.4	0.6	0.7	0.7	0.8	0.8	0.9	0.9	1.0	1.0	1.1	1.1	1.2	1.2	1.3
0.5	0.8	0.8	0.9	0.9	1.0	1.1	1.1	1.2	1.3	1.3	1.4	1.5	1.6	1.6
0.6	0.9	1.0	1.1	1.1	1.2	1.3	1.4	1.4	1.5	1.6	1.7	1.8	1.9	2.0
0.7	1.1	1.2	1.2	1.3	1.4	1.5	1.6	1.7	1.8	1.9	2.0	2.1	2.2	2.3
0.8	1.2	1.3	1.4	1.5	1.6	1.7	1.8	1.9	2.0	2.1	2.3	2.4	2.5	2.6
0.9	1.4	1.5	1.6	1.7	1.8	1.9	2.0	2.2	2.3	2.4	2.5	2.7	2.8	2.9
1.0	1.6	1.7	1.8	1.9	2.0	2.1	2.3	2.4	2.5	2.7	2.8	3.0	3.1	3.3
2.0	3.1	3.3	3.6	3.8	4.0	4.3	4.5	4.8	5.1	5.3	5.6	5.9	6.2	6.5
3.0	4.7	5.0	5.3	5.7	6.0	6.4	6.8	7.2	7.6	8.0	8.4	8.9	9.3	9.8
4.0	6.2	6.7	7.1	7.6	8.1	8.6	9.1	9.6	10.1	10.7	11.3	11.9	12.5	13.1
5.0	7.8	8.3	8.9	9.5	10.1	10.7	11.3	12.0	12.7	13.4	14.1	14.8	15.6	16.3
6.0	9.3	10.0	10.7	11.4	12.1	12.8	13.6	14.4	15.2	16.0	16.9	17.8	18.7	19.6
7.0	10.9	11.7	12.5	13.3	14.1	15.0	15.9	16.8	17.7	18.7	19.7	20.7	21.8	22.9
8.0	12.5	13.3	14.2	15.2	16.1	17.1	18.1	19.2	20.3	21.4	22.5	23.7	24.9	26.1
9.0	14.0	15.0	16.0	17.1	18.2	19.3	20.4	21.6	22.8	24.1	25.4	26.7	28.0	29.4
10.0	15.6	16.7	17.8	19.0	20.2	21.4	22.7	24.0	25.4	26.7	28.2	29.6		
11.0	17.1	18.3	19.6	20.9	22.2	23.5	25.0	26.4	27.9	29.4				
12.0	18.7	20.0	21.4	22.8	24.2	25.7	27.2	28.8						
13.0	20.2	21.7	23.1	24.7	26.2	27.8	29.5							
14.0	21.8	23.3	24.9	26.5	28.2	30.0								
15.0	23.4	25.0	26.7	28.4	30.2									
16.0	24.9	26.7	28.5	30.3										
17.0	26.5	28.3	30.3											

a (table 29)	3°35'	3°40'	3°45'	3°50'	3°55'	4°00'	4°05'	4°10'	4°15'	4°20'	4°25'	4°30'	4°35'	4°40'
	14m20s	14m40s	15m00s	15m20s	15m40s	16m00s	16m20s	16m40s	17m00s	17m20s	17m40s	18m00s	18m20s	18m40s
0.1	0.3	0.4	0.4	0.4	0.4	0.4	0.4	0.5	0.5	0.5	0.5	0.5	0.6	0.6
0.2	0.7	0.7	0.8	0.8	0.8	0.9	0.9	0.9	1.0	1.0	1.0	1.1	1.1	1.2
0.3	1.0	1.1	1.1	1.2	1.2	1.3	1.3	1.4	1.4	1.5	1.6	1.6	1.7	1.7
0.4	1.4	1.4	1.5	1.6	1.6	1.7	1.8	1.9	1.9	2.0	2.1	2.2	2.2	2.3
0.5	1.7	1.8	1.9	2.0	2.0	2.1	2.2	2.3	2.4	2.5	2.6	2.7	2.8	2.9
0.6	2.1	2.2	2.2	2.4	2.5	2.6	2.7	2.8	2.9	3.0	3.1	3.2	3.4	3.5
0.7	2.4	2.5	2.6	2.7	2.9	3.0	3.1	3.2	3.4	3.5	3.6	3.8	3.9	4.1
0.8	2.7	2.9	3.0	3.1	3.3	3.4	3.6	3.7	3.9	4.0	4.2	4.3	4.5	4.6
0.9	3.1	3.2	3.4	3.5	3.7	3.8	4.0	4.2	4.3	4.5	4.7	4.9	5.0	5.2
1.0	3.4	3.6	3.8	3.9	4.1	4.3	4.4	4.6	4.8	5.0	5.2	5.4	5.6	5.8
2.0	6.8	7.2	7.5	7.8	8.2	8.5	8.9	9.3	9.6	10.0	10.4	10.8	11.2	11.6
3.0	10.3	10.8	11.2	11.8	12.3	12.8	13.3	13.9	14.4	15.0	15.6	16.2	16.8	17.4
4.0	13.7	14.3	15.0	15.7	16.4	17.1	17.8	18.5	19.3	20.0	20.8	21.6	22.4	23.2
5.0	17.1	17.9	18.8	19.6	20.5	21.3	22.2	23.1	24.1	25.0	26.0	27.0	28.0	29.0
6.0	20.5	21.5	22.5	23.5	24.5	25.6	26.7	27.8						
7.0	24.0	25.1	26.2	27.4										
8.0	27.4	28.7	30.0											

TABLE 30

Change of Altitude in Given Time from Meridian Transit

a (table 29)	4°45'	4°50'	4°55'	5°00'	5°05'	5°10'	5°15'	5°20'	5°25'	5°30'	5°35'	5°40'	5°45'	5°50'
	19m00s	19m20s	19m40s	20m00s	20m20s	20m40s	21m00s	21m20s	21m40s	22m00s	22m20s	22m40s	23m00s	23m20s
0.1	0.6	0.6	0.6	0.7	0.7	0.7	0.7	0.8	0.8	0.8	0.8	0.9	0.9	0.9
0.2	1.2	1.2	1.3	1.3	1.4	1.4	1.5	1.5	1.6	1.6	1.7	1.7	1.8	1.8
0.3	1.8	1.9	1.9	2.0	2.1	2.1	2.2	2.3	2.3	2.4	2.5	2.6	2.6	2.7
0.4	2.4	2.5	2.6	2.7	2.8	2.8	2.9	3.0	3.1	3.2	3.3	3.4	3.5	3.6
0.5	3.0	3.1	3.2	3.3	3.4	3.6	3.7	3.8	3.9	4.0	4.2	4.3	4.4	4.5
0.6	3.6	3.7	3.9	4.0	4.1	4.3	4.4	4.6	4.7	4.8	5.0	5.1	5.3	5.4
0.7	4.2	4.4	4.5	4.7	4.8	5.0	5.1	5.3	5.5	5.6	5.8	6.0	6.2	6.4
0.8	4.8	5.0	5.2	5.3	5.5	5.7	5.9	6.1	6.3	6.5	6.7	6.9	7.1	7.3
0.9	5.4	5.6	5.8	6.0	6.2	6.4	6.6	6.8	7.0	7.3	7.5	7.7	7.9	8.2
1.0	6.0	6.2	6.4	6.7	6.9	7.1	7.4	7.6	7.8	8.1	8.3	8.6	8.8	9.1
2.0	12.0	12.5	12.9	13.3	13.8	14.2	14.7	15.2	15.6	16.1	16.6	17.1	17.6	18.1
3.0	18.0	18.7	19.3	20.0	20.7	21.4	22.0	22.8	23.5	24.2	24.9	25.7	26.4	27.2
4.0	24.1	24.9	25.8	26.7	27.6	28.5	29.4	30.3	31.3					

t, meridian angle

a (table 29)	5°55'	6°00'	6°05'	6°10'	6°15'	6°20'	6°25'	6°30'	6°35'	6°40'	6°45'	6°50'	6°55'	7°00'
	23m40s	24m00s	24m20s	24m40s	25m00s	25m20s	25m40s	26m00s	26m20s	26m40s	27m00s	27m20s	27m40s	28m00s
0.1	0.9	1.0	1.0	1.0	1.0	1.1	1.1	1.1	1.2	1.2	1.2	1.2	1.3	1.3
0.2	1.9	1.9	2.0	2.0	2.1	2.1	2.2	2.3	2.3	2.4	2.4	2.5	2.6	2.6
0.3	2.8	2.9	3.0	3.0	3.1	3.2	3.3	3.4	3.5	3.6	3.6	3.7	3.8	3.9
0.4	3.7	3.8	3.9	4.1	4.2	4.3	4.4	4.5	4.6	4.7	4.9	5.0	5.1	5.2
0.5	4.7	4.8	4.9	5.1	5.2	5.3	5.5	5.6	5.8	5.9	6.1	6.2	6.4	6.5
0.6	5.6	5.8	5.9	6.1	6.2	6.4	6.6	6.8	6.9	7.1	7.3	7.5	7.7	7.8
0.7	6.5	6.7	6.9	7.1	7.3	7.5	7.7	7.9	8.1	8.3	8.5	8.7	8.9	9.1
0.8	7.5	7.7	7.9	8.1	8.3	8.6	8.8	9.0	9.2	9.5	9.7	10.0	10.2	10.5
0.9	8.4	8.6	8.9	9.1	9.4	9.6	9.9	10.1	10.4	10.7	10.9	11.2	11.5	11.8
1.0	9.3	9.6	9.9	10.1	10.4	10.7	11.0	11.3	11.6	11.9	12.2	12.5	12.8	13.1
2.0	18.7	19.2	19.7	20.3	20.8	21.4	22.0	22.5	23.1	23.7	24.3	24.9	25.5	26.1
3.0	28.0	28.8	29.6	30.4										

Part III

HIGH ALTITUDE SUN SIGHTS

High Altitude Sun Sights

AS YOUR latitude and the declination of the sun become numerically more equal, it gets more difficult to work up noon sights for latitude, because the sun is passing nearly overhead. It is very difficult to keep the sun in the sextant mirror and swing the arc when it is that high overhead.

The way around this difficulty is provided by the fact that the GP of the sun is passing very near you. So it is possible to mark it down on a chart at two separate times and then, with a pair of dividers, mark off the zenith distance directly on the chart.

What you do is take a sight when the sun's GP lies a few degrees of longitude *east* of you, plot that GP on your chart, wait until the sun's GP is a few degrees of longitude *west* of you, take another sight, plot that GP, and then mark out the zenith distances from each. Where they cross is your position—latitude *and* longitude. (These arcs are shown in Figure 43.)

Example: your DR has you at 2°30′N and 5°16′E on March 29, 1981. You get a sight of 85°30′ at 11-25-24 GMT and at 12-11-00, a sight of 82°57′.

This example is worked out in Figure 42 and Figure 43, and

EXAMPLE: HIGH ALTITUDE SUN SIGHT

GMT = 11-25-24

1. GHA = 350°09'   Longitude = 360° – GHA = 9°51'E
   Dec = N3°26'   Latitude = Declination = 3°26'N

   This position (GP of sun @ 1125) is plotted on the chart near right margin.

2. Ho = 85°23'   Zenith Distance = 90° – Ho = 4°37'

   4°37' is measured from the latitude scale with a bow compass and arc of that radius swung using GP as center.

GMT = 12-11-00

3. GHA = 1°35'   Longitude = GHA = 1°35'W
   Dec = 3°27'N   Latitude = Declination = 3°27'N

   This GP is plotted on the chart South of Cape Three Points.

4. Ho = 82°48'   Zenith Distance = 90° – Ho = 7°12'

   7°12' is measured from the latitude scale on the right vertical chart margin and arc of that radius swung using GP as center.

5. Your position is at the intersection of the two arcs nearest the DR.

Figure 42. The result of this sample problem is laid out graphically in the chart shown in Figure 43.

is a good example of the usefulness of the DR in selecting between two theoretically possible positions.

In locating the *latitude* of the GP at times *between* the hourly intervals given in the almanac, interpolation by inspection is simple, due to the fact that the declination change is always small. In fact, in this example (March 29) it is as rapid as it ever gets—1' of arc per hour.

Interpolating for the change in GHA, however, is not so easy. While it is possible to work out the GHA for the time of the sight by knowing that the sun covers 15° of longitude (i.e. GHA increases 15°) each hour, there is a table for this shown in Figure 44, on page 104.

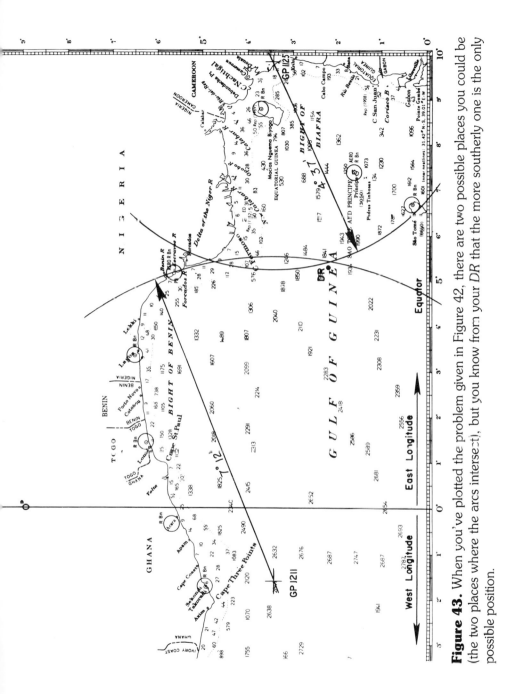

Figure 43. When you've plotted the problem given in Figure 42, there are two possible places you could be (the two places where the arcs intersect), but you know from your *DR* that the more southerly one is the only possible position.

GREENWICH HOUR ANGLE INTERPOLATION

Minutes m (° ')		Minutes m (° ')		Seconds s (')		Seconds s (')	
0	0 00	30	7 30	0	0.00	30	7.50
1	0 15	31	7 45	1	0.25	31	7.75
2	0 30	32	8 00	2	0.50	32	8.00
3	0 45	33	8 15	3	0.75	33	8.25
4	1 00	34	8 30	4	1.00	34	8.50
5	1 15	35	8 45	5	1.25	35	8.75
6	1 30	36	9 00	6	1.50	36	9.00
7	1 45	37	9 15	7	1.75	37	9.25
8	2 00	38	9 30	8	2.00	38	9.50
9	2 15	39	9 45	9	2.25	39	9.75
10	2 30	40	10 00	10	2.50	40	10.00
11	2 45	41	10 15	11	2.75	41	10.25
12	3 00	42	10 30	12	3.00	42	10.50
13	3 15	43	10 45	13	3.25	43	10.75
14	3 30	44	11 00	14	3.50	44	11.00
15	3 45	45	11 15	15	3.75	45	11.25
16	4 00	46	11 30	16	4.00	46	11.50
17	4 15	47	11 45	17	4.25	47	11.75
18	4 30	48	12 00	18	4.50	48	12.00
19	4 45	49	12 15	19	4.75	49	12.25
20	5 00	50	12 30	20	5.00	50	12.50
21	5 15	51	12 45	21	5.25	51	12.75
22	5 30	52	13 00	22	5.50	52	13.00
23	5 45	53	13 15	23	5.75	53	13.25
24	6 00	54	13 30	24	6.00	54	13.50
25	6 15	55	13 45	25	6.25	55	13.75
26	6 30	56	14 00	26	6.50	56	14.00
27	6 45	57	14 15	27	6.75	57	14.25
28	7 00	58	14 30	28	7.00	58	14.50
29	7 15	59	14 45	29	7.25	59	14.75
		60	15 00			60	15.00

Figure 44. To use this table, you look down the left-hand column headed minutes (m) and read the Greenwich Hour Angle increase shown in the right-hand column. For example, 7 minutes in time (arrow) equals 1°45' of GHA.

In the seconds column(s), GHA is given in minutes (') and *tenths of minutes of GHA*. Thus, in 5 seconds of time, GHA increases 1.25 minutes of arc (see arrows in right-hand column).

Part IV

LONGITUDE BY NOON SIGHT

11

A.M.-P.M. Equal Altitudes

SINCE, AS we've said before, the sun is on your longitude when it reaches its maximum altitude, it would seem that if you noted the time at which you got the maximum reading on your sextant, you could simply go to the almanac, look up the GHA for that time and find your longitude—GHA and longitude being equivalent.

Well, as you will discover once you have taken a few noon sights, that's not possible because as the sun approaches its maximum altitude, the *rate of change* in the altitude decreases dramatically, and there is an appreciable interval of time during which the sextant *can detect no change*. This interval (called the "hang") varies directly with the zenith distance—the greater the zenith distance, the longer the hang—and results from the fact that the sun's path through the sky is a curve (Figure 45) and the top of that curve is essentially flat. Since the sun's longitude is changing at the rate of 15′ per minute of time and "hang" times are typically two to five minutes, it's easy to see that accurate longitudes cannot be determined by the simple method of following the sun up to its highest altitude.

If, however, you take and time a sight sometime *before* noon

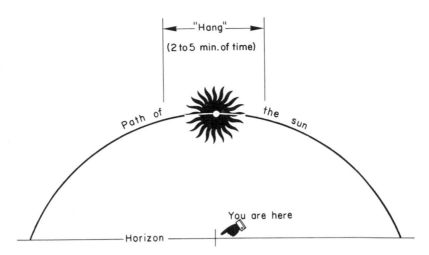

Figure 45. As the sun arcs across the sky, its altitude *does not change* at a constant rate. As it approaches its maximum altitude (just *before* noon) and then declines from its maximum altitude (just after noon), there is a brief interval (2 to 5 minutes) when no change is discernible. This interval is called the "Hang".

and then, *after* noon, reset your sextant to the AM altitude and take the time in the PM when the sun again reaches that *same* altitude (see Figure 46), you will have located two points on the curved path of the sun across the sky that lie equidistant from the time it reached its maximum altitude. Adding the times of these two sights and dividing by two will give you the time of maximum altitude and, therefore, the time the sun was *on your longitude.* Going to your almanac and looking up the GHA for that time gives you your longitude.

100 MINUTES DO NOT AN HOUR MAKE

In adding times, remember that seconds total to 60 (not 100) and become a minute; minutes total to 60 (not 100) to become an hour; hours total to 24 to become a day.

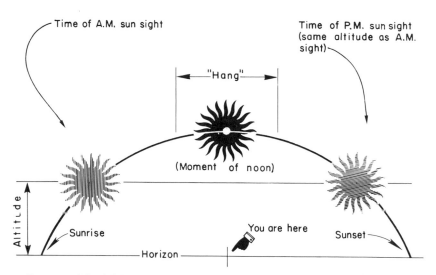

Figure 46. A.M.-P.M. equal altitudes. By taking and timing a sight sometime *before* meridian passage (noon), and then, *after* noon, by resetting your sextant to the A.M. altitude, and taking the time in the P.M. when the sun again reaches that same altitude, the moment of noon is determined by adding the two times and dividing by two.

In taking these observations, you don't have to apply any corrections to your raw sextant reading, since these altitudes are being used only to locate points equidistant on a curve. It *is* necessary, however, to take the sights at a time when the sun's rate of change in altitude is rapid enough to ensure an accurate

WHEN TO TAKE THE LONGITUDE SIGHTS:

No corrections to sextant readings are used for longitude by noon sight, since only time is needed.

For best results, take AM sight an appreciable time before noon —about one minute of time for each degree of zenith distance gives reliable results.

timing. The usual rule of thumb is to allow *four minutes* of time for the total interval between your AM and PM sights for each 1° of zenith distance. Thus, if you were going to use this method on a day when you were at latitude 20°S and the sun's declination was 20°N, the zenith distance would be 40° and you should take your AM sight about 80 minutes *before* noon. In practice, I have found that cutting this rule in half produces reliable results—i.e., two minutes of time for each degree of zenith distance.

12

Correction for Motion of the Vessel

NOW, IT happens that if you are traveling directly toward or away from the sun (i.e., due north or south) during the time you are taking your AM-PM sights for longitude, your vessel's motion has an effect on your measurement of the sun's altitude.

If you are moving directly *toward* the sun, your motion *raises* the sun's altitude and, thus, makes it take a little longer to reach its maximum altitude. This, effectively, makes the time of noon *later* and so shifts your longitude *westward*.

If you are moving directly *away* from the sun, your motion *lowers* the altitude of the sun, makes noon occur *earlier* and shifts the longitude *eastward*. Motion directly east or west has no effect.

The graph in Figure 47 illustrates this effect upon three imaginary navigators on three different vessels. They begin taking sights at 24 minutes before meridian passage of the sun. One vessel—the top curve—is standing still. The middle curve is for a vessel moving directly away from the sun at a speed of six knots. The bottom curve is for a vessel moving directly away from the sun at 15 knots. The middle vertical line represents the longitude on which all three vessels are located. As you can see

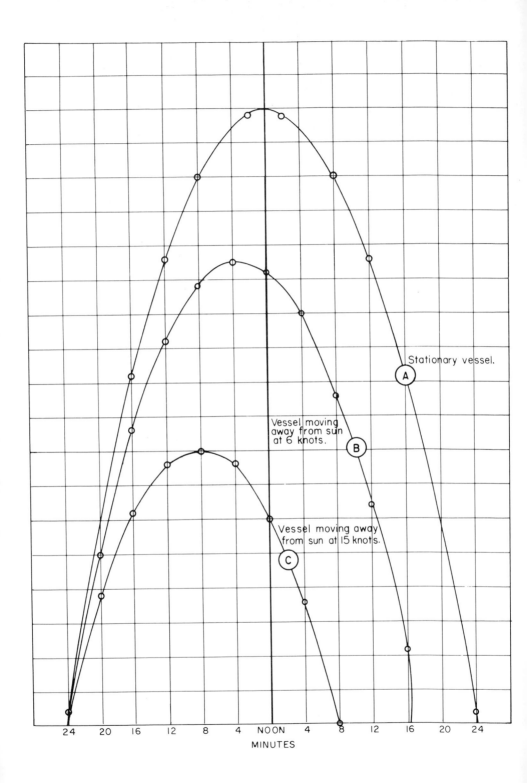

Stationary vessel.

Ⓐ

Vessel moving away from sun at 6 knots.

Ⓑ

Vessel moving away from sun at 15 knots.

Ⓒ

24 20 16 12 8 4 NOON 4 8 12 16 20 24

MINUTES

on the graph, the maximum altitudes for the two vessels moving away from the sun are shifted to earlier times than those of the stationary vessel. And, so, longitudes calculated from AM-PM sights would be in error.

The table in Figure 48 shows how much the change in longitude would be for a vessel moving six knots directly north or south. To use it for other speeds, it is only necessary to multiply by the ratio of the two speeds—i.e. for a vessel traveling 12 knots, the change would be double the value shown in the table; for 3 knots, half the value; and for 9 knots, 1½ times the value.

If you are not proceeding directly north or south, use a protractor to scale off the amount of the north or south component (See Figure 49 and complete example in following chapter.)

LONGITUDE CHECKLIST

1. GMT of AM Altitude
   + GMT of PM Same Altitude
   2 / 00-00-00

   GMT of Noon

2. From Almanac get:
   GHA of Sun at Time of Noon

3. GHA less than 180°
   GHA = Longitude West

   GHA greater than 180°
   360° − GHA = Longitude East

4. Estimate effective speed directly toward or away from sun and apply correction from Table 9.

5. Result: Your Longitude.

Shift in Longitude at Speed of 6 Knots

Declination / Latitude	24°N	21°N	18°N	15°N	12°N	9°N	6°N	3°N	0°	3°S	6°S	9°S	12°S	15°S	18°S	21°S	24°S
65°N	39'	40'	42'	43'	44'	46'	47'	48'	49'	50'	52'	53'	54'	55'	57'	58'	59'
60°N	29'	31'	32'	34'	35'	36'	37'	38'	40'	41'	42'	43'	45'	46'	47'	48'	50'
55°N	23'	24'	25'	27'	28'	29'	30'	31'	33'	34'	35'	36'	38'	39'	40'	42'	43'
50°N	17'	19'	20'	21'	22'	24'	25'	26'	27'	29'	30'	31'	32'	33'	35'	36'	38'
45°N	13'	14'	15'	17'	18'	19'	21'	22'	23'	24'	25'	27'	28'	29'	30'	32'	33'
40°N	9'	10'	12'	13'	14'	16'	17'	18'	19'	20'	22'	23'	24'	25'	27'	28'	29'
35°N	6'	7'	9'	10'	11'	12'	14'	15'	16'	17'	18'	20'	21'	22'	23'	25'	26'
30°N	3'	4'	6'	7'	8'	10'	11'	12'	13'	14'	16'	17'	18'	19'	21'	22'	23'
25°N	0'	2'	3'	5'	6'	7'	8'	9'	11'	12'	13'	14'	16'	17'	18'	19'	21'
20°N	2'	0'	1'	2'	3'	5'	6'	7'	8'	10'	11'	12'	13'	14'	16'	17'	19'
15°N	4'	3'	1'	0'	1'	3'	4'	5'	6'	7'	9'	10'	11'	12'	14'	15'	16'
10°N	6'	5'	3'	2'	1'	0'	2'	3'	4'	5'	6'	8'	9'	10'	11'	13'	14'
5°N	8'	7'	5'	4'	3'	2'	0'	1'	2'	3'	4'	6'	7'	8'	9'	11'	12'
0°	10'	9'	7'	6'	5'	4'	2'	1'	0'	1'	2'	4'	5'	6'	7'	9'	10'
5°S	12'	11'	9'	8'	7'	6'	4'	3'	2'	1'	0'	2'	3'	4'	5'	7'	8'
10°S	14'	13'	11'	10'	9'	8'	6'	5'	4'	3'	2'	0'	1'	2'	3'	5'	6'
15°S	16'	15'	14'	12'	11'	10'	9'	7'	6'	5'	4'	3'	1'	0'	1'	3'	4'
20°S	19'	17'	16'	14'	13'	12'	11'	10'	8'	7'	6'	5'	3'	2'	1'	0'	2'
25°S	21'	19'	18'	17'	16'	14'	13'	12'	11'	9'	8'	7'	6'	5'	3'	2'	0'
30°S	23'	22'	21'	19'	18'	17'	16'	14'	13'	12'	11'	10'	8'	7'	6'	4'	3'
35°S	26'	25'	23'	22'	21'	20'	18'	17'	16'	15'	14'	12'	11'	10'	9'	7'	6'
40°S	29'	28'	27'	25'	24'	23'	22'	20'	19'	18'	17'	16'	14'	13'	12'	10'	9'
45°S	33'	32'	30'	29'	28'	27'	25'	24'	23'	22'	21'	19'	18'	17'	15'	14'	13'
50°S	38'	36'	35'	33'	32'	31'	30'	29'	27'	26'	25'	24'	22'	21'	20'	19'	17'
55°S	43'	42'	40'	39'	38'	36'	35'	34'	33'	31'	30'	29'	28'	27'	25'	24'	23'
60°S	50'	48'	47'	46'	45'	43'	42'	41'	40'	38'	37'	36'	35'	34'	32'	31'	29'
65°S	59'	58'	57'	55'	54'	53'	52'	50'	49'	48'	47'	46'	44'	43'	42'	40'	39'

Figure 48. To find the correction to be applied to the GHA derived from the a.m.-p.m. equal altitude sights, find the meeting point of your DR latitude (left column) and the sun's declination (top line). This is the number of minutes to add to *or* subtract from the GHA. *Add* to GHA if vessel is moving *away* from the sun; *subtract* from GHA if vessel is moving *toward* the sun. In the example shown above, you have a DR latitude of 40° north on the day when the sun's declination is 3° south. Where they meet shows a correction of 20'. This correction is only applied fully if your vessel is going a *full* six knots directly toward or away from the sun, i.e. true north or true south. (For *speeds* other than six knots, the correction is multiplied by the ratio of the two speeds. For courses other than true north or true south, you can readily find the north or south component of such courses. See example in Figure 49.)

Figure 49. This diagram shows how to find the effective speed of a vessel moving directly toward or away from the sun when its course is *not* true north or true south. Let's assume your vessel is going six knots on a course of 60° true. Effectively, then, you are moving north-ward at three knots.

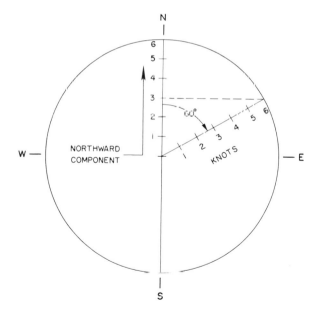

Part V

COMPLETE EXAMPLE: LONGITUDE & LATITUDE BY NOON SIGHT

LONGITUDE AT NOON

Date: 11-4-81
DR: N33°07'   W77°51'
Course: 240° True
Speed: 6 Knots

Chronometer
(Digital Watch)
on Eastern
Standard Time

1. AM Sight: Sextant Reads: 40°47' @ 11-30-47
    PM Sight: Sextant Again Reads: 40°47' @ 12-20-35
                                             2 / 23-50-82
                 Chronometer Time of Noon: 11-55-41
                 Convert EST to GMT: +   5-00-00
                 GMT of Noon: 16-55-41

2. Almanac: GHA @ 1600:   64°05!9
    Table 6:   4-Year Change: −   !07
                 GHA @ 1600:   64°05!83

    Table 3:   GHA increase in 55 minutes.   13°45'
                 GHA increase in 41 seconds: 10!25
                                                77°61!08

3. GHA @ Noon: 78°01'

4. Effect of boat's motion
    - Draw any convenient-sized circle.
    - Lay out course in degrees (True = 240° in this example).
    - Draw a line from the edge of the circle in the direction of sun's bearing at noon (has to be either N or S).
    - Compare length of (N S) line to length of course line. This is speed toward or away from sun.
    - With this speed, go to Table 4.

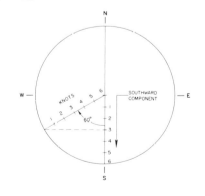

5. Table 4:
    - Nearest latitude and declination in table are 35°N and 15°S, respectively. Correction (for 6 knots directly toward or away from the sun) is 22'.
    - From Step 4 (boat in example going roughly southwest) south component—that part of speed directly toward sun—is 3 knots; so, take 3/6 (½) of correction. 22'/2 = 11'. Subtract this from uncorrected GHA at noon.

$$78°01' = 77°61'$$
$$- \quad 11'$$

Longitude @ Noon	77°50'W
Latitude @ Noon	33°08'N

I took these sights during a delivery trip from New York to Florida; and, as you can see, I took the sights for longitude well below the rule of "one minute of time for each degree of zenith distance." The results, though, were very good.

DR	Noon Position
N33°07'	N33°08'
W77°51'	W77°50'

My experience with this particular technique, and with celestial navigation in general, is that you can expect DR and fixes to be within 5% of the distance run between sights. This is in average going. In rough weather, expect a deterioration.

Interestingly, the Loran C at the time of this sight had us on shore; the depth finder agreed with the DR.

Navigation is an art and consists of, first and foremost, keeping the DR and, second, comparing other information to the DR to verify, update, confirm or refine it.

In this case, it was easy to tell that the Loran was wrong—we weren't aground. But, in any case, when three out of four items agree (DR, noon sight, depth finder), it's unlikely that the fourth (the Loran) is correct. Always check one thing against another.

LATITUDE AT NOON

Date: 11-4-81
DR: N33°07′   W77°51′
Course: 240° True
Speed: 6 Knots
Watch Reading at Maximum Altitude: 11-56

Chronometer
(Digital Watch)
on Eastern
Standard Time

1.	Maximum Sextant Reading (Hs)	41°08′
	Index Correction	+ 2′
	Dip (always −)	− 3′
	Apparent Altitude (ha)	41°07′
	Semi-Diameter + ☉ − ☉	+ 16′
	Refraction (always −)	1′
	Observed (True) Altitude (Ho)	41°22′

2.	90° Written so ′ can be subtracted	89°60′
	Observed Altitude (always subtracted)	−41°22′
	Zenith Distance	48°38′

Approximate GMT of Noon = 11-56 + 5 hours = 1700

	From Almanac, Declination =	S15°29′3
	4-Year Change (Table 6):	+ ′42
		S15°29′72
	Rounding Off:	S15°30′

3.	Zenith Distance	48°38′
	Declination	15°30′S
	Latitude	33°08′N

While there are obviously lots of cute rules about how to combine declination and zenith distance, the most practical approach is to simply find the combination that comes out closest to the DR.

Another way is to look at the situation from the outside, above the surface of the earth, as it were, and draw it.

Complete Example: Longitude & Latitude by Noon Sight   _121_

For this example:

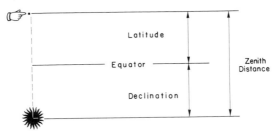

*Latitude = Zenith Distance − Declination

*The other possible arrangements are:

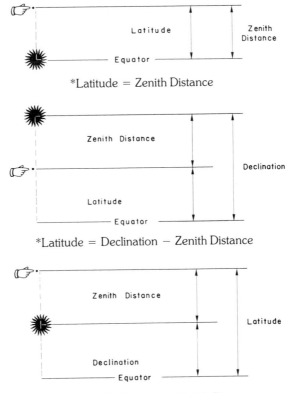

*Latitude = Zenith Distance

*Latitude = Declination − Zenith Distance

*Latitude = Declination + Zenith Distance

*For positions in south latitudes, the possible alignments are shown by turning the drawings upside down.

The important thing, though, is to remember the three great principles and Bedrock's Rule:

I. Latitude is the length of the arc (in degrees and minutes) on the surface of the earth between the equator and you. This is what you want to find.
II. Declination is the length of the arc on the surface of the earth between the equator and the GP of the sun. This info is in the almanac.
III. Zenith distance is the length of the arc on the surface of the earth between the GP of the sun and you. You use your sextant to find this: ZD = 90° − Ho.

Bedrock's Rule: If you know two out of three, that's not bad.

Corollary: If they're related, you can *always* figure out the third.

Part VI

TABLES

1. Dip
2. Refraction
3. GHA Interpolation
4. Shift in Longitude at Speed of 6 knots
5. Long Term Almanac—Base year 1976
6. Long Term Almanac—Base year 1977
7. Long Term Almanac—Base year 1978
8. Long Term Almanac—Base year 1979
9. Four-year change in GHA and Declination

Table 1

Dip

Height of eye in meters (m)—left column
Height of eye in feet (ft.)—right column
Dip correction (always subtractive)—middle column

Ht. of Eye	Corrn	Ht. of Eye		Ht. of Eye	Corrn	Ht. of Eye
m	′	ft.		m	′	ft.
2·4	−2·8	8·0		9·9	−5·6	32·7
2·6	−2·9	8·6		10·3	−5·7	33·9
2·8	−3·0	9·2		10·6	−5·8	35·1
3·0	−3·1	9·8		11·0	−5·9	36·3
3·2	−3·2	10·5		11·4	−6·0	37·6
3·4	−3·3	11·2		11·8	−6·1	38·9
3·6	−3·4	11·9		12·2	−6·2	40·1
3·8	−3·5	12·6		12·6	−6·3	41·5
4·0	−3·6	13·3		13·0	−6·4	42·8
4·3	−3·7	14·1		13·4	−6·5	44·2
4·5	−3·8	14·9		13·8	−6·6	45·5
4·7	−3·9	15·7		14·2	−6·7	46·9
5·0	−4·0	16·5		14·7	−6·8	48·4
5·2	−4·1	17·4		15·1	−6·9	49·8
5·5	−4·2	18·3		15·5	−7·0	51·3
5·8	−4·3	19·1		16·0	−7·1	52·8
6·1	−4·4	20·1		16·5	−7·2	54·3
6·3	−4·5	21·0		16·9	−7·3	55·8
6·6	−4·6	22·0		17·4	−7·4	57·4
6·9	−4·7	22·9		17·9	−7·5	58·9
7·2	−4·8	23·9		18·4	−7·6	60·5
7·5	−4·9	24·9		18·8	−7·7	62·1
7·9	−5·0	26·0		19·3	−7·8	63·8
8·2	−5·1	27·1		19·8	−7·9	65·4
8·5	−5·2	28·1		20·4	−8·0	67·1
8·8	−5·3	29·2		20·9	−8·1	68·8
9·2	−5·4	30·4		21·4		70·5
9·5	−5·5	31·5				

Table 2

REFRACTION

App. Alt.	Corrn	App. Alt.	Corrn
° ′	′	° ′	′
9 56		20 42	
	−5·3		−2·5
10 08		21 28	
	−5·2		−2·4
10 20		22 19	
	−5·1		−2·3
10 33		23 13	
	−5·0		−2·2
10 46		24 11	
	−4·9		−2·1
11 00		25 14	
	−4·8		−2·0
11 14		26 22	
	−4·7		−1·9
11 29		27 36	
	−4·6		−1·8
11 45		28 56	
	−4·5		−1·7
12 01		30 24	
	−4·4		−1·6
12 18		32 00	
	−4·3		−1·5
12 35		33 45	
	−4·2		−1·4
12 54		35 40	
	−4·1		−1·3
13 13		37 48	
	−4·0		−1·2
13 33		40 08	
	−3·9		−1·1
13 54		42 44	
	−3·8		−1·0
14 16		45 36	
	−3·7		−0·9
14 40		48 47	
	−3·6		−0·8
15 04		52 18	
	−3·5		−0·7
15 30		56 11	
	−3·4		−0·6
15 57		60 28	
	−3·3		−0·5
16 26		65 08	
	−3·2		−0·4
16 56		70 11	
	−3·1		−0·3
17 28		75 34	
	−3·0		−0·2
18 02		81 13	
	−2·9		−0·1
18 38		87 03	
	−2·8		0·0
19 17		90 00	
	−2·7		
19 58			
	−2·6		

App. Alt. = Apparent altitude = Sextant altitude corrected for index error and dip.

Table 3

GHA INTERPOLATION

m	° ′	m	° ′		s	′ ″	s	′ ″
0	0 00	30	7 30		0	0 00	30	7 30
1	0 15	31	7 45		1	0 15	31	7 45
2	0 30	32	8 00		2	0 30	32	8 00
3	0 45	33	8 15		3	0 45	33	8 15
4	1 00	34	8 30		4	1 00	34	8 30
5	1 15	35	8 45		5	1 15	35	8 45
6	1 30	36	9 00		6	1 30	36	9 00
7	1 45	37	9 15		7	1 45	37	9 15
8	2 00	38	9 30		8	2 00	38	9 30
9	2 15	39	9 45		9	2 15	39	9 45
10	2 30	40	10 00		10	2 30	40	10 00
11	2 45	41	10 15		11	2 45	41	10 15
12	3 00	42	10 30		12	3 00	42	10 30
13	3 15	43	10 45		13	3 15	43	10 45
14	3 30	44	11 00		14	3 30	44	11 00
15	3 45	45	11 15		15	3 45	45	11 15
16	4 00	46	11 30		16	4 00	46	11 30
17	4 15	47	11 45		17	4 15	47	11 45
18	4 30	48	12 00		18	4 30	48	12 00
19	4 45	49	12 15		19	4 45	49	12 15
20	5 00	50	12 30		20	5 00	50	12 30
21	5 15	51	12 45		21	5 15	51	12 45
22	5 30	52	13 00		22	5 30	52	13 00
23	5 45	53	13 15		23	5 45	53	13 15
24	6 00	54	13 30		24	6 00	54	13 30
25	6 15	55	13 45		25	6 15	55	13 45
26	6 30	56	14 00		26	6 30	56	14 00
27	6 45	57	14 15		27	6 45	57	14 15
28	7 00	58	14 30		28	7 00	58	14 30
29	7 15	59	14 45		29	7 15	59	14 45

Table 4

Shift in Longitude at Speed of 6 Knots

Declination Latitude	24°N	21°N	18°N	15°N	12°N	9°N	6°N	3°N	0°	3°S	6°S	9°S	12°S	15°S	18°S	21°S	24°S
65°N	39'	40'	42'	43'	44'	46'	47'	48'	49'	50'	52'	53'	54'	55'	57'	58'	59'
60°N	29'	31'	32'	34'	35'	36'	37'	38'	40'	41'	42'	43'	45'	46'	47'	48'	50'
55°N	23'	24'	25'	27'	28'	29'	30'	31'	33'	34'	35'	36'	38'	39'	40'	42'	43'
50°N	17'	19'	20'	21'	22'	24'	25'	26'	27'	29'	30'	31'	32'	33'	35'	36'	38'
45°N	13'	14'	15'	17'	18'	19'	21'	22'	23'	24'	25'	27'	28'	29'	30'	32'	33'
40°N	9'	10'	12'	13'	14'	15'	17'	18'	19'	20'	22'	23'	24'	25'	27'	28'	29'
35°N	6'	7'	9'	10'	11'	12'	14'	15'	16'	17'	18'	20'	21'	22'	23'	25'	26'
30°N	3'	4'	6'	7'	8'	9'	11'	12'	13'	14'	16'	17'	18'	19'	21'	22'	23'
25°N	0'	2'	3'	4'	6'	7'	8'	9'	11'	12'	13'	14'	16'	17'	18'	19'	21'
20°N	2'	0'	1'	2'	3'	5'	6'	7'	8'	10'	11'	12'	13'	14'	16'	17'	19'
15°N	4'	3'	1'	0'	1'	2'	4'	5'	6'	7'	9'	10'	11'	12'	14'	15'	16'
10°N	6'	5'	3'	2'	1'	0'	2'	3'	4'	5'	6'	8'	9'	10'	11'	13'	14'
5°N	8'	7'	5'	4'	3'	2'	0'	1'	2'	3'	4'	6'	7'	8'	9'	11'	12'
0°	10'	9'	7'	6'	5'	4'	2'	1'	0'	1'	2'	4'	5'	6'	7'	9'	10'
5°S	12'	11'	9'	8'	7'	6'	4'	3'	2'	1'	0'	2'	3'	4'	5'	7'	8'
10°S	14'	13'	11'	10'	9'	8'	6'	5'	4'	3'	2'	0'	1'	2'	3'	5'	6'
15°S	16'	15'	14'	12'	11'	10'	9'	7'	6'	5'	4'	2'	1'	0'	1'	3'	4'
20°S	19'	17'	16'	14'	13'	12'	11'	10'	8'	7'	6'	5'	3'	2'	1'	1'	2'
25°S	21'	19'	18'	17'	16'	14'	13'	12'	11'	9'	8'	7'	6'	4'	3'	2'	0'
30°S	23'	22'	21'	19'	18'	17'	16'	14'	13'	12'	11'	9'	8'	7'	6'	4'	3'
35°S	26'	25'	23'	22'	21'	20'	18'	17'	16'	15'	14'	12'	11'	10'	9'	7'	6'
40°S	29'	28'	27'	25'	24'	23'	22'	20'	19'	18'	17'	15'	14'	13'	12'	10'	9'
45°S	33'	32'	30'	29'	28'	27'	25'	24'	23'	22'	21'	19'	18'	17'	15'	14'	13'
50°S	38'	36'	35'	33'	32'	31'	30'	29'	27'	26'	25'	24'	22'	21'	20'	19'	17'
55°S	43'	42'	40'	39'	38'	36'	35'	34'	33'	31'	30'	29'	28'	27'	25'	24'	23'
60°S	50'	48'	47'	46'	45'	43'	42'	41'	40'	38'	37'	36'	35'	34'	32'	31'	29'
65°S	59'	58'	57'	55'	54'	53'	52'	50'	49'	48'	47'	46'	44'	43'	42'	40'	39'

Table 5

DAILY LONG-TERM ALMANAC OF THE SUN FOR THE LEAP YEARS 1976, 1980, 1984, 1988, 1992, 1996, 2000, ETC.*

This table is a compilation of the columns for Greenwich Hour Angle and declination of the sun from the 1976 *Nautical Almanac.* For use beyond 1976, a correction may be necessary, depending upon the degree of precision desired. The amount of change in *4 years* in the GHA and declination (DEC) is given for each 3-day interval in Table 9.

Since the correction factors are changing slowly with the passage of time, this table is *not* a perpetual almanac, but an almanac designed for use in practical navigation for a period of about 30 years.

*See note at end of table on use of 4-year correction factors.

JANUARY

Days 1, 4, 7

d h	G.H.A.	Dec.	G.H.A.	Dec.	G.H.A.	Dec.
1 00	179 14.4	S23 05.2	**4** 178 53.2	S22 49.8	**7** 178 32.9	S22 30.2
01	194 14.1	05.0	193 52.9	49.5	193 32.6	29.9
02	209 13.8	04.8	208 52.6	49.3	208 32.3	29.6
03	224 13.5	·· 04.6	223 52.3	·· 49.0	223 32.0	·· 29.3
04	239 13.2	04.4	238 52.0	48.8	238 31.8	29.0
05	254 12.9	04.2	253 51.7	48.5	253 31.5	28.7
06	269 12.6	S23 04.1	268 51.5	S22 48.3	268 31.2	S22 28.4
07	284 12.3	03.9	283 51.2	48.0	283 30.9	28.1
08	299 12.0	03.7	298 50.9	47.8	298 30.7	27.8
09	314 11.7	·· 03.5	313 50.6	·· 47.5	313 30.4	·· 27.5
10	329 11.4	03.3	328 50.3	47.3	328 30.1	27.2
11	344 11.1	03.1	343 50.0	47.0	343 29.8	26.9
12	359 10.8	S23 02.9	358 49.7	S22 46.8	358 29.6	S22 26.6
13	14 10.5	02.7	13 49.4	46.5	13 29.3	26.3
14	29 10.2	02.5	28 49.2	46.3	28 29.0	26.0
15	44 09.9	·· 02.3	43 48.9	·· 46.0	43 28.7	·· 25.7
16	59 09.6	02.1	58 48.6	45.8	58 28.5	25.4
17	74 09.3	01.9	73 48.3	45.5	73 28.2	25.1
18	89 09.0	S23 01.7	88 48.0	S22 45.3	88 27.9	S22 24.7
19	104 08.7	01.5	103 47.7	45.0	103 27.7	24.4
20	119 08.4	01.3	118 47.4	44.7	118 27.4	24.1
21	134 08.1	·· 01.1	133 47.2	·· 44.5	133 27.1	·· 23.8
22	149 07.8	00.9	148 46.9	44.2	148 26.8	23.5
23	164 07.5	00.7	163 46.6	44.0	163 26.6	23.2

Days 2, 5, 8

d h	G.H.A.	Dec.	G.H.A.	Dec.	G.H.A.	Dec.
2 00	179 07.2	S23 00.5	**5** 178 46.3	S22 43.7	**8** 178 26.3	S22 22.9
01	194 06.9	00.3	193 46.0	43.4	193 26.0	22.5
02	209 06.6	23 00.1	208 45.7	43.2	208 25.8	22.2
03	224 06.3	22 59.9	223 45.4	·· 42.9	223 25.5	·· 21.9
04	239 06.0	59.7	238 45.2	42.6	238 25.2	21.6
05	254 05.8	59.5	253 44.9	42.4	253 25.0	21.3
06	269 05.5	S22 59.3	268 44.6	S22 42.1	268 24.7	S22 20.9
07	284 05.2	59.0	283 44.3	41.8	283 24.4	20.6
08	299 04.9	58.8	298 44.0	41.6	298 24.2	20.3
09	314 04.6	·· 58.6	313 43.7	·· 41.3	313 23.9	·· 20.0
10	329 04.3	58.4	328 43.5	41.0	328 23.6	19.6
11	344 04.0	58.2	343 43.2	40.8	343 23.4	19.3
12	359 03.7	S22 58.0	358 42.9	S22 40.5	358 23.1	S22 19.0
13	14 03.4	57.8	13 42.6	40.2	13 22.8	18.7
14	29 03.1	57.6	28 42.3	40.0	28 22.5	18.3
15	44 02.8	·· 57.3	43 42.0	·· 39.7	43 22.3	·· 18.0
16	59 02.5	57.1	58 41.8	39.4	58 22.0	17.7
17	74 02.2	56.9	73 41.5	39.1	73 21.7	17.3
18	89 01.9	S22 56.7	88 41.2	S22 38.9	88 21.5	S22 17.0
19	104 01.6	56.5	103 40.9	38.6	103 21.2	16.7
20	119 01.3	56.2	118 40.6	38.3	118 21.0	16.4
21	134 01.0	·· 56.0	133 40.4	·· 38.0	133 20.7	·· 16.0
22	149 00.7	55.8	148 40.1	37.8	148 20.4	15.7
23	164 00.5	55.6	163 39.8	37.5	163 20.2	15.4

Days 3, 6, 9

d h	G.H.A.	Dec.	G.H.A.	Dec.	G.H.A.	Dec.
3 00	179 00.2	S22 55.4	**6** 178 39.5	S22 37.2	**9** 178 19.9	S22 15.0
01	193 59.9	55.1	193 39.2	36.9	193 19.6	14.7
02	208 59.6	54.9	208 39.0	36.6	208 19.4	14.3
03	223 59.3	·· 54.7	223 38.7	·· 36.4	223 19.1	·· 14.0
04	238 59.0	54.5	238 38.4	36.1	238 18.8	13.7
05	253 58.7	54.2	253 38.1	35.8	253 18.6	13.3
06	268 58.4	S22 54.0	268 37.8	S22 35.5	268 18.3	S22 13.0
07	283 58.1	53.8	283 37.6	35.2	283 18.0	12.6
08	298 57.8	53.5	298 37.3	34.9	298 17.8	12.3
09	313 57.5	·· 53.3	313 37.0	·· 34.6	313 17.5	·· 12.0
10	328 57.2	53.1	328 36.7	34.4	328 17.3	11.6
11	343 57.0	52.8	343 36.4	34.1	343 17.0	11.3
12	358 56.7	S22 52.6	358 36.2	S22 33.8	358 16.7	S22 10.9
13	13 56.4	52.4	13 35.9	33.5	13 16.5	10.6
14	28 56.1	52.1	28 35.6	33.2	28 16.2	10.2
15	43 55.8	·· 51.9	43 35.3	·· 32.9	43 15.9	·· 09.9
16	58 55.5	51.7	58 35.1	32.6	58 15.7	09.6
17	73 55.2	51.4	73 34.8	32.3	73 15.4	09.2
18	88 54.9	S22 51.2	88 34.5	S22 32.0	88 15.2	S22 08.9
19	103 54.6	51.0	103 34.2	31.7	103 14.9	08.5
20	118 54.3	50.7	118 34.0	31.4	118 14.6	08.2
21	133 54.0	·· 50.5	133 33.7	·· 31.1	133 14.4	·· 07.8
22	148 53.8	50.2	148 33.4	30.8	148 14.1	07.5
23	163 53.5	50.0	163 33.1	30.5	163 13.9	07.1

	G.H.A.	Dec.
10 00	178 13.6	S22 06.7
01	193 13.3	06.4
02	208 13.1	06.0
03	223 12.8 ··	05.7
04	238 12.6	05.3
05	253 12.3	05.0
06	268 12.1	S22 04.6
07	283 11.8	04.3
08	298 11.5	03.9
09	313 11.3 ··	03.5
10	328 11.0	03.2
11	343 10.8	02.8
12	358 10.5	S22 02.4
13	13 10.3	02.1
14	28 10.0	01.7
15	43 09.7 ··	01.4
16	58 09.5	01.0
17	73 09.2	00.6
18	88 09.0	S22 00.3
19	103 08.7	21 59.9
20	118 08.5	59.5
21	133 08.2 ··	59.2
22	148 08.0	58.8
23	163 07.7	58.4
11 00	178 07.5	S21 58.0
01	193 07.2	57.7
02	208 07.0	57.3
03	223 06.7 ··	56.9
04	238 06.4	56.5
05	253 06.2	56.2
06	268 05.9	S21 55.8
07	283 05.7	55.4
08	298 05.4	55.0
09	313 05.2 ··	54.7
10	328 04.9	54.3
11	343 04.7	53.9
12	358 04.4	S21 53.5
13	13 04.2	53.1
14	28 03.9	52.8
15	43 03.7 ··	52.4
16	58 03.4	52.0
17	73 03.2	51.6
18	88 02.9	S21 51.2
19	103 02.7	50.8
20	118 02.4	50.5
21	133 02.2 ··	50.1
22	148 02.0	49.7
23	163 01.7	49.3
12 00	178 01.5	S21 48.9
01	193 01.2	48.5
02	208 01.0	48.1
03	223 00.7 ··	47.7
04	238 00.5	47.3
05	253 00.2	47.0
06	268 00.0	S21 46.6
07	282 59.7	46.2
08	297 59.5	45.8
09	312 59.2 ··	45.4
10	327 59.0	45.0
11	342 58.8	44.6
12	357 58.5	S21 44.2
13	12 58.3	43.8
14	27 58.0	43.4
15	42 57.8 ··	43.0
16	57 57.5	42.6
17	72 57.3	42.2
18	87 57.1	S21 41.8
19	102 56.8	41.4
20	117 56.6	41.0
21	132 56.3 ··	40.6
22	147 56.1	40.2
23	162 55.9	39.8

	G.H.A.	Dec.
13 00	177 55.6	S21 39.4
01	192 55.4	38.9
02	207 55.1	38.5
03	222 54.9 ··	38.1
04	237 54.7	37.7
05	252 54.4	37.3
06	267 54.2	S21 36.9
07	282 53.9	36.5
08	297 53.7	36.1
09	312 53.5 ··	35.7
10	327 53.2	35.3
11	342 53.0	34.8
12	357 52.7	S21 34.4
13	12 52.5	34.0
14	27 52.3	33.6
15	42 52.0 ··	33.2
16	57 51.8	32.8
17	72 51.6	32.3
18	87 51.3	S21 31.9
19	102 51.1	31.5
20	117 50.9	31.1
21	132 50.6 ··	30.7
22	147 50.4	30.2
23	162 50.2	29.8
14 00	177 49.9	S21 29.4
01	192 49.7	29.0
02	207 49.5	28.5
03	222 49.2 ··	28.1
04	237 49.0	27.7
05	252 48.8	27.3
06	267 48.5	S21 26.8
07	282 48.3	26.4
08	297 48.1	26.0
09	312 47.8 ··	25.5
10	327 47.6	25.1
11	342 47.4	24.7
12	357 47.1	S21 24.2
13	12 46.9	23.8
14	27 46.7	23.4
15	42 46.4 ··	22.9
16	57 46.2	22.5
17	72 46.0	22.1
18	87 45.8	S21 21.6
19	102 45.5	21.2
20	117 45.3	20.8
21	132 45.1 ··	20.3
22	147 44.8	19.9
23	162 44.6	19.4
15 00	177 44.4	S21 19.0
01	192 44.2	18.6
02	207 43.9	18.1
03	222 43.7 ··	17.7
04	237 43.5	17.2
05	252 43.3	16.8
06	267 43.0	S21 16.3
07	282 42.8	15.9
08	297 42.6	15.4
09	312 42.4 ··	15.0
10	327 42.1	14.5
11	342 41.9	14.1
12	357 41.7	S21 13.6
13	12 41.5	13.2
14	27 41.2	12.7
15	42 41.0 ··	12.3
16	57 40.8	11.8
17	72 40.6	11.4
18	87 40.4	S21 10.9
19	102 40.1	10.5
20	117 39.9	10.0
21	132 39.7 ··	09.6
22	147 39.5	09.1
23	162 39.2	08.7

	G.H.A.	Dec.
16 00	177 39.0	S21 08.2
01	192 38.8	07.7
02	207 38.6	07.3
03	222 38.4 ··	06.8
04	237 38.1	06.4
05	252 37.9	05.9
06	267 37.7	S21 05.4
07	282 37.5	05.0
08	297 37.3	04.5
09	312 37.1 ··	04.0
10	327 36.8	03.6
11	342 36.6	03.1
12	357 36.4	S21 02.6
13	12 36.2	02.2
14	27 36.0	01.7
15	42 35.8 ··	01.2
16	57 35.5	00.8
17	72 35.3	21 00.3
18	87 35.1	S20 59.8
19	102 34.9	59.4
20	117 34.7	58.9
21	132 34.5 ··	58.4
22	147 34.3	57.9
23	162 34.0	57.5
17 00	177 33.8	S20 57.0
01	192 33.6	56.5
02	207 33.4	56.0
03	222 33.2 ··	55.6
04	237 33.0	55.1
05	252 32.8	54.6
06	267 32.6	S20 54.1
07	282 32.3	53.7
08	297 32.1	53.2
09	312 31.9 ··	52.7
10	327 31.7	52.2
11	342 31.5	51.7
12	357 31.3	S20 51.2
13	12 31.1	50.8
14	27 30.9	50.3
15	42 30.7 ··	49.8
16	57 30.5	49.3
17	72 30.3	48.8
18	87 30.0	S20 48.3
19	102 29.8	47.8
20	117 29.6	47.4
21	132 29.4 ··	46.9
22	147 29.2	46.4
23	162 29.0	45.9
18 00	177 28.8	S20 45.4
01	192 28.6	44.9
02	207 28.4	44.4
03	222 28.2 ··	43.9
04	237 28.0	43.4
05	252 27.8	42.9
06	267 27.6	S20 42.4
07	282 27.4	41.9
08	297 27.2	41.4
09	312 27.0 ··	40.9
10	327 26.8	40.4
11	342 26.6	39.9
12	357 26.4	S20 39.4
13	12 26.2	38.9
14	27 26.0	38.4
15	42 25.8 ··	37.9
16	57 25.6	37.4
17	72 25.4	36.9
18	87 25.2	S20 36.4
19	102 25.0	35.9
20	117 24.8	35.4
21	132 24.6 ··	34.9
22	147 24.4	34.4
23	162 24.2	33.9

	G.H.A.	Dec.		G.H.A.	Dec.		G.H.A.	Dec.
d h	° ′	° ′	d h	° ′	° ′	d h	° ′	° ′
19 00	177 24.0	S20 33.4	22 00	177 10.5	S19 55.1	25 00	176 58.7	S19 13.4
01	192 23.8	32.9	01	192 10.3	54.5	01	191 58.5	12.8
02	207 23.6	32.4	02	207 10.1	54.0	02	206 58.4	12.2
03	222 23.4	·· 31.9	03	222 10.0	·· 53.4	03	221 58.2	·· 11.6
04	237 23.2	31.4	04	237 09.8	52.9	04	236 58.1	11.0
05	252 23.0	30.9	05	252 09.6	52.3	05	251 57.9	10.4
06	267 22.8	S20 30.3	06	267 09.4	S19 51.8	06	266 57.8	S19 09.8
07	282 22.6	29.8	07	282 09.3	51.2	07	281 57.6	09.2
08	297 22.4	29.3	08	297 09.1	50.6	08	296 57.5	08.6
09	312 22.2	·· 28.8	09	312 08.9	·· 50.1	09	311 57.3	·· 08.0
10	327 22.0	28.3	10	327 08.8	49.5	10	326 57.2	07.4
11	342 21.8	27.8	11	342 08.6	48.9	11	341 57.0	06.8
12	357 21.6	S20 27.3	12	357 08.4	S19 48.4	12	356 56.9	S19 06.2
13	12 21.4	26.7	13	12 08.2	47.8	13	11 56.7	05.6
14	27 21.2	26.2	14	27 08.1	47.3	14	26 56.6	05.0
15	42 21.0	·· 25.7	15	42 07.9	·· 46.7	15	41 56.4	·· 04.4
16	57 20.8	25.2	16	57 07.7	46.1	16	56 56.3	03.7
17	72 20.6	24.7	17	72 07.6	45.6	17	71 56.2	03.1
18	87 20.4	S20 24.1	18	87 07.4	S19 45.0	18	86 56.0	S19 02.5
19	102 20.2	23.6	19	102 07.2	44.4	19	101 55.9	01.9
20	117 20.1	23.1	20	117 07.0	43.9	20	116 55.7	01.3
21	132 19.9	·· 22.6	21	132 06.9	·· 43.3	21	131 55.6	·· 00.7
22	147 19.7	22.1	22	147 06.7	42.7	22	146 55.4	19 00.1
23	162 19.5	21.5	23	162 06.5	42.1	23	161 55.3	18 59.4
20 00	177 19.3	S20 21.0	23 00	177 06.4	S19 41.6	26 00	176 55.1	S18 58.8
01	192 19.1	20.5	01	192 06.2	41.0	01	191 55.0	58.2
02	207 18.9	20.0	02	207 06.0	40.4	02	206 54.9	57.6
03	222 18.7	·· 19.4	03	222 05.9	·· 39.9	03	221 54.7	·· 57.0
04	237 18.5	18.9	04	237 05.7	39.3	04	236 54.6	56.4
05	252 18.3	18.4	05	252 05.5	38.7	05	251 54.4	55.7
06	267 18.2	S20 17.9	06	267 05.4	S19 38.1	06	266 54.3	S18 55.1
07	282 18.0	17.3	07	282 05.2	37.6	07	281 54.1	54.5
08	297 17.8	16.8	08	297 05.0	37.0	08	296 54.0	53.9
09	312 17.6	·· 16.3	09	312 04.9	·· 36.4	09	311 53.9	·· 53.3
10	327 17.4	15.7	10	327 04.7	35.8	10	326 53.7	52.6
11	342 17.2	15.2	11	342 04.5	35.3	11	341 53.6	52.0
12	357 17.0	S20 14.7	12	357 04.4	S19 34.7	12	356 53.4	S18 51.4
13	12 16.8	14.1	13	12 04.2	34.1	13	11 53.3	50.8
14	27 16.6	13.6	14	27 04.1	33.5	14	26 53.2	50.2
15	42 16.5	·· 13.1	15	42 03.9	·· 32.9	15	41 53.0	·· 49.5
16	57 16.3	12.5	16	57 03.7	32.4	16	56 52.9	48.9
17	72 16.1	12.0	17	72 03.6	31.8	17	71 52.7	48.3
18	87 15.9	S20 11.5	18	87 03.4	S19 31.2	18	86 52.6	S18 47.7
19	102 15.7	10.9	19	102 03.2	30.6	19	101 52.5	47.0
20	117 15.5	10.4	20	117 03.1	30.0	20	116 52.3	46.4
21	132 15.4	·· 09.9	21	132 02.9	·· 29.4	21	131 52.2	·· 45.8
22	147 15.2	09.3	22	147 02.8	28.9	22	146 52.1	45.1
23	162 15.0	08.8	23	162 02.6	28.3	23	161 51.9	44.5
21 00	177 14.8	S20 08.2	24 00	177 02.4	S19 27.7	27 00	176 51.8	S18 43.9
01	192 14.6	07.7	01	192 02.3	27.1	01	191 51.6	43.2
02	207 14.4	07.2	02	207 02.1	26.5	02	206 51.5	42.6
03	222 14.3	·· 06.6	03	222 02.0	·· 25.9	03	221 51.4	·· 42.0
04	237 14.1	06.1	04	237 01.8	25.3	04	236 51.2	41.4
05	252 13.9	05.5	05	252 01.6	24.7	05	251 51.1	40.7
06	267 13.7	S20 05.0	06	267 01.5	S19 24.2	06	266 51.0	S18 40.1
07	282 13.5	04.5	07	282 01.3	23.6	07	281 50.8	39.5
08	297 13.3	03.9	08	297 01.2	23.0	08	296 50.7	38.8
09	312 13.2	·· 03.4	09	312 01.0	·· 22.4	09	311 50.6	·· 38.2
10	327 13.0	02.8	10	327 00.9	21.8	10	326 50.4	37.5
11	342 12.8	02.3	11	342 00.7	21.2	11	341 50.3	36.9
12	357 12.6	S20 01.7	12	357 00.5	S19 20.6	12	356 50.2	S18 36.3
13	12 12.4	01.2	13	12 00.4	20.0	13	11 50.0	35.6
14	27 12.3	00.6	14	27 00.2	19.4	14	26 49.9	35.0
15	42 12.1	20 00.1	15	42 00.1	·· 18.8	15	41 49.8	·· 34.4
16	57 11.9	19 59.5	16	56 59.9	18.2	16	56 49.7	33.8
17	72 11.7	59.0	17	71 59.8	17.6	17	71 49.5	33.1
18	87 11.6	S19 58.4	18	86 59.6	S19 17.0	18	86 49.4	S18 32.4
19	102 11.4	57.9	19	101 59.5	16.4	19	101 49.3	31.8
20	117 11.2	57.3	20	116 59.3	15.8	20	116 49.1	31.2
21	132 11.0	·· 56.8	21	131 59.1	·· 15.2	21	131 49.0	·· 30.5
22	147 10.8	56.2	22	146 59.0	14.6	22	146 48.9	29.9
23	162 10.7	55.7	23	161 58.8	14.0	23	161 48.7	29.2

134

d	h	G.H.A.		Dec.	
		°	′	°	′
28	00	176	48.6	S18	28.6
	01	191	48.5		27.9
	02	206	48.4		27.3
	03	221	48.2	··	26.6
	04	236	48.1		26.0
	05	251	48.0		25.4
	06	266	47.9	S18	24.7
	07	281	47.7		24.1
	08	296	47.6		23.4
	09	311	47.5	··	22.8
	10	326	47.4		22.1
	11	341	47.2		21.5
	12	356	47.1	S18	20.8
	13	11	47.0		20.2
	14	26	46.9		19.5
	15	41	46.7	··	18.9
	16	56	46.6		18.2
	17	71	46.5		17.5
	18	86	46.4	S18	16.9
	19	101	46.3		16.2
	20	116	46.1		15.6
	21	131	46.0	··	14.9
	22	146	45.9		14.3
	23	161	45.8		13.6
29	00	176	45.7	S18	13.0
	01	191	45.5		12.3
	02	206	45.4		11.6
	03	221	45.3	··	11.0
	04	236	45.2		10.3
	05	251	45.1		09.7
	06	266	44.9	S18	09.0
	07	281	44.8		08.3
	08	296	44.7		07.7
	09	311	44.6	··	07.0
	10	326	44.5		06.3
	11	341	44.4		05.7
	12	356	44.3	S18	05.0
	13	11	44.1		04.4
	14	26	44.0		03.7
	15	41	43.9	··	03.0
	16	56	43.8		02.4
	17	71	43.7		01.7
	18	86	43.6	S18	01.0
	19	101	43.5	18	00.3
	20	116	43.3	17	59.7
	21	131	43.2	··	59.0
	22	146	43.1		58.3
	23	161	43.0		57.7
30	00	176	42.9	S17	57.0
	01	191	42.8		56.3
	02	206	42.7		55.7
	03	221	42.6	··	55.0
	04	236	42.5		54.3
	05	251	42.3		53.6
	06	266	42.2	S17	53.0
	07	281	42.1		52.3
	08	296	42.0		51.6
	09	311	41.9	··	50.9
	10	326	41.8		50.3
	11	341	41.7		49.6
	12	356	41.6	S17	48.9
	13	11	41.5		48.2
	14	26	41.4		47.5
	15	41	41.3	··	46.9
	16	56	41.2		46.2
	17	71	41.1		45.5
	18	86	41.0	S17	44.8
	19	101	40.9		44.1
	20	116	40.7		43.5
	21	131	40.6	··	42.8
	22	146	40.5		42.1
	23	161	40.4		41.4

FEBRUARY

d	h	G.H.A.		Dec.	
		°	′	°	′
31	00	176	40.3	S17	40.7
	01	191	40.2		40.0
	02	206	40.1		39.3
	03	221	40.0	··	38.7
	04	236	39.9		38.0
	05	251	39.8		37.3
	06	266	39.7	S17	36.6
	07	281	39.6		35.9
	08	296	39.5		35.2
	09	311	39.4	··	34.5
	10	326	39.3		33.8
	11	341	39.2		33.2
	12	356	39.1	S17	32.5
	13	11	39.0		31.8
	14	26	38.9		31.1
	15	41	38.8	··	30.4
	16	56	38.7		29.7
	17	71	38.6		29.0
	18	86	38.5	S17	28.3
	19	101	38.5		27.6
	20	116	38.4		26.9
	21	131	38.3	··	26.2
	22	146	38.2		25.5
	23	161	38.1		24.8
1	00	176	38.0	S17	24.1
	01	191	37.9		23.4
	02	206	37.8		22.7
	03	221	37.7	··	22.0
	04	236	37.6		21.3
	05	251	37.5		20.6
	06	266	37.4	S17	19.9
	07	281	37.3		19.2
	08	296	37.2		18.5
	09	311	37.1	··	17.8
	10	326	37.1		17.1
	11	341	37.0		16.4
	12	356	36.9	S17	15.7
	13	11	36.8		15.0
	14	26	36.7		14.3
	15	41	36.6	··	13.6
	16	56	36.5		12.9
	17	71	36.4		12.2
	18	86	36.3	S17	11.5
	19	101	36.3		10.8
	20	116	36.2		10.1
	21	131	36.1	··	09.4
	22	146	36.0		08.6
	23	161	35.9		07.9
2	00	176	35.8	S17	07.2
	01	191	35.7		06.5
	02	206	35.7		05.8
	03	221	35.6	··	05.1
	04	236	35.5		04.4
	05	251	35.4		03.7
	06	266	35.3	S17	03.0
	07	281	35.2		02.2
	08	296	35.2		01.5
	09	311	35.1	··	00.8
	10	326	35.0	17	00.1
	11	341	34.9	16	59.4
	12	356	34.8	S16	58.7
	13	11	34.8		57.9
	14	26	34.7		57.2
	15	41	34.6	··	56.5
	16	56	34.5		55.8
	17	71	34.4		55.1
	18	86	34.4	S16	54.4
	19	101	34.3		53.6
	20	116	34.2		52.9
	21	131	34.1	··	52.2
	22	146	34.0		51.5
	23	161	34.0		50.7

d	h	G.H.A.		Dec.	
		°	′	°	′
3	00	176	33.9	S16	50.0
	01	191	33.8		49.3
	02	206	33.7		48.6
	03	221	33.7	··	47.9
	04	236	33.6		47.1
	05	251	33.5		46.4
	06	266	33.4	S16	45.7
	07	281	33.4		45.0
	08	296	33.3		44.2
	09	311	33.2	··	43.5
	10	326	33.1		42.8
	11	341	33.1		42.0
	12	356	33.0	S16	41.3
	13	11	32.9		40.6
	14	26	32.8		39.9
	15	41	32.8	··	39.1
	16	56	32.7		38.4
	17	71	32.6		37.7
	18	86	32.6	S16	36.9
	19	101	32.5		36.2
	20	116	32.4		35.5
	21	131	32.4	··	34.7
	22	146	32.3		34.0
	23	161	32.2		33.3
4	00	176	32.1	S16	32.5
	01	191	32.1		31.8
	02	206	32.0		31.1
	03	221	31.9	··	30.3
	04	236	31.9		29.6
	05	251	31.8		28.8
	06	266	31.7	S16	28.1
	07	281	31.7		27.4
	08	296	31.6		26.6
	09	311	31.6	··	25.9
	10	326	31.5		25.2
	11	341	31.4		24.4
	12	356	31.4	S16	23.7
	13	11	31.3		22.9
	14	26	31.2		22.2
	15	41	31.2	··	21.5
	16	56	31.1		20.7
	17	71	31.0		20.0
	18	86	31.0	S16	19.2
	19	101	30.9		18.5
	20	116	30.9		17.7
	21	131	30.8	··	17.0
	22	146	30.7		16.2
	23	161	30.7		15.5
5	00	176	30.6	S16	14.7
	01	191	30.6		14.0
	02	206	30.5		13.3
	03	221	30.4	··	12.5
	04	236	30.4		11.8
	05	251	30.3		11.0
	06	266	30.3	S16	10.3
	07	281	30.2		09.5
	08	296	30.2		08.8
	09	311	30.1	··	08.0
	10	326	30.0		07.3
	11	341	30.0		06.5
	12	356	29.9	S16	05.8
	13	11	29.9		05.0
	14	26	29.8		04.2
	15	41	29.8	··	03.5
	16	56	29.7		02.7
	17	71	29.7		02.0
	18	86	29.6	S16	01.2
	19	101	29.6	16	00.5
	20	116	29.5	15	59.7
	21	131	29.5	··	59.0
	22	146	29.4		58.2
	23	161	29.3		57.4

G.H.A. / Dec. tables

d h	G.H.A. ° ′	Dec. ° ′
6 00	176 29.3	S15 56.7
01	191 29.2	55.9
02	206 29.2	55.2
03	221 29.1	·· 54.4
04	236 29.1	53.7
05	251 29.0	52.9
06	266 29.0	S15 52.1
07	281 28.9	51.4
08	296 28.9	50.6
09	311 28.9	·· 49.8
10	326 28.8	49.1
11	341 28.8	48.3
12	356 28.7	S15 47.6
13	11 28.7	46.8
14	26 28.6	46.0
15	41 28.6	·· 45.3
16	56 28.5	44.5
17	71 28.5	43.7
18	86 28.4	S15 43.0
19	101 28.4	42.2
20	116 28.3	41.4
21	131 28.3	·· 40.7
22	146 28.3	39.9
23	161 28.2	39.1
7 00	176 28.2	S15 38.4
01	191 28.1	37.6
02	206 28.1	36.8
03	221 28.0	·· 36.0
04	236 28.0	35.3
05	251 28.0	34.5
06	266 27.9	S15 33.7
07	281 27.9	33.0
08	296 27.8	32.2
09	311 27.8	·· 31.4
10	326 27.8	30.6
11	341 27.7	29.9
12	356 27.7	S15 29.1
13	11 27.7	28.3
14	26 27.6	27.5
15	41 27.6	·· 26.8
16	56 27.5	26.0
17	71 27.5	25.2
18	86 27.5	S15 24.4
19	101 27.4	23.6
20	116 27.4	22.9
21	131 27.4	·· 22.1
22	146 27.3	21.3
23	161 27.3	20.5
8 00	176 27.3	S15 19.8
01	191 27.2	19.0
02	206 27.2	18.2
03	221 27.2	·· 17.4
04	236 27.1	16.6
05	251 27.1	15.8
06	266 27.1	S15 15.1
07	281 27.0	14.3
08	296 27.0	13.5
09	311 27.0	·· 12.7
10	326 26.9	11.9
11	341 26.9	11.1
12	356 26.9	S15 10.4
13	11 26.9	09.6
14	26 26.8	08.8
15	41 26.8	·· 08.0
16	56 26.8	07.2
17	71 26.7	06.4
18	86 26.7	S15 05.6
19	101 26.7	04.8
20	116 26.7	04.1
21	131 26.6	·· 03.3
22	146 26.6	02.5
23	161 26.6	01.7

d h	G.H.A. ° ′	Dec. ° ′
9 00	176 26.6	S15 00.9
01	191 26.5	15 00.1
02	206 26.5	14 59.3
03	221 26.5	·· 58.5
04	236 26.5	57.7
05	251 26.4	56.9
06	266 26.4	S14 56.1
07	281 26.4	55.3
08	296 26.4	54.6
09	311 26.3	·· 53.8
10	326 26.3	53.0
11	341 26.3	52.2
12	356 26.3	S14 51.4
13	11 26.3	50.6
.4	26 26.2	49.8
15	41 26.2	·· 49.0
1)	56 26.2	48.2
17	71 26.2	47.4
18	86 26.2	S14 46.6
19	101 26.1	45.8
20	116 26.1	45.0
2)	131 26.1	·· 44.2
22	146 26.1	43.4
23	161 26.1	42.6
10 00	176 26.0	S14 41.8
01	191 26.0	41.0
02	206 26.0	40.2
03	221 26.0	·· 39.4
04	236 26.0	38.6
05	251 26.0	37.8
06	266 26.0	S14 37.0
07	281 25.9	36.2
08	296 25.9	35.4
09	311 25.9	·· 34.6
10	326 25.9	33.7
11	341 25.9	32.9
12	356 25.9	S14 32.1
13	11 25.9	31.3
14	26 25.8	30.5
15	41 25.8	·· 29.7
16	56 25.8	28.9
17	71 25.8	28.1
18	86 25.8	S14 27.3
19	101 25.8	26.5
20	116 25.8	25.7
21	131 25.8	·· 24.9
22	146 25.8	24.0
23	161 25.7	23.2
11 00	176 25.7	S14 22.4
01	191 25.7	21.6
02	206 25.7	20.8
03	221 25.7	·· 20.0
04	236 25.7	19.2
05	251 25.7	18.4
06	266 25.7	S14 17.5
07	281 25.7	16.7
08	296 25.7	15.9
09	311 25.7	·· 15.1
10	326 25.7	14.3
11	341 25.7	13.5
12	356 25.7	S14 12.7
13	11 25.7	11.8
14	26 25.7	11.0
15	41 25.6	·· 10.2
16	56 25.6	09.4
17	71 25.6	08.6
18	86 25.6	S14 07.8
19	101 25.6	06.9
20	116 25.6	06.1
21	131 25.6	·· 05.3
22	146 25.6	04.5
23	161 25.6	03.7

d h	G.H.A. ° ′	Dec. ° ′
12 00	176 25.6	S14 02.8
01	191 25.6	02.0
02	206 25.6	01.2
03	221 25.6	14 00.4
04	236 25.6	13 59.5
05	251 25.6	58.7
06	266 25.6	S13 57.9
07	281 25.6	57.1
08	296 25.6	56.2
09	311 25.6	·· 55.4
10	326 25.6	54.6
11	341 25.6	53.8
12	356 25.7	S13 52.9
13	11 25.7	52.1
14	26 25.7	51.3
15	41 25.7	·· 50.5
16	56 25.7	49.6
17	71 25.7	48.8
18	86 25.7	S13 48.0
19	101 25.7	47.2
20	116 25.7	46.3
21	131 25.7	·· 45.5
22	146 25.7	44.7
23	161 25.7	43.8
13 00	176 25.7	S13 43.0
01	191 25.7	42.2
02	206 25.7	41.3
03	221 25.7	·· 40.5
04	236 25.8	39.7
05	251 25.8	38.8
06	266 25.8	S13 38.0
07	281 25.8	37.2
08	296 25.8	36.3
09	311 25.8	·· 35.5
10	326 25.8	34.7
11	341 25.8	33.8
12	356 25.8	S13 33.0
13	11 25.8	32.2
14	26 25.9	31.3
15	41 25.9	·· 30.5
16	56 25.9	29.7
17	71 25.9	28.8
18	86 25.9	S13 28.0
19	101 25.9	27.2
20	116 25.9	26.3
21	131 26.0	·· 25.5
22	146 26.0	24.6
23	161 26.0	23.8
14 00	176 26.0	S13 23.0
01	191 26.0	22.1
02	206 26.0	21.3
03	221 26.0	·· 20.4
04	236 26.1	19.6
05	251 26.1	18.8
06	266 26.1	S13 17.9
07	281 26.1	17.1
08	296 26.1	16.2
09	311 26.2	·· 15.4
10	326 26.2	14.5
11	341 26.2	13.7
12	356 26.2	S13 12.8
13	11 26.2	12.0
14	26 26.2	11.2
15	41 26.3	·· 10.3
16	56 26.3	09.5
17	71 26.3	08.6
18	86 26.3	S13 07.8
19	101 26.4	06.9
20	116 26.4	06.1
21	131 26.4	·· 05.2
22	146 26.4	04.4
23	161 26.4	03.5

d	h	G.H.A. °	'	Dec.
15	00	176	26.5	S13 02.7
	01	191	26.5	01.8
	02	206	26.5	01.0
	03	221	26.5	13 00.1
	04	236	26.6	12 59.3
	05	251	26.6	58.4
	06	266	26.6	S12 57.6
	07	281	26.6	56.7
	08	296	26.7	55.9
	09	311	26.7	·· 55.0
	10	326	26.7	54.2
	11	341	26.7	53.3
	12	356	26.8	S12 52.5
	13	11	26.8	51.6
	14	26	26.8	50.8
	15	41	26.9	·· 49.9
	16	56	26.9	49.1
	17	71	26.9	48.2
	18	86	26.9	S12 47.3
	19	101	27.0	46.5
	20	116	27.0	45.6
	21	131	27.0	·· 44.8
	22	146	27.1	43.9
	23	161	27.1	43.1
16	00	176	27.1	S12 42.2
	01	191	27.2	41.3
	02	206	27.2	40.5
	03	221	27.2	·· 39.6
	04	236	27.2	38.8
	05	251	27.3	37.9
	06	266	27.3	S12 37.1
	07	281	27.3	36.2
	08	296	27.4	35.3
	09	311	27.4	·· 34.5
	10	326	27.4	33.6
	11	341	27.5	32.8
	12	356	27.5	S12 31.9
	13	11	27.5	31.0
	14	26	27.6	30.2
	15	41	27.6	·· 29.3
	16	56	27.7	28.4
	17	71	27.7	27.6
	18	86	27.7	S12 26.7
	19	101	27.8	25.8
	20	116	27.8	25.0
	21	131	27.8	·· 24.1
	22	146	27.9	23.3
	23	161	27.9	22.4
17	00	176	28.0	S12 21.5
	01	191	28.0	20.7
	02	206	28.0	19.8
	03	221	28.1	·· 18.9
	04	236	28.1	18.1
	05	251	28.1	17.2
	06	266	28.2	S12 16.3
	07	281	28.2	15.5
	08	296	28.3	14.6
	09	311	28.3	·· 13.7
	10	326	28.4	12.8
	11	341	28.4	12.0
	12	356	28.4	S12 11.1
	13	11	28.5	10.2
	14	26	28.5	09.4
	15	41	28.6	·· 08.5
	16	56	28.6	07.6
	17	71	28.7	06.8
	18	86	28.7	S12 05.9
	19	101	28.7	05.0
	20	116	28.8	04.1
	21	131	28.8	·· 03.3
	22	146	28.9	02.4
	23	161	28.9	01.5

d	h	G.H.A. °	'	Dec.
18	00	176	29.0	S12 00.6
	01	191	29.0	11 59.8
	02	206	29.1	58.9
	03	221	29.1	·· 58.0
	04	236	29.1	57.1
	05	251	29.2	56.3
	06	266	29.2	S11 55.4
	07	281	29.3	54.5
	08	296	29.3	53.6
	09	311	29.4	·· 52.8
	10	326	29.4	51.9
	11	341	29.5	51.0
	12	356	29.5	S11 50.1
	13	11	29.6	49.2
	14	26	29.6	48.4
	15	41	29.7	·· 47.5
	16	56	29.7	46.6
	17	71	29.8	45.7
	18	86	29.8	S11 44.9
	19	101	29.9	44 0
	20	116	29.9	43.1
	21	131	30.0	·· 42.2
	22	146	30.0	41.3
	23	161	30.1	40.4
19	00	176	30.1	S11 39.6
	01	191	30.2	38.7
	02	206	30.3	37.8
	03	221	30.3	·· 36.9
	04	236	30.4	36.0
	05	251	30.4	35.1
	06	266	30.5	S11 34.3
	07	281	30.5	33.4
	08	296	30.6	32.5
	09	311	30.6	·· 31.6
	10	326	30.7	30.7
	11	341	30.7	29.8
	12	356	30.8	S11 29.0
	13	11	30.9	28.1
	14	26	30.9	27.2
	15	41	31.0	·· 26.3
	16	56	31.0	25.4
	17	71	31.1	24.5
	18	86	31.1	S11 23.6
	19	101	31.2	22.7
	20	116	31.3	21.9
	21	131	31.3	·· 21.0
	22	146	31.4	20.1
	23	161	31.4	19.2
20	00	176	31.5	S11 18.3
	01	191	31.6	17.4
	02	206	31.6	16.5
	03	221	31.7	·· 15.6
	04	236	31.7	14.7
	05	251	31.8	13.9
	06	266	31.9	S11 13.0
	07	281	31.9	12.1
	08	296	32.0	11.2
	09	311	32.0	·· 10.3
	10	326	32.1	09.4
	11	341	32.2	08.5
	12	356	32.2	S11 07.6
	13	11	32.3	06.7
	14	26	32.4	05.8
	15	41	32.4	·· 04.9
	16	56	32.5	04.0
	17	71	32.6	03.1
	18	86	32.6	S11 02.2
	19	101	32.7	01.3
	20	116	32.8	11 00.4
	21	131	32.8	10 59.6
	22	146	32.9	58.7
	23	161	32.9	57.8

d	h	G.H.A. °	'	Dec.
21	00	176	33.0	S10 56.9
	01	191	33.1	56.0
	02	206	33.1	55.1
	03	221	33.2	·· 54.2
	04	236	33.3	53.3
	05	251	33.4	52.4
	06	266	33.4	S10 51.5
	07	281	33.5	50.6
	08	296	33.6	49.7
	09	311	33.6	·· 48.8
	10	326	33.7	47.9
	11	341	33.8	47.0
	12	356	33.8	S10 46.1
	13	11	33.9	45.2
	14	26	34.0	44.3
	15	41	34.0	·· 43.4
	16	56	34.1	42.5
	17	71	34.2	41.6
	18	86	34.3	S10 40.7
	19	101	34.3	39.8
	20	116	34.4	38.9
	21	131	34.5	·· 38.0
	22	146	34.5	37.1
	23	161	34.6	36.2
22	00	176	34.7	S10 35.3
	01	191	34.8	34.3
	02	206	34.8	33.4
	03	221	34.9	·· 32.5
	04	236	35.0	31.6
	05	251	35.1	30.7
	06	266	35.1	S10 29.8
	07	281	35.2	28.9
	08	296	35.3	28.0
	09	311	35.4	·· 27.1
	10	326	35.4	26.2
	11	341	35.5	25.3
	12	356	35.6	S10 24.4
	13	11	35.7	23.5
	14	26	35.7	22.6
	15	41	35.8	·· 21.7
	16	56	35.9	20.7
	17	71	36.0	19.8
	18	86	36.0	S10 18.9
	19	101	36.1	18.0
	20	116	36.2	17.1
	21	131	36.3	·· 16.2
	22	146	36.4	15.3
	23	161	36.4	14.4
23	00	176	36.5	S10 13.5
	01	191	36.6	12.6
	02	206	36.7	11.6
	03	221	36.8	·· 10.7
	04	236	36.8	09.8
	05	251	36.9	08.9
	06	266	37.0	S10 08.0
	07	281	37.1	07.1
	08	296	37.2	06.2
	09	311	37.2	·· 05.3
	10	326	37.3	04.4
	11	341	37.4	03.4
	12	356	37.5	S10 02.5
	13	11	37.6	01.6
	14	26	37.7	10 00.7
	15	41	37.7	9 59.8
	16	56	37.8	58.9
	17	71	37.9	57.9
	18	86	38.0	S 9 57.0
	19	101	38.1	56.1
	20	116	38.2	55.2
	21	131	38.2	·· 54.3
	22	146	38.3	53.4
	23	161	38.4	52.5

d h	G.H.A.	Dec.
24 00	176 38.5	S 9 51.5
01	191 38.6	50.6
02	206 38.7	49.7
03	221 38.8	·· 48.8
04	236 38.8	47.9
05	251 38.9	46.9
06	266 39.0	S 9 46.0
07	281 39.1	45.1
08	296 39.2	44.2
09	311 39.3	·· 43.3
10	326 39.4	42.4
11	341 39.5	41.4
12	356 39.5	S 9 40.5
13	11 39.6	39.6
14	26 39.7	38.7
15	41 39.8	·· 37.7
16	56 39.9	36.8
17	71 40.0	35.9
18	86 40.1	S 9 35.0
19	101 40.2	34.1
20	116 40.3	33.1
21	131 40.4	·· 32.2
22	146 40.4	31.3
23	161 40.5	30.4
25 00	176 40.6	S 9 29.4
01	191 40.7	28.5
02	206 40.8	27.6
03	221 40.9	·· 26.7
04	236 41.0	25.8
05	251 41.1	24.8
06	266 41.2	S 9 23.9
07	281 41.3	23.0
08	296 41.4	22.1
09	311 41.5	·· 21.1
10	326 41.6	20.2
11	341 41.7	19.3
12	356 41.8	S 9 18.3
13	11 41.8	17.4
14	26 41.9	16.5
15	41 42.0	·· 15.6
16	56 42.1	14.6
17	71 42.2	13.7
18	86 42.3	S 9 12.8
19	101 42.4	11.9
20	116 42.5	10.9
21	131 42.6	·· 10.0
22	146 42.7	09.1
23	161 42.8	08.1
26 00	176 42.9	S 9 07.2
01	191 43.0	06.3
02	206 43.1	05.4
03	221 43.2	·· 04.4
04	236 43.3	03.5
05	251 43.4	02.6
06	266 43.5	S 9 01.6
07	281 43.6	9 00.7
08	296 43.7	8 59.8
09	311 43.8	·· 58.8
10	326 43.9	57.9
11	341 44.0	57.0
12	356 44.1	S 8 56.0
13	11 44.2	55.1
14	26 44.3	54.2
15	41 44.4	·· 53.3
16	56 44.5	52.3
17	71 44.6	51.4
18	86 44.7	S 8 50.5
19	101 44.8	49.5
20	116 44.9	48.6
21	131 45.0	·· 47.7
22	146 45.1	46.7
23	161 45.2	45.8

d h	G.H.A.	Dec.
27 00	176 45.3	S 8 44.8
01	191 45.4	43.9
02	206 45.5	43.0
03	221 45.6	·· 42.0
04	236 45.7	41.1
05	251 45.8	40.2
06	266 46.0	S 8 39.2
07	281 46.1	38.3
08	296 46.2	37.4
09	311 46.3	·· 36.4
10	326 46.4	35.5
11	341 46.5	34.6
12	356 46.6	S 8 33.6
13	11 46.7	32.7
14	26 46.8	31.7
15	41 46.9	·· 30.8
16	56 47.0	29.9
17	71 47.1	28.9
18	86 47.2	S 8 28.0
19	101 47.3	27.0
20	116 47.4	26.1
21	131 47.6	·· 25.2
22	146 47.7	24.2
23	161 47.8	23.3
28 00	176 47.9	S 8 22.3
01	191 48.0	21.4
02	206 48.1	20.5
03	221 48.2	·· 19.5
04	236 48.3	18.6
05	251 48.4	17.6
06	266 48.5	S 8 16.7
07	281 48.7	15.8
08	296 48.8	14.8
09	311 48.9	·· 13.9
10	326 49.0	12.9
11	341 49.1	12.0
12	356 49.2	S 8 11.1
13	11 49.3	10.1
14	26 49.4	09.2
15	41 49.5	·· 08.2
16	56 49.7	07.3
17	71 49.8	06.3
18	86 49.9	S 8 05.4
19	101 50.0	04.5
20	116 50.1	03.5
21	131 50.2	·· 02.6
22	146 50.3	01.6
23	161 50.5	8 00.7
29 00	176 50.6	S 7 59.7
01	191 50.7	58.8
02	206 50.8	57.8
03	221 50.9	·· 56.9
04	236 51.0	55.9
05	251 51.2	55.0
06	266 51.3	S 7 54.1
07	281 51.4	53.1
08	296 51.5	52.2
09	311 51.6	·· 51.2
10	326 51.7	50.3
11	341 51.9	49.3
12	356 52.0	S 7 48.4
13	11 52.1	47.4
14	26 52.2	46.5
15	41 52.3	·· 45.5
16	56 52.4	44.6
17	71 52.6	43.6
18	86 52.7	S 7 42.7
19	101 52.8	41.7
20	116 52.9	40.8
21	131 53.0	·· 39.8
22	146 53.2	38.9
23	161 53.3	37.9

MARCH

d h	G.H.A.	Dec.
1 00	176 53.4	S 7 37.0
01	191 53.5	36.0
02	206 53.6	35.1
03	221 53.8	·· 34.1
04	236 53.9	33.2
05	251 54.0	32.2
06	266 54.1	S 7 31.3
07	281 54.2	30.3
08	296 54.4	29.4
09	311 54.5	·· 28.4
10	326 54.6	27.5
11	341 54.7	26.5
12	356 54.9	S 7 25.6
13	11 55.0	24.6
14	26 55.1	23.7
15	41 55.2	·· 22.7
16	56 55.3	21.8
17	71 55.5	20.8
18	86 55.6	S 7 19.9
19	101 55.7	18.9
20	116 55.8	18.0
21	131 56.0	·· 17.0
22	146 56.1	16.1
23	161 56.2	15.1
2 00	176 56.3	S 7 14.1
01	191 56.5	13.2
02	206 56.6	12.2
03	221 56.7	·· 11.3
04	236 56.8	10.3
05	251 57.0	09.4
06	266 57.1	S 7 08.4
07	281 57.2	07.5
08	296 57.4	06.5
09	311 57.5	·· 05.6
10	326 57.6	04.6
11	341 57.7	03.6
12	356 57.9	S 7 02.7
13	11 58.0	01.7
14	26 58.1	7 00.8
15	41 58.2	6 59.8
16	56 58.4	58.9
17	71 58.5	57.9
18	86 58.6	S 6 56.9
19	101 58.8	56.0
20	116 58.9	55.0
21	131 59.0	·· 54.1
22	146 59.2	53.1
23	161 59.3	52.2
3 00	176 59.4	S 6 51.2
01	191 59.5	50.2
02	206 59.7	49.3
03	221 59.8	·· 48.3
04	236 59.9	47.4
05	252 00.1	46.4
06	267 00.2	S 6 45.4
07	282 00.3	44.5
08	297 00.5	43.5
09	312 00.6	·· 42.6
10	327 00.7	41.6
11	342 00.9	40.6
12	357 01.0	S 6 39.7
13	12 01.1	38.7
14	27 01.3	37.8
15	42 01.4	·· 36.8
16	57 01.5	35.8
17	72 01.7	34.9
18	87 01.8	S 6 33.9
19	102 01.9	33.0
20	117 02.1	32.0
21	132 02.2	·· 31.0
22	147 02.3	30.0
23	162 02.5	29.1

d h	G.H.A. ° '	Dec. ° '
4 00	177 02.6	S 6 28.2
01	192 02.7	27.2
02	207 02.9	26.2
03	222 03.0	·· 25.3
04	237 03.1	24.3
05	252 03.3	23.3
06	267 03.4	S 6 22.4
07	282 03.6	21.4
08	297 03.7	20.4
09	312 03.8	·· 19.5
10	327 04.0	18.5
11	342 04.1	17.6
12	357 04.2	S 6 16.6
13	12 04.4	15.6
14	27 04.5	14.7
15	42 04.7	·· 13.7
16	57 04.8	12.7
17	72 04.9	11.8
18	87 05.1	S 6 10.8
19	102 05.2	09.8
20	117 05.4	08.9
21	132 05.5	·· 07.9
22	147 05.6	06.9
23	162 05.8	06.0
5 00	177 05.9	S 6 05.0
01	192 06.1	04.1
02	207 06.2	03.1
03	222 06.3	·· 02.1
04	237 06.5	01.2
05	252 06.6	6 00.2
06	267 06.8	S 5 59.2
07	282 06.9	58.3
08	297 07.0	57.3
09	312 07.2	·· 56.3
10	327 07.3	55.4
11	342 07.5	54.4
12	357 07.6	S 5 53.4
13	12 07.7	52.4
14	27 07.9	51.5
15	42 08.0	·· 50.5
16	57 08.2	49.5
17	72 08.3	48.6
18	07 08.5	S 5 47.6
19	102 08.6	46.6
20	117 08.8	45.7
21	132 08.9	·· 44.7
22	147 09.0	43.7
23	162 09.2	42.8
6 00	177 09.3	S 5 41.8
01	192 09.5	40.8
02	207 09.6	39.9
03	222 09.8	·· 38.9
04	237 09.9	37.9
05	252 10.1	37.0
06	267 10.2	S 5 36.0
07	282 10.3	35.0
08	297 10.5	34.0
09	312 10.6	·· 33.1
10	327 10.8	32.1
11	342 10.9	31.1
12	357 11.1	S 5 30.2
13	12 11.2	29.2
14	27 11.4	28.2
15	42 11.5	·· 27.3
16	57 11.7	26.3
17	72 11.8	25.3
18	87 12.0	S 5 24.3
19	102 12.1	23.4
20	117 12.3	22.4
21	132 12.4	·· 21.4
22	147 12.6	20.5
23	162 12.7	19.5

d h	G.H.A. ° '	Dec. ° '
7 00	177 12.9	S 5 18.5
01	192 13.0	17.5
02	207 13.2	16.6
03	222 13.3	·· 15.6
04	237 13.5	14.6
05	252 13.6	13.6
06	267 13.8	S 5 12.7
07	282 13.9	11.7
08	297 14.1	10.7
09	312 14.2	·· 09.8
10	327 14.4	08.8
11	342 14.5	07.8
12	357 14.7	S 5 06.8
13	12 14.8	05.9
14	27 15.0	04.9
15	42 15.1	·· 03.9
16	57 15.3	02.9
17	72 15.4	02.0
18	87 15.6	S 5 01.0
19	102 15.7	5 00.0
20	117 15.9	4 59.0
21	132 16.0	·· 58.1
22	147 16.2	57.1
23	162 16.3	56.1
8 00	177 16.5	S 4 55.1
01	192 16.6	54.2
02	207 16.8	53.2
03	222 16.9	·· 52.2
04	237 17.1	51.2
05	252 17.3	50.3
06	267 17.4	S 4 49.3
07	282 17.6	48.3
08	297 17.7	47.3
09	312 17.9	·· 46.4
10	327 18.0	45.4
11	342 18.2	44.4
12	357 18.3	S 4 43.4
13	12 18.5	42.5
14	27 18.6	41.5
15	42 18.8	·· 40.5
16	57 19.0	39.5
17	72 19.1	38.6
18	07 19.3	S 4 37.6
19	102 19.4	36.6
20	117 19.6	35.6
21	132 19.7	·· 34.7
22	147 19.9	33.7
23	162 20.0	32.7
9 00	177 20.2	S 4 31.7
01	192 20.4	30.7
02	207 20.5	29.8
03	222 20.7	·· 28.8
04	237 20.8	27.8
05	252 21.0	26.8
06	267 21.2	S 4 25.9
07	282 21.3	24.9
08	297 21.5	23.9
09	312 21.6	·· 22.9
10	327 21.8	21.9
11	342 21.9	21.0
12	357 22.1	S 4 20.0
13	12 22.3	19.0
14	27 22.4	18.0
15	42 22.6	·· 17.1
16	57 22.7	16.1
17	72 22.9	15.1
18	87 23.1	S 4 14.1
19	102 23.2	13.1
20	117 23.4	12.2
21	132 23.5	·· 11.2
22	147 23.7	10.2
23	162 23.9	09.2

d h	G.H.A. ° '	Dec. ° '
10 00	177 24.0	S 4 08.2
01	192 24.2	07.3
02	207 24.3	06.3
03	222 24.5	·· 05.3
04	237 24.7	04.3
05	252 24.8	03.3
06	267 25.0	S 4 02.4
07	282 25.2	01.4
08	297 25.3	4 00.4
09	312 25.5	3 59.4
10	327 25.6	58.4
11	342 25.8	57.5
12	357 26.0	S 3 56.5
13	12 26.1	55.5
14	27 26.3	54.5
15	42 26.5	·· 53.5
16	57 26.6	52.6
17	72 26.8	51.6
18	87 26.9	S 3 50.6
19	102 27.1	49.6
20	117 27.3	48.6
21	132 27.4	·· 47.7
22	147 27.6	46.7
23	162 27.8	45.7
11 00	177 27.9	S 3 44.7
01	192 28.1	43.7
02	207 28.3	42.7
03	222 28.4	·· 41.8
04	237 28.6	40.8
05	252 28.8	39.8
06	267 28.9	S 3 38.8
07	282 29.1	37.8
08	297 29.3	36.9
09	312 29.4	·· 35.9
10	327 29.6	34.9
11	342 29.8	33.9
12	357 29.9	S 3 32.9
13	12 30.1	32.0
14	27 30.3	31.0
15	42 30.4	·· 30.0
16	57 30.6	29.0
17	72 30.8	28.0
18	87 30.9	S 3 27.0
19	102 31.1	26.1
20	117 31.3	25.1
21	132 31.4	·· 24.1
22	147 31.6	23.1
23	162 31.8	22.1
12 00	177 31.9	S 3 21.1
01	192 32.1	20.2
02	207 32.3	19.2
03	222 32.4	·· 18.2
04	237 32.6	17.2
05	252 32.8	16.2
06	267 32.9	S 3 15.2
07	282 33.1	14.3
08	297 33.3	13.3
09	312 33.4	·· 12.3
10	327 33.6	11.3
11	342 33.8	10.3
12	357 34.0	S 3 09.3
13	12 34.1	08.4
14	27 34.3	07.4
15	42 34.5	·· 06.4
16	57 34.6	05.4
17	72 34.8	04.4
18	87 35.0	S 3 03.4
19	102 35.1	02.4
20	117 35.3	01.5
21	132 35.5	3 00.5
22	147 35.7	2 59.5
23	162 35.8	58.5

d h	G.H.A. ° ′	Dec. ° ′
13 00	177 36.0	S 2 57.5
01	192 36.2	56.5
02	207 36.3	55.6
03	222 36.5	.. 54.6
04	237 36.7	53.6
05	252 36.9	52.6
06	267 37.0	S 2 51.6
07	282 37.2	50.6
08	297 37.4	49.6
09	312 37.5	.. 48.7
10	327 37.7	47.7
11	342 37.9	46.7
12	357 38.1	S 2 45.7
13	12 38.2	44.7
14	27 38.4	43.7
15	42 38.6	.. 42.8
16	57 38.7	41.8
17	72 38.9	40.8
18	87 39.1	S 2 39.8
19	102 39.3	38.8
20	117 39.4	37.8
21	132 39.6	.. 36.8
22	147 39.8	35.9
23	162 40.0	34.9
14 00	177 40.1	S 2 33.9
01	192 40.3	32.9
02	207 40.5	31.9
03	222 40.7	.. 30.9
04	237 40.8	29.9
05	252 41.0	29.0
06	267 41.2	S 2 28.0
07	282 41.4	27.0
08	297 41.5	26.0
09	312 41.7	.. 25.0
10	327 41.9	24.0
11	342 42.1	23.0
12	357 42.2	S 2 22.1
13	12 42.4	21.1
14	27 42.6	20.1
15	42 42.8	.. 19.1
16	57 42.9	18.1
17	72 43.1	17.1
18	87 43.3	S 2 16.1
19	102 43.5	15.1
20	117 43.6	14.2
21	132 43.8	.. 13.2
22	147 44.0	12.2
23	162 44.2	11.2
15 00	177 44.3	S 2 10.2
01	192 44.5	09.2
02	207 44.7	08.2
03	222 44.9	.. 07.3
04	237 45.1	06.3
05	252 45.2	05.3
06	267 45.4	S 2 04.3
07	282 45.6	03.3
08	297 45.8	02.3
09	312 45.9	.. 01.3
10	327 46.1	2 00.3
11	342 46.3	1 59.4
12	357 46.5	S 1 58.4
13	12 46.7	57.4
14	27 46.8	56.4
15	42 47.0	.. 55.4
16	57 47.2	54.4
17	72 47.4	53.4
18	87 47.5	S 1 52.4
19	102 47.7	51.5
20	117 47.9	50.5
21	132 48.1	.. 49.5
22	147 48.3	48.5
23	162 48.4	47.5

d h	G.H.A. ° ′	Dec. ° ′
16 00	177 48.6	S 1 46.5
01	192 48.8	45.5
02	207 49.0	44.6
03	222 49.2	.. 43.6
04	237 49.3	42.6
05	252 49.5	41.6
06	267 49.7	S 1 40.6
07	282 49.9	39.6
08	297 50.0	38.6
09	312 50.2	.. 37.6
10	327 50.4	36.7
11	342 50.6	35.7
12	357 50.8	S 1 34.7
13	12 50.9	33.7
14	27 51.1	32.7
15	42 51.3	.. 31.7
16	57 51.5	30.7
17	72 51.7	29.7
18	87 51.9	S 1 28.8
19	102 52.0	27.8
20	117 52.2	26.8
21	132 52.4	.. 25.8
22	147 52.6	24.8
23	162 52.8	23.8
17 00	177 52.9	S 1 22.8
01	192 53.1	21.8
02	207 53.3	20.8
03	222 53.5	.. 19.9
04	237 53.7	18.9
05	252 53.8	17.9
06	267 54.0	S 1 16.9
07	282 54.2	15.9
08	297 54.4	14.9
09	312 54.6	.. 13.9
10	327 54.8	12.9
11	342 54.9	12.0
12	357 55.1	S 1 11.0
13	12 55.3	10.0
14	27 55.5	09.0
15	42 55.7	.. 08.0
16	57 55.8	07.0
17	72 56.0	06.0
18	87 56.2	S 1 05.0
19	102 56.4	04.1
20	117 56.6	03.1
21	132 56.8	.. 02.1
22	147 56.9	01.1
23	162 57.1	1 00.1
18 00	177 57.3	S 0 59.1
01	192 57.5	58.1
02	207 57.7	57.1
03	222 57.9	.. 56.1
04	237 58.0	55.2
05	252 58.2	54.2
06	267 58.4	S 0 53.2
07	282 58.6	52.2
08	297 58.8	51.2
09	312 59.0	.. 50.2
10	327 59.1	49.2
11	342 59.3	48.2
12	357 59.5	S 0 47.3
13	12 59.7	46.3
14	27 59.9	45.3
15	43 00.1	.. 44.3
16	58 00.2	43.3
17	73 00.4	42.3
18	88 00.6	S 0 41.3
19	103 00.8	40.3
20	118 01.0	39.3
21	133 01.2	.. 38.4
22	148 01.3	37.4
23	163 01.5	36.4

d h	G.H.A. ° ′	Dec. ° ′
19 00	178 01.7	S 0 35.4
01	193 01.9	34.4
02	208 02.1	33.4
03	223 02.3	.. 32.4
04	238 02.5	31.4
05	253 02.6	30.5
06	268 02.8	S 0 29.5
07	283 03.0	28.5
08	298 03.2	27.5
09	313 03.4	.. 26.5
10	328 03.6	25.5
11	343 03.7	24.5
12	358 03.9	S 0 23.5
13	13 04.1	22.6
14	28 04.3	21.6
15	43 04.5	.. 20.6
16	58 04.7	19.6
17	73 04.9	18.6
18	88 05.0	S 0 17.6
19	103 05.2	16.6
20	118 05.4	15.6
21	133 05.6	.. 14.6
22	148 05.8	13.7
23	163 06.0	12.7
20 00	178 06.2	S 0 11.7
01	193 06.3	10.7
02	208 06.5	09.7
03	223 06.7	.. 08.7
04	238 06.9	07.7
05	253 07.1	06.7
06	268 07.3	S 0 05.8
07	283 07.5	04.8
08	298 07.6	03.8
09	313 07.8	.. 02.8
10	328 08.0	01.8
11	343 08.2	S 0 00.8
12	358 08.4	N 0 00.2
13	13 08.6	01.2
14	28 08.8	02.1
15	43 09.0	.. 03.1
16	58 09.1	04.1
17	73 09.3	05.1
18	88 09.5	N 0 06.1
19	103 09.7	07.1
20	118 09.9	08.1
21	133 10.1	.. 09.1
22	148 10.3	10.0
23	163 10.4	11.0
21 00	178 10.6	N 0 12.0
01	193 10.8	13.0
02	208 11.0	14.0
03	223 11.2	.. 15.0
04	238 11.4	16.0
05	253 11.6	17.0
06	268 11.8	N 0 17.9
07	283 11.9	18.9
08	298 12.1	19.9
09	313 12.3	.. 20.9
10	328 12.5	21.9
11	343 12.7	22.9
12	358 12.9	N 0 23.9
13	13 13.1	24.9
14	28 13.3	25.8
15	43 13.4	.. 26.8
16	58 13.6	27.8
17	73 13.8	28.8
18	88 14.0	N 0 29.8
19	103 14.2	30.8
20	118 14.4	31.8
21	133 14.6	.. 32.7
22	148 14.8	33.7
23	163 14.9	34.7

22

d h	G.H.A.	Dec.
00	178 15.1	N 0 35.7
01	193 15.3	36.7
02	208 15.5	37.7
03	223 15.7	·· 38.7
04	238 15.9	39.7
05	253 16.1	40.6
06	268 16.3	N 0 41.6
07	283 16.5	42.6
08	298 16.6	43.6
09	313 16.8	·· 44.6
10	328 17.0	45.6
11	343 17.2	46.6
12	358 17.4	N 0 47.5
13	13 17.6	48.5
14	28 17.8	49.5
15	43 18.0	·· 50.5
16	58 18.1	51.5
17	73 18.3	52.5
18	88 18.5	N 0 53.5
19	103 18.7	54.4
20	118 18.9	55.4
21	133 19.1	·· 56.4
22	148 19.3	57.4
23	163 19.5	58.4

23

d h	G.H.A.	Dec.
00	178 19.7	N 0 59.4
01	193 19.8	1 00.4
02	208 20.0	01.3
03	223 20.2	·· 02.3
04	238 20.4	03.3
05	253 20.6	04.3
06	268 20.8	N 1 05.3
07	283 21.0	06.3
08	298 21.2	07.3
09	313 21.4	·· 08.2
10	328 21.5	09.2
11	343 21.7	10.2
12	358 21.9	N 1 11.2
13	13 22.1	12.2
14	28 22.3	13.2
15	43 22.5	·· 14.2
16	58 22.7	15.1
17	73 22.9	16.1
18	88 23.0	N 1 17.1
19	103 23.2	18.1
20	118 23.4	19.1
21	133 23.6	·· 20.1
22	148 23.8	21.0
23	163 24.0	22.0

24

d h	G.H.A.	Dec.
00	178 24.2	N 1 23.0
01	193 24.4	24.0
02	208 24.6	25.0
03	223 24.7	·· 26.0
04	238 24.9	27.0
05	253 25.1	27.9
06	268 25.3	N 1 28.9
07	283 25.5	29.9
08	298 25.7	30.9
09	313 25.9	·· 31.9
10	328 26.1	32.9
11	343 26.3	33.8
12	358 26.5	N 1 34.8
13	13 26.6	35.8
14	28 26.8	36.8
15	43 27.0	·· 37.8
16	58 27.2	38.8
17	73 27.4	39.7
18	88 27.6	N 1 40.7
19	103 27.8	41.7
20	118 28.0	42.7
21	133 28.2	·· 43.7
22	148 28.3	44.7
23	163 28.5	45.6

25

d h	G.H.A.	Dec.
00	178 28.7	N 1 46.6
01	193 28.9	47.6
02	208 29.1	48.6
03	223 29.3	·· 49.6
04	238 29.5	50.6
05	253 29.7	51.5
06	268 29.9	N 1 52.5
07	283 30.0	53.5
08	298 30.2	54.5
09	313 30.4	·· 55.5
10	328 30.6	56.5
11	343 30.8	57.4
12	358 31.0	N 1 58.4
13	13 31.2	1 59.4
14	28 31.4	2 00.4
15	43 31.6	·· 01.4
16	58 31.8	02.3
17	73 31.9	03.3
18	88 32.1	N 2 04.3
19	103 32.3	05.3
20	118 32.5	06.3
21	133 32.7	·· 07.3
22	148 32.9	08.2
23	163 33.1	09.2

26

d h	G.H.A.	Dec.
00	178 33.3	N 2 10.2
01	193 33.5	11.2
02	208 33.6	12.2
03	223 33.8	·· 13.1
04	238 34.0	14.1
05	253 34.2	15.1
06	268 34.4	N 2 16.1
07	283 34.6	17.1
08	298 34.8	18.0
09	313 35.0	·· 19.0
10	328 35.2	20.0
11	343 35.3	21.0
12	358 35.5	N 2 22.0
13	13 35.7	22.9
14	28 35.9	23.9
15	43 36.1	·· 24.9
16	58 36.3	25.9
17	73 36.5	26.9
18	88 36.7	N 2 27.8
19	103 36.9	28.8
20	118 37.0	29.8
21	133 37.2	·· 30.8
22	148 37.4	31.8
23	163 37.6	32.7

27

d h	G.H.A.	Dec.
00	178 37.8	N 2 33.7
01	193 38.0	34.7
02	208 38.2	35.7
03	223 38.4	·· 36.7
04	238 38.6	37.6
05	253 38.8	38.6
06	268 38.9	N 2 39.6
07	283 39.1	40.6
08	298 39.3	41.5
09	313 39.5	·· 42.5
10	328 39.7	43.5
11	343 39.9	44.5
12	358 40.1	N 2 45.5
13	13 40.3	46.4
14	28 40.5	47.4
15	43 40.6	·· 48.4
16	58 40.8	49.4
17	73 41.0	50.3
18	88 41.2	N 2 51.3
19	103 41.4	52.3
20	118 41.6	53.3
21	133 41.8	·· 54.3
22	148 42.0	55.2
23	163 42.2	56.2

28

d h	G.H.A.	Dec.
00	178 42.3	N 2 57.2
01	193 42.5	58.2
02	208 42.7	2 59.1
03	223 42.9	3 00.1
04	238 43.1	01.1
05	253 43.3	02.1
06	268 43.5	N 3 03.0
07	283 43.7	04.0
08	298 43.9	05.0
09	313 44.0	·· 06.0
10	328 44.2	07.0
11	343 44.4	07.9
12	358 44.6	N 3 08.9
13	13 44.8	09.9
14	28 45.0	10.9
15	43 45.2	·· 11.8
16	58 45.4	12.8
17	73 45.5	13.8
18	88 45.7	N 3 14.8
19	103 45.9	15.7
20	118 46.1	16.7
21	133 46.3	·· 17.7
22	148 46.5	18.7
23	163 46.7	19.6

29

d h	G.H.A.	Dec.
00	178 46.9	N 3 20.6
01	193 47.1	21.6
02	208 47.2	22.6
03	223 47.4	·· 23.5
04	238 47.6	24.5
05	253 47.8	25.5
06	268 48.0	N 3 26.4
07	283 48.2	27.4
08	298 48.4	28.4
09	313 48.6	·· 29.4
10	328 48.8	30.3
11	343 48.9	31.3
12	358 49.1	N 3 32.3
13	13 49.3	33.3
14	28 49.5	34.2
15	43 49.7	·· 35.2
16	58 49.9	36.2
17	73 50.1	37.2
18	88 50.3	N 3 38.1
19	103 50.4	39.1
20	118 50.6	40.1
21	133 50.8	·· 41.0
22	148 51.0	42.0
23	163 51.2	43.0

30

d h	G.H.A.	Dec.
00	178 51.4	N 3 44.0
01	193 51.6	44.9
02	208 51.8	45.9
03	223 51.9	·· 46.9
04	238 52.1	47.8
05	253 52.3	48.8
06	268 52.5	N 3 49.8
07	283 52.7	50.8
08	298 52.9	51.7
09	313 53.1	·· 52.7
10	328 53.3	53.7
11	343 53.4	54.6
12	358 53.6	N 3 55.6
13	13 53.8	56.6
14	28 54.0	57.5
15	43 54.2	·· 58.5
16	58 54.4	3 59.5
17	73 54.6	4 00.5
18	88 54.8	N 4 01.4
19	103 54.9	02.4
20	118 55.1	03.4
21	133 55.3	·· 04.3
22	148 55.5	05.3
23	163 55.7	06.3

	G.H.A.	Dec.		G.H.A.	Dec.		G.H.A.	Dec.
d h	° ′	° ′	d h	° ′	° ′	d h	° ′	° ′
31 00	178 55.9	N 4 07.2	3 00	179 09.2	N 5 16.6	6 00	179 22.3	N 6 25.1
01	193 56.1	08.2	01	194 09.4	17.6	01	194 22.4	26.1
02	208 56.3	09.2	02	209 09.6	18.5	02	209 22.6	27.0
03	223 56.4	·· 10.1	03	224 09.8	·· 19.5	03	224 22.8	·· 27.9
04	238 56.6	11.1	04	239 10.0	20.4	04	239 23.0	28.9
05	253 56.8	12.1	05	254 10.1	21.4	05	254 23.1	29.8
06	268 57.0	N 4 13.0	06	269 10.3	N 5 22.3	06	269 23.3	N 6 30.8
07	283 57.2	14.0	07	284 10.5	23.3	07	284 23.5	31.7
08	298 57.4	15.0	08	299 10.7	24.3	08	299 23.7	32.7
09	313 57.6	·· 15.9	09	314 10.9	·· 25.2	09	314 23.9	·· 33.6
10	328 57.7	16.9	10	329 11.1	26.2	10	329 24.0	34.6
11	343 57.9	17.9	11	344 11.2	27.1	11	344 24.2	35.5
12	358 58.1	N 4 18.8	12	359 11.4	N 5 28.1	12	359 24.4	N 6 36.4
13	13 58.3	19.8	13	14 11.6	29.0	13	14 24.6	37.4
14	28 58.5	20.8	14	29 11.8	30.0	14	29 24.7	38.3
15	43 58.7	·· 21.7	15	44 12.0	·· 30.9	15	44 24.9	·· 39.3
16	58 58.9	22.7	16	59 12.1	31.9	16	59 25.1	40.2
17	73 59.1	23.7	17	74 12.3	32.9	17	74 25.3	41.2
18	88 59.2	N 4 24.6	18	89 12.5	N 5 33.8	18	89 25.5	N 6 42.1
19	103 59.4	25.6	19	104 12.7	34.8	19	104 25.6	43.0
20	118 59.6	26.6	20	119 12.9	35.7	20	119 25.8	44.0
21	133 59.8	·· 27.5	21	134 13.1	·· 36.7	21	134 26.0	·· 44.9
22	149 00.0	28.5	22	149 13.2	37.6	22	149 26.2	45.9
23	164 00.2	29.5	23	164 13.4	38.6	23	164 26.3	46.8
1 00	179 00.4	N 4 30.4	4 00	179 13.6	N 5 39.5	7 00	179 26.5	N 6 47.7
01	194 00.5	31.4	01	194 13.8	40.5	01	194 26.7	48.7
02	209 00.7	32.4	02	209 14.0	41.4	02	209 26.9	49.6
03	224 00.9	·· 33.3	03	224 14.1	·· 42.4	03	224 27.0	·· 50.6
04	239 01.1	34.3	04	239 14.3	43.4	04	239 27.2	51.5
05	254 01.3	35.3	05	254 14.5	44.3	05	254 27.4	52.4
06	269 01.5	N 4 36.2	06	269 14.7	N 5 45.3	06	269 27.6	N 6 53.4
07	284 01.7	37.2	07	284 14.9	46.2	07	284 27.7	54.3
08	299 01.8	38.2	08	299 15.1	47.2	08	299 27.9	55.3
09	314 02.0	·· 39.1	09	314 15.2	·· 48.1	09	314 28.1	·· 56.2
10	329 02.2	40.1	10	329 15.4	49.1	10	329 28.3	57.1
11	344 02.4	41.0	11	344 15.6	50.0	11	344 28.4	58.1
12	359 02.6	N 4 42.0	12	359 15.8	N 5 51.0	12	359 28.6	N 6 59.0
13	14 02.8	43.0	13	14 16.0	51.9	13	14 28.8	6 59.9
14	29 03.0	43.9	14	29 16.1	52.9	14	29 29.0	7 00.9
15	44 03.1	·· 44.9	15	44 16.3	·· 53.8	15	44 29.2	·· 01.8
16	59 03.3	45.9	16	59 16.5	54.8	16	59 29.3	02.8
17	74 03.5	46.8	17	74 16.7	55.7	17	74 29.5	03.7
18	89 03.7	N 4 47.8	18	89 16.9	N 5 56.7	18	89 29.7	N 7 04.6
19	104 03.9	48.8	19	104 17.0	57.6	19	104 29.9	05.6
20	119 04.1	49.7	20	119 17.2	58.6	20	119 30.0	06.5
21	134 04.2	·· 50.7	21	134 17.4	5 59.5	21	134 30.2	·· 07.4
22	149 04.4	51.6	22	149 17.6	6 00.5	22	149 30.4	08.4
23	164 04.6	52.6	23	164 17.8	01.4	23	164 30.5	09.3
2 00	179 04.8	N 4 53.6	5 00	179 17.9	N 6 02.4	8 00	179 30.7	N 7 10.2
01	194 05.0	54.5	01	194 18.1	03.3	01	194 30.9	11.2
02	209 05.2	55.5	02	209 18.3	04.3	02	209 31.1	12.1
03	224 05.4	·· 56.4	03	224 18.5	·· 05.2	03	224 31.2	·· 13.1
04	239 05.5	57.4	04	239 18.7	06.2	04	239 31.4	14.0
05	254 05.7	58.4	05	254 18.8	07.1	05	254 31.6	14.9
06	269 05.9	N 4 59.3	06	269 19.0	N 6 08.1	06	269 31.8	N 7 15.9
07	284 06.1	5 00.3	07	284 19.2	09.0	07	284 31.9	16.8
08	299 06.3	01.2	08	299 19.4	10.0	08	299 32.1	17.7
09	314 06.5	·· 02.2	09	314 19.6	·· 10.9	09	314 32.3	·· 18.7
10	329 06.6	03.2	10	329 19.7	11.9	10	329 32.5	19.6
11	344 06.8	04.1	11	344 19.9	12.8	11	344 32.6	20.5
12	359 07.0	N 5 05.1	12	359 20.1	N 6 13.8	12	359 32.8	N 7 21.5
13	14 07.2	06.0	13	14 20.3	14.7	13	14 33.0	22.4
14	29 07.4	07.0	14	29 20.5	15.7	14	29 33.2	23.3
15	44 07.6	·· 08.0	15	44 20.6	·· 16.6	15	44 33.3	·· 24.2
16	59 07.8	08.9	16	59 20.8	17.5	16	59 33.5	25.2
17	74 07.9	09.9	17	74 21.0	18.5	17	74 33.7	26.1
18	89 08.1	N 5 10.8	18	89 21.2	N 6 19.4	18	89 33.8	N 7 27.0
19	104 08.3	11.8	19	104 21.4	20.4	19	104 34.0	28.0
20	119 08.5	12.8	20	119 21.5	21.3	20	119 34.2	28.9
21	134 08.7	·· 13.7	21	134 21.7	·· 22.3	21	134 34.4	·· 29.8
22	149 08.9	14.7	22	149 21.9	23.2	22	149 34.5	30.8
23	164 09.0	15.6	23	164 22.1	24.2	23	164 34.7	31.7

142

9, 10, 11

d h	G.H.A.	Dec.
9 00	179 34.9	N 7 32.6
01	194 35.1	33.6
02	209 35.2	34.5
03	224 35.4	·· 35.4
04	239 35.6	36.3
05	254 35.7	37.3
06	269 35.9	N 7 38.2
07	284 36.1	39.1
08	299 36.3	40.1
09	314 36.4	·· 41.0
10	329 36.6	41.9
11	344 36.8	42.8
12	359 36.9	N 7 43.8
13	14 37.1	44.7
14	29 37.3	45.6
15	44 37.4	·· 46.6
16	59 37.6	47.5
17	74 37.8	48.4
18	89 38.0	N 7 49.3
19	104 30.1	50.3
20	119 38.3	51.2
21	134 38.5	·· 52.1
22	149 38.6	53.0
23	164 38.8	54.0
10 00	179 39.0	N 7 54.9
01	194 39.1	55.8
02	209 39.3	56.7
03	224 39.5	·· 57.7
04	239 39.6	58.6
05	254 39.8	7 59.5
06	269 40.0	N 8 00.4
07	284 40.2	01.3
08	299 40.3	02.3
09	314 40.5	·· 03.2
10	329 40.7	04.1
11	344 40.8	05.0
12	359 41.0	N 8 06.0
13	14 41.2	06.9
14	29 41.3	07.8
15	44 41.5	·· 08.7
16	59 41.7	09.6
17	74 41.8	10.6
18	89 42.0	N 8 11.5
19	104 42.2	12.4
20	119 42.3	13.3
21	134 42.5	·· 14.2
22	149 42.7	15.2
23	164 42.8	16.1
11 00	179 43.0	N 8 17.0
01	194 43.2	17.9
02	209 43.3	18.8
03	224 43.5	·· 19.8
04	239 43.7	20.7
05	254 43.8	21.6
06	269 44.0	N 8 22.5
07	284 44.2	23.4
08	299 44.3	24.3
09	314 44.5	·· 25.3
10	329 44.7	26.2
11	344 44.8	27.1
12	359 45.0	N 8 28.0
13	14 45.2	28.9
14	29 45.3	29.8
15	44 45.5	·· 30.8
16	59 45.7	31.7
17	74 45.8	32.6
18	89 46.0	N 8 33.5
19	104 46.2	34.4
20	119 46.3	35.3
21	134 46.5	·· 36.2
22	149 46.6	37.2
23	164 46.8	38.1

12, 13, 14

d h	G.H.A.	Dec.
12 00	179 47.0	N 8 39.0
01	194 47.1	39.9
02	209 47.3	40.8
03	224 47.5	·· 41.7
04	239 47.6	42.6
05	254 47.8	43.5
06	269 48.0	N 8 44.5
07	284 48.1	45.4
08	299 48.3	46.3
09	314 48.4	·· 47.2
10	329 48.6	48.1
11	344 48.8	49.0
12	359 48.9	N 8 49.9
13	14 49.1	50.8
14	29 49.3	51.7
15	44 49.4	·· 52.7
16	59 49.6	53.6
17	74 49.7	54.5
18	89 49.9	N 8 55.4
19	104 50.1	56.3
20	119 50.2	57.2
21	134 50.4	·· 58.1
22	149 50.5	59.0
23	164 50.7	8 59.9
13 00	179 50.9	N 9 00.8
01	194 51.0	01.7
02	209 51.2	02.6
03	224 51.3	·· 03.5
04	239 51.5	04.4
05	254 51.7	05.4
06	269 51.8	N 9 06.3
07	284 52.0	07.2
08	299 52.1	08.1
09	314 52.3	·· 09.0
10	329 52.5	09.9
11	344 52.6	10.8
12	359 52.8	N 9 11.7
13	14 52.9	12.6
14	29 53.1	13.5
15	44 53.3	·· 14.4
16	59 53.4	15.3
17	74 53.6	16.2
18	89 53.7	N 9 17.1
19	104 53.9	18.0
20	119 54.1	18.9
21	134 54.2	·· 19.8
22	149 54.4	20.7
23	164 54.5	21.6
14 00	179 54.7	N 9 22.5
01	194 54.8	23.4
02	209 55.0	24.3
03	224 55.2	·· 25.2
04	239 55.3	26.1
05	254 55.5	27.0
06	269 55.6	N 9 27.9
07	284 55.8	28.8
08	299 55.9	29.7
09	314 56.1	·· 30.6
10	329 56.2	31.5
11	344 56.4	32.4
12	359 56.6	N 9 33.3
13	14 56.7	34.2
14	29 56.9	35.1
15	44 57.0	·· 36.0
16	59 57.2	36.9
17	74 57.3	37.8
18	89 57.5	N 9 38.7
19	104 57.6	39.6
20	119 57.8	40.5
21	134 58.0	·· 41.4
22	149 58.1	42.2
23	164 58.3	43.1

15, 16, 17

d h	G.H.A.	Dec.
15 00	179 58.4	N 9 44.0
01	194 58.6	44.9
02	209 58.7	45.8
03	224 58.9	·· 46.7
04	239 59.0	47.6
05	254 59.2	48.5
06	269 59.3	N 9 49.4
07	284 59.5	50.3
08	299 59.6	51.2
09	314 59.8	·· 52.1
10	329 59.9	53.0
11	345 00.1	53.9
12	0 00.2	N 9 54.7
13	15 00.4	55.6
14	30 00.5	56.5
15	45 00.7	·· 57.4
16	60 00.8	58.3
17	75 01.0	9 59.2
18	90 01.2	N10 00.1
19	105 01.3	01.0
20	120 01.5	01.9
21	135 01.6	·· 02.7
22	150 01.8	03.6
23	165 01.9	04.5
16 00	180 02.1	N10 05.4
01	195 02.2	06.3
02	210 02.4	07.2
03	225 02.5	·· 08.1
04	240 02.7	09.0
05	255 02.8	09.8
06	270 02.9	N10 10.7
07	285 03.1	11.6
08	300 03.2	12.5
09	315 03.4	·· 13.4
10	330 03.5	14.3
11	345 03.7	15.1
12	0 03.8	N10 16.0
13	15 04.0	16.9
14	30 04.1	17.8
15	45 04.3	·· 18.7
16	60 04.4	19.6
17	75 04.6	20.4
18	90 04.7	N10 21.3
19	105 04.9	22.2
20	120 05.0	23.1
21	135 05.2	·· 24.0
22	150 05.3	24.8
23	165 05.5	25.7
17 00	180 05.6	N10 26.6
01	195 05.7	27.5
02	210 05.9	28.4
03	225 06.0	·· 29.2
04	240 06.2	30.1
05	255 06.3	31.0
06	270 06.5	N10 31.9
07	285 06.6	32.8
08	300 06.8	33.6
09	315 06.9	·· 34.5
10	330 07.1	35.4
11	345 07.2	36.3
12	0 07.3	N10 37.1
13	15 07.5	38.0
14	30 07.6	38.9
15	45 07.8	·· 39.8
16	60 07.9	40.6
17	75 08.1	41.5
18	90 08.2	N10 42.4
19	105 08.3	43.3
20	120 08.5	44.1
21	135 08.6	·· 45.0
22	150 08.8	45.9
23	165 08.9	46.8

		G.H.A.	Dec.
d h		° ′	° ′
18 00		180 09.1	N10 47.6
01		195 09.2	48.5
02		210 09.3	49.4
03		225 09.5 ··	50.3
04		240 09.6	51.1
05		255 09.8	52.0
06		270 09.9	N10 52.9
07		285 10.0	53.7
08		300 10.2	54.6
09		315 10.3 ··	55.5
10		330 10.5	56.4
11		345 10.6	57.2
12		0 10.7	N10 58.1
13		15 10.9	59.0
14		30 11.0	.10 59.8
15		45 11.2	11 00.7
16		60 11.3	01.6
17		75 11.4	02.4
18		90 11.6	N11 03.3
19		105 11.7	04.2
20		120 11.9	05.0
21		135 12.0 ··	05.9
22		150 12.1	06.8
23		165 12.3	07.6
19 00		180 12.4	N11 08.5
01		195 12.5	09.4
02		210 12.7	10.2
03		225 12.8 ··	11.1
04		240 13.0	12.0
05		255 13.1	12.8
06		270 13.2	N11 13.7
07		285 13.4	14.5
08		300 13.5	15.4
09		315 13.6 ··	16.3
10		330 13.8	17.1
11		345 13.9	18.0
12		0 14.0	N11 18.9
13		15 14.2	19.7
14		30 14.3	20.6
15		45 14.4 ··	21.4
16		60 14.6	22.3
17		75 14.7	23.2
18		90 14.8	N11 24.0
19		105 15.0	24.9
20		120 15.1	25.7
21		135 15.2 ··	26.6
22		150 15.4	27.5
23		165 15.5	28.3
20 00		180 15.6	N11 29.2
01		195 15.8	30.0
02		210 15.9	30.9
03		225 16.0 ··	31.7
04		240 16.2	32.6
05		255 16.3	33.5
06		270 16.4	N11 34.3
07		285 16.6	35.2
08		300 16.7	36.0
09		315 16.8 ··	36.9
10		330 17.0	37.7
11		345 17.1	38.6
12		0 17.2	N11 39.4
13		15 17.4	40.3
14		30 17.5	41.1
15		45 17.6 ··	42.0
16		60 17.7	42.9
17		75 17.9	43.7
18		90 18.0	N11 44.6
19		105 18.1	45.4
20		120 18.3	46.3
21		135 18.4 ··	47.1
22		150 18.5	48.0
23		165 18.6	48.8

		G.H.A.	Dec.
d h		° ′	° ′
21 00		180 18.8	N11 49.7
01		195 18.9	50.5
02		210 19.0	51.4
03		225 19.2 ··	52.2
04		240 19.3	53.1
05		255 19.4	53.9
06		270 19.5	N11 54.8
07		285 19.7	55.6
08		300 19.8	56.4
09		315 19.9 ··	57.3
10		330 20.0	58.1
11		345 20.2	59.0
12		0 20.3	N11 59.8
13		15 20.4	12 00.7
14		30 20.5	01.5
15		45 20.7 ··	02.4
16		60 20.8	03.2
17		75 20.9	04.1
18		90 21.0	N12 04.9
19		105 21.2	05.7
20		120 21.3	06.6
21		135 21.4 ··	07.4
22		150 21.5	08.3
23		165 21.7	09.1
22 00		180 21.8	N12 10.0
01		195 21.9	10.8
02		210 22.0	11.6
03		225 22.2 ··	12.5
04		240 22.3	13.3
05		255 22.4	14.2
06		270 22.5	N12 15.0
07		285 22.7	15.8
08		300 22.8	16.7
09		315 22.9 ··	17.5
10		330 23.0	18.3
11		345 23.1	19.2
12		0 23.3	N12 20.0
13		15 23.4	20.9
14		30 23.5	21.7
15		45 23.6 ··	22.5
16		60 23.7	23.4
17		75 23.9	24.2
18		90 24.0	N12 25.0
19		105 24.1	25.9
20		120 24.2	26.7
21		135 24.3 ··	27.5
22		150 24.5	28.4
23		165 24.6	29.2
23 00		180 24.7	N12 30.0
01		195 24.8	30.9
02		210 24.9	31.7
03		225 25.0 ··	32.5
04		240 25.2	33.4
05		255 25.3	34.2
06		270 25.4	N12 35.0
07		285 25.5	35.9
08		300 25.6	36.7
09		315 25.8 ··	37.5
10		330 25.9	38.4
11		345 26.0	39.2
12		0 26.1	N12 40.0
13		15 26.2	40.8
14		30 26.3	41.7
15		45 26.4 ··	42.5
16		60 26.6	43.3
17		75 26.7	44.2
18		90 26.8	N12 45.0
19		105 26.9	45.8
20		120 27.0	46.6
21		135 27.1 ··	47.5
22		150 27.2	48.3
23		165 27.4	49.1

		G.H.A.	Dec.
d h		° ′	° ′
24 00		180 27.5	N12 49.9
01		195 27.6	50.8
02		210 27.7	51.6
03		225 27.8 ··	52.4
04		240 27.9	53.2
05		255 28.0	54.1
06		270 28.2	N12 54.9
07		285 28.3	55.7
08		300 28.4	56.5
09		315 28.5 ··	57.3
10		330 28.6	58.2
11		345 28.7	59.0
12		0 28.8	N12 59.8
13		15 28.9	13 00.6
14		30 29.0	01.4
15		45 29.2 ··	02.3
16		60 29.3	03.1
17		75 29.4	03.9
18		90 29.5	N13 04.7
19		105 29.6	05.5
20		120 29.7	06.3
21		135 29.8 ··	07.2
22		150 29.9	08.0
23		165 30.0	08.8
25 00		180 30.1	N13 09.6
01		195 30.2	10.4
02		210 30.4	11.2
03		225 30.5 ··	12.1
04		240 30.6	12.9
05		255 30.7	13.7
06		270 30.8	N13 14.5
07		285 30.9	15.3
08		300 31.0	16.1
09		315 31.1 ··	16.9
10		330 31.2	17.8
11		345 31.3	18.6
12		0 31.4	N13 19.4
13		15 31.5	20.2
14		30 31.6	21.0
15		45 31.7 ··	21.8
16		60 31.8	22.6
17		75 31.9	23.4
18		90 32.1	N13 24.2
19		105 32.2	25.0
20		120 32.3	25.8
21		135 32.4 ··	26.7
22		150 32.5	27.5
23		165 32.6	28.3
26 00		180 32.7	N13 29.1
01		195 32.8	29.9
02		210 32.9	30.7
03		225 33.0 ··	31.5
04		240 33.1	32.3
05		255 33.2	33.1
06		270 33.3	N13 33.9
07		285 33.4	34.7
08		300 33.5	35.5
09		315 33.6 ··	36.3
10		330 33.7	37.1
11		345 33.8	37.9
12		0 33.9	N13 38.7
13		15 34.0	39.5
14		30 34.1	40.3
15		45 34.2 ··	41.1
16		60 34.3	41.9
17		75 34.4	42.7
18		90 34.5	N13 43.5
19		105 34.6	44.3
20		120 34.7	45.1
21		135 34.8 ··	45.9
22		150 34.9	46.7
23		165 35.0	47.5

SUN — G.H.A. and Dec.

d h	G.H.A. ° '	Dec. ° '
27 00	180 35.1	N13 48.3
01	195 35.2	49.1
02	210 35.3	49.9
03	225 35.4 ··	50.7
04	240 35.5	51.5
05	255 35.6	52.3
06	270 35.7	N13 53.1
07	285 35.8	53.9
08	300 35.9	54.7
09	315 36.0 ··	55.5
10	330 36.0	56.3
11	345 36.1	57.1
12	0 36.2	N13 57.9
13	15 36.3	58.6
14	30 36.4	13 59.4
15	45 36.5	14 00.2
16	60 36.6	01.0
17	75 36.7	01.8
18	90 36.8	N14 02.6
19	105 36.9	03.4
20	120 37.0	04.2
21	135 37.1 ··	05.0
22	150 37.2	05.8
23	165 37.3	06.5
28 00	180 37.4	N14 07.3
01	195 37.5	08.1
02	210 37.5	08.9
03	225 37.6 ··	09.7
04	240 37.7	10.5
05	255 37.8	11.3
06	270 37.9	N14 12.0
07	285 38.0	12.8
08	300 38.1	13.6
09	315 38.2 ··	14.4
10	330 38.3	15.2
11	345 38.4	16.0
12	0 38.5	N14 16.8
13	15 38.5	17.5
14	30 38.6	18.3
15	45 38.7 ··	19.1
16	60 38.8	19.9
17	75 38.9	20.7
18	90 39.0	N14 21.4
19	105 39.1	22.2
20	120 39.2	23.0
21	135 39.3 ··	23.8
22	150 39.3	24.6
23	165 39.4	25.3
29 00	180 39.5	N14 26.1
01	195 39.6	26.9
02	210 39.7	27.7
03	225 39.8 ··	28.4
04	240 39.9	29.2
05	255 40.0	30.0
06	270 40.0	N14 30.8
07	285 40.1	31.5
08	300 40.2	32.3
09	315 40.3 ··	33.1
10	330 40.4	33.9
11	345 40.5	34.6
12	0 40.5	N14 35.4
13	15 40.6	36.2
14	30 40.7	37.0
15	45 40.8 ··	37.7
16	60 40.9	38.5
17	75 41.0	39.3
18	90 41.1	N14 40.0
19	105 41.1	40.8
20	120 41.2	41.6
21	135 41.3 ··	42.4
22	150 41.4	43.1
23	165 41.5	43.9

MAY

d h	G.H.A. ° '	Dec. ° '
30 00	180 41.5	N14 44.7
01	195 41.6	45.4
02	210 41.7	46.2
03	225 41.8 ··	47.0
04	240 41.9	47.7
05	255 41.9	48.5
06	270 42.0	N14 49.3
07	285 42.1	50.0
08	300 42.2	50.8
09	315 42.3 ··	51.5
10	330 42.3	52.3
11	345 42.4	53.1
12	0 42.5	N14 53.8
13	15 42.6	54.6
14	30 42.7	55.4
15	45 42.7 ··	56.1
16	60 42.8	56.9
17	75 42.9	57.6
18	90 43.0	N14 58.4
19	105 43.0	59.2
20	120 43.1	14 59.9
21	135 43.2	15 00.7
22	150 43.3	01.4
23	165 43.4	02.2
1 00	180 43.4	N15 03.0
01	195 43.5	03.7
02	210 43.6	04.5
03	225 43.7 ··	05.2
04	240 43.7	06.0
05	255 43.8	06.7
06	270 43.9	N15 07.5
07	285 44.0	08.2
08	300 44.0	09.0
09	315 44.1 ··	09.8
10	330 44.2	10.5
11	345 44.3	11.3
12	0 44.3	N15 12.0
13	15 44.4	12.8
14	30 44.5	13.5
15	45 44.5 ··	14.3
16	60 44.6	15.0
17	75 44.7	15.8
18	90 44.8	N15 16.5
19	105 44.8	17.3
20	120 44.9	18.0
21	135 45.0 ··	18.8
22	150 45.0	19.5
23	165 45.1	20.3
2 00	180 45.2	N15 21.0
01	195 45.3	21.8
02	210 45.3	22.5
03	225 45.4 ··	23.2
04	240 45.5	24.0
05	255 45.5	24.7
06	270 45.6	N15 25.5
07	285 45.7	26.2
08	300 45.7	27.0
09	315 45.8 ··	27.7
10	330 45.9	28.5
11	345 46.0	29.2
12	0 46.0	N15 29.9
13	15 46.1	30.7
14	30 46.2	31.4
15	45 46.2 ··	32.2
16	60 46.3	32.9
17	75 46.4	33.6
18	90 46.4	N15 34.4
19	105 46.5	35.1
20	120 46.6	35.9
21	135 46.6 ··	36.6
22	150 46.7	37.3
23	165 46.7	38.1

d h	G.H.A. ° '	Dec. ° '
3 00	180 46.8	N15 38.8
01	195 46.9	39.5
02	210 46.9	40.3
03	225 47.0 ··	41.0
04	240 47.1	41.7
05	255 47.1	42.5
06	270 47.2	N15 43.2
07	285 47.3	44.0
08	300 47.3	44.7
09	315 47.4 ··	45.4
10	330 47.5	46.1
11	345 47.5	46.9
12	0 47.6	N15 47.6
13	15 47.6	48.3
14	30 47.7	49.1
15	45 47.8 ··	49.8
16	60 47.8	50.5
17	75 47.9	51.3
18	90 47.9	N15 52.0
19	105 48.0	52.7
20	120 48.1	53.4
21	135 48.1 ··	54.2
22	150 48.2	54.9
23	165 48.2	55.6
4 00	180 48.3	N15 56.3
01	195 48.4	57.1
02	210 48.4	57.8
03	225 48.5 ··	58.5
04	240 48.5	15 59.2
05	255 48.6	16 00.0
06	270 48.7	N16 00.7
07	285 48.7	01.4
08	300 48.8	02.1
09	315 48.8 ··	02.9
10	330 48.9	03.6
11	345 48.9	04.3
12	0 49.0	N16 05.0
13	15 49.1	05.7
14	30 49.1	06.5
15	45 49.2 ··	07.2
16	60 49.2	07.9
17	75 49.3	08.6
18	90 49.3	N16 09.3
19	105 49.4	10.0
20	120 49.4	10.8
21	135 49.5 ··	11.5
22	150 49.5	12.2
23	165 49.6	12.9
5 00	180 49.7	N16 13.6
01	195 49.7	14.3
02	210 49.8	15.0
03	225 49.8 ··	15.8
04	240 49.9	16.5
05	255 49.9	17.2
06	270 50.0	N16 17.9
07	285 50.0	18.6
08	300 50.1	19.3
09	315 50.1 ··	20.0
10	330 50.2	20.7
11	345 50.2	21.4
12	0 50.3	N16 22.2
13	15 50.3	22.9
14	30 50.4	23.6
15	45 50.4 ··	24.3
16	60 50.5	25.0
17	75 50.5	25.7
18	90 50.6	N16 26.4
19	105 50.6	27.1
20	120 50.7	27.8
21	135 50.7 ··	28.5
22	150 50.8	29.2
23	165 50.8	29.9

d h	G.H.A. ° '	Dec. ° '
6 00	180 50.9	N16 30.6
01	195 50.9	31.3
02	210 51.0	32.0
03	225 51.0 ··	32.7
04	240 51.1	33.4
05	255 51.1	34.1
06	270 51.2	N16 34.8
07	285 51.2	35.5
08	300 51.3	36.2
09	315 51.3 ··	36.9
10	330 51.3	37.6
11	345 51.4	38.3
12	0 51.4	N16 39.0
13	15 51.5	39.7
14	30 51.5	40.4
15	45 51.6 ··	41.1
16	60 51.6	41.8
17	75 51.7	42.5
18	90 51.7	N16 43.2
19	105 51.7	43.9
20	120 51.8	44.6
21	135 51.8 ··	45.3
22	150 51.9	46.0
23	165 51.9	46.7
7 00	180 52.0	N16 47.3
01	195 52.0	48.0
02	210 52.0	48.7
03	225 52.1 ··	49.4
04	240 52.1	50.1
05	255 52.2	50.8
06	270 52.2	N16 51.5
07	285 52.3	52.2
08	300 52.3	52.9
09	315 52.3 ··	53.5
10	330 52.4	54.2
11	345 52.4	54.9
12	0 52.5	N16 55.6
13	15 52.5	56.3
14	30 52.5	57.0
15	45 52.6 ··	57.7
16	60 52.6	58.3
17	75 52.6	59.0
18	90 52.7	N16 59.7
19	105 52.7	17 00.4
20	120 52.8	01.1
21	135 52.8 ··	01.8
22	150 52.8	02.4
23	165 52.9	03.1
8 00	180 52.9	N17 03.8
01	195 52.9	04.5
02	210 53.0	05.2
03	225 53.0 ··	05.8
04	240 53.1	06.5
05	255 53.1	07.2
06	270 53.1	N17 07.9
07	285 53.2	08.5
08	300 53.2	09.2
09	315 53.2 ··	09.9
10	330 53.3	10.6
11	345 53.3	11.2
12	0 53.3	N17 11.9
13	15 53.4	12.6
14	30 53.4	13.3
15	45 53.4 ··	13.9
16	60 53.5	14.6
17	75 53.5	15.3
18	90 53.5	N17 15.9
19	105 53.6	16.6
20	120 53.6	17.3
21	135 53.6 ··	18.0
22	150 53.7	18.6
23	165 53.7	19.3

d h	G.H.A. ° '	Dec. ° '
9 00	180 53.7	N17 20.0
01	195 53.7	20.6
02	210 53.8	21.3
03	225 53.8 ··	22.0
04	240 53.8	22.6
05	255 53.9	23.3
06	270 53.9	N17 24.0
07	285 53.9	24.6
08	300 54.0	25.3
09	315 54.0 ··	25.9
10	330 54.0	26.6
11	345 54.0	27.3
12	0 54.1	N17 27.9
13	15 54.1	28.6
14	30 54.1	29.2
15	45 54.2 ··	29.9
16	60 54.2	30.6
17	75 54.2	31.2
18	90 54.2	N17 31.9
19	105 54.3	32.5
20	120 54.3	33.2
21	135 54.3 ··	33.9
22	150 54.3	34.5
23	165 54.4	35.2
10 00	180 54.4	N17 35.8
01	195 54.4	36.5
02	210 54.4	37.1
03	225 54.5 ··	37.8
04	240 54.5	38.4
05	255 54.5	39.1
06	270 54.5	N17 39.7
07	285 54.6	40.4
08	300 54.6	41.1
09	315 54.6 ··	41.7
10	330 54.6	42.4
11	345 54.7	43.0
12	0 54.7	N17 43.7
13	15 54.7	44.3
14	30 54.7	44.9
15	45 54.7 ··	45.6
16	60 54.8	46.2
17	75 54.8	46.9
18	90 54.8	N17 47.5
19	105 54.8	48.2
20	120 54.8	48.8
21	135 54.9 ··	49.5
22	150 54.9	50.1
23	165 54.9	50.8
11 00	180 54.9	N17 51.4
01	195 54.9	52.0
02	210 55.0	52.7
03	225 55.0 ··	53.3
04	240 55.0	54.0
05	255 55.0	54.6
06	270 55.0	N17 55.3
07	285 55.1	55.9
08	300 55.1	56.5
09	315 55.1 ··	57.2
10	330 55.1	57.8
11	345 55.1	58.4
12	0 55.1	N17 59.1
13	15 55.2	17 59.7
14	30 55.2	18 00.4
15	45 55.2 ··	01.0
16	60 55.2	01.6
17	75 55.2	02.3
18	90 55.2	N18 02.9
19	105 55.2	03.5
20	120 55.3	04.2
21	135 55.3 ··	04.8
22	150 55.3	05.4
23	165 55.3	06.0

d h	G.H.A. ° '	Dec. ° '
12 00	180 55.3	N18 06.7
01	195 55.3	07.3
02	210 55.3	07.9
03	225 55.4 ··	08.6
04	240 55.4	09.2
05	255 55.4	09.8
06	270 55.4	N18 10.4
07	285 55.4	11.1
08	300 55.4	11.7
09	315 55.4 ··	12.3
10	330 55.4	13.0
11	345 55.5	13.6
12	0 55.5	N18 14.2
13	15 55.5	14.8
14	30 55.5	15.4
15	45 55.5 ··	16.1
16	60 55.5	16.7
17	75 55.5	17.3
18	90 55.5	N18 17.9
19	105 55.5	18.6
20	120 55.5	19.2
21	135 55.5 ··	19.8
22	150 55.6	20.4
23	165 55.6	21.0
13 00	180 55.6	N18 21.6
01	195 55.6	22.3
02	210 55.6	22.9
03	225 55.6 ··	23.5
04	240 55.6	24.1
05	255 55.6	24.7
06	270 55.6	N18 25.3
07	285 55.6	26.0
08	300 55.6	26.6
09	315 55.6 ··	27.2
10	330 55.6	27.8
11	345 55.6	28.4
12	0 55.6	N18 29.0
13	15 55.6	29.6
14	30 55.6	30.2
15	45 55.7 ··	30.8
16	60 55.7	31.5
17	75 55.7	32.1
18	90 55.7	N18 32.7
19	105 55.7	33.3
20	120 55.7	33.9
21	135 55.7 ··	34.5
22	150 55.7	35.1
23	165 55.7	35.7
14 00	180 55.7	N18 36.3
01	195 55.7	36.9
02	210 55.7	37.5
03	225 55.7 ··	38.1
04	240 55.7	38.7
05	255 55.7	39.3
06	270 55.7	N18 39.9
07	285 55.7	40.5
08	300 55.7	41.1
09	315 55.7 ··	41.7
10	330 55.7	42.3
11	345 55.7	42.9
12	0 55.7	N18 43.5
13	15 55.7	44.1
14	30 55.7	44.7
15	45 55.7 ··	45.3
16	60 55.7	45.9
17	75 55.7	46.5
18	90 55.7	N18 47.1
19	105 55.7	47.7
20	120 55.7	48.3
21	135 55.7 ··	48.9
22	150 55.7	49.5
23	165 55.7	50.1

Day 15

d h	G.H.A.	Dec.
15 00	180 55.6	N18 50.7
01	195 55.6	51.2
02	210 55.6	51.8
03	225 55.6 ··	52.4
04	240 55.6	53.0
05	255 55.6	53.6
06	270 55.6	N18 54.2
07	285 55.6	54.8
08	300 55.6	55.4
09	315 55.6 ··	56.0
10	330 55.6	56.5
11	345 55.6	57.1
12	0 55.6	N18 57.7
13	15 55.6	58.3
14	30 55.6	58.9
15	45 55.6	18 59.5
16	60 55.5	19 00.0
17	75 55.5	00.6
18	90 55.5	N19 01.2
19	105 55.5	01.8
20	120 55.5	02.4
21	135 55.5 ··	03.0
22	150 55.5	03.5
23	165 55.5	04.1

Day 16

d h	G.H.A.	Dec.
16 00	180 55.5	N19 04.7
01	195 55.5	05.3
02	210 55.5	05.8
03	225 55.4 ··	06.4
04	240 55.4	07.0
05	255 55.4	07.6
06	270 55.4	N19 08.1
07	285 55.4	08.7
08	300 55.4	09.3
09	315 55.4 ··	09.9
10	330 55.4	10.4
11	345 55.3	11.0
12	0 55.3	N19 11.6
13	15 55.3	12.2
14	30 55.3	12.7
15	45 55.3 ··	13.3
16	60 55.3	13.9
17	75 55.3	14.4
18	90 55.2	N19 15.0
19	105 55.2	15.6
20	120 55.2	16.1
21	135 55.2 ··	16.7
22	150 55.2	17.3
23	165 55.2	17.8

Day 17

d h	G.H.A.	Dec.
17 00	180 55.2	N19 18.4
01	195 55.1	19.0
02	210 55.1	19.5
03	225 55.1 ··	20.1
04	240 55.1	20.7
05	255 55.1	21.2
06	270 55.1	N19 21.8
07	285 55.0	22.3
08	300 55.0	22.9
09	315 55.0 ··	23.5
10	330 55.0	24.0
11	345 55.0	24.6
12	0 54.9	N19 25.1
13	15 54.9	25.7
14	30 54.9	26.2
15	45 54.9 ··	26.8
16	60 54.9	27.4
17	75 54.8	27.9
18	90 54.8	N19 28.5
19	105 54.8	29.0
20	120 54.8	29.6
21	135 54.8 ··	30.1
22	150 54.7	30.7
23	165 54.7	31.2

Day 18

d h	G.H.A.	Dec.
18 00	180 54.7	N19 31.8
01	195 54.7	32.3
02	210 54.6	32.9
03	225 54.6 ··	33.4
04	240 54.6	34.0
05	255 54.6	34.5
06	270 54.6	N19 35.1
07	285 54.5	35.6
08	300 54.5	36.2
09	315 54.5 ··	36.7
10	330 54.5	37.3
11	345 54.4	37.8
12	0 54.4	N19 38.3
13	15 54.4	38.9
14	30 54.4	39.4
15	45 54.3 ··	40.0
16	60 54.3	40.5
17	75 54.3	41.1
18	90 54.3	N19 41.6
19	105 54.2	42.1
20	120 54.2	42.7
21	135 54.2 ··	43.2
22	150 54.1	43.8
23	165 54.1	44.3

Day 19

d h	G.H.A.	Dec.
19 00	180 54.1	N19 44.8
01	195 54.1	45.4
02	210 54.0	45.9
03	225 54.0 ··	46.4
04	240 54.0	47.0
05	255 53.9	47.5
06	270 53.9	N19 48.0
07	285 53.9	48.6
08	300 53.9	49.1
09	315 53.8 ··	49.6
10	330 53.8	50.2
11	345 53.8	50.7
12	0 53.7	N19 51.2
13	15 53.7	51.8
14	30 53.7	52.3
15	45 53.6 ··	52.8
16	60 53.6	53.3
17	75 53.6	53.9
18	90 53.5	N19 54.4
19	105 53.5	54.9
20	120 53.5	55.5
21	135 53.4 ··	56.0
22	150 53.4	56.5
23	165 53.4	57.0

Day 20

d h	G.H.A.	Dec.
20 00	180 53.3	N19 57.5
01	195 53.3	58.1
02	210 53.3	58.6
03	225 53.2 ··	59.1
04	240 53.2	19 59.6
05	255 53.2	20 00.2
06	270 53.1	N20 00.7
07	285 53.1	01.2
08	300 53.1	01.7
09	315 53.0 ··	02.2
10	330 53.0	02.7
11	345 53.0	03.3
12	0 52.9	N20 03.8
13	15 52.9	04.3
14	30 52.8	04.8
15	45 52.8 ··	05.3
16	60 52.8	05.8
17	75 52.7	06.3
18	90 52.7	N20 06.9
19	105 52.6	07.4
20	120 52.6	07.9
21	135 52.6 ··	08.4
22	150 52.5	08.9
23	165 52.5	09.4

Day 21

d h	G.H.A.	Dec.
21 00	180 52.5	N20 09.9
01	195 52.4	10.4
02	210 52.4	10.9
03	225 52.3 ··	11.4
04	240 52.3	12.0
05	255 52.3	12.5
06	270 52.2	N20 13.0
07	285 52.2	13.5
08	300 52.1	14.0
09	315 52.1 ··	14.5
10	330 52.0	15.0
11	345 52.0	15.5
12	0 52.0	N20 16.0
13	15 51.9	16.5
14	30 51.9	17.0
15	45 51.8 ··	17.5
16	60 51.8	18.0
17	75 51.7	18.5
18	90 51.7	N20 19.0
19	105 51.7	19.5
20	120 51.6	20.0
21	135 51.6 ··	20.5
22	150 51.5	21.0
23	165 51.5	21.5

Day 22

d h	G.H.A.	Dec.
22 00	180 51.4	N20 22.0
01	195 51.4	22.4
02	210 51.3	22.9
03	225 51.3 ··	23.4
04	240 51.2	23.9
05	255 51.2	24.4
06	270 51.1	N20 24.9
07	285 51.1	25.4
08	300 51.1	25.9
09	315 51.0 ··	26.4
10	330 51.0	26.9
11	345 50.9	27.4
12	0 50.9	N20 27.8
13	15 50.8	28.3
14	30 50.8	28.8
15	45 50.7 ··	29.3
16	60 50.7	29.8
17	75 50.6	30.3
18	90 50.6	N20 30.7
19	105 50.5	31.2
20	120 50.5	31.7
21	135 50.4 ··	32.2
22	150 50.4	32.7
23	165 50.3	33.2

Day 23

d h	G.H.A.	Dec.
23 00	180 50.3	N20 33.6
01	195 50.2	34.1
02	210 50.2	34.6
03	225 50.1 ··	35.1
04	240 50.1	35.5
05	255 50.0	36.0
06	270 50.0	N20 36.5
07	285 49.9	37.0
08	300 49.8	37.5
09	315 49.8 ··	37.9
10	330 49.7	38.4
11	345 49.7	38.9
12	0 49.6	N20 39.3
13	15 49.6	39.8
14	30 49.5	40.3
15	45 49.5 ··	40.8
16	60 49.4	41.2
17	75 49.4	41.7
18	90 49.3	N20 42.2
19	105 49.3	42.6
20	120 49.2	43.1
21	135 49.1 ··	43.6
22	150 49.1	44.0
23	165 49.0	44.5

d h	G.H.A.	Dec.
24 00	180 49.0	N20 45.0
01	195 48.9	45.4
02	210 48.9	45.9
03	225 48.8 ··	46.4
04	240 48.7	46.8
05	255 48.7	47.3
06	270 48.6 N20	47.7
07	285 48.6	48.2
08	300 48.5	48.7
09	315 48.5 ··	49.1
10	330 48.4	49.6
11	345 48.3	50.0
12	0 48.3 N20	50.5
13	15 48.2	51.0
14	30 48.2	51.4
15	45 48.1 ··	51.9
16	60 48.0	52.3
17	75 48.0	52.8
18	90 47.9 N20	53.2
19	105 47.9	53.7
20	120 47.8	54.1
21	135 47.7 ··	54.6
22	150 47.7	55.0
23	165 47.6	55.5
25 00	180 47.5 N20	55.9
01	195 47.5	56.4
02	210 47.4	56.8
03	225 47.4 ··	57.3
04	240 47.3	57.7
05	255 47.2	58.2
06	270 47.2 N20	58.6
07	285 47.1	59.1
08	300 47.0 20	59.5
09	315 47.0 21	00.0
10	330 46.9	00.4
11	345 46.8	00.8
12	0 46.8 N21	01.3
13	15 46.7	01.7
14	30 46.7	02.2
15	45 46.6 ··	02.6
16	60 46.5	03.1
17	75 46.5	03.5
18	90 46.4 N21	03.9
19	105 46.3	04.4
20	120 46.3	04.8
21	135 46.2 ··	05.2
22	150 46.1	05.7
23	165 46.1	06.1
26 00	180 46.0 N21	06.6
01	195 45.9	07.0
02	210 45.9	07.4
03	225 45.8 ··	07.9
04	240 45.7	08.3
05	255 45.7	08.7
06	270 45.6 N21	09.1
07	285 45.5	09.6
08	300 45.4	10.0
09	315 45.4 ··	10.4
10	330 45.3	10.9
11	345 45.2	11.3
12	0 45.2 N21	11.7
13	15 45.1	12.1
14	30 45.0	12.6
15	45 45.0 ··	13.0
16	60 44.9	13.4
17	75 44.8	13.8
18	90 44.7 N21	14.3
19	105 44.7	14.7
20	120 44.6	15.1
21	135 44.5 ··	15.5
22	150 44.5	16.0
23	165 44.4	16.4

d h	G.H.A.	Dec.
27 00	180 44.3 N21	16.8
01	195 44.2	17.2
02	210 44.2	17.6
03	225 44.1 ··	18.1
04	240 44.0	18.5
05	255 44.0	18.9
06	270 43.9 N21	19.3
07	285 43.8	19.7
08	300 43.7	20.1
09	315 43.7 ··	20.5
10	330 43.6	21.0
11	345 43.5	21.4
12	0 43.4 N21	21.8
13	15 43.4	22.2
14	30 43.3	22.6
15	45 43.2 ··	23.0
16	60 43.1	23.4
17	75 43.1	23.8
18	90 43.0 N21	24.2
19	105 42.9	24.7
20	120 42.8	25.1
21	135 42.8 ··	25.5
22	150 42.7	25.9
23	165 42.6	26.3
28 00	180 42.5 N21	26.7
01	195 42.4	27.1
02	210 42.4	27.5
03	225 42.3 ··	27.9
04	240 42.2	28.3
05	255 42.1	28.7
06	270 42.1 N21	29.1
07	285 42.0	29.5
08	300 41.9	29.9
09	315 41.8 ··	30.3
10	330 41.7	30.7
11	345 41.7	31.1
12	0 41.6 N21	31.5
13	15 41.5	31.9
14	30 41.4	32.3
15	45 41.3 ··	32.7
16	60 41.3	33.1
17	75 41.2	33.5
18	90 41.1 N21	33.8
19	105 41.0	34.2
20	120 40.9	34.6
21	135 40.9 ··	35.0
22	150 40.8	35.4
23	165 40.7	35.8
29 00	180 40.6 N21	36.2
01	195 40.5	36.6
02	210 40.4	37.0
03	225 40.4 ··	37.4
04	240 40.3	37.7
05	255 40.2	38.1
06	270 40.1 N21	38.5
07	285 40.0	38.9
08	300 39.9	39.3
09	315 39.9 ··	39.7
10	330 39.8	40.0
11	345 39.7	40.4
12	0 39.6 N21	40.8
13	15 39.5	41.2
14	30 39.4	41.6
15	45 39.4 ··	41.9
16	60 39.3	42.3
17	75 39.2	42.7
18	90 39.1 N21	43.1
19	105 39.0	43.5
20	120 38.9	43.8
21	135 38.8 ··	44.2
22	150 38.8	44.6
23	165 38.7	45.0

d h	G.H.A.	Dec.
30 00	180 38.6 N21	45.3
01	195 38.5	45.7
02	210 38.4	46.1
03	225 38.3 ··	46.4
04	240 38.2	46.8
05	255 38.1	47.2
06	270 38.1 N21	47.6
07	285 38.0	47.9
08	300 37.9	48.3
09	315 37.8 ··	48.7
10	330 37.7	49.0
11	345 37.6	49.4
12	0 37.5 N21	49.8
13	15 37.4	50.1
14	30 37.4	50.5
15	45 37.3 ··	50.8
16	60 37.2	51.2
17	75 37.1	51.6
18	90 37.0 N21	51.9
19	105 36.9	52.3
20	120 36.8	52.7
21	135 36.7 ··	53.0
22	150 36.6	53.4
23	165 36.5	53.7
31 00	180 36.5 N21	54.1
01	195 36.4	54.4
02	210 36.3	54.8
03	225 36.2 ··	55.1
04	240 36.1	55.5
05	255 36.0	55.9
06	270 35.9 N21	56.2
07	285 35.8	56.6
08	300 35.7	56.9
09	₃15 35.6 ··	57.3
10	330 35.5	57.6
11	345 35.4	58.0
12	0 35.3 N21	58.3
13	15 35.3	58.7
14	30 35.2	59.0
15	45 35.1 ··	59.4
16	60 35.0 21	59.7
17	75 34.9 22	00.1
18	90 34.8 N22	00.4
19	105 34.7	00.8
20	120 34.6	01.1
21	135 34.5 ··	01.4
22	150 34.4	01.8
23	165 34.3	02.1
JUNE 1 00	180 34.2 N22	02.5
01	195 34.1	02.8
02	210 34.0	03.1
03	225 33.9 ··	03.5
04	240 33.8	03.8
05	255 33.7	04.2
06	270 33.6 N22	04.5
07	285 33.6	04.8
08	300 33.5	05.2
09	315 33.4	05.5
10	330 33.3	05.8
11	345 33.2	06.2
12	0 33.1 N22	06.5
13	15 33.0	06.8
14	30 32.9	07.2
15	45 32.8 ··	07.5
16	60 32.7	07.8
17	75 32.6	08.2
18	90 32.5 N22	08.5
19	105 32.4	08.8
20	120 32.3	09.2
21	135 32.2 ··	09.5
22	150 32.1	09.8
23	165 32.0	10.1

d h	G.H.A.	Dec.		d h	G.H.A.	Dec.		d h	G.H.A.	Dec.
2 00	180 31.9	N22 10.5		**5** 00	180 24.4	N22 32.1		**8** 00	180 16.2	N22 50.2
01	195 31.8	10.8		01	195 24.3	32.4		01	195 16.0	50.4
02	210 31.7	11.1		02	210 24.2	32.7		02	210 15.9	50.7
03	225 31.6 ··	11.4		03	225 24.0 ··	32.9		03	225 15.8 ··	50.9
04	240 31.5	11.8		04	240 23.9	33.2		04	240 15.7	51.1
05	255 31.4	12.1		05	255 23.8	33.5		05	255 15.6	51.3
06	270 31.3	N22 12.4		06	270 23.7	N22 33.8		06	270 15.5	N22 51.5
07	285 31.2	12.7		07	285 23.6	34.0		07	285 15.3	51.8
08	300 31.1	13.0		08	300 23.5	34.3		08	300 15.2	52.0
09	315 31.0 ··	13.4		09	315 23.4 ··	34.6		09	315 15.1 ··	52.2
10	330 30.9	13.7		10	330 23.3	34.8		10	330 15.0	52.4
11	345 30.8	14.0		11	345 23.2	35.1		11	345 14.9	52.7
12	0 30.7	N22 14.3		12	0 23.1	N22 35.4		12	0 14.7	N22 52.9
13	15 30.6	14.6		13	15 22.9	35.6		13	15 14.6	53.1
14	30 30.5	14.9		14	30 22.8	35.9		14	30 14.5	53.3
15	45 30.4 ··	15.3		15	45 22.7 ··	36.2		15	45 14.4 ··	53.5
16	60 30.3	15.6		16	60 22.6	36.4		16	60 14.3	53.7
17	75 30.2	15.9		17	75 22.5	36.7		17	75 14.1	54.0
18	90 30.1	N22 16.2		18	90 22.4	N22 37.0		18	90 14.0	N22 54.2
19	105 30.0	16.5		19	105 22.3	37.2		19	105 13.9	54.4
20	120 29.9	16.8		20	120 22.2	37.5		20	120 13.8	54.6
21	135 29.8 ··	17.1		21	135 22.0 ··	37.8		21	135 13.7 ··	54.8
22	150 29.7	17.4		22	150 21.9	38.0		22	150 13.5	55.0
23	165 29.6	17.8		23	165 21.8	38.3		23	165 13.4	55.2
3 00	180 29.5	N22 18.1		**6** 00	180 21.7	N22 38.5		**9** 00	180 13.3	N22 55.4
01	195 29.4	18.4		01	195 21.6	38.8		01	195 13.2	55.6
02	210 29.3	18.7		02	210 21.5	39.1		02	210 13.1	55.9
03	225 29.2 ··	19.0		03	225 21.4 ··	39.3		03	225 12.9 ··	56.1
04	240 29.1	19.3		04	240 21.3	39.6		04	240 12.8	56.3
05	255 29.0	19.6		05	255 21.1	39.8		05	255 12.7	56.5
06	270 28.9	N22 19.9		06	270 21.0	N22 40.1		06	270 12.6	N22 56.7
07	285 28.7	20.2		07	285 20.9	40.3		07	285 12.5	56.9
08	300 28.6	20.5		08	300 20.8	40.6		08	300 12.3	57.1
09	315 28.5 ··	20.8		09	315 20.7 ··	40.8		09	315 12.2 ··	57.3
10	330 28.4	21.1		10	330 20.6	41.1		10	330 12.1	57.5
11	345 28.3	21.4		11	345 20.5	41.4		11	345 12.0	57.7
12	0 28.2	N22 21.7		12	0 20.3	N22 41.6		12	0 11.8	N22 57.9
13	15 28.1	22.0		13	15 20.2	41.9		13	15 11.7	58.1
14	30 28.0	22.3		14	30 20.1	42.1		14	30 11.6	58.3
15	45 27.9 ··	22.6		15	45 20.0 ··	42.4		15	45 11.5 ··	58.5
16	60 27.8	22.9		16	60 19.9	42.6		16	60 11.4	58.7
17	75 27.7	23.2		17	75 19.8	42.9		17	75 11.2	58.9
18	90 27.6	N22 23.5		18	90 19.7	N22 43.1		18	90 11.1	N22 59.1
19	105 27.5	23.8		19	105 19.5	43.3		19	105 11.0	59.3
20	120 27.4	24.1		20	120 19.4	43.6		20	120 10.9	59.5
21	135 27.3 ··	24.4		21	135 19.3 ··	43.8		21	135 10.7 ··	59.7
22	150 27.2	24.7		22	150 19.2	44.1		22	150 10.6	22 59.9
23	165 27.1	25.0		23	165 19.1	44.3		23	165 10.5	23 00.1
4 00	180 27.0	N22 25.3		**7** 00	180 19.0	N22 44.6		**10** 00	180 10.4	N23 00.3
01	195 26.9	25.6		01	195 18.9	44.8		01	195 10.3	00.5
02	210 26.8	25.9		02	210 18.7	45.1		02	210 10.1	00.6
03	225 26.6 ··	26.2		03	225 18.6 ··	45.3		03	225 10.0 ··	00.8
04	240 26.5	26.4		04	240 18.5	45.5		04	240 09.9	01.0
05	255 26.4	26.7		05	255 18.4	45.8		05	255 09.8	01.2
06	270 26.3	N22 27.0		06	270 18.3	N22 46.0		06	270 09.6	N23 01.4
07	285 26.2	27.3		07	285 18.2	46.3		07	285 09.5	01.6
08	300 26.1	27.6		08	300 18.0	46.5		08	300 09.4	01.8
09	315 26.0 ··	27.9		09	315 17.9 ··	46.7		09	315 09.3 ··	02.0
10	330 25.9	28.2		10	330 17.8	47.0		10	330 09.1	02.2
11	345 25.8	28.5		11	345 17.7	47.2		11	345 09.0	02.3
12	0 25.7	N22 28.7		12	0 17.6	N22 47.4		12	0 08.9	N23 02.5
13	15 25.6	29.0		13	15 17.5	47.7		13	15 08.8	02.7
14	30 25.5	29.3		14	30 17.3	47.9		14	30 08.6	02.9
15	45 25.4 ··	29.6		15	45 17.2 ··	48.1		15	45 08.5 ··	03.1
16	60 25.2	29.9		16	60 17.1	48.4		16	60 08.4	03.3
17	75 25.1	30.2		17	75 17.0	48.6		17	75 08.3	03.4
18	90 25.0	N22 30.4		18	90 16.9	N22 48.8		18	90 08.1	N23 03.6
19	105 24.9	30.7		19	105 16.8	49.1		19	105 08.0	03.8
20	120 24.8	31.0		20	120 16.6	49.3		20	120 07.9	04.0
21	135 24.7 ··	31.3		21	135 16.5 ··	49.5		21	135 07.8 ··	04.2
22	150 24.6	31.6		22	150 16.4	49.7		22	150 07.6	04.3
23	165 24.5	31.8		23	165 16.3	50.0		23	165 07.5	04.5

	G.H.A.	Dec.

d h	G.H.A. ° ′	Dec. ° ′
11 00	180 07.4 N23	04.7
01	195 07.3	04.9
02	210 07.1	05.0
03	225 07.0 ··	05.2
04	240 06.9	05.4
05	255 06.8	05.6
06	270 06.6 N23	05.7
07	285 06.5	05.9
08	300 06.4	06.1
09	315 06.3 ··	06.2
10	330 06.1	06.4
11	345 06.0	06.6
12	0 05.9 N23	06.7
13	15 05.8	06.9
14	30 05.6	07.1
15	45 05.5 ··	07.2
16	60 05.4	07.4
17	75 05.2	07.6
18	90 05.1 N23	07.7
19	105 05.0	07.9
20	120 04.9	08.1
21	135 04.7 ··	08.2
22	150 04.6	08.4
23	165 04.5	08.5
12 00	180 04.4 N23	08.7
01	195 04.2	08.9
02	210 04.1	09.0
03	225 04.0 ··	09.2
04	240 03.9	09.3
05	255 03.7	09.5
06	270 03.6 N23	09.6
07	285 03.5	09.8
08	300 03.3	10.0
09	315 03.2 ··	10.1
10	330 03.1	10.3
11	345 03.0	10.4
12	0 02.8 N23	10.6
13	15 02.7	10.7
14	30 02.6	10.9
15	45 02.4 ··	11.0
16	60 02.3	11.2
17	75 02.2	11.3
18	90 02.1 N23	11.5
19	105 01.9	11.6
20	120 01.8	11.7
21	135 01.7 ··	11.9
22	150 01.5	12.0
23	165 01.4	12.2
13 00	180 01.3 N23	12.3
01	195 01.2	12.5
02	210 01.0	12.6
03	225 00.9 ··	12.7
04	240 00.8	12.9
05	255 00.6	13.0
06	270 00.5 N23	13.2
07	285 00.4	13.3
08	300 00.3	13.4
09	315 00.1 ··	13.6
10	330 00.0	13.7
11	344 59.9	13.8
12	359 59.7 N23	14.0
13	14 59.6	14.1
14	29 59.5	14.2
15	44 59.3 ··	14.4
16	59 59.2	14.5
17	74 59.1	14.6
18	89 59.0 N23	14.8
19	104 58.8	14.9
20	119 58.7	15.0
21	134 58.6 ··	15.1
22	149 58.4	15.3
23	164 58.3	15.4

d h	G.H.A. ° ′	Dec. ° ′
14 00	179 58.2 N23	15.5
01	194 58.0	15.6
02	209 57.9	15.8
03	224 57.8 ··	15.9
04	239 57.6	16.0
05	254 57.5	16.1
06	269 57.4 N23	16.3
07	284 57.3	16.4
08	299 57.1	16.5
09	314 57.0 ··	16.6
10	329 56.9	16.7
11	344 56.7	16.8
12	359 56.6 N23	17.0
13	14 56.5	17.1
14	29 56.3	17.2
15	44 56.2 ··	17.3
16	59 56.1	17.4
17	74 55.9	17.5
18	89 55.8 N23	17.7
19	104 55.7	17.8
20	119 55.5	17.9
21	134 55.4 ··	18.0
22	149 55.3	18.1
23	164 55.1	18.2
15 00	179 55.0 N23	18.3
01	194 54.9	18.4
02	209 54.8	18.5
03	224 54.6 ··	18.6
04	239 54.5	18.7
05	254 54.4	18.8
06	269 54.2 N23	18.9
07	284 54.1	19.0
08	299 54.0	19.1
09	314 53.8 ··	19.3
10	329 53.7	19.4
11	344 53.6	19.5
12	359 53.4 N23	19.6
13	14 53.3	19.7
14	29 53.2	19.7
15	44 53.0 ··	19.8
16	59 52.9	19.9
17	74 52.8	20.0
18	89 52.6 N23	20.1
19	104 52.5	20.2
20	119 52.4	20.3
21	134 52.2 ··	20.4
22	149 52.1	20.5
23	164 52.0	20.6
16 00	179 51.8 N23	20.7
01	194 51.7	20.8
02	209 51.6	20.9
03	224 51.4 ··	21.0
04	239 51.3	21.0
05	254 51.2	21.1
06	269 51.0 N23	21.2
07	284 50.9	21.3
08	299 50.8	21.4
09	314 50.6 ··	21.5
10	329 50.5	21.6
11	344 50.4	21.6
12	359 50.2 N23	21.7
13	14 50.1	21.8
14	29 50.0	21.9
15	44 49.8 ··	22.0
16	59 49.7	22.0
17	74 49.6	22.1
18	89 49.4 N23	22.2
19	104 49.3	22.3
20	119 49.2	22.4
21	134 49.0 ··	22.4
22	149 48.9	22.5
23	164 48.8	22.6

d h	G.H.A. ° ′	Dec. ° ′
17 00	179 48.6 N23	22.7
01	194 48.5	22.7
02	209 48.3	22.8
03	224 48.2 ··	22.9
04	239 48.1	22.9
05	254 47.9	23.0
06	269 47.8 N23	23.1
07	284 47.7	23.2
08	299 47.5	23.2
09	314 47.4 ··	23.3
10	329 47.3	23.4
11	344 47.1	23.4
12	359 47.0 N23	23.5
13	14 46.9	23.6
14	29 46.7	23.6
15	44 46.6 ··	23.7
16	59 46.5	23.7
17	74 46.3	23.8
18	89 46.2 N23	23.9
19	104 46.1	23.9
20	119 45.9	24.0
21	134 45.8 ··	24.0
22	149 45.7	24.1
23	164 45.5	24.2
18 00	179 45.4 N23	24.2
01	194 45.2	24.3
02	209 45.1	24.3
03	224 45.0 ··	24.4
04	239 44.8	24.4
05	254 44.7	24.5
06	269 44.6 N23	24.5
07	284 44.4	24.6
08	299 44.3	24.6
09	314 44.2 ··	24.7
10	329 44.0	24.7
11	344 43.9	24.8
12	359 43.8 N23	24.8
13	14 43.6	24.9
14	29 43.5	24.9
15	44 43.4 ··	25.0
16	59 43.2	25.0
17	74 43.1	25.1
18	89 42.9 N23	25.1
19	104 42.8	25.2
20	119 42.7	25.2
21	134 42.5 ··	25.2
22	149 42.4	25.3
23	164 42.3	25.3
19 00	179 42.1 N23	25.4
01	194 42.0	25.4
02	209 41.9	25.4
03	224 41.7 ··	25.5
04	239 41.6	25.5
05	254 41.5	25.5
06	269 41.3 N23	25.6
07	284 41.2	25.6
08	299 41.0	25.6
09	314 40.9 ··	25.7
10	329 40.8	25.7
11	344 40.6	25.7
12	359 40.5 N23	25.8
13	14 40.4	25.8
14	29 40.2	25.8
15	44 40.1 ··	25.9
16	59 40.0	25.9
17	74 39.8	25.9
18	89 39.7 N23	25.9
19	104 39.5	26.0
20	119 39.4	26.0
21	134 39.3 ··	26.0
22	149 39.1	26.0
23	164 39.0	26.1

20

d h	G.H.A.		Dec.
00	179 38.9	N23	26.1
01	194 38.7		26.1
02	209 38.6		26.1
03	224 38.5 ··		26.1
04	239 38.3		26.2
05	254 38.2		26.2
06	269 38.1	N23	26.2
07	284 37.9		26.2
08	299 37.8		26.2
09	314 37.6 ··		26.3
10	329 37.5		26.3
11	344 37.4		26.3
12	359 37.2	N23	26.3
13	14 37.1		26.3
14	29 37.0		26.3
15	44 36.8 ··		26.3
16	59 36.7		26.3
17	74 36.6		26.4
18	89 36.4	N23	26.4
19	104 36.3		26.4
20	119 36.1		26.4
21	134 36.0 ··		26.4
22	149 35.9		26.4
23	164 35.7		26.4

21

d h	G.H.A.		Dec.
00	179 35.6	N23	26.4
01	194 35.5		26.4
02	209 35.3		26.4
03	224 35.2 ··		26.4
04	239 35.1		26.4
05	254 34.9		26.4
06	269 34.8	N23	26.4
07	284 34.7		26.4
08	299 34.5		26.4
09	314 34.4 ··		26.4
10	329 34.2		26.4
11	344 34.1		26.4
12	359 34.0	N23	26.4
13	14 33.8		26.4
14	29 33.7		26.4
15	44 33.6 ··		26.4
16	59 33.4		26.4
17	74 33.3		26.4
18	89 33.2	N23	26.4
19	104 33.0		26.4
20	119 32.9		26.3
21	134 32.7 ··		26.3
22	149 32.6		26.3
23	164 32.5		26.3

22

d h	G.H.A.		Dec.
00	179 32.3	N23	26.3
01	194 32.2		26.3
02	209 32.1		26.3
03	224 31.9 ··		26.3
04	239 31.8		26.2
05	254 31.7		26.2
06	269 31.5	N23	26.2
07	284 31.4		26.2
08	299 31.3		26.2
09	314 31.1 ··		26.2
10	329 31.0		26.1
11	344 30.8		26.1
12	359 30.7	N23	26.1
13	14 30.6		26.1
14	29 30.4		26.1
15	44 30.3 ··		26.0
16	59 30.2		26.0
17	74 30.0		26.0
18	89 29.9	N23	26.0
19	104 29.8		25.9
20	119 29.6		25.9
21	134 29.5 ··		25.9
22	149 29.4		25.8
23	164 29.2		25.8

23

d h	G.H.A.		Dec.
00	179 29.1	N23	25.8
01	194 28.9		25.8
02	209 28.8		25.7
03	224 28.7 ··		25.7
04	239 28.5		25.7
05	254 28.4		25.6
06	269 28.3	N23	25.6
07	284 28.1		25.6
08	299 28.0		25.5
09	314 27.9 ··		25.5
10	329 27.7		25.5
11	344 27.6		25.4
12	359 27.5	N23	25.4
13	14 27.3		25.3
14	29 27.2		25.3
15	44 27.1 ··		25.3
16	59 26.9		25.2
17	74 26.8		25.2
18	89 26.6	N23	25.1
19	104 26.5		25.1
20	119 26.4		25.0
21	134 26.2 ··		25.0
22	149 26.1		25.0
23	164 26.0		24.9

24

d h	G.H.A.		Dec.
00	179 25.8	N23	24.9
01	194 25.7		24.8
02	209 25.6		24.8
03	224 25.4 ··		24.7
04	239 25.3		24.7
05	254 25.2		24.6
06	269 25.0	N23	24.6
07	284 24.9		24.5
08	299 24.8		24.5
09	314 24.6 ··		24.4
10	329 24.5		24.4
11	344 24.4		24.3
12	359 24.2	N23	24.2
13	14 24.1		24.2
14	29 24.0		24.1
15	44 23.8 ··		24.1
16	59 23.7		24.0
17	74 23.6		24.0
18	89 23.4	N23	23.9
19	104 23.3		23.8
20	119 23.2		23.8
21	134 23.0 ··		23.7
22	149 22.9		23.7
23	164 22.8		23.6

25

d h	G.H.A.		Dec.
00	179 22.6	N23	23.5
01	194 22.5		23.5
02	209 22.3		23.4
03	224 22.2 ··		23.3
04	239 22.1		23.3
05	254 21.9		23.2
06	269 21.8	N23	23.1
07	284 21.7		23.1
08	299 21.5		23.0
09	314 21.4 ··		22.9
10	329 21.3		22.8
11	344 21.1		22.8
12	359 21.0	N23	22.7
13	14 20.9		22.6
14	29 20.7		22.6
15	44 20.6 ··		22.5
16	59 20.5		22.4
17	74 20.4		22.3
18	89 20.2	N23	22.3
19	104 20.1		22.2
20	119 20.0		22.1
21	134 19.8 ··		22.0
22	149 19.7		21.9
23	164 19.6		21.9

26

d h	G.H.A.		Dec.
00	179 19.4	N23	21.8
01	194 19.3		21.7
02	209 19.2		21.6
03	224 19.0 ··		21.5
04	239 18.9		21.4
05	254 18.8		21.4
06	269 18.6	N23	21.3
07	284 18.5		21.2
08	299 18.4		21.1
09	314 18.2 ··		21.0
10	329 18.1		20.9
11	344 18.0		20.8
12	359 17.8	N23	20.7
13	14 17.7		20.7
14	29 17.6		20.6
15	44 17.4 ··		20.5
16	59 17.3		20.4
17	74 17.2		20.3
18	89 17.0	N23	20.2
19	104 16.9		20.1
20	119 16.8		20.0
21	134 16.7 ··		19.9
22	149 16.5		19.8
23	164 16.4		19.7

27

d h	G.H.A.		Dec.
00	179 16.3	N23	19.6
01	194 16.1		19.5
02	209 16.0		19.4
03	224 15.9 ··		19.3
04	239 15.7		19.2
05	254 15.6		19.1
06	269 15.5	N23	19.0
07	284 15.3		18.9
08	299 15.2		18.8
09	314 15.1 ··		18.7
10	329 15.0		18.6
11	344 14.8		18.5
12	359 14.7	N23	18.4
13	14 14.6		18.3
14	29 14.4		18.2
15	44 14.3 ··		18.1
16	59 14.2		18.0
17	74 14.0		17.8
18	89 13.9	N23	17.7
19	104 13.8		17.6
20	119 13.7		17.5
21	134 13.5 ··		17.4
22	149 13.4		17.3
23	164 13.3		17.2

28

d h	G.H.A.		Dec.
00	179 13.1	N23	17.0
01	194 13.0		16.9
02	209 12.9		16.8
03	224 12.8 ··		16.7
04	239 12.6		16.6
05	254 12.5		16.5
06	269 12.4	N23	16.3
07	284 12.2		16.2
08	299 12.1		16.1
09	314 12.0 ··		15.9
10	329 11.9		15.9
11	344 11.7		15.7
12	359 11.6	N23	15.6
13	14 11.5		15.5
14	29 11.3		15.4
15	44 11.2 ··		15.2
16	59 11.1		15.1
17	74 11.0		15.0
18	89 10.8	N23	14.9
19	104 10.7		14.7
20	119 10.6		14.6
21	134 10.5 ··		14.5
22	149 10.3		14.3
23	164 10.2		14.2

JULY

d h	G.H.A. ° '	Dec. ° '		d h	G.H.A. ° '	Dec. ° '		d h	G.H.A. ° '	Dec. ° '
29 00	179 10.1	N23 14.1		**2** 00	179 01.2	N23 02.7		**5** 00	178 53.0	N22 47.7
01	194 09.9	13.9		01	194 01.1	02.5		01	193 52.9	47.5
02	209 09.8	13.8		02	209 01.0	02.3		02	208 52.8	47.2
03	224 09.7	·· 13.7		03	224 00.9	·· 02.1		03	223 52.7	·· 47.0
04	239 09.6	13.5		04	239 00.7	02.0		04	238 52.6	46.8
05	254 09.4	13.4		05	254 00.6	01.8		05	253 52.5	46.5
06	269 09.3	N23 13.3		06	269 00.5	N23 01.6		06	268 52.4	N22 46.3
07	284 09.2	13.1		07	284 00.4	01.4		07	283 52.3	46.1
08	299 09.1	13.0		08	299 00.3	01.2		08	298 52.1	45.8
09	314 08.9	·· 12.8		09	314 00.1	·· 01.0		09	313 52.0	·· 45.6
10	329 08.8	12.7		10	329 00.0	00.8		10	328 51.9	45.4
11	344 08.7	12.6		11	343 59.9	00.6		11	343 51.8	45.1
12	359 08.6	N23 12.4		12	358 59.8	N23 00.5		12	358 51.7	N22 44.9
13	14 08.4	12.3		13	13 59.7	00.3		13	13 51.6	44.6
14	29 08.3	12.1		14	28 59.6	23 00.1		14	28 51.5	44.4
15	44 08.2	·· 12.0		15	43 59.4	22 59.9		15	43 51.4	·· 44.1
16	59 08.1	11.9		16	58 59.3	59.7		16	58 51.3	43.9
17	74 07.9	11.7		17	73 59.2	59.5		17	73 51.2	43.7
18	89 07.8	N23 11.6		18	88 59.1	N22 59.3		18	88 51.1	N22 43.4
19	104 07.7	11.4		19	103 59.0	59.1		19	103 51.0	43.2
20	119 07.6	11.3		20	118 58.9	58.9		20	118 50.9	42.9
21	134 07.4	·· 11.1		21	133 58.7	·· 58.7		21	133 50.8	·· 42.7
22	149 07.3	11.0		22	148 58.6	58.5		22	148 50.7	42.4
23	164 07.2	10.8		23	163 58.5	58.3		23	163 50.6	42.2
30 00	179 07.1	N23 10.7		**3** 00	178 58.4	N22 58.1		**6** 00	178 50.4	N22 41.9
01	194 06.9	10.5		01	193 58.3	57.9		01	193 50.3	41.7
02	209 06.8	10.4		02	208 58.2	57.7		02	208 50.2	41.4
03	224 06.7	·· 10.2		03	223 58.0	·· 57.5		03	223 50.1	·· 41.2
04	239 06.6	10.1		04	238 57.9	57.3		04	238 50.0	40.9
05	254 06.4	09.9		05	253 57.8	57.1		05	253 49.9	40.7
06	269 06.3	N23 09.8		06	268 57.7	N22 56.9		06	268 49.8	N22 40.4
07	284 06.2	09.6		07	283 57.6	56.7		07	283 49.7	40.2
08	299 06.1	09.5		08	298 57.5	56.5		08	298 49.6	39.9
09	314 05.9	·· 09.3		09	313 57.4	·· 56.3		09	313 49.5	·· 39.7
10	329 05.8	09.2		10	328 57.2	56.1		10	328 49.4	39.4
11	344 05.7	09.0		11	343 57.1	55.9		11	343 49.3	39.1
12	359 05.6	N23 08.8		12	358 57.0	N22 55.7		12	358 49.2	N22 38.9
13	14 05.4	08.7		13	13 56.9	55.4		13	13 49.1	38.6
14	29 05.3	08.5		14	28 56.8	55.2		14	28 49.0	38.4
15	44 05.2	·· 08.4		15	43 56.7	·· 55.0		15	43 48.9	·· 38.1
16	59 05.1	08.2		16	58 56.6	54.8		16	58 48.8	37.8
17	74 05.0	08.0		17	73 56.4	54.6		17	73 48.7	37.6
18	89 04.8	N23 07.9		18	88 56.3	N22 54.4		18	88 48.6	N22 37.3
19	104 04.7	07.7		19	103 56.2	54.2		19	103 48.5	37.1
20	119 04.6	07.6		20	118 56.1	54.0		20	118 48.4	36.8
21	134 04.5	·· 07.4		21	133 56.0	·· 53.8		21	133 48.3	·· 36.5
22	149 04.3	07.2		22	148 55.9	53.5		22	148 48.2	36.3
23	164 04.2	07.1		23	163 55.8	53.3		23	163 48.1	36.0
1 00	179 04.1	N23 06.9		**4** 00	178 55.7	N22 53.1		**7** 00	178 48.0	N22 35.7
01	194 04.0	06.7		01	193 55.5	52.9		01	193 47.9	35.5
02	209 03.9	06.6		02	208 55.4	52.7		02	208 47.8	35.2
03	224 03.7	·· 06.4		03	223 55.3	·· 52.5		03	223 47.7	·· 34.9
04	239 03.6	06.2		04	238 55.2	52.2		04	238 47.6	34.7
05	254 03.5	06.1		05	253 55.1	52.0		05	253 47.5	34.4
06	269 03.4	N23 05.9		06	268 55.0	N22 51.8		06	268 47.4	N22 34.1
07	284 03.3	05.7		07	283 54.9	51.6		07	283 47.3	33.9
08	299 03.1	05.5		08	298 54.8	51.4		08	298 47.2	33.6
09	314 03.0	·· 05.4		09	313 54.7	·· 51.1		09	313 47.1	·· 33.3
10	329 02.9	05.2		10	328 54.5	50.9		10	328 47.0	33.1
11	344 02.8	05.0		11	343 54.4	50.7		11	343 46.9	32.8
12	359 02.6	N23 04.8		12	358 54.3	N22 50.5		12	358 46.8	N22 32.5
13	14 02.5	04.7		13	13 54.2	50.2		13	13 46.7	32.2
14	29 02.4	04.5		14	28 54.1	50.0		14	28 46.6	32.0
15	44 02.3	·· 04.3		15	43 54.0	·· 49.8		15	43 46.5	·· 31.7
16	59 02.2	04.1		16	58 53.9	49.6		16	58 46.4	31.4
17	74 02.0	04.0		17	73 53.8	49.3		17	73 46.3	31.1
18	89 01.9	N23 03.8		18	88 53.7	N22 49.1		18	88 46.2	N22 30.8
19	104 01.8	03.6		19	103 53.6	48.9		19	103 46.1	30.6
20	119 01.7	03.4		20	118 53.4	48.6		20	118 46.0	30.3
21	134 01.6	·· 03.2		21	133 53.3	·· 48.4		21	133 45.9	·· 30.0
22	149 01.4	03.1		22	148 53.2	48.2		22	148 45.8	29.7
23	164 01.3	02.9		23	163 53.1	47.9		23	163 45.7	29.5

d h	G.H.A.	Dec.		d h	G.H.A.	Dec.		d h	G.H.A.	Dec.
8 00	178 45.6	N22 29.2		**11** 00	178 39.1	N22 07.1		**14** 00	178 33.7	N21 41.7
01	193 45.5	28.9		01	193 39.0	06.8		01	193 33.6	41.3
02	208 45.4	28.6		02	208 39.0	06.5		02	208 33.5	40.9
03	223 45.3 ··	28.3		03	223 38.9 ··	06.1		03	223 33.4 ··	40.5
04	238 45.2	28.0		04	238 38.8	05.8		04	238 33.4	40.2
05	253 45.1	27.8		05	253 38.7	05.5		05	253 33.3	39.8
06	268 45.0	N22 27.5		06	268 38.6	N22 05.1		06	268 33.2	N21 39.4
07	283 44.9	27.2		07	283 38.6	04.8		07	283 33.2	39.0
08	298 44.8	26.9		08	298 38.5	04.5		08	298 33.1	38.6
09	313 44.7 ··	26.6		09	313 38.4 ··	04.1		09	313 33.0 ··	38.3
10	328 44.7	26.3		10	328 38.3	03.8		10	328 33.0	37.9
11	343 44.6	26.0		11	343 38.2	03.5		11	343 32.9	37.5
12	358 44.5	N22 25.7		12	358 38.1	N22 03.1		12	358 32.8	N21 37.1
13	13 44.4	25.4		13	13 38.1	02.8		13	13 32.8	36.7
14	28 44.3	25.2		14	28 38.0	02.4		14	28 32.7	36.3
15	43 44.2 ··	24.9		15	43 37.9 ··	02.1		15	43 32.6 ··	35.9
16	58 44.1	24.6		16	58 37.8	01.8		16	58 32.6	35.6
17	73 44.0	24.3		17	73 37.7	01.4		17	73 32.5	35.2
18	88 43.9	N22 24.0		18	88 37.7	N22 01.1		18	88 32.5	N21 34.8
19	103 43.8	23.7		19	103 37.6	00.7		19	103 32.4	34.4
20	118 43.7	23.4		20	118 37.5	00.4		20	118 32.3	34.0
21	133 43.6 ··	23.1		21	133 37.4	22 00.1		21	133 32.3 ··	33.6
22	148 43.5	22.8		22	148 37.3	21 59.7		22	148 32.2	33.2
23	163 43.4	22.5		23	163 37.3	59.4		23	163 32.1	32.8
9 00	178 43.3	N22 22.2		**12** 00	178 37.2	N21 59.0		**15** 00	178 32.1	N21 32.4
01	193 43.3	21.9		01	193 37.1	58.7		01	193 32.0	32.1
02	208 43.2	21.6		02	208 37.0	58.3		02	208 31.9	31.7
03	223 43.1 ··	21.3		03	223 37.0 ··	58.0		03	223 31.9 ··	31.3
04	238 43.0	21.0		04	238 36.9	57.6		04	238 31.8	30.9
05	253 42.9	20.7		05	253 36.8	57.3		05	253 31.7	30.5
06	268 42.8	N22 20.4		06	268 36.7	N21 56.9		06	268 31.7	N21 30.1
07	283 42.7	20.1		07	283 36.6	56.6		07	283 31.6	29.7
08	298 42.6	19.8		08	298 36.6	56.2		08	298 31.6	29.3
09	313 42.5 ··	19.5		09	313 36.5 ··	55.9		09	313 31.5 ··	28.9
10	328 42.4	19.2		10	328 36.4	55.5		10	328 31.4	28.5
11	343 42.3	18.9		11	343 36.3	55.2		11	343 31.4	28.1
12	358 42.2	N22 18.6		12	358 36.3	N21 54.8		12	358 31.3	N21 27.7
13	13 42.2	18.3		13	13 36.2	54.5		13	13 31.3	27.3
14	28 42.1	18.0		14	28 36.1	54.1		14	28 31.2	26.9
15	43 42.0 ··	17.7		15	43 36.0 ··	53.8		15	43 31.1 ··	26.5
16	58 41.9	17.4		16	58 36.0	53.4		16	58 31.1	26.1
17	73 41.8	17.0		17	73 35.9	53.0		17	73 31.0	25.7
18	88 41.7	N22 16.7		18	88 35.8	N21 52.7		18	88 31.0	N21 25.3
19	103 41.6	16.4		19	103 35.7	52.3		19	103 30.9	24.9
20	118 41.5	16.1		20	118 35.7	52.0		20	118 30.8	24.5
21	133 41.4 ··	15.8		21	133 35.6 ··	51.6		21	133 30.8 ··	24.1
22	148 41.4	15.5		22	148 35.5	51.3		22	148 30.7	23.7
23	163 41.3	15.2		23	163 35.4	50.9		23	163 30.7	23.3
10 00	178 41.2	N22 14.9		**13** 00	178 35.4	N21 50.5		**16** 00	178 30.6	N21 22.8
01	193 41.1	14.5		01	193 35.3	50.2		01	193 30.5	22.4
02	208 41.0	14.2		02	208 35.2	49.8		02	208 30.5	22.0
03	223 40.9 ··	13.9		03	223 35.1 ··	49.4		03	223 30.4 ··	21.6
04	238 40.8	13.6		04	238 35.1	49.1		04	238 30.4	21.2
05	253 40.7	13.3		05	253 35.0	48.7		05	253 30.3	20.8
06	268 40.7	N22 13.0		06	268 34.9	N21 48.4		06	268 30.2	N21 20.4
07	283 40.6	12.6		07	283 34.9	48.0		07	283 30.2	20.0
08	298 40.5	12.3		08	298 34.8	47.6		08	298 30.1	19.6
09	313 40.4 ··	12.0		09	313 34.7 ··	47.3		09	313 30.1 ··	19.2
10	328 40.3	11.7		10	328 34.6	46.9		10	328 30.0	18.7
11	343 40.2	11.4		11	343 34.6	46.5		11	343 30.0	18.3
12	358 40.1	N22 11.0		12	358 34.5	N21 46.2		12	358 29.9	N21 17.9
13	13 40.1	10.7		13	13 34.4	45.8		13	13 29.9	17.5
14	28 40.0	10.4		14	28 34.4	45.4		14	28 29.8	17.1
15	43 39.9 ··	10.1		15	43 34.3 ··	45.0		15	43 29.7 ··	16.7
16	58 39.8	09.8		16	58 34.2	44.7		16	58 29.7	16.2
17	73 39.7	09.4		17	73 34.1	44.3		17	73 29.6	15.8
18	88 39.6	N22 09.1		18	88 34.1	N21 43.9		18	88 29.6	N21 15.4
19	103 39.5	08.8		19	103 34.0	43.6		19	103 29.5	15.0
20	118 39.5	08.4		20	118 33.9	43.2		20	118 29.5	14.6
21	133 39.4 ··	08.1		21	133 33.9 ··	42.8		21	133 29.4 ··	14.1
22	148 39.3	07.8		22	148 33.8	42.4		22	148 29.4	13.7
23	163 39.2	07.5		23	163 33.7	42.1		23	163 29.3	13.3

d h	G.H.A. ° '	Dec. ° '
17 00	178 29.3	N21 12.9
01	193 29.2	12.5
02	208 29.1	12.0
03	223 29.1 · ·	11.6
04	238 29.0	11.2
05	253 29.0	10.8
06	268 28.9	N21 10.3
07	283 28.9	09.9
08	298 28.8	09.5
09	313 28.8 · ·	09.1
10	328 28.7	08.6
11	343 28.7	08.2
12	358 28.6	N21 07.8
13	13 28.6	07.3
14	28 28.5	06.9
15	43 28.5 · ·	06.5
16	58 28.4	06.0
17	73 28.4	05.6
18	88 28.3	N21 05.2
19	103 28.3	04.7
20	118 28.2	04.3
21	133 28.2 · ·	03.9
22	148 28.1	03.4
23	163 28.1	03.0
18 00	178 28.0	N21 02.6
01	193 28.0	02.1
02	208 27.9	01.7
03	223 27.9 · ·	01.2
04	238 27.8	00.8
05	253 27.8	21 00.4
06	268 27.8	N20 59.9
07	283 27.7	59.5
08	298 27.7	59.0
09	313 27.6 · ·	58.6
10	328 27.6	58.2
11	343 27.5	57.7
12	358 27.5	N20 57.3
13	13 27.4	56.8
14	28 27.4	56.4
15	43 27.3 · ·	55.9
16	58 27.3	55.5
17	73 27.3	55.0
18	88 27.2	N20 54.6
19	103 27.2	54.1
20	118 27.1	53.7
21	133 27.1 · ·	53.2
22	148 27.0	52.8
23	163 27.0	52.3
19 00	178 26.9	N20 51.9
01	193 26.9	51.4
02	208 26.9	51.0
03	223 26.8 · ·	50.5
04	238 26.8	50.1
05	253 26.7	49.6
06	268 26.7	N20 49.2
07	283 26.7	48.7
08	298 26.6	48.2
09	313 26.6 · ·	47.8
10	328 26.5	47.3
11	343 26.5	46.9
12	358 26.5	N20 46.4
13	13 26.4	45.9
14	28 26.4	45.5
15	43 26.3 · ·	45.0
16	58 26.3	44.6
17	73 26.3	44.1
18	88 26.2	N20 43.6
19	103 26.2	43.2
20	118 26.1	42.7
21	133 26.1 · ·	42.2
22	148 26.1	41.8
23	163 26.0	41.3

d h	G.H.A. ° '	Dec. ° '
20 00	178 26.0	N20 40.8
01	193 26.0	40.4
02	208 25.9	39.9
03	223 25.9 · ·	39.4
04	238 25.8	39.0
05	253 25.8	38.5
06	268 25.8	N20 38.0
07	283 25.7	37.6
08	298 25.7	37.1
09	313 25.7 · ·	36.6
10	328 25.6	36.1
11	343 25.6	35.7
12	358 25.6	N20 35.2
13	13 25.5	34.7
14	28 25.5	34.2
15	43 25.5 · ·	33.8
16	58 25.4	33.3
17	73 25.4	32.8
18	88 25.4	N20 32.3
19	103 25.3	31.9
20	118 25.3	31.4
21	133 25.3 · ·	30.9
22	148 25.2	30.4
23	163 25.2	29.9
21 00	178 25.2	N20 29.5
01	193 25.1	29.0
02	208 25.1	28.5
03	223 25.1 · ·	28.0
04	238 25.1	27.5
05	253 25.0	27.0
06	268 25.0	N20 26.6
07	283 25.0	26.1
08	298 24.9	25.6
09	313 24.9 · ·	25.1
10	328 24.9	24.6
11	343 24.8	24.1
12	358 24.8	N20 23.6
13	13 24.8	23.2
14	28 24.8	22.7
15	43 24.7 · ·	22.2
16	58 24.7	21.7
17	73 24.7	21.2
18	88 24.7	N20 20.7
19	103 24.6	20.2
20	118 24.6	19.7
21	133 24.6 · ·	19.2
22	148 24.5	18.7
23	163 24.5	18.2
22 00	178 24.5	N20 17.7
01	193 24.5	17.2
02	208 24.4	16.7
03	223 24.4 · ·	16.2
04	238 24.4	15.7
05	253 24.4	15.2
06	268 24.3	N20 14.7
07	283 24.3	14.2
08	298 24.3	13.7
09	313 24.3 · ·	13.2
10	328 24.2	12.7
11	343 24.2	12.2
12	358 24.2	N20 11.7
13	13 24.2	11.2
14	28 24.2	10.7
15	43 24.1 · ·	10.2
16	58 24.1	09.7
17	73 24.1	09.2
18	88 24.1	N20 08.7
19	103 24.1	08.2
20	118 24.0	07.7
21	133 24.0 · ·	07.2
22	148 24.0	06.7
23	163 24.0	06.2

d h	G.H.A. ° '	Dec. ° '
23 00	178 23.9	N20 05.7
01	193 23.9	05.2
02	208 23.9	04.6
03	223 23.9 · ·	04.1
04	238 23.9	03.6
05	253 23.9	03.1
06	268 23.8	N20 02.6
07	283 23.8	02.1
08	298 23.8	01.6
09	313 23.8 · ·	01.1
10	328 23.8	00.5
11	343 23.7	20 00.0
12	358 23.7	N19 59.5
13	13 23.7	59.0
14	28 23.7	58.5
15	43 23.7 · ·	58.0
16	58 23.7	57.4
17	73 23.6	56.9
18	88 23.6	N19 56.4
19	103 23.6	55.9
20	118 23.6	55.4
21	133 23.6 · ·	54.8
22	148 23.6	54.3
23	163 23.6	53.8
24 00	178 23.5	N19 53.3
01	193 23.5	52.7
02	208 23.5	52.2
03	223 23.5 · ·	51.7
04	238 23.5	51.2
05	253 23.5	50.6
06	268 23.5	N19 50.1
07	283 23.5	49.6
08	298 23.4	49.1
09	313 23.4 · ·	48.5
10	328 23.4	48.0
11	343 23.4	47.5
12	358 23.4	N19 46.9
13	13 23.4	46.4
14	28 23.4	45.9
15	43 23.4 · ·	45.3
16	58 23.4	44.8
17	73 23.3	44.3
18	88 23.3	N19 43.7
19	103 23.3	43.2
20	118 23.3	42.7
21	133 23.3 · ·	42.1
22	148 23.3	41.6
23	163 23.3	41.1
25 00	178 23.3	N19 40.5
01	193 23.3	40.0
02	208 23.3	39.5
03	223 23.3 · ·	38.9
04	238 23.3	38.4
05	253 23.3	37.8
06	268 23.2	N19 37.3
07	283 23.2	36.8
08	298 23.2	36.2
09	313 23.2 · ·	35.7
10	328 23.2	35.1
11	343 23.2	34.6
12	358 23.2	N19 34.0
13	13 23.2	33.5
14	28 23.2	32.9
15	43 23.2 · ·	32.4
16	58 23.2	31.9
17	73 23.2	31.3
18	88 23.2	N19 30.8
19	103 23.2	30.2
20	118 23.2	29.7
21	133 23.2 · ·	29.1
22	148 23.2	28.6
23	163 23.2	28.0

Table 1

d h	G.H.A.	Dec.
26 00	178 23.2	N19 27.5
01	193 23.2	26.9
02	208 23.2	26.4
03	223 23.2	·· 25.8
04	238 23.2	25.3
05	253 23.2	24.7
06	268 23.2	N19 24.1
07	283 23.2	23.6
08	298 23.2	23.0
09	313 23.2	·· 22.5
10	328 23.2	21.9
11	343 23.2	21.4
12	358 23.2	N19 20.8
13	13 23.2	20.3
14	28 23.2	19.7
15	43 23.2	·· 19.1
16	58 23.2	18.6
17	73 23.2	18.0
18	88 23.2	N19 17.5
19	103 23.2	16.9
20	118 23.2	16.3
21	133 23.2	·· 15.8
22	148 23.2	15.2
23	163 23.2	14.6
27 00	178 23.2	N19 14.1
01	193 23.2	13.5
02	208 23.2	13.0
03	223 23.2	·· 12.4
04	238 23.2	11.8
05	253 23.2	11.3
06	268 23.2	N19 10.7
07	283 23.2	10.1
08	298 23.3	09.5
09	313 23.3	·· 09.0
10	328 23.3	08.4
11	343 23.3	07.8
12	358 23.3	N19 07.3
13	13 23.3	06.7
14	28 23.3	06.1
15	43 23.3	·· 05.6
16	58 23.3	05.0
17	73 23.3	04.4
18	88 23.3	N19 03.8
19	103 23.3	03.3
20	118 23.4	02.7
21	133 23.4	·· 02.1
22	148 23.4	01.5
23	163 23.4	01.0
28 00	178 23.4	N19 00.4
01	193 23.4	18 59.8
02	208 23.4	59.2
03	223 23.4	·· 58.6
04	238 23.4	58.1
05	253 23.5	57.5
06	268 23.5	N18 56.9
07	283 23.5	56.3
08	298 23.5	55.7
09	313 23.5	·· 55.2
10	328 23.5	54.6
11	343 23.5	54.0
12	358 23.5	N18 53.4
13	13 23.6	52.8
14	28 23.6	52.2
15	43 23.6	·· 51.7
16	58 23.6	51.1
17	73 23.6	50.5
18	88 23.6	N18 49.9
19	103 23.7	49.3
20	118 23.7	48.7
21	133 23.7	·· 48.1
22	148 23.7	47.5
23	163 23.7	47.0

Table 2

d h	G.H.A.	Dec.
29 00	178 23.7	N18 46.4
01	193 23.8	45.8
02	208 23.8	45.2
03	223 23.8	·· 44.6
04	238 23.8	44.0
05	253 23.8	43.4
06	268 23.8	N18 42.8
07	283 23.9	42.2
08	298 23.9	41.6
09	313 23.9	·· 41.0
10	328 23.9	40.4
11	343 23.9	39.8
12	358 24.0	N18 39.2
13	13 24.0	38.6
14	28 24.0	38.0
15	43 24.0	·· 37.4
16	58 24.0	36.9
17	73 24.1	36.3
18	88 24.1	N18 35.7
19	103 24.1	35.1
20	118 24.1	34.5
21	133 24.2	·· 33.8
22	148 24.2	33.2
23	163 24.2	32.6
30 00	178 24.2	N18 32.0
01	193 24.2	31.4
02	208 24.3	30.8
03	223 24.3	·· 30.2
04	238 24.3	29.6
05	253 24.3	29.0
06	268 24.4	N18 28.4
07	283 24.4	27.8
08	298 24.4	27.2
09	313 24.4	·· 26.6
10	328 24.5	26.0
11	343 24.5	25.4
12	358 24.5	N18 24.8
13	13 24.6	24.2
14	28 24.6	23.5
15	43 24.6	·· 22.9
16	58 24.6	22.3
17	73 24.7	21.7
18	88 24.7	N18 21.1
19	103 24.7	20.5
20	118 24.8	19.9
21	133 24.8	·· 19.3
22	148 24.8	18.6
23	163 24.8	18.0
31 00	178 24.9	N18 17.4
01	193 24.9	16.8
02	208 24.9	16.2
03	223 25.0	·· 15.6
04	238 25.0	14.9
05	253 25.0	14.3
06	268 25.1	N18 13.7
07	283 25.1	13.1
08	298 25.1	12.5
09	313 25.2	·· 11.9
10	328 25.2	11.2
11	343 25.2	10.6
12	358 25.3	N18 10.0
13	13 25.3	09.4
14	28 25.3	08.7
15	43 25.4	·· 08.1
16	58 25.4	07.5
17	73 25.4	06.9
18	88 25.5	N18 06.3
19	103 25.5	05.6
20	118 25.5	05.0
21	133 25.6	·· 04.4
22	148 25.6	03.7
23	163 25.6	03.1

Table 3 — AUGUST

d h	G.H.A.	Dec.
1 00	178 25.7	N18 02.5
01	193 25.7	01.9
02	208 25.7	01.2
03	223 25.8	·· 00.6
04	238 25.8	18 00.0
05	253 25.9	17 59.3
06	268 25.9	N17 58.7
07	283 25.9	58.1
08	298 26.0	57.5
09	313 26.0	·· 56.8
10	328 26.1	56.2
11	343 26.1	55.6
12	358 26.1	N17 54.9
13	13 26.2	54.3
14	28 26.2	53.7
15	43 26.3	·· 53.0
16	58 26.3	52.4
17	73 26.3	51.7
18	88 26.4	N17 51.1
19	103 26.4	50.5
20	118 26.5	49.8
21	133 26.5	49.2
22	148 26.5	48.6
23	163 26.6	47.9
2 00	178 26.6	N17 47.3
01	193 26.7	46.6
02	208 26.7	46.0
03	223 26.8	·· 45.4
04	238 26.8	44.7
05	253 26.8	44.1
06	268 26.9	N17 43.4
07	283 26.9	42.8
08	298 27.0	42.1
09	313 27.0	·· 41.5
10	328 27.1	40.8
11	343 27.1	40.2
12	358 27.2	N17 39.6
13	13 27.2	38.9
14	28 27.3	38.3
15	43 27.3	·· 37.6
16	58 27.4	37.0
17	73 27.4	36.3
18	88 27.4	N17 35.7
19	103 27.5	35.0
20	118 27.5	34.4
21	133 27.6	·· 33.7
22	148 27.6	33.1
23	163 27.7	32.4
3 00	178 27.7	N17 31.8
01	193 27.8	31.1
02	208 27.8	30.5
03	223 27.9	·· 29.8
04	238 27.9	29.2
05	253 28.0	28.5
06	268 28.0	N17 27.8
07	283 28.1	27.2
08	298 28.1	26.5
09	313 28.2	·· 25.9
10	328 28.3	25.2
11	343 28.3	24.6
12	358 28.4	N17 23.9
13	13 28.4	23.3
14	28 28.5	22.6
15	43 28.5	·· 21.9
16	58 28.6	21.3
17	73 28.6	20.6
18	88 28.7	N17 20.0
19	103 28.7	19.3
20	118 28.8	18.6
21	133 28.8	·· 18.0
22	148 28.9	17.3
23	163 29.0	16.6

d h	G.H.A.	Dec.
4 00	178 29.0	N17 16.0
01	193 29.1	15.3
02	208 29.1	14.7
03	223 29.2 ··	14.0
04	238 29.2	13.3
05	253 29.3	12.7
06	268 29.4 N17	12.0
07	283 29.4	11.3
08	298 29.5	10.7
09	313 29.5 ··	10.0
10	328 29.6	09.3
11	343 29.6	08.7
12	358 29.7 N17	08.0
13	13 29.8	07.3
14	28 29.8	06.6
15	43 29.9 ··	06.0
16	58 29.9	05.3
17	73 30.0	04.6
18	88 30.1 N17	04.0
19	103 30.1	03.3
20	118 30.2	02.6
21	133 30.2 ··	01.9
22	148 30.3	01.3
23	163 30.4	17 00.6
5 00	178 30.4 N16	59.9
01	193 30.5	59.2
02	208 30.6	58.6
03	223 30.6 ··	57.9
04	238 30.7	57.2
05	253 30.8	56.5
06	268 30.8 N16	55.9
07	283 30.9	55.2
08	298 30.9	54.5
09	313 31.0 ··	53.8
10	328 31.1	53.1
11	343 31.1	52.5
12	358 31.2 N16	51.8
13	13 31.3	51.1
14	28 31.3	50.4
15	43 31.4 ··	49.7
16	58 31.5	49.0
17	73 31.5	48.4
18	88 31.6 N16	47.7
19	103 31.7	47.0
20	118 31.7	46.3
21	133 31.8 ··	45.6
22	148 31.9	44.9
23	163 31.9	44.3
6 00	178 32.0 N16	43.6
01	193 32.1	42.9
02	208 32.2	42.2
03	223 32.2 ··	41.5
04	238 32.3	40.8
05	253 32.4	40.1
06	268 32.4 N16	39.4
07	283 32.5	38.8
08	298 32.6	38.1
09	313 32.6 ··	37.4
10	328 32.7	36.7
11	343 32.8	36.0
12	358 32.9 N16	35.3
13	13 32.9	34.6
14	28 33.0	33.9
15	43 33.1 ··	33.2
16	58 33.2	32.5
17	73 33.2	31.8
18	88 33.3 N16	31.1
19	103 33.4	30.4
20	118 33.5	29.7
21	133 33.5 ··	29.1
22	148 33.6	28.4
23	163 33.7	27.7

d h	G.H.A.	Dec.
7 00	178 33.8 N16	27.0
01	193 33.8	26.3
02	208 33.9	25.6
03	223 34.0 ··	24.9
04	238 34.1	24.2
05	253 34.1	23.5
06	268 34.2 N16	22.8
07	283 34.3	22.1
08	298 34.4	21.4
09	313 34.4 ··	20.7
10	328 34.5	20.0
11	343 34.6	19.3
12	358 34.7 N16	18.6
13	13 34.8	17.9
14	28 34.8	17.1
15	43 34.9 ··	16.4
16	58 35.0	15.7
17	73 35.1	15.0
18	88 35.1 N16	14.3
19	103 35.2	13.6
20	118 35.3	12.9
21	133 35.4 ··	12.2
22	148 35.5	11.5
23	163 35.6	10.8
8 00	178 35.6 N16	10.1
01	193 35.7	09.4
02	208 35.8	08.7
03	223 35.9 ··	08.0
04	238 36.0	07.2
05	253 36.0	06.5
06	268 36.1 N16	05.8
07	283 36.2	05.1
08	298 36.3	04.4
09	313 36.4 ··	03.7
10	328 36.5	03.0
11	343 36.5	02.3
12	358 36.6 N16	01.6
13	13 36.7	00.8
14	28 36.8	16 00.1
15	43 36.9	15 59.4
16	58 37.0	58.7
17	73 37.1	58.0
18	88 37.1 N15	57.3
19	103 37.2	56.5
20	118 37.3	55.8
21	133 37.4 ··	55.1
22	148 37.5	54.4
23	163 37.6	53.7
9 00	178 37.7 N15	53.0
01	193 37.8	52.2
02	208 37.8	51.5
03	223 37.9 ··	50.8
04	238 38.0	50.1
05	253 38.1	49.4
06	268 38.2 N15	48.6
07	283 38.3	47.9
08	298 38.4	47.2
09	313 38.5 ··	46.5
10	328 38.6	45.7
11	343 38.6	45.0
12	358 38.7 N15	44.3
13	13 38.8	43.6
14	28 38.9	42.8
15	43 39.0 ··	42.1
16	58 39.1	41.4
17	73 39.2	40.7
18	88 39.3 N15	39.9
19	103 39.4	39.2
20	118 39.5	38.5
21	133 39.6 ··	37.8
22	148 39.7	37.0
23	163 39.7	36.3

d h	G.H.A.	Dec.
10 00	178 39.8 N15	35.6
01	193 39.9	34.8
02	208 40.0	34.1
03	223 40.1 ··	33.4
04	238 40.2	32.6
05	253 40.3	31.9
06	268 40.4 N15	31.2
07	283 40.5	30.4
08	298 40.6	29.7
09	313 40.7 ··	29.0
10	328 40.8	28.2
11	343 40.9	27.5
12	358 41.0 N15	26.8
13	13 41.1	26.0
14	28 41.2	25.3
15	43 41.3 ··	24.6
16	58 41.4	23.8
17	73 41.5	23.1
18	88 41.6 N15	22.4
19	103 41.7	21.6
20	118 41.8	20.9
21	133 41.9 ··	20.1
22	148 42.0	19.4
23	163 42.1	18.7
11 00	178 42.2 N15	17.9
01	193 42.3	17.2
02	208 42.4	16.4
03	223 42.5 ··	15.7
04	238 42.6	15.0
05	253 42.7	14.2
06	268 42.8 N15	13.5
07	283 42.9	12.7
08	298 43.0	12.0
09	313 43.1 ··	11.2
10	328 43.2	10.5
11	343 43.3	09.8
12	358 43.4 N15	09.0
13	13 43.5	08.3
14	28 43.6	07.5
15	43 43.7 ··	06.8
16	58 43.8	06.0
17	73 43.9	05.3
18	88 44.0 N15	04.5
19	103 44.1	03.8
20	118 44.2	03.0
21	133 44.3 ··	02.3
22	148 44.4	01.5
23	163 44.5	00.8
12 00	178 44.6 N15	00.0
01	193 44.7	14 59.3
02	208 44.8	58.5
03	223 44.9 ··	57.8
04	238 45.1	57.0
05	253 45.2	56.3
06	268 45.3 N14	55.5
07	283 45.4	54.8
08	298 45.5	54.0
09	313 45.6 ··	53.3
10	328 45.7	52.5
11	343 45.8	51.8
12	358 45.9 N14	51.0
13	13 46.0	50.3
14	28 46.1	49.5
15	43 46.2 ··	48.7
16	58 46.3	48.0
17	73 46.5	47.2
18	88 46.6 N14	46.5
19	103 46.7	45.7
20	118 46.8	45.0
21	133 46.9 ··	44.2
22	148 47.0	43.4
23	163 47.1	42.7

G.H.A. and Dec. tables

d h	G.H.A.	Dec.
13 00	178 47.2	N14 41.9
01	193 47.3	41.2
02	208 47.4	40.4
03	223 47.6 ··	39.6
04	238 47.7	38.9
05	253 47.8	38.1
06	268 47.9	N14 37.4
07	283 48.0	36.6
08	298 48.1	35.8
09	313 48.2 ··	35.1
10	328 48.4	34.3
11	343 48.5	33.5
12	358 48.6	N14 32.8
13	13 48.7	32.0
14	28 48.8	31.2
15	43 48.9 ··	30.5
16	58 49.0	29.7
17	73 49.2	28.9
18	88 49.3	N14 28.2
19	103 49.4	27.4
20	118 49.5	26.6
21	133 49.6 ··	25.9
22	148 49.7	25.1
23	163 49.8	24.3
14 00	178 50.0	N14 23.6
01	193 50.1	22.8
02	208 50.2	22.0
03	223 50.3 ··	21.3
04	238 50.4	20.5
05	253 50.5	19.7
06	268 50.7	N14 18.9
07	283 50.8	18.2
08	298 50.9	17.4
09	313 51.0 ··	16.6
10	328 51.1	15.8
11	343 51.3	15.1
12	358 51.4	N14 14.3
13	13 51.5	13.5
14	28 51.6	12.7
15	43 51.7 ··	12.0
16	58 51.9	11.2
17	73 52.0	10.4
18	88 52.1	N14 09.6
19	103 52.2	08.9
20	118 52.3	08.1
21	133 52.5 ··	07.3
22	148 52.6	06.5
23	163 52.7	05.8
15 00	178 52.8	N14 05.0
01	193 53.0	04.2
02	208 53.1	03.4
03	223 53.2 ··	02.6
04	238 53.3	01.9
05	253 53.4	01.1
06	268 53.6	N14 00.3
07	283 53.7	13 59.5
08	298 53.8	58.7
09	313 53.9 ··	57.9
10	328 54.1	57.2
11	343 54.2	56.4
12	358 54.3	N13 55.6
13	13 54.4	54.8
14	28 54.6	54.0
15	43 54.7 ··	53.2
16	58 54.8	52.5
17	73 54.9	51.7
18	88 55.1	N13 50.9
19	103 55.2	50.1
20	118 55.3	49.3
21	133 55.4 ··	48.5
22	148 55.6	47.7
23	163 55.7	46.9

d h	G.H.A.	Dec.
16 00	178 55.8	N13 46.2
01	193 56.0	45.4
02	208 56.1	44.6
03	223 56.2 ··	43.8
04	238 56.3	43.0
05	253 56.5	42.2
06	268 56.6	N13 41.4
07	283 56.7	40.6
08	298 56.9	39.8
09	313 57.0 ··	39.0
10	328 57.1	38.3
11	343 57.2	37.5
12	358 57.4	N13 36.7
13	13 57.5	35.9
14	28 57.6	35.1
15	43 57.8 ··	34.3
16	58 57.9	33.5
17	73 58.0	32.7
18	88 58.2	N13 31.9
19	103 58.3	31.1
20	118 58.4	30.3
21	133 58.6 ··	29.5
22	148 58.7	28.7
23	163 58.8	27.9
17 00	178 59.0	N13 27.1
01	193 59.1	26.3
02	208 59.2	25.5
03	223 59.4 ··	24.7
04	238 59.5	23.9
05	253 59.6	23.1
06	268 59.8	N13 22.3
07	283 59.9	21.5
08	299 00.0	20.7
09	314 00.2 ··	19.9
10	329 00.3	19.1
11	344 00.4	18.3
12	359 00.6	N13 17.5
13	14 00.7	16.7
14	29 00.8	15.9
15	44 01.0 ··	15.1
16	59 01.1	14.3
17	74 01.2	13.5
18	89 01.4	N13 12.7
19	104 01.5	11.9
20	119 01.6	11.1
21	134 01.8 ··	10.3
22	149 01.9	09.5
23	164 02.1	08.7
18 00	179 02.2	N13 07.9
01	194 02.3	07.1
02	209 02.5	06.3
03	224 02.6 ··	05.5
04	239 02.7	04.6
05	254 02.9	03.8
06	269 03.0	N13 03.0
07	284 03.2	02.2
08	299 03.3	01.4
09	314 03.4	13 00.6
10	329 03.6	12 59.8
11	344 03.7	59.0
12	359 03.9	N12 58.2
13	14 04.0	57.4
14	29 04.1	56.6
15	44 04.3 ··	55.7
16	59 04.4	54.9
17	74 04.6	54.1
18	89 04.7	N12 53.3
19	104 04.9	52.5
20	119 05.0	51.7
21	134 05.1 ··	50.9
22	149 05.3	50.0
23	164 05.4	49.2

d h	G.H.A.	Dec.
19 00	179 05.6	N12 48.4
01	194 05.7	47.6
02	209 05.8	46.8
03	224 06.0 ··	46.0
04	239 06.1	45.2
05	254 06.3	44.3
06	269 06.4	N12 43.5
07	284 06.6	42.7
08	299 06.7	41.9
09	314 06.9 ··	41.1
10	329 07.0	40.3
11	344 07.1	39.4
12	359 07.3	N12 38.6
13	14 07.4	37.8
14	29 07.6	37.0
15	44 07.7 ··	36.2
16	59 07.9	35.3
17	74 08.0	34.5
18	89 08.2	N12 33.7
19	104 08.3	32.9
20	119 08.5	32.0
21	134 08.6 ··	31.2
22	149 08.8	30.4
23	164 08.9	29.6
20 00	179 09.1	N12 28.8
01	194 09.2	27.9
02	209 09.3	27.1
03	224 09.5 ··	26.3
04	239 09.6	25.5
05	254 09.8	24.6
06	269 09.9	N12 23.8
07	284 10.1	23.0
08	299 10.2	22.2
09	314 10.4 ··	21.3
10	329 10.5	20.5
11	344 10.7	19.7
12	359 10.8	N12 18.9
13	14 11.0	18.0
14	29 11.1	17.2
15	44 11.3 ··	16.4
16	59 11.4	15.5
17	74 11.6	14.7
18	89 11.7	N12 13.9
19	104 11.9	13.1
20	119 12.0	12.2
21	134 12.2 ··	11.4
22	149 12.3	10.6
23	164 12.5	09.7
21 00	179 12.6	N12 08.9
01	194 12.8	08.1
02	209 13.0	07.2
03	224 13.1 ··	06.4
04	239 13.3	05.6
05	254 13.4	04.7
06	269 13.6	N12 03.9
07	284 13.7	03.1
08	299 13.9	02.2
09	314 14.0 ··	01.4
10	329 14.2	12 00.6
11	344 14.3	11 59.7
12	359 14.5	N11 58.9
13	14 14.6	58.1
14	29 14.8	57.2
15	44 15.0 ··	56.4
16	59 15.1	55.5
17	74 15.3	54.7
18	89 15.4	N11 53.9
19	104 15.6	53.0
20	119 15.7	52.2
21	134 15.9 ··	51.4
22	149 16.0	50.5
23	164 16.2	49.7

22

d h	G.H.A.	Dec.
00	179 16.4	N11 48.8
01	194 16.5	48.0
02	209 16.7	47.2
03	224 16.8 ··	46.3
04	239 17.0	45.5
05	254 17.2	44.6
06	269 17.3	N11 43.8
07	284 17.5	43.0
08	299 17.6	42.1
09	314 17.8 ··	41.3
10	329 17.9	40.4
11	344 18.1	39.6
12	359 18.3	N11 38.7
13	14 18.4	37.9
14	29 18.6	37.1
15	44 18.7 ··	36.2
16	59 18.9	35.4
17	74 19.1	34.5
18	89 19.2	N11 33.7
19	104 19.4	32.8
20	119 19.5	32.0
21	134 19.7 ··	31.1
22	149 19.9	30.3
23	164 20.0	29.4

23

d h	G.H.A.	Dec.
00	179 20.2	N11 28.6
01	194 20.3	27.8
02	209 20.5	26.9
03	224 20.7 ··	26.1
04	239 20.8	25.2
05	254 21.0	24.4
06	269 21.2	N11 23.5
07	284 21.3	22.7
08	299 21.5	21.8
09	314 21.6 ··	21.0
10	329 21.8	20.1
11	344 22.0	19.3
12	359 22.1	N11 18.4
13	14 22.3	17.6
14	29 22.5	16.7
15	44 22.6 ··	15.9
16	59 22.8	15.0
17	74 23.0	14.2
18	89 23.1	N11 13.3
19	104 23.3	12.4
20	119 23.5	11.6
21	134 23.6 ··	10.7
22	149 23.8	09.9
23	164 23.9	09.0

24

d h	G.H.A.	Dec.
00	179 24.1	N11 08.2
01	194 24.3	07.3
02	209 24.4	06.5
03	224 24.6 ··	05.6
04	239 24.8	04.8
05	254 24.9	03.9
06	269 25.1	N11 03.0
07	284 25.3	02.2
08	299 25.4	01.3
09	314 25.6	11 00.5
10	329 25.8	10 59.6
11	344 25.9	58.8
12	359 26.1	N10 57.9
13	14 26.3	57.0
14	29 26.5	56.2
15	44 26.6 ··	55.3
16	59 26.8	54.5
17	74 27.0	53.6
18	89 27.1	N10 52.7
19	104 27.3	51.9
20	119 27.5	51.0
21	134 27.6 ··	50.2
22	149 27.8	49.3
23	164 28.0	48.4

25

d h	G.H.A.	Dec.
00	179 28.1	N10 47.6
01	194 28.3	46.7
02	209 28.5	45.9
03	224 28.7 ··	45.0
04	239 28.8	44.1
05	254 29.0	43.3
06	269 29.2	N10 42.4
07	284 29.3	41.5
08	299 29.5	40.7
09	314 29.7 ··	39.8
10	329 29.9	38.9
11	344 30.0	38.1
12	359 30.2	N10 37.2
13	14 30.4	36.3
14	29 30.5	35.5
15	44 30.7 ··	34.6
16	59 30.9	33.7
17	74 31.1	32.9
18	89 31.2	N10 32.0
19	104 31.4	31.1
20	119 31.6	30.3
21	134 31.8 ··	29.4
22	149 31.9	28.5
23	164 32.1	27.7

26

d h	G.H.A.	Dec.
00	179 32.3	N10 26.8
01	194 32.5	25.9
02	209 32.6	25.1
03	224 32.8 ··	24.2
04	239 33.0	23.3
05	254 33.2	22.5
06	269 33.3	N10 21.6
07	284 33.5	20.7
08	299 33.7	19.8
09	314 33.9 ··	19.0
10	329 34.0	18.1
11	344 34.2	17.2
12	359 34.4	N10 16.4
13	14 34.6	15.5
14	29 34.7	14.6
15	44 34.9 ··	13.7
16	59 35.1	12.9
17	74 35.3	12.0
18	89 35.5	N10 11.1
19	104 35.6	10.2
20	119 35.8	09.4
21	134 36.0 ··	08.5
22	149 36.2	07.6
23	164 36.3	06.7

27

d h	G.H.A.	Dec.
00	179 36.5	N10 05.9
01	194 36.7	05.0
02	209 36.9	04.1
03	224 37.1 ··	03.2
04	239 37.2	02.4
05	254 37.4	01.5
06	269 37.6	N10 00.6
07	284 37.8	9 59.7
08	299 38.0	58.9
09	314 38.1 ··	58.0
10	329 38.3	57.1
11	344 38.5	56.2
12	359 38.7	N 9 55.3
13	14 38.9	54.5
14	29 39.0	53.6
15	44 39.2 ··	52.7
16	59 39.4	51.8
17	74 39.6	50.9
18	89 39.8	N 9 50.1
19	104 39.9	49.2
20	119 40.1	48.3
21	134 40.3 ··	47.4
22	149 40.5	46.5
23	164 40.7	45.7

28

d h	G.H.A.	Dec.
00	179 40.8	N 9 44.8
01	194 41.0	43.9
02	209 41.2	43.0
03	224 41.4 ··	42.1
04	239 41.6	41.2
05	254 41.8	40.4
06	269 41.9	N 9 39.5
07	284 42.1	38.6
08	299 42.3	37.7
09	314 42.5 ··	36.8
10	329 42.7	35.9
11	344 42.9	35.1
12	359 43.1	N 9 34.2
13	14 43.2	33.3
14	29 43.4	32.4
15	44 43.6 ··	31.5
16	59 43.8	30.6
17	74 44.0	29.7
18	89 44.2	N 9 28.9
19	104 44.3	28.0
20	119 44.5	27.1
21	134 44.7 ··	26.2
22	149 44.9	25.3
23	164 45.1	24.4

29

d h	G.H.A.	Dec.
00	179 45.3	N 9 23.5
01	194 45.5	22.6
02	209 45.6	21.7
03	224 45.8 ··	20.9
04	239 46.0	20.0
05	254 46.2	19.1
06	269 46.4	N 9 18.2
07	284 46.6	17.3
08	299 46.8	16.4
09	314 47.0 ··	15.5
10	329 47.1	14.6
11	344 47.3	13.7
12	359 47.5	N 9 12.8
13	14 47.7	12.0
14	29 47.9	11.1
15	44 48.1 ··	10.2
16	59 48.3	09.3
17	74 48.5	08.4
18	89 48.7	N 9 07.5
19	104 48.8	06.6
20	119 49.0	05.7
21	134 49.2 ··	04.8
22	149 49.4	03.9
23	164 49.6	03.0

30

d h	G.H.A.	Dec.
00	179 49.8	N 9 02.1
01	194 50.0	01.2
02	209 50.2	9 00.3
03	224 50.4	8 59.4
04	239 50.5	58.5
05	254 50.7	57.6
06	269 50.9	N 8 56.8
07	284 51.1	55.9
08	299 51.3	55.0
09	314 51.5 ··	54.1
10	329 51.7	53.2
11	344 51.9	52.3
12	359 52.1	N 8 51.4
13	14 52.3	50.5
14	29 52.5	49.6
15	44 52.7 ··	48.7
16	59 52.8	47.8
17	74 53.0	46.9
18	89 53.2	N 8 46.0
19	104 53.4	45.1
20	119 53.6	44.2
21	134 53.8 ··	43.3
22	149 54.0	42.4
23	164 54.2	41.5

SEPTEMBER

d h	G.H.A.	Dec.		d h	G.H.A.	Dec.		d h	G.H.A.	Dec.
31 00	179 54.4	N 8 40.6		**3** 00	180 08.7	N 7 35.2		**6** 00	180 23.6	N 6 28.6
01	194 54.6	39.7		01	195 08.9	34.2		01	195 23.8	27.7
02	209 54.8	38.8		02	210 09.1	33.3		02	210 24.0	26.8
03	224 55.0	·· 37.9		03	225 09.3	·· 32.4		03	225 24.2	·· 25.9
04	239 55.2	37.0		04	240 09.5	31.5		04	240 24.5	24.9
05	254 55.4	36.1		05	255 09.7	30.6		05	255 24.7	24.0
06	269 55.6	N 8 35.2		06	270 09.9	N 7 29.7		06	270 24.9	N 6 23.1
07	284 55.8	34.3		07	285 10.1	28.7		07	285 25.1	22.1
08	299 55.9	33.4		08	300 10.3	27.8		08	300 25.3	21.2
09	314 56.1	·· 32.5		09	315 10.5	·· 26.9		09	315 25.5	·· 20.3
10	329 56.3	31.6		10	330 10.7	26.0		10	330 25.7	19.3
11	344 56.5	30.7		11	345 10.9	25.1		11	345 25.9	18.4
12	359 56.7	N 8 29.8		12	0 11.1	N 7 24.1		12	0 26.2	N 6 17.5
13	14 56.9	28.9		13	15 11.3	23.2		13	15 26.4	16.5
14	29 57.1	28.0		14	30 11.5	22.3		14	30 26.6	15.6
15	44 57.3	·· 27.1		15	45 11.7	·· 21.4		15	45 26.8	·· 14.7
16	59 57.5	26.1		16	60 12.0	20.5		16	60 27.0	13.7
17	74 57.7	25.2		17	75 12.2	19.6		17	75 27.2	12.8
18	89 57.9	N 8 24.3		18	90 12.4	N 7 18.6		18	90 27.4	N 6 11.9
19	104 50.1	23.4		19	105 12.6	17.7		19	105 27.6	10.9
20	119 58.3	22.5		20	120 12.8	16.8		20	120 27.9	10.0
21	134 58.5	·· 21.6		21	135 13.0	·· 15.9		21	135 28.1	·· 09.1
22	149 58.7	20.7		22	150 13.2	14.9		22	150 28.3	08.1
23	164 58.9	19.8		23	165 13.4	14.0		23	165 28.5	07.2
1 00	179 59.1	N 8 18.9		**4** 00	180 13.6	N 7 13.1		**7** 00	180 28.7	N 6 06.3
01	194 59.3	18.0		01	195 13.8	12.2		01	195 28.9	05.3
02	209 59.5	17.1		02	210 14.0	11.3		02	210 29.1	04.4
03	224 59.7	·· 16.2		03	225 14.2	·· 10.3		03	225 29.3	·· 03.4
04	239 59.9	15.3		04	240 14.4	09.4		04	240 29.6	02.5
05	255 00.1	14.4		05	255 14.6	08.5		05	255 29.8	01.6
06	270 00.3	N 8 13.5		06	270 14.8	N 7 07.6		06	270 30.0	N 6 00.6
07	285 00.5	12.6		07	285 15.0	06.6		07	285 30.2	5 59.7
08	300 00.7	11.7		08	300 15.3	05.7		08	300 30.4	58.8
09	315 00.9	·· 10.7		09	315 15.5	·· 04.8		09	315 30.6	·· 57.8
10	330 01.1	09.8		10	330 15.7	03.9		10	330 30.8	56.9
11	345 01.3	08.9		11	345 15.9	03.0		11	345 31.1	56.0
12	0 01.5	N 8 08.0		12	0 16.1	N 7 02.0		12	0 31.3	N 5 55.0
13	15 01.7	07.1		13	15 16.3	01.1		13	15 31.5	54.1
14	30 01.9	06.2		14	30 16.5	7 00.2		14	30 31.7	53.1
15	45 02.0	·· 05.3		15	45 16.7	6 59.3		15	45 31.9	·· 52.2
16	60 02.2	04.4		16	60 16.9	58.3		16	60 32.1	51.3
17	75 02.4	03.5		17	75 17.1	57.4		17	75 32.3	50.3
18	90 02.6	N 8 02.6		18	90 17.3	N 6 56.5		18	90 32.6	N 5 49.4
19	105 02.8	01.7		19	105 17.5	55.6		19	105 32.8	48.5
20	120 03.0	8 00.7		20	120 17.7	54.6		20	120 33.0	47.5
21	135 03.2	7 59.8		21	135 18.0	·· 53.7		21	135 33.2	·· 46.6
22	150 03.4	58.9		22	150 18.2	52.8		22	150 33.4	45.6
23	165 03.6	58.0		23	165 18.4	51.9		23	165 33.6	44.7
2 00	180 03.8	N 7 57.1		**5** 00	180 18.6	N 6 50.9		**8** 00	180 33.9	N 5 43.8
01	195 04.0	56.2		01	195 18.8	50.0		01	195 34.1	42.8
02	210 04.2	55.3		02	210 19.0	49.1		02	210 34.3	41.9
03	225 04.4	·· 54.4		03	225 19.2	·· 48.2		03	225 34.5	·· 40.9
04	240 04.6	53.5		04	240 19.4	47.2		04	240 34.7	40.0
05	255 04.8	52.5		05	255 19.6	46.3		05	255 34.9	39.1
06	270 05.0	N 7 51.6		06	270 19.8	N 6 45.4		06	270 35.1	N 5 38.1
07	285 05.3	50.7		07	285 20.0	44.4		07	285 35.4	37.2
08	300 05.5	49.8		08	300 20.3	43.5		08	300 35.6	36.2
09	315 05.7	·· 48.9		09	315 20.5	·· 42.6		09	315 35.8	·· 35.3
10	330 05.9	48.0		10	330 20.7	41.7		10	330 36.0	34.4
11	345 06.1	47.1		11	345 20.9	40.7		11	345 36.2	33.4
12	0 06.3	N 7 46.1		12	0 21.1	N 6 39.8		12	0 36.4	N 5 32.5
13	15 06.5	45.2		13	15 21.3	38.9		13	15 36.7	31.5
14	30 06.7	44.3		14	30 21.5	37.9		14	30 36.9	30.6
15	45 06.9	·· 43.4		15	45 21.7	·· 37.0		15	45 37.1	·· 29.7
16	60 07.1	42.5		16	60 21.9	36.1		16	60 37.3	28.7
17	75 07.3	41.6		17	75 22.1	35.2		17	75 37.5	27.8
18	90 07.5	N 7 40.7		18	90 22.4	N 6 34.2		18	90 37.7	N 5 26.8
19	105 07.7	39.7		19	105 22.6	33.3		19	105 38.0	25.9
20	120 07.9	38.8		20	120 22.8	32.4		20	120 38.2	24.9
21	135 08.1	·· 37.9		21	135 23.0	·· 31.4		21	135 38.4	·· 24.0
22	150 08.3	37.0		22	150 23.2	30.5		22	150 38.6	23.1
23	165 08.5	36.1		23	165 23.4	29.6		23	165 38.8	22.1

d h	G.H.A.	Dec.
9 00	180 39.0	N 5 21.2
01	195 39.3	20.2
02	210 39.5	19.3
03	225 39.7 ··	18.3
04	240 39.9	17.4
05	255 40.1	16.5
06	270 40.3	N 5 15.5
07	285 40.6	14.6
08	300 40.8	13.6
09	315 41.0 ··	12.7
10	330 41.2	11.7
11	345 41.4	10.8
12	0 41.6	N 5 09.8
13	15 41.9	08.9
14	30 42.1	08.0
15	45 42.3 ··	07.0
16	60 42.5	06.1
17	75 42.7	05.1
18	90 43.0	N 5 04.2
19	105 43.2	03.2
20	120 43.4	02.3
21	135 43.6 ··	01.3
22	150 43.8	5 00.4
23	165 44.0	4 59.4
10 00	180 44.3	N 4 58.5
01	195 44.5	57.6
02	210 44.7	56.6
03	225 44.9 ··	55.7
04	240 45.1	54.7
05	255 45.4	53.8
06	270 45.6	N 4 52.8
07	285 45.8	51.9
08	300 46.0	50.9
09	315 46.2 ··	50.0
10	330 46.5	49.0
11	345 46.7	48.1
12	0 46.9	N 4 47.1
13	15 47.1	46.2
14	30 47.3	45.2
15	45 47.5 ··	44.3
16	60 47.8	43.3
17	75 48.0	42.4
18	90 48.2	N 4 41.4
19	105 48.4	40.5
20	120 48.6	39.5
21	135 48.9 ··	38.6
22	150 49.1	37.6
23	165 49.3	36.7
11 00	180 49.5	N 4 35.7
01	195 49.7	34.8
02	210 50.0	33.8
03	225 50.2 ··	32.9
04	240 50.4	31.9
05	255 50.6	31.0
06	270 50.8	N 4 30.0
07	285 51.1	29.1
08	300 51.3	28.1
09	315 51.5 ··	27.2
10	330 51.7	26.2
11	345 51.9	25.3
12	0 52.2	N 4 24.3
13	15 52.4	23.4
14	30 52.6	22.4
15	45 52.8 ··	21.5
16	60 53.0	20.5
17	75 53.3	19.6
18	90 53.5	N 4 18.6
19	105 53.7	17.7
20	120 53.9	16.7
21	135 54.1 ··	15.8
22	150 54.4	14.8
23	165 54.6	13.8

d h	G.H.A.	Dec.
12 00	180 54.8	N 4 12.9
01	195 55.0	11.9
02	210 55.2	11.0
03	225 55.5 ··	10.0
04	240 55.7	09.1
05	255 55.9	08.1
06	270 56.1	N 4 07.2
07	285 56.4	06.2
08	300 56.6	05.3
09	315 56.8 ··	04.3
10	330 57.0	03.4
11	345 57.2	02.4
12	0 57.5	N 4 01.4
13	15 57.7	4 00.5
14	30 57.9	3 59.5
15	45 58.1 ··	58.6
16	60 58.3	57.6
17	75 58.6	56.7
18	90 58.8	N 3 55.7
19	105 59.0	54.8
20	120 59.2	53.8
21	135 59.5 ··	52.8
22	150 59.7	51.9
23	165 59.9	50.9
13 00	181 00.1	N 3 50.0
01	196 00.3	49.0
02	211 00.6	48.1
03	226 00.8 ··	47.1
04	241 01.0	46.1
05	256 01.2	45.2
06	271 01.4	N 3 44.2
07	286 01.7	43.3
08	301 01.9	42.3
09	316 02.1 ··	41.4
10	331 02.3	40.4
11	346 02.6	39.4
12	1 02.8	N 3 38.5
13	16 03.0	37.5
14	31 03.2	36.6
15	46 03.4 ··	35.6
16	61 03.7	34.7
17	76 03.9	33.7
18	91 04.1	N 3 32.7
19	106 04.3	31.8
20	121 04.6	30.8
21	136 04.8 ··	29.9
22	151 05.0	28.9
23	166 05.2	27.9
14 00	181 05.4	N 3 27.0
01	196 05.7	26.0
02	211 05.9	25.1
03	226 06.1 ··	24.1
04	241 06.3	23.1
05	256 06.5	22.2
06	271 06.8	N 3 21.2
07	286 07.0	20.3
08	301 07.2	19.3
09	316 07.4 ··	18.3
10	331 07.7	17.4
11	346 07.9	16.4
12	1 08.1	N 3 15.5
13	16 08.3	14.5
14	31 08.6	13.5
15	46 08.8 ··	12.6
16	61 09.0	11.6
17	76 09.2	10.7
18	91 09.4	N 3 09.7
19	106 09.7	08.7
20	121 09.9	07.8
21	136 10.1 ··	06.8
22	151 10.3	05.9
23	166 10.6	04.9

d h	G.H.A.	Dec.
15 00	181 10.8	N 3 03.9
01	196 11.0	03.0
02	211 11.2	02.0
03	226 11.4 ··	01.0
04	241 11.7	3 00.1
05	256 11.9	2 59.1
06	271 12.1	N 2 58.2
07	286 12.3	57.2
08	301 12.6	56.2
09	316 12.8 ··	55.3
10	331 13.0	54.3
11	346 13.2	53.3
12	1 13.4	N 2 52.4
13	16 13.7	51.4
14	31 13.9	50.5
15	46 14.1 ··	49.5
16	61 14.3	48.5
17	76 14.6	47.6
18	91 14.8	N 2 46.6
19	106 15.0	45.6
20	121 15.2	44.7
21	136 15.4 ··	43.7
22	151 15.7	42.7
23	166 15.9	41.8
16 00	181 16.1	N 2 40.8
01	196 16.3	39.9
02	211 16.6	38.9
03	226 16.8 ··	37.9
04	241 17.0	37.0
05	256 17.2	36.0
06	271 17.4	N 2 35.0
07	286 17.7	34.1
08	301 17.9	33.1
09	316 18.1 ··	32.1
10	331 18.3	31.2
11	346 18.6	30.2
12	1 18.8	N 2 29.2
13	16 19.0	28.3
14	31 19.2	27.3
15	46 19.5 ··	26.3
16	61 19.7	25.4
17	76 19.9	24.4
18	91 20.1	N 2 23.5
19	106 20.3	22.5
20	121 20.6	21.5
21	136 20.8 ··	20.6
22	151 21.0	19.6
23	166 21.2	18.6
17 00	181 21.5	N 2 17.7
01	196 21.7	16.7
02	211 21.9	15.7
03	226 22.1 ··	14.8
04	241 22.3	13.8
05	256 22.6	12.8
06	271 22.8	N 2 11.9
07	286 23.0	10.9
08	301 23.2	09.9
09	316 23.5 ··	09.0
10	331 23.7	08.0
11	346 23.9	07.0
12	1 24.1	N 2 06.1
13	16 24.3	05.1
14	31 24.6	04.1
15	46 24.8 ··	03.2
16	61 25.0	02.2
17	76 25.2	01.2
18	91 25.5	N 2 00.2
19	106 25.7	1 59.3
20	121 25.9	58.3
21	136 26.1 ··	57.3
22	151 26.3	56.4
23	166 26.6	55.4

d	h	G.H.A. ° ′	Dec. ° ′
18	00	181 26.8	N 1 54.4
	01	196 27.0	53.5
	02	211 27.2	52.5
	03	226 27.5	·· 51.5
	04	241 27.7	50.6
	05	256 27.9	49.6
	06	271 28.1	N 1 48.6
	07	286 28.3	47.7
	08	301 28.6	46.7
	09	316 28.8	·· 45.7
	10	331 29.0	44.8
	11	346 29.2	43.8
	12	1 29.5	N 1 42.8
	13	16 29.7	41.8
	14	31 29.9	40.9
	15	46 30.1	·· 39.9
	16	61 30.3	38.9
	17	76 30.6	38.0
	18	91 30.8	N 1 37.0
	19	106 31.0	36.0
	20	121 31.2	35.1
	21	136 31.5	·· 34.1
	22	151 31.7	33.1
	23	166 31.9	32.2
19	00	181 32.1	N 1 31.2
	01	196 32.3	30.2
	02	211 32.6	29.2
	03	226 32.8	·· 28.3
	04	241 33.0	27.3
	05	256 33.2	26.3
	06	271 33.4	N 1 25.4
	07	286 33.7	24.4
	08	301 33.9	23.4
	09	316 34.1	·· 22.5
	10	331 34.3	21.5
	11	346 34.6	20.5
	12	1 34.8	N 1 19.5
	13	16 35.0	18.6
	14	31 35.2	17.6
	15	46 35.4	·· 16.6
	16	61 35.7	15.7
	17	76 35.9	14.7
	18	91 36.1	N 1 13.7
	19	106 36.3	12.7
	20	121 36.5	11.8
	21	136 36.8	·· 10.8
	22	151 37.0	09.8
	23	166 37.2	08.9
20	00	181 37.4	N 1 07.9
	01	196 37.7	06.9
	02	211 37.9	06.0
	03	226 38.1	·· 05.0
	04	241 38.3	04.0
	05	256 38.5	03.0
	06	271 38.8	N 1 02.1
	07	286 39.0	01.1
	08	301 39.2	1 00.1
	09	316 39.4	0 59.2
	10	331 39.6	58.2
	11	346 39.9	57.2
	12	1 40.1	N 0 56.2
	13	16 40.3	55.3
	14	31 40.5	54.3
	15	46 40.7	·· 53.3
	16	61 41.0	52.3
	17	76 41.2	51.4
	18	91 41.4	N 0 50.4
	19	106 41.6	49.4
	20	121 41.8	48.5
	21	136 42.1	·· 47.5
	22	151 42.3	46.5
	23	166 42.5	45.5
21	00	181 42.7	N 0 44.6
	01	196 42.9	43.6
	02	211 43.2	42.6
	03	226 43.4	·· 41.7
	04	241 43.6	40.7
	05	256 43.8	39.7
	06	271 44.0	N 0 38.7
	07	286 44.3	37.8
	08	301 44.5	36.8
	09	316 44.7	·· 35.8
	10	331 44.9	34.8
	11	346 45.1	33.9
	12	1 45.4	N 0 32.9
	13	16 45.6	31.9
	14	31 45.8	31.0
	15	46 46.0	·· 30.0
	16	61 46.2	29.0
	17	76 46.5	28.0
	18	91 46.7	N 0 27.1
	19	106 46.9	26.1
	20	121 47.1	25.1
	21	136 47.3	·· 24.1
	22	151 47.6	23.2
	23	166 47.8	22.2
22	00	181 48.0	N 0 21.2
	01	196 48.2	20.3
	02	211 48.4	19.3
	03	226 48.7	·· 18.3
	04	241 48.9	17.3
	05	256 49.1	16.4
	06	271 49.3	N 0 15.4
	07	286 49.5	14.4
	08	301 49.7	13.4
	09	316 50.0	·· 12.5
	10	331 50.2	11.5
	11	346 50.4	10.5
	12	1 50.6	N 0 09.5
	13	16 50.8	08.6
	14	31 51.1	07.6
	15	46 51.3	·· 06.6
	16	61 51.5	05.7
	17	76 51.7	04.7
	18	91 51.9	N 0 03.7
	19	106 52.1	02.7
	20	121 52.4	01.8
	21	136 52.6	N 0 00.8
	22	151 52.8	S 0 00.2
	23	166 53.0	01.2
23	00	181 53.2	S 0 02.1
	01	196 53.5	03.1
	02	211 53.7	04.1
	03	226 53.9	·· 05.1
	04	241 54.1	06.0
	05	256 54.3	07.0
	06	271 54.5	S 0 08.0
	07	286 54.8	09.0
	08	301 55.0	09.9
	09	316 55.2	·· 10.9
	10	331 55.4	11.9
	11	346 55.6	12.9
	12	1 55.8	S 0 13.8
	13	16 56.1	14.8
	14	31 56.3	15.8
	15	46 56.5	·· 16.7
	16	61 56.7	17.7
	17	76 56.9	18.7
	18	91 57.1	S 0 19.7
	19	106 57.4	20.6
	20	121 57.6	21.6
	21	136 57.8	·· 22.6
	22	151 58.0	23.6
	23	166 58.2	24.5
24	00	181 58.4	S 0 25.5
	01	196 58.7	26.5
	02	211 58.9	27.5
	03	226 59.1	·· 28.4
	04	241 59.3	29.4
	05	256 59.5	30.4
	06	271 59.7	S 0 31.4
	07	287 00.0	32.3
	08	302 00.2	33.3
	09	317 00.4	·· 34.3
	10	332 00.6	35.3
	11	347 00.8	36.2
	12	2 01.0	S 0 37.2
	13	17 01.3	38.2
	14	32 01.5	39.2
	15	47 01.7	·· 40.1
	16	62 01.9	41.1
	17	77 02.1	42.1
	18	92 02.3	S 0 43.1
	19	107 02.5	44.0
	20	122 02.8	45.0
	21	137 03.0	·· 46.0
	22	152 03.2	47.0
	23	167 03.4	47.9
25	00	182 03.6	S 0 48.9
	01	197 03.8	49.9
	02	212 04.0	50.8
	03	227 04.3	·· 51.8
	04	242 04.5	52.8
	05	257 04.7	53.8
	06	272 04.9	S 0 54.7
	07	287 05.1	55.7
	08	302 05.3	56.7
	09	317 05.6	·· 57.7
	10	332 05.8	58.6
	11	347 06.0	0 59.6
	12	2 06.2	S 1 00.6
	13	17 06.4	01.6
	14	32 06.6	02.5
	15	47 06.8	·· 03.5
	16	62 07.0	04.5
	17	77 07.3	05.5
	18	92 07.5	S 1 06.4
	19	107 07.7	07.4
	20	122 07.9	08.4
	21	137 08.1	·· 09.4
	22	152 08.3	10.3
	23	167 08.5	11.3
26	00	182 08.8	S 1 12.3
	01	197 09.0	13.3
	02	212 09.2	14.2
	03	227 09.4	·· 15.2
	04	242 09.6	16.2
	05	257 09.8	17.2
	06	272 10.0	S 1 18.1
	07	287 10.2	19.1
	08	302 10.5	20.1
	09	317 10.7	·· 21.1
	10	332 10.9	22.0
	11	347 11.1	23.0
	12	2 11.3	S 1 24.0
	13	17 11.5	24.9
	14	32 11.7	25.9
	15	47 11.9	·· 26.9
	16	62 12.2	27.9
	17	77 12.4	28.8
	18	92 12.6	S 1 29.8
	19	107 12.8	30.8
	20	122 13.0	31.8
	21	137 13.2	·· 32.7
	22	152 13.4	33.7
	23	167 13.6	34.7

d h	G.H.A.	Dec.		d h	G.H.A.	Dec.		d h	G.H.A.	Dec.
27 00	182 13.8	S 1 35.7		30 00	182 28.8	S 2 45.7		3 00	182 43.2	S 3 55.5
01	197 14.1	36.6		01	197 29.0	46.7		01	197 43.4	56.5
02	212 14.3	37.6		02	212 29.2	47.7		02	212 43.6	57.4
03	227 14.5	·· 38.6		03	227 29.4	·· 48.6		03	227 43.8	·· 58.4
04	242 14.7	39.6		04	242 29.6	49.6		04	242 44.0	3 59.4
05	257 14.9	40.5		05	257 29.8	50.6		05	257 44.2	4 00.3
06	272 15.1	S 1 41.5		06	272 30.0	S 2 51.5		06	272 44.4	S 4 01.3
07	287 15.3	42.5		07	287 30.2	52.5		07	287 44.6	02.3
08	302 15.5	43.5		08	302 30.4	53.5		08	302 44.8	03.2
09	317 15.7	·· 44.4		09	317 30.6	·· 54.4		09	317 45.0	·· 04.2
10	332 16.0	45.4		10	332 30.8	55.4		10	332 45.2	05.1
11	347 16.2	46.4		11	347 31.1	56.4		11	347 45.4	06.1
12	2 16.4	S 1 47.3		12	2 31.3	S 2 57.4		12	2 45.6	S 4 07.1
13	17 16.6	48.3		13	17 31.5	58.3		13	17 45.8	08.0
14	32 16.8	49.3		14	32 31.7	2 59.3		14	32 45.9	09.0
15	47 17.0	·· 50.3		15	47 31.9	3 00.3		15	47 46.1	·· 10.0
16	62 17.2	51.2		16	62 32.1	01.2		16	62 46.3	10.9
17	77 17.4	52.2		17	77 32.3	02.2		17	77 46.5	11.9
18	92 17.6	S 1 53.2		18	92 32.5	S 3 03.2		18	92 46.7	S 4 12.9
19	107 17.8	54.2		19	107 32.7	04.2		19	107 46.9	13.8
20	122 18.1	55.1		20	122 32.9	05.1		20	122 47.1	14.8
21	137 18.3	·· 56.1		21	137 33.1	·· 06.1		21	137 47.3	·· 15.8
22	152 18.5	57.1		22	152 33.3	07.1		22	152 47.5	16.7
23	167 18.7	58.1		23	167 33.5	08.0		23	167 47.7	17.7
28 00	182 18.9	S 1 59.0		1 00	182 33.7	S 3 09.0		4 00	182 47.9	S 4 18.7
01	197 19.1	2 00.0		01	197 33.9	10.0		01	197 48.1	19.6
02	212 19.3	01.0		02	212 34.1	10.9		02	212 48.3	20.6
03	227 19.5	·· 02.0		03	227 34.3	·· 11.9		03	227 48.5	·· 21.6
04	242 19.7	02.9		04	242 34.5	12.9		04	242 48.6	22.5
05	257 19.9	03.9		05	257 34.7	13.9		05	257 48.8	23.5
06	272 20.1	S 2 04.9		06	272 34.9	S 3 14.8		06	272 49.0	S 4 24.4
07	287 20.3	05.8		07	287 35.1	15.8		07	287 49.2	25.4
08	302 20.6	06.8		08	302 35.3	16.8		08	302 49.4	26.4
09	317 20.8	·· 07.8		09	317 35.5	·· 17.7		09	317 49.6	·· 27.3
10	332 21.0	08.8		10	332 35.7	18.7		10	332 49.8	28.3
11	347 21.2	09.7		11	347 35.9	19.7		11	347 50.0	29.3
12	2 21.4	S 2 10.7		12	2 36.1	S 3 20.6		12	2 50.2	S 4 30.2
13	17 21.6	11.7		13	17 36.3	21.6		13	17 50.4	31.2
14	32 21.8	12.7		14	32 36.5	22.6		14	32 50.6	32.2
15	47 22.0	·· 13.6		15	47 36.7	·· 23.5		15	47 50.7	·· 33.1
16	62 22.2	14.6		16	62 36.9	24.5		16	62 50.9	34.1
17	77 22.4	15.6		17	77 37.1	25.5		17	77 51.1	35.0
18	92 22.6	S 2 16.5		18	92 37.3	S 3 26.5		18	92 51.3	S 4 36.0
19	107 22.8	17.5		19	107 37.5	27.4		19	107 51.5	37.0
20	122 23.0	18.5		20	122 37.7	28.4		20	122 51.7	37.9
21	137 23.3	·· 19.5		21	137 37.9	·· 29.4		21	137 51.9	·· 38.9
22	152 23.5	20.4		22	152 38.1	30.3		22	152 52.1	39.9
23	167 23.7	21.4		23	167 38.3	31.3		23	167 52.3	40.8
29 00	182 23.9	S 2 22.4		2 00	182 38.5	S 3 32.3		5 00	182 52.4	S 4 41.8
01	197 24.1	23.4		01	197 38.7	33.2		01	197 52.6	42.7
02	212 24.3	24.3		02	212 38.9	34.2		02	212 52.8	43.7
03	227 24.5	·· 25.3		03	227 39.1	·· 35.2		03	227 53.0	·· 44.7
04	242 24.7	26.3		04	242 39.3	36.1		04	242 53.2	45.6
05	257 24.9	27.2		05	257 39.5	37.1		05	257 53.4	46.6
06	272 25.1	S 2 28.2		06	272 39.7	S 3 38.1		06	272 53.6	S 4 47.6
07	287 25.3	29.2		07	287 39.9	39.0		07	287 53.8	48.5
08	302 25.5	30.2		08	302 40.1	40.0		08	302 54.0	49.5
09	317 25.7	·· 31.1		09	317 40.3	·· 41.0		09	317 54.1	·· 50.4
10	332 25.9	32.1		10	332 40.5	41.9		10	332 54.3	51.4
11	347 26.1	33.1		11	347 40.7	42.9		11	347 54.5	52.4
12	2 26.4	S 2 34.0		12	2 40.9	S 3 43.9		12	2 54.7	S 4 53.3
13	17 26.6	35.0		13	17 41.1	44.9		13	17 54.9	54.3
14	32 26.8	36.0		14	32 41.3	45.8		14	32 55.1	55.2
15	47 27.0	·· 37.0		15	47 41.5	·· 46.8		15	47 55.3	·· 56.2
16	62 27.2	37.9		16	62 41.7	47.8		16	62 55.4	57.2
17	77 27.4	38.9		17	77 41.8	48.7		17	77 55.6	58.1
18	92 27.6	S 2 39.9		18	92 42.0	S 3 49.7		18	92 55.8	S 4 59.1
19	107 27.8	40.9		19	107 42.2	50.7		19	107 56.0	5 00.0
20	122 28.0	41.8		20	122 42.4	51.6		20	122 56.2	01.0
21	137 28.2	·· 42.8		21	137 42.6	·· 52.6		21	137 56.4	·· 02.0
22	152 28.4	43.8		22	152 42.8	53.6		22	152 56.6	02.9
23	167 28.6	44.7		23	167 43.0	54.5		23	167 56.7	03.9

d	h	G.H.A. ° ′	Dec. ° ′
6	00	182 56.9	S 5 04.8
	01	197 57.1	05.8
	02	212 57.3	06.8
	03	227 57.5	·· 07.7
	04	242 57.7	08.7
	05	257 57.9	09.6
	06	272 58.0	S 5 10.6
	07	287 58.2	11.6
	08	302 58.4	12.5
	09	317 58.6	·· 13.5
	10	332 58.8	14.4
	11	347 59.0	15.4
	12	2 59.1	S 5 16.4
	13	17 59.3	17.3
	14	32 59.5	18.3
	15	47 59.7	·· 19.2
	16	62 59.9	20.2
	17	78 00.0	21.1
	18	93 00.2	S 5 22.1
	19	108 00.4	23.1
	20	123 00.6	24.0
	21	138 00.8	·· 25.0
	22	153 01.0	25.9
	23	168 01.1	26.9
7	00	183 01.3	S 5 27.8
	01	198 01.5	28.8
	02	213 01.7	29.8
	03	228 01.9	·· 30.7
	04	243 02.0	31.7
	05	258 02.2	32.6
	06	273 02.4	S 5 33.6
	07	288 02.6	34.5
	08	303 02.8	35.5
	09	318 02.9	·· 36.5
	10	333 03.1	37.4
	11	348 03.3	38.4
	12	3 03.5	S 5 39.3
	13	18 03.7	40.3
	14	33 03.8	41.2
	15	48 04.0	·· 42.2
	16	63 04.2	43.1
	17	78 04.4	44.1
	18	93 04.5	S 5 45.0
	19	108 04.7	46.0
	20	123 04.9	47.0
	21	138 05.1	·· 47.9
	22	153 05.2	48.9
	23	168 05.4	49.8
8	00	183 05.6	S 5 50.8
	01	198 05.8	51.7
	02	213 06.0	52.7
	03	228 06.1	·· 53.6
	04	243 06.3	54.6
	05	258 06.5	55.5
	06	273 06.7	S 5 56.5
	07	288 06.8	57.4
	08	303 07.0	58.4
	09	318 07.2	5 59.4
	10	333 07.4	6 00.3
	11	348 07.5	01.3
	12	3 07.7	S 6 02.2
	13	18 07.9	03.2
	14	33 08.1	04.1
	15	48 08.2	·· 05.1
	16	63 08.4	06.0
	17	78 08.6	07.0
	18	93 08.7	S 6 07.9
	19	108 08.9	08.9
	20	123 09.1	09.8
	21	138 09.3	·· 10.8
	22	153 09.4	11.7
	23	168 09.6	12.7

d	h	G.H.A. ° ′	Dec. ° ′
9	00	183 09.8	S 6 13.6
	01	198 10.0	14.6
	02	213 10.1	15.5
	03	228 10.3	·· 16.5
	04	243 10.5	17.4
	05	258 10.6	18.4
	06	273 10.8	S 6 19.3
	07	288 11.0	20.3
	08	303 11.2	21.2
	09	318 11.3	·· 22.2
	10	333 11.5	23.1
	11	348 11.7	24.1
	12	3 11.8	S 6 25.0
	13	18 12.0	26.0
	14	33 12.2	26.9
	15	48 12.3	·· 27.9
	16	63 12.5	28.8
	17	78 12.7	29.8
	18	93 12.8	S 6 30.7
	19	108 13.0	31.7
	20	123 13.2	32.6
	21	138 13.3	·· 33.6
	22	153 13.5	34.5
	23	168 13.7	35.5
10	00	183 13.9	S 6 36.4
	01	198 14.0	37.3
	02	213 14.2	38.3
	03	228 14.4	·· 39.2
	04	243 14.5	40.2
	05	258 14.7	41.1
	06	273 14.8	S 6 42.1
	07	288 15.0	43.0
	08	303 15.2	44.0
	09	318 15.3	·· 44.9
	10	333 15.5	45.9
	11	348 15.7	46.8
	12	3 15.8	S 6 47.8
	13	18 16.0	48.7
	14	33 16.2	49.6
	15	48 16.3	·· 50.6
	16	63 16.5	51.5
	17	78 16.7	52.5
	18	93 16.8	S 6 53.4
	19	108 17.0	54.4
	20	123 17.1	55.3
	21	138 17.3	·· 56.3
	22	153 17.5	57.2
	23	168 17.6	58.1
11	00	183 17.8	S 6 59.1
	01	198 18.0	7 00.0
	02	213 18.1	01.0
	03	228 18.3	·· 01.9
	04	243 18.4	02.9
	05	258 18.6	03.8
	06	273 18.8	S 7 04.7
	07	288 18.9	05.7
	08	303 19.1	06.6
	09	318 19.2	·· 07.6
	10	333 19.4	08.5
	11	348 19.6	09.5
	12	3 19.7	S 7 10.4
	13	18 19.9	11.3
	14	33 20.0	12.3
	15	48 20.2	·· 13.2
	16	63 20.4	14.2
	17	78 20.5	15.1
	18	93 20.7	S 7 16.0
	19	108 20.8	17.0
	20	123 21.0	17.9
	21	138 21.2	·· 18.9
	22	153 21.3	19.8
	23	168 21.5	20.7

d	h	G.H.A. ° ′	Dec. ° ′
12	00	183 21.6	S 7 21.7
	01	198 21.8	22.6
	02	213 21.9	23.6
	03	228 22.1	·· 24.5
	04	243 22.3	25.4
	05	258 22.4	26.4
	06	273 22.6	S 7 27.3
	07	288 22.7	28.2
	08	303 22.9	29.2
	09	318 23.0	·· 30.1
	10	333 23.2	31.1
	11	348 23.3	32.0
	12	3 23.5	S 7 32.9
	13	18 23.6	33.9
	14	33 23.8	34.8
	15	48 24.0	·· 35.7
	16	63 24.1	36.7
	17	78 24.3	37.6
	18	93 24.4	S 7 38.6
	19	108 24.6	39.5
	20	123 24.7	40.4
	21	138 24.9	·· 41.4
	22	153 25.0	42.3
	23	168 25.2	43.2
13	00	183 25.3	S 7 44.2
	01	198 25.5	45.1
	02	213 25.6	46.0
	03	228 25.8	·· 47.0
	04	243 25.9	47.9
	05	258 26.1	48.8
	06	273 26.2	S 7 49.8
	07	288 26.4	50.7
	08	303 26.5	51.6
	09	318 26.7	·· 52.6
	10	333 26.8	53.5
	11	348 27.0	54.4
	12	3 27.1	S 7 55.4
	13	18 27.3	56.3
	14	33 27.4	57.2
	15	48 27.6	·· 58.2
	16	63 27.7	7 59.1
	17	78 27.9	8 00.0
	18	93 28.0	S 8 01.0
	19	108 28.2	01.9
	20	123 28.3	02.8
	21	138 28.5	·· 03.8
	22	153 28.6	04.7
	23	168 28.7	05.6
14	00	183 28.9	S 8 06.6
	01	198 29.0	07.5
	02	213 29.2	08.4
	03	228 29.3	·· 09.3
	04	243 29.5	10.3
	05	258 29.6	11.2
	06	273 29.8	S 8 12.1
	07	288 29.9	13.1
	08	303 30.0	14.0
	09	318 30.2	·· 14.9
	10	333 30.3	15.9
	11	348 30.5	16.8
	12	3 30.6	S 8 17.7
	13	18 30.8	18.6
	14	33 30.9	19.6
	15	48 31.0	·· 20.5
	16	63 31.2	21.4
	17	78 31.3	22.3
	18	93 31.5	S 8 23.3
	19	108 31.6	24.2
	20	123 31.8	25.1
	21	138 31.9	·· 26.1
	22	153 32.0	27.0
	23	168 32.2	27.9

d h	G.H.A.	Dec.
15 00	183 32.3	S 8 28.8
01	198 32.5	29.8
02	213 32.6	30.7
03	228 32.7	·· 31.6
04	243 32.9	32.5
05	258 33.0	33.5
06	273 33.2	S 8 34.4
07	288 33.3	35.3
08	303 33.4	36.2
09	318 33.6	·· 37.2
10	333 33.7	38.1
11	348 33.8	39.0
12	3 34.0	S 8 39.9
13	18 34.1	40.8
14	33 34.2	41.8
15	48 34.4	·· 42.7
16	63 34.5	43.6
17	78 34.7	44.5
18	93 34.8	S 8 45.5
19	108 34.9	46.4
20	123 35.1	47.3
21	138 35.2	·· 48.2
22	153 35.3	49.1
23	168 35.5	50.1
16 00	183 35.6	S 8 51.0
01	198 35.7	51.9
02	213 35.9	52.8
03	228 36.0	·· 53.7
04	243 36.1	54.7
05	258 36.3	55.6
06	273 36.4	S 8 56.5
07	288 36.5	57.4
08	303 36.7	58.3
09	318 36.8	8 59.3
10	333 36.9	9 00.2
11	348 37.1	01.1
12	3 37.2	S 9 02.0
13	18 37.3	02.9
14	33 37.4	03.9
15	48 37.6	·· 04.8
16	63 37.7	05.7
17	78 37.8	06.6
18	93 38.0	S 9 07.5
19	108 38.1	08.4
20	123 38.2	09.4
21	138 38.4	·· 10.3
22	153 38.5	11.2
23	168 38.6	12.1
17 00	183 38.7	S 9 13.0
01	198 38.9	13.9
02	213 39.0	14.9
03	228 39.1	·· 15.8
04	243 39.2	16.7
05	258 39.4	17.6
06	273 39.5	S 9 18.5
07	288 39.6	19.4
08	303 39.7	20.3
09	318 39.9	·· 21.2
10	333 40.0	22.2
11	348 40.1	23.1
12	3 40.2	S 9 24.0
13	18 40.4	24.9
14	33 40.5	25.8
15	48 40.6	·· 26.7
16	63 40.7	27.6
17	78 40.9	28.5
18	93 41.0	S 9 29.5
19	108 41.1	30.4
20	123 41.2	31.3
21	138 41.4	·· 32.2
22	153 41.5	33.1
23	168 41.6	34.0

d h	G.H.A.	Dec.
18 00	183 41.7	S 9 34.9
01	198 41.8	35.8
02	213 42.0	36.7
03	228 42.1	·· 37.6
04	243 42.2	38.6
05	258 42.3	39.5
06	273 42.4	S 9 40.4
07	288 42.6	41.3
08	303 42.7	42.2
09	318 42.8	·· 43.1
10	333 42.9	44.0
11	348 43.0	44.9
12	3 43.1	S 9 45.8
13	18 43.3	46.7
14	33 43.4	47.6
15	48 43.5	·· 48.5
16	63 43.6	49.4
17	78 43.7	50.4
18	93 43.8	S 9 51.3
19	108 44.0	52.2
20	123 44.1	53.1
21	138 44.2	·· 54.0
22	153 44.3	54.9
23	168 44.4	55.8
19 00	183 44.5	S 9 56.7
01	198 44.7	57.6
02	213 44.8	58.5
03	228 44.9	9 59.4
04	243 45.0	10 00.3
05	258 45.1	01.2
06	273 45.2	S10 02.1
07	288 45.3	03.0
08	303 45.4	03.9
09	318 45.6	·· 04.8
10	333 45.7	05.7
11	348 45.8	06.6
12	3 45.9	S10 07.5
13	18 46.0	08.4
14	33 46.1	09.3
15	48 46.2	·· 10.2
16	63 46.3	11.1
17	78 46.4	12.0
18	93 46.6	S10 12.9
19	108 46.7	13.8
20	123 46.8	14.7
21	138 46.9	·· 15.6
22	153 47.0	16.5
23	168 47.1	17.4
20 00	183 47.2	S10 18.3
01	198 47.3	19.2
02	213 47.4	20.1
03	228 47.5	·· 21.0
04	243 47.6	21.9
05	258 47.7	22.8
06	273 47.8	S10 23.7
07	288 47.9	24.6
08	303 48.1	25.5
09	318 48.2	·· 26.4
10	333 48.3	27.3
11	348 48.4	28.2
12	3 48.5	S10 29.1
13	18 48.6	30.0
14	33 48.7	30.8
15	48 48.8	·· 31.7
16	63 48.9	32.6
17	78 49.0	33.5
18	93 49.1	S10 34.4
19	108 49.2	35.3
20	123 49.3	36.2
21	138 49.4	·· 37.1
22	153 49.5	38.0
23	168 49.6	38.9

d h	G.H.A.	Dec.
21 00	183 49.7	S10 39.8
01	198 49.8	40.7
02	213 49.9	41.6
03	228 50.0	·· 42.4
04	243 50.1	43.3
05	258 50.2	44.2
06	273 50.3	S10 45.1
07	288 50.4	46.0
08	303 50.5	46.9
09	318 50.6	·· 47.8
10	333 50.7	48.7
11	348 50.8	49.6
12	3 50.9	S10 50.4
13	18 51.0	51.3
14	33 51.1	52.2
15	48 51.2	·· 53.1
16	63 51.3	54.0
17	78 51.4	54.9
18	93 51.5	S10 55.8
19	108 51.6	56.7
20	123 51.7	57.7
21	138 51.8	·· 58.4
22	153 51.8	10 59.3
23	168 51.9	11 00.2
22 00	183 52.0	S11 01.1
01	198 52.1	02.0
02	213 52.2	02.9
03	228 52.3	·· 03.7
04	243 52.4	04.6
05	258 52.5	05.5
06	273 52.6	S11 06.4
07	288 52.7	07.3
08	303 52.8	08.2
09	318 52.9	·· 09.0
10	333 53.0	09.9
11	348 53.0	10.8
12	3 53.1	S11 11.7
13	18 53.2	12.6
14	33 53.3	13.4
15	48 53.4	·· 14.3
16	63 53.5	15.2
17	78 53.6	16.1
18	93 53.7	S11 17.0
19	108 53.8	17.8
20	123 53.8	18.7
21	138 53.9	·· 19.6
22	153 54.0	20.5
23	168 54.1	21.4
23 00	183 54.2	S11 22.2
01	198 54.3	23.1
02	213 54.4	24.0
03	228 54.5	·· 24.9
04	243 54.5	25.7
05	258 54.6	26.6
06	273 54.7	S11 27.5
07	288 54.8	28.4
08	303 54.9	29.2
09	318 55.0	·· 30.1
10	333 55.0	31.0
11	348 55.1	31.9
12	3 55.2	S11 32.7
13	18 55.3	33.6
14	33 55.4	34.5
15	48 55.5	·· 35.4
16	63 55.5	36.2
17	78 55.6	37.1
18	93 55.7	S11 38.0
19	108 55.8	38.9
20	123 55.9	39.7
21	138 55.9	·· 40.6
22	153 56.0	41.5
23	168 56.1	42.3

24

d h	G.H.A.	Dec.
24 00	183 56.2	S11 43.2
01	198 56.3	44.1
02	213 56.3	44.9
03	228 56.4	·· 45.8
04	243 56.5	46.7
05	258 56.6	47.6
06	273 56.7	S11 48.4
07	288 56.7	49.3
08	303 56.8	50.2
09	318 56.9	·· 51.0
10	333 57.0	51.9
11	348 57.0	52.8
12	3 57.1	S11 53.6
13	18 57.2	54.5
14	33 57.3	55.4
15	48 57.3	·· 56.2
16	63 57.4	57.1
17	78 57.5	58.0
18	93 57.6	S11 58.8
19	108 57.6	11 59.7
20	123 57.7	12 00.6
21	138 57.8	·· 01.4
22	153 57.9	02.3
23	168 57.9	03.1
25 00	183 58.0	S12 04.0
01	198 58.1	04.9
02	213 58.1	05.7
03	228 58.2	·· 06.6
04	243 58.3	07.5
05	258 58.4	08.3
06	273 58.4	S12 09.2
07	288 58.5	10.0
08	303 58.6	10.9
09	318 58.6	·· 11.8
10	333 58.7	12.6
11	348 58.8	13.5
12	3 58.8	S12 14.3
13	18 58.9	15.2
14	33 59.0	16.1
15	48 59.0	·· 16.9
16	63 59.1	17.8
17	78 59.2	18.6
18	93 59.2	S12 19.5
19	108 59.3	20.3
20	123 59.4	21.2
21	138 59.4	·· 22.0
22	153 59.5	22.9
23	168 59.6	23.8
26 00	183 59.6	S12 24.6
01	198 59.7	25.5
02	213 59.8	26.3
03	228 59.8	·· 27.2
04	243 59.9	28.0
05	258 59.9	28.9
06	274 00.0	S12 29.7
07	289 00.1	30.6
08	304 00.1	31.4
09	319 00.2	·· 32.3
10	334 00.3	33.1
11	349 00.3	34.0
12	4 00.4	S12 34.9
13	19 00.4	35.7
14	34 00.5	36.6
15	49 00.6	·· 37.4
16	64 00.6	38.3
17	79 00.7	39.1
18	94 00.7	S12 40.0
19	109 00.8	40.8
20	124 00.9	41.6
21	139 00.9	·· 42.5
22	154 01.0	43.3
23	169 01.0	44.2

27

d h	G.H.A.	Dec.
27 00	184 01.1	S12 45.0
01	199 01.1	45.9
02	214 01.2	46.7
03	229 01.3	·· 47.6
04	244 01.3	48.4
05	259 01.4	49.3
06	274 01.4	S12 50.1
07	289 01.5	51.0
08	304 01.5	51.8
09	319 01.6	·· 52.6
10	334 01.6	53.5
11	349 01.7	54.3
12	4 01.7	S12 55.2
13	19 01.8	56.0
14	34 01.9	56.9
15	49 01.9	·· 57.7
16	64 02.0	58.5
17	79 02.0	12 59.4
18	94 02.1	S13 00.2
19	109 02.1	01.1
20	124 02.2	01.9
21	139 02.2	·· 02.7
22	154 02.3	03.6
23	169 02.3	04.4
28 00	184 02.4	S13 05.3
01	199 02.4	06.1
02	214 02.5	06.9
03	229 02.5	·· 07.8
04	244 02.6	08.6
05	259 02.6	09.4
06	274 02.7	S13 10.3
07	289 02.7	11.1
08	304 02.7	12.0
09	319 02.8	·· 12.8
10	334 02.8	13.6
11	349 02.9	14.5
12	4 02.9	S13 15.3
13	19 03.0	16.1
14	34 03.0	17.0
15	49 03.1	·· 17.8
16	64 03.1	18.6
17	79 03.2	19.5
18	94 03.2	S13 20.3
19	109 03.2	21.1
20	124 03.3	22.0
21	139 03.3	·· 22.8
22	154 03.4	23.6
23	169 03.4	24.4
29 00	184 03.5	S13 25.3
01	199 03.5	26.1
02	214 03.5	26.9
03	229 03.6	·· 27.8
04	244 03.6	28.6
05	259 03.7	29.4
06	274 03.7	S13 30.2
07	289 03.7	31.1
08	304 03.8	31.9
09	319 03.8	·· 32.7
10	334 03.9	33.6
11	349 03.9	34.4
12	4 03.9	S13 35.2
13	19 04.0	36.0
14	34 04.0	36.9
15	49 04.0	·· 37.7
16	64 04.1	38.5
17	79 04.1	39.3
18	94 04.2	S13 40.1
19	109 04.2	41.0
20	124 04.2	41.8
21	139 04.3	·· 42.6
22	154 04.3	43.4
23	169 04.3	44.3

30

d h	G.H.A.	Dec.
30 00	184 04.4	S13 45.1
01	199 04.4	45.9
02	214 04.4	46.7
03	229 04.5	·· 47.5
04	244 04.5	48.4
05	259 04.5	49.2
06	274 04.6	S13 50.0
07	289 04.6	50.8
08	304 04.6	51.6
09	319 04.7	·· 52.4
10	334 04.7	53.3
11	349 04.7	54.1
12	4 04.7	S13 54.9
13	19 04.8	55.7
14	34 04.8	56.5
15	49 04.8	·· 57.3
16	64 04.9	58.2
17	79 04.9	59.0
18	94 04.9	S13 59.8
19	109 04.9	14 00.6
20	124 05.0	01.4
21	139 05.0	·· 02.2
22	154 05.0	03.0
23	169 05.1	03.9
31 00	184 05.1	S14 04.7
01	199 05.1	05.5
02	214 05.1	06.3
03	229 05.2	·· 07.1
04	244 05.2	07.9
05	259 05.2	08.7
06	274 05.2	S14 09.5
07	289 05.3	10.3
08	304 05.3	11.1
09	319 05.3	·· 11.9
10	334 05.3	12.8
11	349 05.3	13.6
12	4 05.4	S14 14.4
13	19 05.4	15.2
14	34 05.4	16.0
15	49 05.4	·· 16.8
16	64 05.4	17.6
17	79 05.5	18.4
18	94 05.5	S14 19.2
19	109 05.5	20.0
20	124 05.5	20.8
21	139 05.5	·· 21.6
22	154 05.6	22.4
23	169 05.6	23.2

NOVEMBER

d h	G.H.A.	Dec.
1 00	184 05.6	S14 24.0
01	199 05.6	24.8
02	214 05.6	25.6
03	229 05.7	·· 26.4
04	244 05.7	27.2
05	259 05.7	28.0
06	274 05.7	S14 28.8
07	289 05.7	29.6
08	304 05.7	30.4
09	319 05.7	·· 31.2
10	334 05.8	32.0
11	349 05.8	32.8
12	4 05.8	S14 33.6
13	19 05.8	34.4
14	34 05.8	35.2
15	49 05.8	·· 36.0
16	64 05.8	36.8
17	79 05.9	37.6
18	94 05.9	S14 38.4
19	109 05.9	39.2
20	124 05.9	40.0
21	139 05.9	·· 40.8
22	154 05.9	41.6
23	169 05.9	42.4

d	h	G.H.A. ° ′	Dec. ° ′
2	00	184 05.9	S14 43.1
	01	199 05.9	43.9
	02	214 05.9	44.7
	03	229 06.0	·· 45.5
	04	244 06.0	46.3
	05	259 06.0	47.1
	06	274 06.0	S14 47.9
	07	289 06.0	48.7
	08	304 06.0	49.5
	09	319 06.0	·· 50.3
	10	334 06.0	51.0
	11	349 06.0	51.8
	12	4 06.0	S14 52.6
	13	19 06.0	53.4
	14	34 06.0	54.2
	15	49 06.0	·· 55.0
	16	64 06.0	55.8
	17	79 06.0	56.5
	18	94 06.0	S14 57.3
	19	109 06.1	58.1
	20	124 06.1	58.9
	21	139 06.1	14 59.7
	22	154 06.1	15 00.5
	23	169 06.1	01.2
3	00	184 06.1	S15 02.0
	01	199 06.1	02.8
	02	214 06.1	03.6
	03	229 06.1	·· 04.4
	04	244 06.1	05.2
	05	259 06.1	05.9
	06	274 06.1	S15 06.7
	07	289 06.1	07.5
	08	304 06.1	08.3
	09	319 06.1	·· 09.0
	10	334 06.1	09.8
	11	349 06.1	10.6
	12	4 06.1	S15 11.4
	13	19 06.0	12.2
	14	34 06.0	12.9
	15	49 06.0	·· 13.7
	16	64 06.0	14.5
	17	79 06.0	15.3
	18	94 06.0	S15 16.0
	19	109 06.0	16.8
	20	124 06.0	17.6
	21	139 06.0	·· 18.4
	22	154 06.0	19.1
	23	169 06.0	19.9
4	00	184 06.0	S15 20.7
	01	199 06.0	21.4
	02	214 06.0	22.2
	03	229 06.0	·· 23.0
	04	244 06.0	23.8
	05	259 06.0	24.5
	06	274 05.9	S15 25.3
	07	289 05.9	26.1
	08	304 05.9	26.8
	09	319 05.9	·· 27.6
	10	334 05.9	28.4
	11	349 05.9	29.1
	12	4 05.9	S15 29.9
	13	19 05.9	30.7
	14	34 05.9	31.4
	15	49 05.8	·· 32.2
	16	64 05.8	33.0
	17	79 05.8	33.7
	18	94 05.8	S15 34.5
	19	109 05.8	35.2
	20	124 05.8	36.0
	21	139 05.8	·· 36.8
	22	154 05.8	37.5
	23	169 05.7	38.3

d	h	G.H.A. ° ′	Dec. ° ′
5	00	184 05.7	S15 39.1
	01	199 05.7	39.8
	02	214 05.7	40.6
	03	229 05.7	·· 41.3
	04	244 05.7	42.1
	05	259 05.6	42.9
	06	274 05.6	S15 43.6
	07	289 05.6	44.4
	08	304 05.6	45.1
	09	319 05.6	·· 45.9
	10	334 05.5	46.6
	11	349 05.5	47.4
	12	4 05.5	S15 48.2
	13	19 05.5	48.9
	14	34 05.5	49.7
	15	49 05.4	·· 50.4
	16	64 05.4	51.2
	17	79 05.4	51.9
	18	94 05.4	S15 52.7
	19	109 05.4	53.4
	20	124 05.3	54.2
	21	139 05.3	·· 54.9
	22	154 05.3	55.7
	23	169 05.3	56.4
6	00	184 05.2	S15 57.2
	01	199 05.2	57.9
	02	214 05.2	58.7
	03	229 05.2	15 59.4
	04	244 05.1	16 00.2
	05	259 05.1	00.9
	06	274 05.1	S16 01.7
	07	289 05.1	02.4
	08	304 05.0	03.2
	09	319 05.0	·· 03.9
	10	334 05.0	04.7
	11	349 05.0	05.4
	12	4 04.9	S16 06.2
	13	19 04.9	06.9
	14	34 04.9	07.6
	15	49 04.8	·· 08.4
	16	64 04.8	09.1
	17	79 04.8	09.9
	18	94 04.7	S16 10.6
	19	109 04.7	11.4
	20	124 04.7	12.1
	21	139 04.7	·· 12.8
	22	154 04.6	13.6
	23	169 04.6	14.3
7	00	184 04.6	S16 15.1
	01	199 04.5	15.8
	02	214 04.5	16.5
	03	229 04.5	·· 17.3
	04	244 04.4	18.0
	05	259 04.4	18.7
	06	274 04.3	S16 19.5
	07	289 04.3	20.2
	08	304 04.3	20.9
	09	319 04.2	·· 21.7
	10	334 04.2	22.4
	11	349 04.2	23.1
	12	4 04.1	S16 23.9
	13	19 04.1	24.6
	14	34 04.1	25.3
	15	49 04.0	·· 26.1
	16	64 04.0	26.8
	17	79 03.9	27.5
	18	94 03.9	S16 28.3
	19	109 03.9	29.0
	20	124 03.8	29.7
	21	139 03.8	·· 30.5
	22	154 03.7	31.2
	23	169 03.7	31.9

d	h	G.H.A. ° ′	Dec. ° ′
8	00	184 03.7	S16 32.6
	01	199 03.6	33.4
	02	214 03.6	34.1
	03	229 03.5	·· 34.8
	04	244 03.5	35.5
	05	259 03.4	36.3
	06	274 03.4	S16 37.0
	07	289 03.4	37.7
	08	304 03.3	38.4
	09	319 03.3	·· 39.2
	10	334 03.2	39.9
	11	349 03.2	40.6
	12	4 03.1	S16 41.3
	13	19 03.1	42.1
	14	34 03.0	42.8
	15	49 03.0	·· 43.5
	16	64 02.9	44.2
	17	79 02.9	44.9
	18	94 02.8	S16 45.7
	19	109 02.8	46.4
	20	124 02.7	47.1
	21	139 02.7	·· 47.8
	22	154 02.7	48.5
	23	169 02.6	49.2
9	00	184 02.6	S16 50.0
	01	199 02.5	50.7
	02	214 02.4	51.4
	03	229 02.4	·· 52.1
	04	244 02.3	52.8
	05	259 02.3	53.5
	06	274 02.2	S16 54.2
	07	289 02.2	55.0
	08	304 02.1	55.7
	09	319 02.1	·· 56.4
	10	334 02.0	57.1
	11	349 02.0	57.8
	12	4 01.9	S16 58.5
	13	19 01.9	59.2
	14	34 01.8	16 59.9
	15	49 01.7	17 00.6
	16	64 01.7	01.3
	17	79 01.6	02.0
	18	94 01.6	S17 02.8
	19	109 01.5	03.5
	20	124 01.5	04.2
	21	139 01.4	·· 04.9
	22	154 01.3	05.6
	23	169 01.3	06.3
10	00	184 01.2	S17 07.0
	01	199 01.2	07.7
	02	214 01.1	08.4
	03	229 01.1	·· 09.1
	04	244 01.0	09.8
	05	259 00.9	10.5
	06	274 00.9	S17 11.2
	07	289 00.8	11.9
	08	304 00.7	12.6
	09	319 00.7	·· 13.3
	10	334 00.6	14.0
	11	349 00.6	14.7
	12	4 00.5	S17 15.4
	13	19 00.4	16.1
	14	34 00.4	16.8
	15	49 00.3	·· 17.5
	16	64 00.2	18.2
	17	79 00.2	18.9
	18	94 00.1	S17 19.6
	19	109 00.0	20.3
	20	124 00.0	21.0
	21	138 59.9	·· 21.6
	22	153 59.8	22.3
	23	168 59.8	23.0

d h	G.H.A. ° '	Dec. ° '		d h	G.H.A. ° '	Dec. ° '		d h	G.H.A. ° '	Dec. ° '
11 00	183 59.7	S17 23.7		14 00	183 53.8	S18 12.1		17 00	183 45.9	S18 57.7
01	198 59.6	24.4		01	198 53.7	12.8		01	198 45.8	58.3
02	213 59.6	25.1		02	213 53.6	13.4		02	213 45.7	58.9
03	228 59.5 ··	25.8		03	228 53.5 ··	14.1		03	228 45.6	18 59.5
04	243 59.4	26.5		04	243 53.4	14.7		04	243 45.5	19 00.1
05	258 59.3	27.2		05	258 53.3	15.4		05	258 45.3	00.7
06	273 59.3	S17 27.9		06	273 53.2	S18 16.0		06	273 45.2	S19 01.3
07	288 59.2	28.5		07	288 53.1	16.7		07	288 45.1	01.9
08	303 59.1	29.2		08	303 53.0	17.3		08	303 45.0	02.5
09	318 59.1 ··	29.9		09	318 52.9 ··	18.0		09	318 44.8 ··	03.1
10	333 59.0	30.6		10	333 52.8	18.6		10	333 44.7	03.7
11	348 58.9	31.3		11	348 52.7	19.3		11	348 44.6	04.3
12	3 58.8	S17 32.0		12	3 52.6	S18 19.9		12	3 44.5	S19 04.9
13	18 58.8	32.7		13	18 52.5	20.6		13	18 44.3	05.6
14	33 58.7	33.3		14	33 52.4	21.2		14	33 44.2	06.2
15	48 58.6 ··	34.0		15	48 52.3 ··	21.8		15	48 44.1 ··	06.8
16	63 58.6	34.7		16	63 52.2	22.5		16	63 43.9	07.4
17	70 50.5	35.4		17	78 52.1	23.1		17	78 43.8	08.0
18	93 58.4	S17 36.1		18	93 52.0	S18 23.8		18	93 43.7	S19 08.6
19	108 58.3	36.8		19	108 51.9	24.4		19	108 43.6	09.2
20	123 58.2	37.4		20	123 51.8	25.1		20	123 43.4	09.8
21	138 58.2 ··	38.1		21	138 51.7 ··	25.7		21	138 43.3 ··	10.4
22	153 58.1	38.8		22	153 51.6	26.3		22	153 43.2	11.0
23	168 58.0	39.5		23	168 51.5	27.0		23	168 43.0	11.6
12 00	183 57.9	S17 40.2		15 00	183 51.4	S18 27.6		18 00	183 42.9	S19 12.2
01	198 57.9	40.8		01	198 51.3	28.3		01	198 42.8	12.8
02	213 57.8	41.5		02	213 51.2	28.9		02	213 42.6	13.4
03	228 57.7 ··	42.2		03	228 51.1 ··	29.5		03	228 42.5 ··	14.0
04	243 57.6	42.9		04	243 51.0	30.2		04	243 42.4	14.5
05	258 57.6	43.5		05	258 50.9	30.8		05	258 42.2	15.1
06	273 57.5	S17 44.2		06	273 50.8	S18 31.4		06	273 42.1	S19 15.7
07	288 57.4	44.9		07	288 50.7	32.1		07	288 42.0	16.3
08	303 57.3	45.6		08	303 50.5	32.7		08	303 41.8	16.9
09	318 57.2 ··	46.2		09	318 50.4 ··	33.4		09	318 41.7 ··	17.5
10	333 57.2	46.9		10	333 50.3	34.0		10	333 41.6	18.1
11	348 57.1	47.6		11	348 50.2	34.6		11	348 41.4	18.7
12	3 57.0	S17 48.3		12	3 50.1	S18 35.3		12	3 41.3	S19 19.3
13	18 56.9	48.9		13	18 50.0	35.9		13	18 41.2	19.9
14	33 56.8	49.6		14	33 49.9	36.5		14	33 41.0	20.5
15	48 56.7 ··	50.3		15	48 49.8 ··	37.1		15	48 40.9 ··	21.1
16	63 56.7	51.0		16	63 49.7	37.8		16	63 40.8	21.6
17	78 56.6	51.6		17	78 49.6	38.4		17	70 40.6	22.2
18	93 56.5	S17 52.3		18	93 49.5	S18 39.0		18	93 40.5	S19 22.8
19	108 56.4	53.0		19	108 49.3	39.7		19	108 40.3	23.4
20	123 56.3	53.6		20	123 49.2	40.3		20	123 40.2	24.0
21	138 56.2 ··	54.3		21	138 49.1 ··	40.9		21	138 40.1 ··	24.6
22	153 56.1	55.0		22	153 49.0	41.6		22	153 39.9	25.2
23	168 56.1	55.6		23	168 48.9	42.2		23	168 39.8	25.7
13 00	183 56.0	S17 56.3		16 00	183 48.8	S18 42.8		19 00	183 39.6	S19 26.3
01	198 55.9	57.0		01	198 48.7	43.4		01	198 39.5	26.9
02	213 55.8	57.6		02	213 48.5	44.1		02	213 39.4	27.5
03	228 55.7 ··	58.3		03	228 48.4 ··	44.7		03	228 39.2 ··	28.1
04	243 55.6	59.0		04	243 48.3	45.3		04	243 39.1	28.7
05	258 55.5	17 59.6		05	258 48.2	45.9		05	258 38.9	29.2
06	273 55.5	S18 00.3		06	273 48.1	S18 46.5		06	273 38.8	S19 29.8
07	288 55.4	00.9		07	288 48.0	47.2		07	288 38.7	30.4
08	303 55.3	01.6		08	303 47.9	47.8		08	303 38.5	31.0
09	318 55.2 ··	02.3		09	318 47.7 ··	48.4		09	318 38.4 ··	31.6
10	333 55.1	02.9		10	333 47.6	49.0		10	333 38.2	32.1
11	348 55.0	03.6		11	348 47.5	49.6		11	348 38.1	32.7
12	3 54.9	S18 04.2		12	3 47.4	S18 50.3		12	3 37.9	S19 33.3
13	18 54.8	04.9		13	18 47.3	50.9		13	18 37.8	33.9
14	33 54.7	05.6		14	33 47.2	51.5		14	33 37.6	34.4
15	48 54.6 ··	06.2		15	48 47.0 ··	52.1		15	48 37.5 ··	35.0
16	63 54.5	06.9		16	63 46.9	52.7		16	63 37.4	35.6
17	78 54.5	07.5		17	78 46.8	53.4		17	78 37.2	36.2
18	93 54.4	S18 08.2		18	93 46.7	S18 54.0		18	93 37.1	S19 36.7
19	108 54.3	08.8		19	108 46.6	54.6		19	108 36.9	37.3
20	123 54.2	09.5		20	123 46.4	55.2		20	123 36.8	37.9
21	138 54.1 ··	10.2		21	138 46.3 ··	55.8		21	138 36.6 ··	38.4
22	153 54.0	10.8		22	153 46.2	56.4		22	153 36.5	39.0
23	168 53.9	11.5		23	168 46.1	57.0		23	168 36.3	39.6

G.H.A. Dec.

d h	G.H.A. ° '	Dec. ° '
20 00	183 36.2	S19 40.1
01	198 36.0	40.7
02	213 35.9	41.3
03	228 35.7 ··	41.9
04	243 35.6	42.4
05	258 35.4	43.0
06	273 35.3	S19 43.5
07	288 35.1	44.1
08	303 35.0	44.7
09	318 34.8 ··	45.2
10	333 34.7	45.8
11	348 34.5	46.4
12	3 34.4	S19 46.9
13	18 34.2	47.5
14	33 34.1	48.0
15	48 33.9 ··	48.6
16	63 33.7	49.2
17	78 33.6	49.7
18	93 33.4	S19 50.3
19	108 33.3	50.8
20	123 33.1	51.4
21	138 33.0 ··	51.9
22	153 32.8	52.5
23	168 32.7	53.1
21 00	183 32.5	S19 53.6
01	198 32.3	54.2
02	213 32.2	54.7
03	228 32.0 ··	55.3
04	243 31.9	55.8
05	258 31.7	56.4
06	273 31.6	S19 56.9
07	288 31.4	57.5
08	303 31.2	58.0
09	318 31.1 ··	58.6
10	333 30.9	59.1
11	348 30.8	19 59.7
12	3 30.6	S20 00.2
13	18 30.4	00.7
14	33 30.3	01.3
15	48 30.1 ··	01.8
16	63 29.9	02.4
17	78 29.8	02.9
18	93 29.6	S20 03.5
19	108 29.5	04.0
20	123 29.3	04.6
21	138 29.1 ··	05.1
22	153 29.0	05.6
23	168 28.8	06.2
22 00	183 28.6	S20 06.7
01	198 28.5	07.2
02	213 28.3	07.8
03	228 28.1 ··	08.3
04	243 28.0	08.9
05	258 27.8	09.4
06	273 27.6	S20 09.9
07	288 27.5	10.5
08	303 27.3	11.0
09	318 27.1 ··	11.5
10	333 27.0	12.1
11	348 26.8	12.6
12	3 26.6	S20 13.1
13	18 26.4	13.6
14	33 26.3	14.2
15	48 26.1 ··	14.7
16	63 25.9	15.2
17	78 25.8	15.8
18	93 25.6	S20 16.3
19	108 25.4	16.8
20	123 25.2	17.3
21	138 25.1 ··	17.9
22	153 24.9	18.4
23	168 24.7	18.9

d h	G.H.A. ° '	Dec. ° '
23 00	183 24.6	S20 19.4
01	198 24.4	20.0
02	213 24.2	20.5
03	228 24.0 ··	21.0
04	243 23.9	21.5
05	258 23.7	22.0
06	273 23.5	S20 22.6
07	288 23.3	23.1
08	303 23.2	23.6
09	318 23.0 ··	24.1
10	333 22.8	24.6
11	348 22.6	25.1
12	3 22.4	S20 25.7
13	18 22.3	26.2
14	33 22.1	26.7
15	48 21.9 ··	27.2
16	63 21.7	27.7
17	78 21.6	28.2
18	93 21.4	S20 28.7
19	108 21.2	29.3
20	123 21.0	29.8
21	138 20.8 ··	30.3
22	153 20.7	30.8
23	168 20.5	31.3
24 00	183 20.3	S20 31.8
01	198 20.1	32.3
02	213 19.9	32.8
03	228 19.7 ··	33.3
04	243 19.6	33.8
05	258 19.4	34.3
06	273 19.2	S20 34.8
07	288 19.0	35.3
08	303 18.8	35.8
09	318 18.6 ··	36.3
10	333 18.5	36.8
11	348 18.3	37.3
12	3 18.1	S20 37.8
13	18 17.9	38.3
14	33 17.7	38.8
15	48 17.5 ··	39.3
16	63 17.3	39.8
17	78 17.2	40.3
18	93 17.0	S20 40.8
19	108 16.8	41.3
20	123 16.6	41.8
21	138 16.4 ··	42.3
22	153 16.2	42.8
23	168 16.0	43.3
25 00	183 15.8	S20 43.8
01	198 15.6	44.3
02	213 15.5	44.8
03	228 15.3 ··	45.2
04	243 15.1	45.7
05	258 14.9	46.2
06	273 14.7	S20 46.7
07	288 14.5	47.2
08	303 14.3	47.7
09	318 14.1 ··	48.2
10	333 13.9	48.6
11	348 13.7	49.1
12	3 13.5	S20 49.6
13	18 13.3	50.1
14	33 13.1	50.6
15	48 13.0 ··	51.1
16	63 12.8	51.5
17	78 12.6	52.0
18	93 12.4	S20 52.5
19	108 12.2	53.0
20	123 12.0	53.5
21	138 11.8 ··	53.9
22	153 11.6	54.4
23	168 11.4	54.9

d h	G.H.A. ° '	Dec. ° '
26 00	183 11.2	S20 55.4
01	198 11.0	55.8
02	213 10.8	56.3
03	228 10.6 ··	56.8
04	243 10.4	57.3
05	258 10.2	57.7
06	273 10.0	S20 58.2
07	288 09.8	58.7
08	303 09.6	59.1
09	318 09.4	20 59.6
10	333 09.2	21 00.1
11	348 09.0	00.5
12	3 08.8	S21 01.0
13	18 08.6	01.5
14	33 08.4	01.9
15	48 08.2 ··	02.4
16	63 08.0	02.9
17	78 07.8	03.3
18	93 07.6	S21 03.8
19	108 07.4	04.3
20	123 07.2	04.7
21	138 07.0 ··	05.2
22	153 06.8	05.6
23	168 06.6	06.1
27 00	183 06.4	S21 06.6
01	198 06.2	07.0
02	213 06.0	07.5
03	228 05.8 ··	07.9
04	243 05.6	08.4
05	258 05.3	08.8
06	273 05.1	S21 09.3
07	288 04.9	09.7
08	303 04.7	10.2
09	318 04.5 ··	10.7
10	333 04.3	11.1
11	348 04.1	11.6
12	3 03.9	S21 12.0
13	18 03.7	12.5
14	33 03.5	12.9
15	48 03.3 ··	13.4
16	63 03.1	13.8
17	78 02.9	14.2
18	93 02.6	S21 14.7
19	108 02.4	15.1
20	123 02.2	15.6
21	138 02.0 ··	16.0
22	153 01.8	16.5
23	168 01.6	16.9
28 00	183 01.4	S21 17.4
01	198 01.2	17.8
02	213 01.0	18.2
03	228 00.7 ··	18.7
04	243 00.5	19.1
05	258 00.3	19.6
06	273 00.1	S21 20.0
07	287 59.9	20.4
08	302 59.7	20.9
09	317 59.5 ··	21.3
10	332 59.2	21.7
11	347 59.0	22.1
12	2 58.8	S21 22.6
13	17 58.6	23.0
14	32 58.4	23.5
15	47 58.2 ··	23.9
16	62 58.0	24.3
17	77 57.7	24.8
18	92 57.5	S21 25.2
19	107 57.3 .	25.6
20	122 57.1	26.0
21	137 56.9 ··	26.5
22	152 56.7	26.9
23	167 56.4	27.3

DECEMBER

d/h	G.H.A.	Dec.		d/h	G.H.A.	Dec.		d/h	G.H.A.	Dec.
29 00	182 56.2	S21 27.8		2 00	182 39.7	S21 56.5		5 00	182 21.9	S22 21.4
01	197 56.0	28.2		01	197 39.5	56.8		01	197 21.6	21.7
02	212 55.8	28.6		02	212 39.3	57.2		02	212 21.4	22.0
03	227 55.6 ··	29.0		03	227 39.0 ··	57.6		03	227 21.1 ··	22.4
04	242 55.3	29.4		04	242 38.8	58.0		04	242 20.8	22.7
05	257 55.1	29.9		05	257 38.5	58.3		05	257 20.6	23.0
06	272 54.9	S21 30.3		06	272 38.3	S21 58.7		06	272 20.3	S22 23.3
07	287 54.7	30.7		07	287 38.1	59.1		07	287 20.1	23.6
08	302 54.5	31.1		08	302 37.8	59.4		08	302 19.8	23.9
09	317 54.2 ··	31.5		09	317 37.6	21 59.8		09	317 19.6 ··	24.2
10	332 54.0	32.0		10	332 37.3	22 00.2		10	332 19.3	24.6
11	347 53.8	32.4		11	347 37.1	00.5		11	347 19.0	24.9
12	2 53.6	S21 32.8		12	2 36.8	S22 00.9		12	2 18.8	S22 25.2
13	17 53.3	33.2		13	17 36.6	01.3		13	17 18.5	25.5
14	32 53.1	33.6		14	32 36.4	01.6		14	32 18.3	25.8
15	47 52.9 ··	34.0		15	47 36.1 ··	02.0		15	47 18.0 ··	26.1
16	62 52.7	34.5		16	62 35.9	02.3		16	62 17.7	26.4
17	77 52.5	34.9		17	77 35.6	02.7		17	77 17.5	26.7
18	92 52.2	S21 35.3		18	92 35.4	S22 03.1		18	92 17.2	S22 27.0
19	107 52.0	35.7		19	107 35.1	03.4		19	107 17.0	27.3
20	122 51.8	36.1		20	122 34.9	03.8		20	122 16.7	27.6
21	137 51.6 ··	36.5		21	137 34.7 ··	04.1		21	137 16.4 ··	27.9
22	152 51.3	36.9		22	152 34.4	04.5		22	152 16.2	28.2
23	167 51.1	37.3		23	167 34.2	04.9		23	167 15.9	28.6
30 00	182 50.9	S21 37.7		3 00	182 33.9	S22 05.2		6 00	182 15.6	S22 28.9
01	197 50.7	38.1		01	197 33.7	05.6		01	197 15.4	29.2
02	212 50.4	38.6		02	212 33.4	05.9		02	212 15.1	29.5
03	227 50.2 ··	39.0		03	227 33.2 ··	06.3		03	227 14.9 ··	29.8
04	242 50.0	39.4		04	242 32.9	06.6		04	242 14.6	30.1
05	257 49.7	39.8		05	257 32.7	07.0		05	257 14.3	30.3
06	272 49.5	S21 40.2		06	272 32.5	S22 07.3		06	272 14.1	S22 30.6
07	287 49.3	40.6		07	287 32.2	07.7		07	287 13.8	30.9
08	302 49.1	41.0		08	302 32.0	08.0		08	302 13.5	31.2
09	317 48.8 ··	41.4		09	317 31.7 ··	08.4		09	317 13.3 ··	31.5
10	332 48.6	41.8		10	332 31.5	08.7		10	332 13.0	31.8
11	347 48.4	42.2		11	347 31.2	09.1		11	347 12.7	32.1
12	2 48.2	S21 42.6		12	2 31.0	S22 09.4		12	2 12.5	S22 32.4
13	17 47.9	43.0		13	17 30.7	09.8		13	17 12.2	32.7
14	32 47.7	43.4		14	32 30.5	10.1		14	32 11.9	33.0
15	47 47.5 ··	43.8		15	47 30.2 ··	10.5		15	47 11.7 ··	33.3
16	62 47.2	44.2		16	62 30.0	10.8		16	62 11.4	33.6
17	77 47.0	44.6		17	77 29.7	11.1		17	77 11.1	33.9
18	92 46.8	S21 45.0		18	92 29.5	S22 11.5		18	92 10.9	S22 34.1
19	107 46.5	45.4		19	107 29.2	11.8		19	107 10.6	34.4
20	122 46.3	45.8		20	122 29.0	12.2		20	122 10.3	34.7
21	137 46.1 ··	46.1		21	137 28.7 ··	12.5		21	137 10.1 ··	35.0
22	152 45.9	46.5		22	152 28.5	12.8		22	152 09.8	35.3
23	167 45.6	46.9		23	167 28.2	13.2		23	167 09.5	35.6
1 00	182 45.4	S21 47.3		4 00	182 28.0	S22 13.5		7 00	182 09.3	S22 35.9
01	197 45.2	47.7		01	197 27.7	13.9		01	197 09.0	36.1
02	212 44.9	48.1		02	212 27.5	14.2		02	212 08.7	36.4
03	227 44.7 ··	48.5		03	227 27.2 ··	14.5		03	227 08.5 ··	36.7
04	242 44.5	48.9		04	242 27.0	14.9		04	242 08.2	37.0
05	257 44.2	49.3		05	257 26.7	15.2		05	257 07.9	37.3
06	272 44.0	S21 49.6		06	272 26.5	S22 15.5		06	272 07.7	S22 37.5
07	287 43.8	50.0		07	287 26.2	15.9		07	287 07.4	37.8
08	302 43.5	50.4		08	302 26.0	16.2		08	302 07.1	38.1
09	317 43.3 ··	50.8		09	317 25.7 ··	16.5		09	317 06.9 ··	38.4
10	332 43.0	51.2		10	332 25.5	16.9		10	332 06.6	38.7
11	347 42.8	51.6		11	347 25.2	17.2		11	347 06.3	38.9
12	2 42.6	S21 51.9		12	2 24.9	S22 17.5		12	2 06.0	S22 39.2
13	17 42.3	52.3		13	17 24.7	17.8		13	17 05.8	39.5
14	32 42.1	52.7		14	32 24.4	18.2		14	32 05.5	39.7
15	47 41.9 ··	53.1		15	47 24.2 ··	18.5		15	47 05.2 ··	40.0
16	62 41.6	53.5		16	62 23.9	18.8		16	62 05.0	40.3
17	77 41.4	53.8		17	77 23.7	19.2		17	77 04.7	40.6
18	92 41.2	S21 54.2		18	92 23.4	S22 19.5		18	92 04.4	S22 40.8
19	107 40.9	54.6		19	107 23.2	19.8		19	107 04.2	41.1
20	122 40.7	55.0		20	122 22.9	20.1		20	122 03.9	41.4
21	137 40.4 ··	55.4		21	137 22.6 ··	20.4		21	137 03.6 ··	41.6
22	152 40.2	55.7		22	152 22.4	20.8		22	152 03.3	41.9
23	167 40.0	56.1		23	167 22.1	21.1		23	167 03.1	42.2

d h	G.H.A. ° ′	Dec. ° ′
8 00	182 02.8	S22 42.4
01	197 02.5	42.7
02	212 02.2	43.0
03	227 02.0 ··	43.2
04	242 01.7	43.5
05	257 01.4	43.7
06	272 01.1	S22 44.0
07	287 00.9	44.3
08	302 00.6	44.5
09	317 00.3 ··	44.8
10	332 00.1	45.0
11	346 59.8	45.3
12	1 59.5	S22 45.5
13	16 59.2	45.8
14	31 58.9	46.1
15	46 58.7 ··	46.3
16	61 58.4	46.6
17	76 58.1	46.8
18	91 57.8	S22 47.1
19	106 57.6	47.3
20	121 57.3	47.6
21	136 57.0 ··	47.8
22	151 56.7	48.1
23	166 56.5	48.3
9 00	181 56.2	S22 48.5
01	196 55.9	48.8
02	211 55.6	49.0
03	226 55.3 ··	49.3
04	241 55.1	49.5
05	256 54.8	49.8
06	271 54.5	S22 50.0
07	286 54.2	50.2
08	301 54.0	50.5
09	316 53.7 ··	50.7
10	331 53.4	51.0
11	346 53.1	51.2
12	1 52.8	S22 51.4
13	16 52.6	51.7
14	31 52.3	51.9
15	46 52.0 ··	52.1
16	61 51.7	52.4
17	76 51.4	52.6
18	91 51.1	S22 52.8
19	106 50.9	53.1
20	121 50.6	53.3
21	136 50.3 ··	53.5
22	151 50.0	53.8
23	166 49.7	54.0
10 00	181 49.5	S22 54.2
01	196 49.2	54.4
02	211 48.9	54.7
03	226 48.6 ··	54.9
04	241 48.3	55.1
05	256 48.0	55.3
06	271 47.8	S22 55.6
07	286 47.5	55.8
08	301 47.2	56.0
09	316 46.9 ··	56.2
10	331 46.6	56.5
11	346 46.3	56.7
12	1 46.1	S22 56.9
13	16 45.8	57.1
14	31 45.5	57.3
15	46 45.2 ··	57.5
16	61 44.9	57.8
17	76 44.6	58.0
18	91 44.3	S22 58.2
19	106 44.1	58.4
20	121 43.8	58.6
21	136 43.5 ··	58.8
22	151 43.2	59.0
23	166 42.9	59.2

d h	G.H.A. ° ′	Dec. ° ′
11 00	181 42.6	S22 59.4
01	196 42.3	59.6
02	211 42.1	22 59.9
03	226 41.8	23 00.1
04	241 41.5	00.3
05	256 41.2	00.5
06	271 40.9	S23 00.7
07	286 40.6	00.9
08	301 40.3	01.1
09	316 40.0 ··	01.3
10	331 39.8	01.5
11	346 39.5	01.7
12	1 39.2	S23 01.9
13	16 38.9	02.1
14	31 38.6	02.3
15	46 38.3 ··	02.5
16	61 38.0	02.7
17	76 37.7	02.9
18	91 37.4	S23 03.1
19	106 37.2	03.3
20	121 36.9	03.4
21	136 36.6 ··	03.6
22	151 36.3	03.8
23	166 36.0	04.0
12 00	181 35.7	S23 04.2
01	196 35.4	04.4
02	211 35.1	04.6
03	226 34.8 ··	04.8
04	241 34.5	05.0
05	256 34.2	05.1
06	271 34.0	S23 05.3
07	286 33.7	05.5
08	301 33.4	05.7
09	316 33.1 ··	05.9
10	331 32.8	06.1
11	346 32.5	06.2
12	1 32.2	S23 06.4
13	16 31.9	06.6
14	31 31.6	06.8
15	46 31.3 ··	07.0
16	61 31.0	07.1
17	76 30.7	07.3
18	91 30.4	S23 07.5
19	106 30.1	07.7
20	121 29.9	07.8
21	136 29.6 ··	08.0
22	151 29.3	08.2
23	166 29.0	08.3
13 00	181 28.7	S23 08.5
01	196 28.4	08.7
02	211 28.1	08.9
03	226 27.8 ··	09.0
04	241 27.5	09.2
05	256 27.2	09.4
06	271 26.9	S23 09.5
07	286 26.6	09.7
08	301 26.3	09.8
09	316 26.0 ··	10.0
10	331 25.7	10.2
11	346 25.4	10.3
12	1 25.1	S23 10.5
13	16 24.8	10.7
14	31 24.5	10.8
15	46 24.2 ··	11.0
16	61 23.9	11.1
17	76 23.6	11.3
18	91 23.4	S23 11.4
19	106 23.1	11.6
20	121 22.8	11.8
21	136 22.5 ··	11.9
22	151 22.2	12.1
23	166 21.9	12.2

d h	G.H.A. ° ′	Dec. ° ′
14 00	181 21.6	S23 12.4
01	196 21.3	12.5
02	211 21.0	12.7
03	226 20.7 ··	12.8
04	241 20.4	13.0
05	256 20.1	13.1
06	271 19.8	S23 13.3
07	286 19.5	13.4
08	301 19.2	13.5
09	316 18.9 ··	13.7
10	331 18.6	13.8
11	346 18.3	14.0
12	1 18.0	S23 14.1
13	16 17.7	14.3
14	31 17.4	14.4
15	46 17.1 ··	14.5
16	61 16.8	14.7
17	76 16.5	14.8
18	91 16.2	S23 14.9
19	106 15.9	15.1
20	121 15.6	15.2
21	136 15.3 ··	15.3
22	151 15.0	15.5
23	166 14.7	15.6
15 00	181 14.4	S23 15.7
01	196 14.1	15.9
02	211 13.8	16.0
03	226 13.5 ··	16.1
04	241 13.2	16.3
05	256 12.9	16.4
06	271 12.6	S23 16.5
07	286 12.3	16.6
08	301 12.0	16.8
09	316 11.7 ··	16.9
10	331 11.4	17.0
11	346 11.1	17.1
12	1 10.8	S23 17.3
13	16 10.5	17.4
14	31 10.2	17.5
15	46 09.8 ··	17.6
16	61 09.5	17.7
17	76 09.2	17.9
18	91 08.9	S23 18.0
19	106 08.6	18.1
20	121 08.3	18.2
21	136 08.0 ··	18.3
22	151 07.7	18.4
23	166 07.4	18.6
16 00	181 07.1	S23 18.7
01	196 06.8	18.8
02	211 06.5	18.9
03	226 06.2 ··	19.0
04	241 05.9	19.1
05	256 05.6	19.2
06	271 05.3	S23 19.3
07	286 05.0	19.4
08	301 04.7	19.5
09	316 04.4 ··	19.6
10	331 04.1	19.7
11	346 03.8	19.9
12	1 03.5	S23 20.0
13	16 03.2	20.1
14	31 02.9	20.2
15	46 02.6 ··	20.3
16	61 02.2	20.4
17	76 01.9	20.5
18	91 01.6	S23 20.6
19	106 01.3	20.6
20	121 01.0	20.7
21	136 00.7 ··	20.8
22	151 00.4	20.9
23	166 00.1	21.0

d	h	G.H.A. °	′	Dec. °	′
17	00	180	59.8	S23	21.1
	01	195	59.5		21.2
	02	210	59.2		21.3
	03	225	58.9	· ·	21.4
	04	240	58.6		21.5
	05	255	58.3		21.6
	06	270	58.0	S23	21.7
	07	285	57.7		21.8
	08	300	57.3		21.8
	09	315	57.0	· ·	21.9
	10	330	56.7		22.0
	11	345	56.4		22.1
	12	0	56.1	S23	22.2
	13	15	55.8		22.3
	14	30	55.5		22.3
	15	45	55.2	· ·	22.4
	16	60	54.9		22.5
	17	75	54.6		22.6
	18	90	54.3	S23	22.7
	19	105	54.0		22.7
	20	120	53.7		22.8
	21	135	53.3	· ·	22.9
	22	150	53.0		23.0
	23	165	52.7		23.0
18	00	180	52.4	S23	23.1
	01	195	52.1		23.2
	02	210	51.8		23.3
	03	225	51.5	· ·	23.3
	04	240	51.2		23.4
	05	255	50.9		23.5
	06	270	50.6	S23	23.5
	07	285	50.3		23.6
	08	300	50.0		23.7
	09	315	49.6	· ·	23.7
	10	330	49.3		23.8
	11	345	49.0		23.9
	12	0	48.7	S23	23.9
	13	15	48.4		24.0
	14	30	48.1		24.1
	15	45	47.8	· ·	24.1
	16	60	47.5		24.2
	17	75	47.2		24.2
	18	90	46.9	S23	24.3
	19	105	46.6		24.4
	20	120	46.2		24.4
	21	135	45.9	· ·	24.5
	22	150	45.6		24.5
	23	165	45.3		24.6
19	00	180	45.0	S23	24.6
	01	195	44.7		24.7
	02	210	44.4		24.7
	03	225	44.1	· ·	24.8
	04	240	43.8		24.8
	05	255	43.5		24.9
	06	270	43.1	S23	24.9
	07	285	42.8		25.0
	08	300	42.5		25.0
	09	315	42.2	· ·	25.1
	10	330	41.9		25.1
	11	345	41.6		25.2
	12	0	41.3	S23	25.2
	13	15	41.0		25.3
	14	30	40.7		25.3
	15	45	40.3	· ·	25.3
	16	60	40.0		25.4
	17	75	39.7		25.4
	18	90	39.4	S23	25.5
	19	105	39.1		25.5
	20	120	38.8		25.5
	21	135	38.5	· ·	25.6
	22	150	38.2		25.6
	23	165	37.9		25.6
20	00	180	37.6	S23	25.7
	01	195	37.2		25.7
	02	210	36.9		25.7
	03	225	36.6	· ·	25.8
	04	240	36.3		25.8
	05	255	36.0		25.8
	06	270	35.7	S23	25.9
	07	285	35.4		25.9
	08	300	35.1		25.9
	09	315	34.7	· ·	25.9
	10	330	34.4		26.0
	11	345	34.1		26.0
	12	0	33.8	S23	26.0
	13	15	33.5		26.0
	14	30	33.2		26.1
	15	45	32.9	· ·	26.1
	16	60	32.6		26.1
	17	75	32.3		26.1
	18	90	31.9	S23	26.2
	19	105	31.6		26.2
	20	120	31.3		26.2
	21	135	31.0	· ·	26.2
	22	150	30.7		26.2
	23	165	30.4		26.2
21	00	180	30.1	S23	26.3
	01	195	29.8		26.3
	02	210	29.4		26.3
	03	225	29.1	· ·	26.3
	04	240	28.8		26.3
	05	255	28.5		26.3
	06	270	28.2	S23	26.3
	07	285	27.9		26.3
	08	300	27.6		26.3
	09	315	27.3	· ·	26.4
	10	330	27.0		26.4
	11	345	26.6		26.4
	12	0	26.3	S23	26.4
	13	15	26.0		26.4
	14	30	25.7		26.4
	15	45	25.4	· ·	26.4
	16	60	25.1		26.4
	17	75	24.8		26.4
	18	90	24.5	S23	26.4
	19	105	24.1		26.4
	20	120	23.8		26.4
	21	135	23.5	· ·	26.4
	22	150	23.2		26.4
	23	165	22.9		26.4
22	00	180	22.6	S23	26.4
	01	195	22.3		26.4
	02	210	22.0		26.4
	03	225	21.6	· ·	26.3
	04	240	21.3		26.3
	05	255	21.0		26.3
	06	270	20.7	S23	26.3
	07	285	20.4		26.3
	08	300	20.1		26.3
	09	315	19.8	· ·	26.3
	10	330	19.5		26.3
	11	345	19.1		26.3
	12	0	18.8	S23	26.2
	13	15	18.5		26.2
	14	30	18.2		26.2
	15	45	17.9	· ·	26.2
	16	60	17.6		26.2
	17	75	17.3		26.2
	18	90	17.0	S23	26.1
	19	105	16.6		26.1
	20	120	16.3		26.1
	21	135	16.0	· ·	26.1
	22	150	15.7		26.1
	23	165	15.4		26.0
23	00	180	15.1	S23	26.0
	01	195	14.8		26.0
	02	210	14.5		26.0
	03	225	14.1	· ·	25.9
	04	240	13.8		25.9
	05	255	13.5		25.9
	06	270	13.2	S23	25.8
	07	285	12.9		25.8
	08	300	12.6		25.8
	09	315	12.3	· ·	25.7
	10	330	12.0		25.7
	11	345	11.6		25.7
	12	0	11.3	S23	25.6
	13	15	11.0		25.6
	14	30	10.7		25.6
	15	45	10.4	· ·	25.5
	16	60	10.1		25.5
	17	75	09.8		25.5
	18	90	09.5	S23	25.4
	19	105	09.2		25.4
	20	120	08.8		25.3
	21	135	08.5	· ·	25.3
	22	150	08.2		25.3
	23	165	07.9		25.2
24	00	180	07.6	S23	25.2
	01	195	07.3		25.1
	02	210	07.0		25.1
	03	225	06.7	· ·	25.0
	04	240	06.3		25.0
	05	255	06.0		24.9
	06	270	05.7	S23	24.9
	07	285	05.4		24.8
	08	300	05.1		24.8
	09	315	04.8	· ·	24.7
	10	330	04.5		24.7
	11	345	04.2		24.6
	12	0	03.9	S23	24.6
	13	15	03.5		24.5
	14	30	03.2		24.5
	15	45	02.9	· ·	24.4
	16	60	02.6		24.4
	17	75	02.3		24.3
	18	90	02.0	S23	24.2
	19	105	01.7		24.2
	20	120	01.4		24.1
	21	135	01.0	· ·	24.1
	22	150	00.7		24.0
	23	165	00.4		23.9
25	00	180	00.1	S23	23.9
	01	194	59.8		23.8
	02	209	59.5		23.7
	03	224	59.2	· ·	23.7
	04	239	58.9		23.6
	05	254	58.6		23.5
	06	269	58.2	S23	23.5
	07	284	57.9		23.4
	08	299	57.6		23.3
	09	314	57.3	· ·	23.3
	10	329	57.0		23.2
	11	344	56.7		23.1
	12	359	56.4	S23	23.0
	13	14	56.1		23.0
	14	29	55.8		22.9
	15	44	55.5	· ·	22.8
	16	59	55.1		22.7
	17	74	54.8		22.7
	18	89	54.5	S23	22.6
	19	104	54.2		22.5
	20	119	53.9		22.4
	21	134	53.6	· ·	22.3
	22	149	53.3		22.3
	23	164	53.0		22.2

d h	G.H.A.	Dec.		d h	G.H.A.	Dec.
26 00	179 52.7	S23 22.1		29 00	179 30.5	S23 14.0
01	194 52.4	22.0		01	194 30.2	13.8
02	209 52.0	21.9		02	209 29.9	13.7
03	224 51.7 ··	21.8		03	224 29.6 ··	13.5
04	239 51.4	21.8		04	239 29.3	13.4
05	254 51.1	21.7		05	254 29.0	13.2
06	269 50.8	S23 21.6		06	269 28.7	S23 13.1
07	284 50.5	21.5		07	284 28.4	13.0
08	299 50.2	21.4		08	299 28.1	12.8
09	314 49.9 ··	21.3		09	314 27.8 ··	12.7
10	329 49.6	21.2		10	329 27.5	12.5
11	344 49.3	21.1		11	344 27.2	12.4
12	359 48.9	S23 21.0		12	359 26.9	S23 12.2
13	14 48.6	20.9		13	14 26.6	12.1
14	29 48.3	20.8		14	29 26.3	11.9
15	44 48.0 ··	20.7		15	44 26.0 ··	11.7
16	59 47.7	20.7		16	59 25.7	11.6
17	74 47.4	20.6		17	74 25.4	11.4
18	89 47.1	S23 20.5		18	89 25.1	S23 11.3
19	104 46.8	20.4		19	104 24.8	11.1
20	119 46.5	20.3		20	119 24.5	11.0
21	134 46.2 ··	20.2		21	134 24.2 ··	10.8
22	149 45.9	20.1		22	149 23.9	10.6
23	164 45.6	20.0		23	164 23.6	10.5
27 00	179 45.2	S23 19.9		30 00	179 23.3	S23 10.3
01	194 44.9	19.8		01	194 23.0	10.2
02	209 44.6	19.6		02	209 22.7	10.0
03	224 44.3 ··	19.5		03	224 22.4 ··	09.8
04	239 44.0	19.4		04	239 22.1	09.7
05	254 43.7	19.3		05	254 21.8	09.5
06	269 43.4	S23 19.2		06	269 21.5	S23 09.3
07	284 43.1	19.1		07	284 21.2	09.2
08	299 42.8	19.0		08	299 20.9	09.0
09	314 42.5 ··	18.9		09	314 20.6 ··	08.8
10	329 42.2	18.8		10	329 20.3	08.7
11	344 41.9	18.7		11	344 20.0	08.5
12	359 41.5	S23 18.6		12	359 19.7	S23 08.3
13	14 41.2	18.4		13	14 19.4	08.2
14	29 40.9	18.3		14	29 19.1	08.0
15	44 40.6 ··	18.2		15	44 18.8 ··	07.8
16	59 40.3	18.1		16	59 18.5	07.6
17	74 40.0	18.0		17	74 18.2	07.5
18	89 39.7	S23 17.9		18	89 17.9	S23 07.3
19	104 39.4	17.7		19	104 17.6	07.1
20	119 39.1	17.6		20	119 17.3	06.9
21	134 38.8 ··	17.5		21	134 17.0 ··	06.8
22	149 38.5	17.4		22	149 16.7	06.6
23	164 38.2	17.3		23	164 16.4	06.4
28 00	179 37.9	S23 17.1		31 00	179 16.1	S23 06.2
01	194 37.6	17.0		01	194 15.8	06.0
02	209 37.3	16.9		02	209 15.5	05.9
03	224 36.9 ··	16.8		03	224 15.2 ··	05.7
04	239 36.6	16.6		04	239 14.9	05.5
05	254 36.3	16.5		05	254 14.6	05.3
06	269 36.0	S23 16.4		06	269 14.3	S23 05.1
07	284 35.7	16.3		07	284 14.0	04.9
08	299 35.4	16.1		08	299 13.7	04.7
09	314 35.1 ··	16.0		09	314 13.4 ··	04.6
10	329 34.8	15.9		10	329 13.1	04.4
11	344 34.5	15.7		11	344 12.8	04.2
12	359 34.2	S23 15.6		12	359 12.5	S23 04.0
13	14 33.9	15.5		13	14 12.2	03.8
14	29 33.6	15.3		14	29 11.9	03.6
15	44 33.3 ··	15.2		15	44 11.6 ··	03.4
16	59 33.0	15.1		16	59 11.3	03.2
17	74 32.7	14.9		17	74 11.0	03.0
18	89 32.4	S23 14.8		18	89 10.7	S23 02.8
19	104 32.1	14.7		19	104 10.4	02.6
20	119 31.8	14.5		20	119 10.1	02.4
21	134 31.5 ··	14.4		21	134 09.8 ··	02.2
22	149 31.1	14.2		22	149 09.5	02.0
23	164 30.8	14.1		23	164 09.2	01.8

Table 6

DAILY LONG-TERM ALMANAC OF THE SUN FOR THE YEARS 1977, 1981, 1985, 1989, 1993, 1997, 2001, ETC.*

This table is a compilation of the columns for Greenwich Hour Angle and declination of the sun from the 1977 *Nautical Almanac*. For use beyond 1977, a correction may be necessary, depending upon the degree of precision desired. The amount of change in 1 years in the GHA and declination (DEC) is given for each 3-day interval in Table 9.

Since the correction factors are changing slowly with the passage of time, this table is *not* a perpetual almanac, but an almanac designed for use in practical navigation for a period of about 30 years.

*See note at end of table on use of 4-year correction factors.

d h	G.H.A.	Dec.
1 00	179 08.9	S23 01.7
01	194 08.6	01.5
02	209 08.3	01.3
03	224 08.0	·· 01.1
04	239 07.7	00.9
05	254 07.5	00.6
06	269 07.2	S23 00.4
07	284 06.9	00.2
08	299 06.6	23 00.0
09	314 06.3	22 59.8
10	329 06.0	59.6
11	344 05.7	59.4
12	359 05.4	S22 59.2
13	14 05.1	59.0
14	29 04.8	58.8
15	44 04.5	·· 58.6
16	59 04.2	58.4
17	74 03.9	58.1
18	89 03.6	S22 57.9
19	104 03.3	57.7
20	119 03.1	57.5
21	134 02.8	·· 57.3
22	149 02.5	57.1
23	164 02.2	56.9
2 00	179 01.9	S22 56.6
01	194 01.6	56.4
02	209 01.3	56.2
03	224 01.0	·· 56.0
04	239 00.7	55.7
05	254 00.4	55.5
06	269 00.1	S22 55.3
07	283 59.8	55.1
08	298 59.6	54.9
09	313 59.3	·· 54.6
10	328 59.0	54.4
11	343 58.7	54.2
12	358 58.4	S22 53.9
13	13 58.1	53.7
14	28 57.8	53.5
15	43 57.5	·· 53.3
16	58 57.2	53.0
17	73 56.9	52.8
18	88 56.7	S22 52.6
19	103 56.4	52.3
20	118 56.1	52.1
21	133 55.8	·· 51.9
22	148 55.5	51.6
23	163 55.2	51.4
3 00	178 54.9	S22 51.1
01	193 54.6	50.9
02	208 54.4	50.7
03	223 54.1	·· 50.4
04	238 53.8	50.2
05	253 53.5	49.9
06	268 53.2	S22 49.7
07	283 52.9	49.5
08	298 52.6	49.2
09	313 52.3	·· 49.0
10	328 52.1	48.7
11	343 51.8	48.5
12	358 51.5	S22 48.2
13	13 51.2	48.0
14	28 50.9	47.7
15	43 50.6	·· 47.5
16	58 50.3	47.2
17	73 50.1	47.0
18	88 49.8	S22 46.7
19	103 49.5	46.5
20	118 49.2	46.2
21	133 48.9	·· 46.0
22	148 48.6	45.7
23	163 48.3	45.5

d h	G.H.A.	Dec.
4 00	178 48.1	S22 45.2
01	193 47.8	45.0
02	208 47.5	44.7
03	223 47.2	·· 44.4
04	238 46.9	44.2
05	253 46.6	43.9
06	268 46.4	S22 43.7
07	283 46.1	43.4
08	298 45.8	43.1
09	313 45.5	·· 42.9
10	328 45.2	42.6
11	343 45.0	42.3
12	358 44.7	S22 42.1
13	13 44.4	41.8
14	28 44.1	41.5
15	43 43.8	·· 41.3
16	58 43.5	41.0
17	73 43.3	40.7
18	88 43.0	S22 40.5
19	103 42.7	40.2
20	118 42.4	39.9
21	133 42.1	·· 39.6
22	148 41.9	39.4
23	163 41.6	39.1
5 00	178 41.3	S22 38.8
01	193 41.0	38.5
02	208 40.7	38.3
03	223 40.5	·· 38.0
04	238 40.2	37.7
05	253 39.9	37.4
06	268 39.6	S22 37.2
07	283 39.3	36.9
08	298 39.1	36.6
09	313 38.8	·· 36.3
10	328 38.5	36.0
11	343 38.2	35.7
12	358 38.0	S22 35.5
13	13 37.7	35.2
14	28 37.4	34.9
15	43 37.1	·· 34.6
16	58 36.9	34.3
17	73 36.6	34.0
18	88 36.3	S22 33.7
19	103 36.0	33.4
20	118 35.7	33.2
21	133 35.5	·· 32.9
22	148 35.2	32.6
23	163 34.9	32.3
6 00	178 34.6	S22 32.0
01	193 34.4	31.7
02	208 34.1	31.4
03	223 33.8	·· 31.1
04	238 33.5	30.8
05	253 33.3	30.5
06	268 33.0	S22 30.2
07	283 32.7	29.9
08	298 32.5	29.6
09	313 32.2	·· 29.3
10	328 31.9	29.0
11	343 31.6	28.7
12	358 31.4	S22 28.4
13	13 31.1	28.1
14	28 30.8	27.8
15	43 30.5	·· 27.5
16	58 30.3	27.2
17	73 30.0	26.9
18	88 29.7	S22 26.6
19	103 29.5	26.3
20	118 29.2	25.9
21	133 28.9	·· 25.6
22	148 28.6	25.3
23	163 28.4	25.0

d h	G.H.A.	Dec.
7 00	178 28.1	S22 24.7
01	193 27.8	24.4
02	208 27.6	24.1
03	223 27.3	·· 23.8
04	238 27.0	23.4
05	253 26.8	23.1
06	268 26.5	S22 22.8
07	283 26.2	22.5
08	298 25.9	22.2
09	313 25.7	·· 21.8
10	328 25.4	21.5
11	343 25.1	21.2
12	358 24.9	S22 20.9
13	13 24.6	20.6
14	28 24.3	20.2
15	43 24.1	·· 19.9
16	58 23.8	19.6
17	73 23.5	19.3
18	88 23.3	S22 18.9
19	103 23.0	18.6
20	118 22.7	18.3
21	133 22.5	·· 18.0
22	148 22.2	17.6
23	163 21.9	17.3
8 00	178 21.7	S22 17.0
01	193 21.4	16.6
02	208 21.1	16.3
03	223 20.9	·· 16.0
04	238 20.6	15.6
05	253 20.4	15.3
06	268 20.1	S22 15.0
07	283 19.8	14.6
08	298 19.6	14.3
09	313 19.3	·· 14.0
10	328 19.0	13.6
11	343 18.8	13.3
12	358 18.5	S22 12.9
13	13 18.2	12.6
14	28 18.0	12.3
15	43 17.7	·· 11.9
16	58 17.5	11.6
17	73 17.2	11.2
18	88 16.9	S22 10.9
19	103 16.7	10.5
20	118 16.4	10.2
21	133 16.2	·· 09.8
22	148 15.9	09.5
23	163 15.6	09.2
9 00	178 15.4	S22 08.8
01	193 15.1	08.5
02	208 14.9	08.1
03	223 14.6	·· 07.8
04	238 14.3	07.4
05	253 14.1	07.0
06	268 13.8	S22 06.7
07	283 13.6	06.3
08	298 13.3	06.0
09	313 13.0	·· 05.6
10	328 12.8	05.3
11	343 12.5	04.9
12	358 12.3	S22 04.6
13	13 12.0	04.2
14	28 11.8	03.8
15	43 11.5	·· 03.5
16	58 11.2	03.1
17	73 11.0	02.8
18	88 10.7	S22 02.4
19	103 10.5	02.0
20	118 10.2	01.7
21	133 10.0	·· 01.3
22	148 09.7	00.9
23	163 09.5	00.6

d h	G.H.A.	Dec.
10 00	178 09.2	S22 00.2
01	193 08.9	21 59.8
02	208 08.7	59.5
03	223 08.4 ··	59.1
04	238 08.2	58.7
05	253 07.9	58.4
06	268 07.7	S21 58.0
07	283 07.4	57.6
08	298 07.2	57.2
09	313 06.9 ··	56.9
10	328 06.7	56.5
11	343 06.4	56.1
12	358 06.2	S21 55.7
13	13 05.9	55.4
14	28 05.7	55.0
15	43 05.4 ··	54.6
16	58 05.2	54.2
17	73 04.9	53.8
18	88 04.7	S21 53.5
19	103 04.4	53.1
20	118 04.2	52.7
21	133 03.9 ··	52.3
22	148 03.7	51.9
23	163 03.4	51.6
11 00	178 03.2	S21 51.2
01	193 02.9	50.8
02	208 02.7	50.4
03	223 02.4 ··	50.0
04	238 02.2	49.6
05	253 01.9	49.2
06	268 01.7	S21 48.8
07	283 01.4	48.5
08	298 01.2	48.1
09	313 00.9 ··	47.7
10	328 00.7	47.3
11	343 00.4	46.9
12	358 00.2	S21 46.5
13	13 00.0	46.1
14	27 59.7	45.7
15	42 59.5 ··	45.3
16	57 59.2	44.9
17	72 59.0	44.5
18	87 58.7	S21 44.1
19	102 58.5	43.7
20	117 58.2	43.3
21	132 58.0 ··	42.9
22	147 57.8	42.5
23	162 57.5	42.1
12 00	177 57.3	S21 41.7
01	192 57.0	41.3
02	207 56.8	40.9
03	222 56.5 ··	40.5
04	237 56.3	40.1
05	252 56.1	39.7
06	267 55.8	S21 39.3
07	282 55.6	38.9
08	297 55.3	38.5
09	312 55.1 ··	38.1
10	327 54.9	37.6
11	342 54.6	37.2
12	357 54.4	S21 36.8
13	12 54.1	36.4
14	27 53.9	36.0
15	42 53.7 ··	35.6
16	57 53.4	35.2
17	72 53.2	34.8
18	87 52.9	S21 34.3
19	102 52.7	33.9
20	117 52.5	33.5
21	132 52.2 ··	33.1
22	147 52.0	32.7
23	162 51.8	32.3

d h	G.H.A.	Dec.
13 00	177 51.5	S21 31.8
01	192 51.3	31.4
02	207 51.0	31.0
03	222 50.8 ··	30.6
04	237 50.6	30.1
05	252 50.3	29.7
06	267 50.1	S21 29.3
07	282 49.9	28.9
08	297 49.6	28.5
09	312 49.4 ··	28.0
10	327 49.2	27.6
11	342 48.9	27.2
12	357 48.7	S21 26.7
13	12 48.5	26.3
14	27 48.2	25.9
15	42 48.0 ··	25.5
16	57 47.8	25.0
17	72 47.5	24.6
18	07 47.3	S21 24.2
19	102 47.1	23.7
20	117 46.8	23.3
21	132 46.6 ··	22.9
22	147 46.4	22.4
23	162 46.2	22.0
14 00	177 45.9	S21 21.5
01	192 45.7	21.1
02	207 45.5	20.7
03	222 45.2 ··	20.2
04	237 45.0	19.8
05	252 44.8	19.3
06	267 44.5	S21 18.9
07	282 44.3	18.5
08	297 44.1	18.0
09	312 43.9 ··	17.6
10	327 43.6	17.1
11	342 43.4	16.7
12	357 43.2	S21 16.2
13	12 43.0	15.8
14	27 42.7	15.3
15	42 42.5 ··	14.9
16	57 42.3	14.5
17	72 42.1	14.0
18	87 41.8	S21 13.6
19	102 41.6	13.1
20	117 41.4	12.6
21	132 41.2 ··	12.2
22	147 40.9	11.7
23	162 40.7	11.3
15 00	177 40.5	S21 10.8
01	192 40.3	10.4
02	207 40.0	10.0
03	222 39.8 ··	09.5
04	237 39.6	09.0
05	252 39.4	08.6
06	267 39.1	S21 08.1
07	282 38.9	07.6
08	297 38.7	07.2
09	312 38.5 ··	06.7
10	327 38.3	06.3
11	342 38.0	05.8
12	357 37.8	S21 05.3
13	12 37.6	04.9
14	27 37.4	04.4
15	42 37.2 ··	03.9
16	57 36.9	03.5
17	72 36.7	03.0
18	87 36.5	S21 02.5
19	102 36.3	02.1
20	117 36.1	01.6
21	132 35.9 ··	01.1
22	147 35.6	00.7
23	162 35.4	00.2

d h	G.H.A.	Dec.
16 00	177 35.2	S20 59.7
01	192 35.0	59.3
02	207 34.8	58.8
03	222 34.6 ··	58.3
04	237 34.3	57.8
05	252 34.1	57.4
06	267 33.9	S20 56.9
07	282 33.7	56.4
08	297 33.5	55.9
09	312 33.3 ··	55.5
10	327 33.1	55.0
11	342 32.8	54.5
12	357 32.6	S20 54.0
13	12 32.4	53.5
14	27 32.2	53.1
15	42 32.0 ··	52.6
16	57 31.8	52.1
17	72 31.6	51.6
18	87 31.4	S20 51.1
19	102 31.1	50.6
20	117 30.9	50.2
21	132 30.7 ··	49.7
22	147 30.5	49.2
23	162 30.3	48.7
17 00	177 30.1	S20 48.2
01	192 29.9	47.7
02	207 29.7	47.2
03	222 29.5 ··	46.7
04	237 29.3	46.3
05	252 29.1	45.8
06	267 28.8	S20 45.3
07	282 28.6	44.8
08	297 28.4	44.3
09	312 28.2 ··	43.8
10	327 28.0	43.3
11	342 27.8	42.8
12	357 27.6	S20 42.3
13	12 27.4	41.8
14	27 27.2	41.3
15	42 27.0 ··	40.8
16	57 26.8	40.3
17	72 26.6	39.8
18	87 26.4	S20 39.3
19	102 26.2	38.8
20	117 26.0	38.3
21	132 25.8 ··	37.8
22	147 25.6	37.3
23	162 25.4	36.8
18 00	177 25.2	S20 36.3
01	192 25.0	35.8
02	207 24.8	35.3
03	222 24.6 ··	34.8
04	237 24.4	34.3
05	252 24.2	33.8
06	267 24.0	S20 33.3
07	282 23.8	32.8
08	297 23.6	32.2
09	312 23.4 ··	31.7
10	327 23.2	31.2
11	342 23.0	30.7
12	357 22.8	S20 30.2
13	12 22.6	29.7
14	27 22.4	29.2
15	42 22.2 ··	28.7
16	57 22.0	28.1
17	72 21.8	27.6
18	87 21.6	S20 27.1
19	102 21.4	26.6
20	117 21.2	26.1
21	132 21.0 ··	25.6
22	147 20.8	25.0
23	162 20.6	24.5

d h	G.H.A.	Dec.
19 00	177 20.4	S20 24.0
01	192 20.2	23.5
02	207 20.0	23.0
03	222 19.8 ··	22.4
04	237 19.6	21.9
05	252 19.4	21.4
06	267 19.2	S20 20.9
07	282 19.1	20.3
08	297 18.9	19.8
09	312 18.7 ··	19.3
10	327 18.5	18.8
11	342 18.3	18.2
12	357 18.1	S20 17.7
13	12 17.9	17.2
14	27 17.7	16.7
15	42 17.5 ··	16.1
16	57 17.3	15.6
17	72 17.1	15.1
18	87 17.0	S20 14.5
19	102 16.8	14.0
20	117 16.6	13.5
21	132 16.4 ··	12.9
22	147 16.2	12.4
23	162 16.0	11.9
20 00	177 15.8	S20 11.3
01	192 15.6	10.8
02	207 15.5	10.3
03	222 15.3 ··	09.7
04	237 15.1	09.2
05	252 14.9	08.6
06	267 14.7	S20 08.1
07	282 14.5	07.6
08	297 14.3	07.0
09	312 14.2 ··	06.5
10	327 14.0	05.9
11	342 13.8	05.4
12	357 13.6	S20 04.8
13	12 13.4	04.3
14	27 13.3	03.8
15	42 13.1 ··	03.2
16	57 12.9	02.7
17	72 12.7	02.1
18	87 12.5	S20 01.6
19	102 12.3	01.0
20	117 12.2	20 00.5
21	132 12.0	19 59.9
22	147 11.8	59.4
23	162 11.6	58.8
21 00	177 11.4	S19 58.3
01	192 11.3	57.7
02	207 11.1	57.2
03	222 10.9 ··	56.6
04	237 10.7	56.1
05	252 10.6	55.5
06	267 10.4	S19 54.9
07	282 10.2	54.4
08	297 10.0	53.8
09	312 09.8 ··	53.3
10	327 09.7	52.7
11	342 09.5	52.2
12	357 09.3	S19 51.6
13	12 09.1	51.0
14	27 09.0	50.5
15	42 08.8 ··	49.9
16	57 08.6	49.4
17	72 08.5	48.8
18	87 08.3	S19 48.2
19	102 08.1	47.7
20	117 07.9	47.1
21	132 07.8 ··	46.5
22	147 07.6	46.0
23	162 07.4	45.4

d h	G.H.A.	Dec.
22 00	177 07.2	S19 44.8
01	192 07.1	44.3
02	207 06.9	43.7
03	222 06.7 ··	43.1
04	237 06.6	42.6
05	252 06.4	42.0
06	267 06.2	S19 41.4
07	282 06.1	40.8
08	297 05.9	40.3
09	312 05.7 ··	39.7
10	327 05.6	39.1
11	342 05.4	38.6
12	357 05.2	S19 38.0
13	12 05.1	37.4
14	27 04.9	36.8
15	42 04.7 ··	36.3
16	57 04.6	35.7
17	72 04.4	35.1
18	87 04.2	S19 34.5
19	102 04.1	33.9
20	117 03.9	33.4
21	132 03.7 ··	32.8
22	147 03.6	32.2
23	162 03.4	31.6
23 00	177 03.2	S19 31.0
01	192 03.1	30.5
02	207 02.9	29.9
03	222 02.8 ··	29.3
04	237 02.6	28.7
05	252 02.4	28.1
06	267 02.3	S19 27.5
07	282 02.1	26.9
08	297 02.0	26.4
09	312 01.8 ··	25.8
10	327 01.6	25.2
11	342 01.5	24.6
12	357 01.3	S19 24.0
13	12 01.2	23.4
14	27 01.0	22.8
15	42 00.8 ··	22.2
16	57 00.7	21.6
17	72 00.5	21.0
18	87 00.4	S19 20.4
19	102 00.2	19.9
20	117 00.1	19.3
21	131 59.9 ··	18.7
22	146 59.8	18.1
23	161 59.6	17.5
24 00	176 59.4	S19 16.9
01	191 59.3	16.3
02	206 59.1	15.7
03	221 59.0 ··	15.1
04	236 58.8	14.5
05	251 58.7	13.9
06	266 58.5	S19 13.3
07	281 58.4	12.7
08	296 58.2	12.1
09	311 58.1 ··	11.5
10	326 57.9	10.9
11	341 57.8	10.3
12	356 57.6	S19 09.7
13	11 57.5	09.1
14	26 57.3	08.5
15	41 57.2 ··	07.8
16	56 57.0	07.2
17	71 56.9	06.6
18	86 56.7	S19 06.0
19	101 56.6	05.4
20	116 56.4	04.8
21	131 56.3 ··	04.2
22	146 56.1	03.6
23	161 56.0	03.0

d h	G.H.A.	Dec.
25 00	176 55.8	S19 02.4
01	191 55.7	01.7
02	206 55.5	01.1
03	221 55.4	19 00.5
04	236 55.3	18 59.9
05	251 55.1	59.3
06	266 55.0	S18 58.7
07	281 54.8	58.1
08	296 54.7	57.4
09	311 54.5 ··	56.8
10	326 54.4	56.2
11	341 54.3	55.6
12	356 54.1	S18 55.0
13	11 54.0	54.4
14	26 53.8	53.7
15	41 53.7 ··	53.1
16	56 53.5	52.5
17	71 53.4	51.9
18	86 53.3	S18 51.2
19	101 53.1	50.6
20	116 53.0	50.0
21	131 52.9 ··	49.4
22	146 52.7	48.8
23	161 52.6	48.1
26 00	176 52.4	S18 47.5
01	191 52.3	46.9
02	206 52.2	46.2
03	221 52.0 ··	45.6
04	236 51.9	45.0
05	251 51.8	44.4
06	266 51.6	S18 43.7
07	281 51.5	43.1
08	296 51.3	42.5
09	311 51.2 ··	41.8
10	326 51.1	41.2
11	341 50.9	40.6
12	356 50.8	S18 39.9
13	11 50.7	39.3
14	26 50.5	38.7
15	41 50.4 ··	38.0
16	56 50.3	37.4
17	71 50.2	36.8
18	86 50.0	S18 36.1
19	101 49.9	35.5
20	116 49.8	34.9
21	131 49.6 ··	34.2
22	146 49.5	33.6
23	161 49.4	32.9
27 00	176 49.2	S18 32.3
01	191 49.1	31.7
02	206 49.0	31.0
03	221 48.9 ··	30.4
04	236 48.7	29.7
05	251 48.6	29.1
06	266 48.5	S18 28.4
07	281 48.3	27.8
08	296 48.2	27.2
09	311 48.1 ··	26.5
10	326 48.0	25.9
11	341 47.8	25.2
12	356 47.7	S18 24.6
13	11 47.6	23.9
14	26 47.5	23.3
15	41 47.3 ··	22.6
16	56 47.2	22.0
17	71 47.1	21.3
18	86 47.0	S18 20.7
19	101 46.9	20.0
20	116 46.7	19.4
21	131 46.6 ··	18.7
22	146 46.5	18.1
23	161 46.4	17.4

d h	G.H.A. ° '	Dec. ° '
28 00	176 46.3	S18 16.8
01	191 46.1	16.1
02	206 46.0	15.4
03	221 45.9	·· 14.8
04	236 45.8	14.1
05	251 45.7	13.5
06	266 45.5	S18 12.8
07	281 45.4	12.2
08	296 45.3	11.5
09	311 45.2	·· 10.8
10	326 45.1	10.2
11	341 44.9	09.5
12	356 44.8	S18 08.9
13	11 44.7	08.2
14	26 44.6	07.5
15	41 44.5	·· 06.9
16	56 44.4	06.2
17	71 44.3	05.5
18	86 44.1	S18 04.9
19	101 44.0	04.2
20	116 43.9	03.6
21	131 43.8	·· 02.9
22	146 43.7	02.2
23	161 43.6	01.6
29 00	176 43.5	S18 00.9
01	191 43.4	18 00.2
02	206 43.2	17 59.6
03	221 43.1	·· 58.9
04	236 43.0	58.2
05	251 42.9	57.5
06	266 42.8	S17 56.9
07	281 42.7	56.2
08	296 42.6	55.5
09	311 42.5	·· 54.9
10	326 42.4	54.2
11	341 42.3	53.5
12	356 42.2	S17 52.8
13	11 42.0	52.2
14	26 41.9	51.5
15	41 41.8	·· 50.8
16	56 41.7	50.1
17	71 41.6	49.5
18	86 41.5	S17 48.8
19	101 41.4	48.1
20	116 41.3	47.4
21	131 41.2	·· 46.7
22	146 41.1	46.1
23	161 41.0	45.4
30 00	176 40.9	S17 44.7
01	191 40.8	44.0
02	206 40.7	43.3
03	221 40.6	·· 42.7
04	236 40.5	42.0
05	251 40.4	41.3
06	266 40.3	S17 40.6
07	281 40.2	39.9
08	296 40.1	39.2
09	311 40.0	·· 38.5
10	326 39.9	37.9
11	341 39.8	37.2
12	356 39.7	S17 36.5
13	11 39.6	35.8
14	26 39.5	35.1
15	41 39.4	·· 34.4
16	56 39.3	33.7
17	71 39.2	33.0
18	86 39.1	S17 32.3
19	101 39.0	31.7
20	116 38.9	31.0
21	131 38.8	·· 30.3
22	146 38.7	29.6
23	161 38.6	28.9

FEBRUARY

d h	G.H.A. ° '	Dec. ° '
31 00	176 38.5	S17 28.2
01	191 38.4	27.5
02	206 38.3	26.8
03	221 38.2	·· 26.1
04	236 38.1	25.4
05	251 38.1	24.7
06	266 38.0	S17 24.0
07	281 37.9	23.3
08	296 37.8	22.6
09	311 37.7	·· 21.9
10	326 37.6	21.2
11	341 37.5	20.5
12	356 37.4	S17 19.8
13	11 37.3	19.1
14	26 37.2	18.4
15	41 37.1	·· 17.7
16	56 37.1	17.0
17	71 37.0	16.3
18	86 36.9	S17 15.6
19	101 36.8	14.9
20	116 36.7	14.2
21	131 36.6	·· 13.5
22	146 36.5	12.8
23	161 36.4	12.1
1 00	176 36.4	S17 11.4
01	191 36.3	10.7
02	206 36.2	10.0
03	221 36.1	·· 09.3
04	236 36.0	08.5
05	251 35.9	07.8
06	266 35.9	S17 07.1
07	281 35.8	06.4
08	296 35.7	05.7
09	311 35.6	·· 05.0
10	326 35.5	04.3
11	341 35.4	03.6
12	356 35.4	S17 02.9
13	11 35.3	02.1
14	26 35.2	01.4
15	41 35.1	·· 00.7
16	56 35.0	17 00.0
17	71 35.0	16 59.3
18	86 34.9	S16 58.6
19	101 34.8	57.8
20	116 34.7	57.1
21	131 34.6	·· 56.4
22	146 34.6	55.7
23	161 34.5	55.0
2 00	176 34.4	S16 54.3
01	191 34.3	53.5
02	206 34.3	52.8
03	221 34.2	·· 52.1
04	236 34.1	51.4
05	251 34.0	50.7
06	266 33.9	S16 49.9
07	281 33.9	49.2
08	296 33.8	48.5
09	311 33.7	·· 47.8
10	326 33.7	47.0
11	341 33.6	46.3
12	356 33.5	S16 45.6
13	11 33.4	44.9
14	26 33.4	44.1
15	41 33.3	·· 43.4
16	56 33.2	42.7
17	71 33.1	41.9
18	86 33.1	S16 41.2
19	101 33.0	40.5
20	116 32.9	39.8
21	131 32.9	·· 39.0
22	146 32.8	38.3
23	161 32.7	37.6

d h	G.H.A. ° '	Dec. ° '
3 00	176 32.7	S16 36.8
01	191 32.6	36.1
02	206 32.5	35.4
03	221 32.5	·· 34.6
04	236 32.4	33.9
05	251 32.3	33.2
06	266 32.3	S16 32.4
07	281 32.2	31.7
08	296 32.1	31.0
09	311 32.1	·· 30.2
10	326 32.0	29.5
11	341 31.9	28.8
12	356 31.9	S16 28.0
13	11 31.8	27.3
14	26 31.7	26.5
15	41 31.7	·· 25.8
16	56 31.6	25.1
17	71 31.5	24.3
18	86 31.5	S16 23.6
19	101 31.4	22.8
20	116 31.4	22.1
21	131 31.3	·· 21.4
22	146 31.2	20.6
23	161 31.2	19.9
4 00	176 31.1	S16 19.1
01	191 31.1	18.4
02	206 31.0	17.6
03	221 30.9	·· 16.9
04	236 30.9	16.2
05	251 30.8	15.4
06	266 30.8	S16 14.7
07	281 30.7	13.9
08	296 30.6	13.2
09	311 30.6	·· 12.4
10	326 30.5	11.7
11	341 30.5	10.9
12	356 30.4	S16 10.2
13	11 30.4	09.4
14	26 30.3	08.7
15	41 30.3	·· 07.9
16	56 30.2	07.2
17	71 30.1	06.4
18	86 30.1	S16 05.7
19	101 30.0	04.9
20	116 30.0	04.2
21	131 29.9	·· 03.4
22	146 29.9	02.7
23	161 29.8	01.9
5 00	176 29.8	S16 01.1
01	191 29.7	16 00.4
02	206 29.7	15 59.6
03	221 29.6	·· 58.9
04	236 29.6	58.1
05	251 29.5	57.4
06	266 29.5	S15 56.6
07	281 29.4	55.8
08	296 29.4	55.1
09	311 29.3	·· 54.3
10	326 29.3	53.6
11	341 29.2	52.8
12	356 29.2	S15 52.1
13	11 29.1	51.3
14	26 29.1	50.5
15	41 29.0	·· 49.8
16	56 29.0	49.0
17	71 28.9	48.2
18	86 28.9	S15 47.5
19	101 28.9	46.7
20	116 28.8	45.9
21	131 28.8	·· 45.2
22	146 28.7	44.4
23	161 28.7	43.7

d h	G.H.A.	Dec.
6 00	176 28.6	S15 42.9
01	191 28.6	42.1
02	206 28.5	41.4
03	221 28.5 ··	40.6
04	236 28.5	39.8
05	251 28.4	39.0
06	266 28.4	S15 38.3
07	281 28.3	37.5
08	296 28.3	36.7
09	311 28.3 ··	36.0
10	326 28.2	35.2
11	341 28.2	34.4
12	356 28.1	S15 33.7
13	11 28.1	32.9
14	26 28.1	32.1
15	41 28.0 ··	31.3
16	56 28.0	30.6
17	71 27.9	29.8
18	86 27.9	S15 29.0
19	101 27.9	28.2
20	116 27.8	27.5
21	131 27.8 ··	26.7
22	146 27.8	25.9
23	161 27.7	25.1
7 00	176 27.7	S15 24.4
01	191 27.7	23.6
02	206 27.6	22.8
03	221 27.6 ··	22.0
04	236 27.6	21.2
05	251 27.5	20.5
06	266 27.5	S15 19.7
07	281 27.5	18.9
08	296 27.4	18.1
09	311 27.4 ··	17.3
10	326 27.4	16.6
11	341 27.3	15.8
12	356 27.3	S15 15.0
13	11 27.3	14.2
14	26 27.2	13.4
15	41 27.2 ··	12.6
16	56 27.2	11.9
17	71 27.1	11.1
18	86 27.1	S15 10.3
19	101 27.1	09.5
20	116 27.1	08.7
21	131 27.0 ··	07.9
22	146 27.0	07.1
23	161 27.0	06.3
8 00	176 27.0	S15 05.6
01	191 26.9	04.8
02	206 26.9	04.0
03	221 26.9 ··	03.2
04	236 26.8	02.4
05	251 26.8	01.6
06	266 26.8	S15 00.8
07	281 26.8	15 00.0
08	296 26.8	14 59.2
09	311 26.7 ··	58.4
10	326 26.7	57.6
11	341 26.7	56.9
12	356 26.7	S14 56.1
13	11 26.6	55.3
14	26 26.6	54.5
15	41 26.6 ··	53.7
16	56 26.6	52.9
17	71 26.5	52.1
18	86 26.5	S14 51.3
19	101 26.5	50.5
20	116 26.5	49.7
21	131 26.5 ··	48.9
22	146 26.4	48.1
23	161 26.4	47.3

d h	G.H.A.	Dec.
9 00	176 26.4	S14 46.5
01	191 26.4	45.7
02	206 26.4	44.9
03	221 26.4 ··	44.1
04	236 26.3	43.3
05	251 26.3	42.5
06	266 26.3	S14 41.7
07	281 26.3	40.9
08	296 26.3	40.1
09	311 26.3 ··	39.3
10	326 26.2	38.5
11	341 26.2	37.7
12	356 26.2	S14 36.9
13	11 26.2	36.1
14	26 26.2	35.3
15	41 26.2 ··	34.5
16	56 26.2	33.7
17	71 26.1	32.9
18	86 26.1	S14 32.1
19	101 26.1	31.2
20	116 26.1	30.4
21	131 26.1 ··	29.6
22	146 26.1	28.8
23	161 26.1	28.0
10 00	176 26.1	S14 27.2
01	191 26.0	26.4
02	206 26.0	25.6
03	221 26.0 ··	24.8
04	236 26.0	24.0
05	251 26.0	23.2
06	266 26.0	S14 22.3
07	281 26.0	21.5
08	296 26.0	20.7
09	311 26.0 ··	19.9
10	326 26.0	19.1
11	341 26.0	18.3
12	356 25.9	S14 17.5
13	11 25.9	16.7
14	26 25.9	15.8
15	41 25.9 ··	15.0
16	56 25.9	14.2
17	71 25.9	13.4
18	86 25.9	S14 12.6
19	101 25.9	11.8
20	116 25.9	10.9
21	131 25.9 ··	10.1
22	146 25.9	09.3
23	161 25.9	08.5
11 00	176 25.9	S14 07.7
01	191 25.9	06.8
02	206 25.9	06.0
03	221 25.9 ··	05.2
04	236 25.9	04.4
05	251 25.9	03.6
06	266 25.9	S14 02.7
07	281 25.9	01.9
08	296 25.9	01.1
09	311 25.9 ··	14 00.3
10	326 25.9	13 59.5
11	341 25.9	58.6
12	356 25.9	S13 57.8
13	11 25.9	57.0
14	26 25.9	56.2
15	41 25.9 ··	55.3
16	56 25.9	54.5
17	71 25.9	53.7
18	86 25.9	S13 52.9
19	101 25.9	52.0
20	116 25.9	51.2
21	131 25.9 ··	50.4
22	146 25.9	49.5
23	161 25.9	48.7

d h	G.H.A.	Dec.
12 00	176 25.9	S13 47.9
01	191 25.9	47.1
02	206 25.9	46.2
03	221 25.9 ··	45.4
04	236 25.9	44.6
05	251 25.9	43.7
06	266 26.0	S13 42.9
07	281 26.0	42.1
08	296 26.0	41.2
09	311 26.0 ··	40.4
10	326 26.0	39.6
11	341 26.0	38.7
12	356 26.0	S13 37.9
13	11 26.0	37.1
14	26 26.0	36.2
15	41 26.0 ··	35.4
16	56 26.0	34.6
17	71 26.0	33.7
18	86 26.1	S13 32.9
19	101 26.1	32.1
20	116 26.1	31.2
21	131 26.1 ··	30.4
22	146 26.1	29.6
23	161 26.1	28.7
13 00	176 26.1	S13 27.9
01	191 26.1	27.0
02	206 26.2	26.2
03	221 26.2 ··	25.4
04	236 26.2	24.5
05	251 26.2	23.7
06	266 26.2	S13 22.8
07	281 26.2	22.0
08	296 26.2	21.2
09	311 26.3 ··	20.3
10	326 26.3	19.5
11	341 26.3	18.6
12	356 26.3	S13 17.8
13	11 26.3	17.0
14	26 26.3	16.1
15	41 26.4 ··	15.3
16	56 26.4	14.4
17	71 26.4	13.6
18	86 26.4	S13 12.7
19	101 26.4	11.9
20	116 26.4	11.0
21	131 26.5 ··	10.2
22	146 26.5	09.4
23	161 26.5	08.5
14 00	176 26.5	S13 07.7
01	191 26.5	06.8
02	206 26.6	06.0
03	221 26.6 ··	05.1
04	236 26.6	04.3
05	251 26.6	03.4
06	266 26.7	S13 02.6
07	281 26.7	01.7
08	296 26.7	00.9
09	311 26.7 ··	13 00.0
10	326 26.7	12 59.2
11	341 26.8	58.3
12	356 26.8	S12 57.5
13	11 26.8	56.6
14	26 26.8	55.8
15	41 26.9 ··	54.9
16	56 26.9	54.1
17	71 26.9	53.2
18	86 26.9	S12 52.4
19	101 27.0	51.5
20	116 27.0	50.6
21	131 27.0 ··	49.8
22	146 27.0	48.9
23	161 27.1	48.1

GHA / Dec. table

Day 15–17

d h	G.H.A.	Dec.
15 00	176 27.1	S12 47.2
01	191 27.1	46.4
02	206 27.2	45.5
03	221 27.2 ··	44.7
04	236 27.2	43.8
05	251 27.2	42.9
06	266 27.3	S12 42.1
07	281 27.3	41.2
08	296 27.3	40.4
09	311 27.4 ··	39.5
10	326 27.4	38.6
11	341 27.4	37.8
12	356 27.5	S12 36.9
13	11 27.5	36.1
14	26 27.5	35.2
15	41 27.6 ··	34.3
16	56 27.6	33.5
17	71 27.6	32.6
18	86 27.7	S12 31.8
19	101 27.7	30.9
20	116 27.7	30.0
21	131 27.8 ··	29.2
22	146 27.8	28.3
23	161 27.8	27.4
16 00	176 27.9	S12 26.6
01	191 27.9	25.7
02	206 27.9	24.8
03	221 28.0 ··	24.0
04	236 28.0	23.1
05	251 28.0	22.3
06	266 28.1	S12 21.4
07	281 28.1	20.5
08	296 28.1	19.7
09	311 28.2 ··	18.8
10	326 28.2	17.9
11	341 28.3	17.0
12	356 28.3	S12 16.2
13	11 28.3	15.3
14	26 28.4	14.4
15	41 28.4 ··	13.6
16	56 28.5	12.7
17	71 28.5	11.8
18	86 28.5	S12 11.0
19	101 28.6	10.1
20	116 28.6	09.2
21	131 28.7 ··	08.3
22	146 28.7	07.5
23	161 28.7	06.6
17 00	176 28.8	S12 05.7
01	191 28.8	04.9
02	206 28.9	04.0
03	221 28.9 ··	03.1
04	236 29.0	02.2
05	251 29.0	01.4
06	266 29.0	S12 00.5
07	281 29.1	11 59.6
08	296 29.1	58.7
09	311 29.2 ··	57.9
10	326 29.2	57.0
11	341 29.3	56.1
12	356 29.3	S11 55.2
13	11 29.4	54.4
14	26 29.4	53.5
15	41 29.5 ··	52.6
16	56 29.5	51.7
17	71 29.6	50.9
18	86 29.6	S11 50.0
19	101 29.6	49.1
20	116 29.7	48.2
21	131 29.7 ··	47.3
22	146 29.8	46.5
23	161 29.8	45.6

Day 18–20

d h	G.H.A.	Dec.
18 00	176 29.9	S11 44.7
01	191 29.9	43.8
02	206 30.0	42.9
03	221 30.0 ··	42.1
04	236 30.1	41.2
05	251 30.1	40.3
06	266 30.2	S11 39.4
07	281 30.2	38.5
08	296 30.3	37.6
09	311 30.4 ··	36.8
10	326 30.4	35.9
11	341 30.5	35.0
12	356 30.5	S11 34.1
13	11 30.6	33.2
14	26 30.6	32.3
15	41 30.7 ··	31.5
16	56 30.7	30.6
17	71 30.8	29.7
18	86 30.8	S11 28.8
19	101 30.9	27.9
20	116 30.9	27.0
21	131 31.0 ··	26.1
22	146 31.1	25.2
23	161 31.1	24.4
19 00	176 31.2	S11 23.5
01	191 31.2	22.6
02	206 31.3	21.7
03	221 31.3 ··	20.8
04	236 31.4	19.9
05	251 31.5	19.0
06	266 31.5	S11 18.1
07	281 31.6	17.2
08	296 31.6	16.4
09	311 31.7 ··	15.5
10	326 31.7	14.6
11	341 31.8	13.7
12	356 31.9	S11 12.8
13	11 31.9	11.9
14	26 32.0	11.0
15	41 32.1 ··	10.1
16	56 32.1	09.2
17	71 32.2	08.3
18	86 32.2	S11 07.4
19	101 32.3	06.5
20	116 32.4	05.6
21	131 32.4 ··	04.8
22	146 32.5	03.9
23	161 32.5	03.0
20 00	176 32.6	S11 02.1
01	191 32.7	01.2
02	206 32.7	11 00.3
03	221 32.8	10 59.4
04	236 32.9	58.5
05	251 32.9	57.6
06	266 33.0	S10 56.7
07	281 33.1	55.8
08	296 33.1	54.9
09	311 33.2 ··	54.0
10	326 33.3	53.1
11	341 33.3	52.2
12	356 33.4	S10 51.3
13	11 33.5	50.4
14	26 33.5	49.5
15	41 33.6 ··	48.6
16	56 33.7	47.7
17	71 33.7	46.8
18	86 33.8	S10 45.9
19	101 33.9	45.0
20	116 33.9	44.1
21	131 34.0 ··	43.2
22	146 34.1	42.3
23	161 34.2	41.4

Day 21–23

d h	G.H.A.	Dec.
21 00	176 34.2	S10 40.5
01	191 34.3	39.6
02	206 34.4	38.7
03	221 34.4 ··	37.8
04	236 34.5	36.9
05	251 34.6	36.0
06	266 34.7	S10 35.1
07	281 34.7	34.2
08	296 34.8	33.3
09	311 34.9 ··	32.4
10	326 34.9	31.5
11	341 35.0	30.6
12	356 35.1	S10 29.6
13	11 35.2	28.7
14	26 35.2	27.8
15	41 35.3 ··	26.9
16	56 35.4	26.0
17	71 35.5	25.1
18	86 35.5	S10 24.2
19	101 35.6	23.3
20	116 35.7	22.4
21	131 35.8 ··	21.5
22	146 35.8	20.6
23	161 35.9	19.7
22 00	176 36.0	S10 18.8
01	191 36.1	17.8
02	206 36.2	16.9
03	221 36.2 ··	16.0
04	236 36.3	15.1
05	251 36.4	14.2
06	266 36.5	S10 13.3
07	281 36.5	12.4
08	296 36.6	11.5
09	311 36.7 ··	10.6
10	326 36.8	09.7
11	341 36.9	08.7
12	356 36.9	S10 07.8
13	11 37.0	06.9
14	26 37.1	06.0
15	41 37.2 ··	05.1
16	56 37.3	04.2
17	71 37.4	03.3
18	86 37.4	S10 02.3
19	101 37.5	01.4
20	116 37.6	10 00.5
21	131 37.7 ··	9 59.6
22	146 37.8	58.7
23	161 37.8	57.8
23 00	176 37.9	S 9 56.9
01	191 38.0	55.9
02	206 38.1	55.0
03	221 38.2 ··	54.1
04	236 38.3	53.2
05	251 38.4	52.3
06	266 38.4	S 9 51.4
07	281 38.5	50.4
08	296 38.6	49.5
09	311 38.7 ··	48.6
10	326 38.8	47.7
11	341 38.9	46.8
12	356 39.0	S 9 45.9
13	11 39.1	44.9
14	26 39.1	44.0
15	41 39.2 ··	43.1
16	56 39.3	42.2
17	71 39.4	41.3
18	86 39.5	S 9 40.3
19	101 39.6	39.4
20	116 39.7	38.5
21	131 39.8 ··	37.6
22	146 39.8	36.7
23	161 39.9	35.7

d h	G.H.A. ° ′	Dec. ° ′
24 00	176 40.0	S 9 34.8
01	191 40.1	33.9
02	206 40.2	33.0
03	221 40.3	·· 32.0
04	236 40.4	31.1
05	251 40.5	30.2
06	266 40.6	S 9 29.3
07	281 40.7	28.4
08	296 40.8	27.4
09	311 40.8	·· 26.5
10	326 40.9	25.6
11	341 41.0	24.7
12	356 41.1	S 9 23.7
13	11 41.2	22.8
14	26 41.3	21.9
15	41 41.4	·· 21.0
16	56 41.5	20.0
17	71 41.6	19.1
18	86 41.7	S 9 18.2
19	101 41.8	17.3
20	116 41.9	16.3
21	131 42.0	·· 15.4
22	146 42.1	14.5
23	161 42.2	13.5
25 00	176 42.3	S 9 12.6
01	191 42.4	11.7
02	206 42.5	10.8
03	221 42.6	·· 09.8
04	236 42.7	08.9
05	251 42.8	08.0
06	266 42.9	S 9 07.1
07	281 43.0	06.1
08	296 43.1	05.2
09	311 43.1	·· 04.3
10	326 43.2	03.3
11	341 43.3	02.4
12	356 43.4	S 9 01.5
13	11 43.5	9 00.5
14	26 43.6	8 59.6
15	41 43.7	·· 58.7
16	56 43.8	57.8
17	71 43.9	56.8
18	86 44.1	S 8 55.9
19	101 44.2	55.0
20	116 44.3	54.0
21	131 44.4	·· 53.1
22	146 44.5	52.2
23	161 44.6	51.2
26 00	176 44.7	S 8 50.3
01	191 44.8	49.4
02	206 44.9	48.4
03	221 45.0	·· 47.5
04	236 45.1	46.6
05	251 45.2	45.6
06	266 45.3	S 8 44.7
07	281 45.4	43.8
08	296 45.5	42.8
09	311 45.6	·· 41.9
10	326 45.7	41.0
11	341 45.8	40.0
12	356 45.9	S 8 39.1
13	11 46.0	38.1
14	26 46.1	37.2
15	41 46.2	·· 36.3
16	56 46.3	35.3
17	71 46.4	34.4
18	86 46.6	S 8 33.5
19	101 46.7	32.5
20	116 46.8	31.6
21	131 46.9	·· 30.6
22	146 47.0	29.7
23	161 47.1	28.8

d h	G.H.A. ° ′	Dec. ° ′
27 00	176 47.2	S 8 27.8
01	191 47.3	26.9
02	206 47.4	26.0
03	221 47.5	·· 25.0
04	236 47.6	24.1
05	251 47.8	23.1
06	266 47.9	S 8 22.2
07	281 48.0	21.3
08	296 48.1	20.3
09	311 48.2	·· 19.4
10	326 48.3	18.4
11	341 48.4	17.5
12	356 48.5	S 8 16.6
13	11 48.6	15.6
14	26 48.8	14.7
15	41 48.9	·· 13.7
16	56 49.0	12.8
17	71 49.1	11.9
18	86 49.2	S 8 10.9
19	101 49.3	10.0
20	116 49.4	09.0
21	131 49.5	·· 08.1
22	146 49.7	07.1
23	161 49.8	06.2
28 00	176 49.9	S 8 05.3
01	191 50.0	04.3
02	206 50.1	03.4
03	221 50.2	·· 02.4
04	236 50.4	01.5
05	251 50.5	8 00.5
06	266 50.6	S 7 59.6
07	281 50.7	58.6
08	296 50.8	57.7
09	311 50.9	·· 56.8
10	326 51.0	55.8
11	341 51.2	54.9
12	356 51.3	S 7 53.9
13	11 51.4	53.0
14	26 51.5	52.0
15	41 51.6	·· 51.1
16	56 51.8	50.1
17	71 51.9	49.2
18	86 52.0	S 7 48.2
19	101 52.1	47.3
20	116 52.2	46.3
21	131 52.4	·· 45.4
22	146 52.5	44.5
23	161 52.6	43.5
MARCH 1 00	176 52.7	S 7 42.6
01	191 52.8	41.6
02	206 53.0	40.7
03	221 53.1	·· 39.7
04	236 53.2	38.8
05	251 53.3	37.8
06	266 53.4	S 7 36.9
07	281 53.6	35.9
08	296 53.7	35.0
09	311 53.8	·· 34.0
10	326 53.9	33.1
11	341 54.1	32.1
12	356 54.2	S 7 31.2
13	11 54.3	30.2
14	26 54.4	29.3
15	41 54.5	·· 28.3
16	56 54.7	27.4
17	71 54.8	26.4
18	86 54.9	S 7 25.5
19	101 55.0	24.5
20	116 55.2	23.6
21	131 55.3	·· 22.6
22	146 55.4	21.7
23	161 55.5	20.7

d h	G.H.A. ° ′	Dec. ° ′
2 00	176 55.7	S 7 19.7
01	191 55.8	18.8
02	206 55.9	17.8
03	221 56.0	·· 16.9
04	236 56.2	15.9
05	251 56.3	15.0
06	266 56.4	S 7 14.0
07	281 56.6	13.1
08	296 56.7	12.1
09	311 56.8	·· 11.2
10	326 56.9	10.2
11	341 57.1	09.3
12	356 57.2	S 7 08.3
13	11 57.3	07.3
14	26 57.5	06.4
15	41 57.6	·· 05.4
16	56 57.7	04.5
17	71 57.8	03.5
18	86 58.0	S 7 02.6
19	101 58.1	01.6
20	116 58.2	7 00.7
21	131 58.4	6 59.7
22	146 58.5	58.7
23	161 58.6	57.8
3 00	176 58.8	S 6 56.8
01	191 58.9	55.9
02	206 59.0	54.9
03	221 59.1	·· 54.0
04	236 59.3	53.0
05	251 59.4	52.0
06	266 59.5	S 6 51.1
07	281 59.7	50.1
08	296 59.8	49.2
09	311 59.9	·· 48.2
10	327 00.1	47.3
11	342 00.2	46.3
12	357 00.3	S 6 45.3
13	12 00.5	44.4
14	27 00.6	43.4
15	42 00.7	·· 42.5
16	57 00.9	41.5
17	72 01.0	40.5
18	87 01.1	S 6 39.6
19	102 01.3	38.6
20	117 01.4	37.7
21	132 01.5	·· 36.7
22	147 01.7	35.7
23	162 01.8	34.8
4 00	177 02.0	S 6 33.8
01	192 02.1	32.9
02	207 02.2	31.9
03	222 02.4	·· 30.9
04	237 02.5	30.0
05	252 02.6	29.0
06	267 02.8	S 6 28.1
07	282 02.9	27.1
08	297 03.1	26.1
09	312 03.2	·· 25.2
10	327 03.4	24.2
11	342 03.5	23.2
12	357 03.6	S 6 22.3
13	12 03.7	21.3
14	27 03.9	20.4
15	42 04.0	·· 19.4
16	57 04.2	18.4
17	72 04.3	17.5
18	87 04.4	S 6 16.5
19	102 04.6	15.5
20	117 04.7	14.6
21	132 04.9	·· 13.6
22	147 05.0	12.7
23	162 05.1	11.7

d h	G.H.A. ° '	Dec. ° '
5 00	177 05.3	S 6 10.7
01	192 05.4	09.8
02	207 05.6	08.8
03	222 05.7	·· 07.8
04	237 05.8	06.9
05	252 06.0	05.9
06	267 06.1	S 6 04.9
07	282 06.3	04.0
08	297 06.4	03.0
09	312 06.6	·· 02.0
10	327 06.7	01.1
11	342 06.8	6 00.1
12	357 07.0	S 5 59.1
13	12 07.1	58.2
14	27 07.3	57.2
15	42 07.4	·· 56.2
16	57 07.6	55.3
17	72 07.7	54.3
18	87 07.8	S 5 53.3
19	102 08.0	52.4
20	117 08.1	51.4
21	132 08.3	·· 50.4
22	147 08.4	49.5
23	162 08.6	48.5
6 00	177 08.7	S 5 47.5
01	192 08.9	46.6
02	207 09.0	45.6
03	222 09.1	·· 44.6
04	237 09.3	43.7
05	252 09.4	42.7
06	267 09.6	S 5 41.7
07	282 09.7	40.8
08	297 09.9	39.8
09	312 10.0	·· 38.8
10	327 10.2	37.9
11	342 10.3	36.9
12	357 10.5	S 5 35.9
13	12 10.6	34.9
14	27 10.8	34.0
15	42 10.9	·· 33.0
16	57 11.1	32.0
17	72 11.2	31.1
18	87 11.4	S 5 30.1
19	102 11.5	29.1
20	117 11.7	28.2
21	132 11.8	·· 27.2
22	147 12.0	26.2
23	162 12.1	25.2
7 00	177 12.3	S 5 24.3
01	192 12.4	23.3
02	207 12.6	22.3
03	222 12.7	·· 21.4
04	237 12.9	20.4
05	252 13.0	19.4
06	267 13.2	S 5 18.4
07	282 13.3	17.5
08	297 13.5	16.5
09	312 13.6	·· 15.5
10	327 13.8	14.6
11	342 13.9	13.6
12	357 14.1	S 5 12.6
13	12 14.2	11.6
14	27 14.4	10.7
15	42 14.5	·· 09.7
16	57 14.7	08.7
17	72 14.8	07.8
18	87 15.0	S 5 06.8
19	102 15.1	05.8
20	117 15.3	04.8
21	132 15.4	·· 03.9
22	147 15.6	02.9
23	162 15.7	01.9

d h	G.H.A. ° '	Dec. ° '
8 00	177 15.9	S 5 00.9
01	192 16.0	5 00.0
02	207 16.2	4 59.0
03	222 16.4	·· 58.0
04	237 16.5	57.0
05	252 16.7	56.1
06	267 16.8	S 4 55.1
07	282 17.0	54.1
08	297 17.1	53.1
09	312 17.3	·· 52.2
10	327 17.4	51.2
11	342 17.6	50.2
12	357 17.7	S 4 49.2
13	12 17.9	48.3
14	27 18.1	47.3
15	42 18.2	·· 46.3
16	57 18.4	45.3
17	72 18.5	44.4
18	87 18.7	S 4 43.4
19	102 18.8	42.4
20	117 19.0	41.4
21	132 19.1	·· 40.5
22	147 19.3	39.5
23	162 19.5	38.5
9 00	177 19.6	S 4 37.5
01	192 19.8	36.6
02	207 19.9	35.6
03	222 20.1	·· 34.6
04	237 20.2	33.6
05	252 20.4	32.7
06	267 20.6	S 4 31.7
07	282 20.7	30.7
08	297 20.9	29.7
09	312 21.0	·· 28.7
10	327 21.2	27.8
11	342 21.4	26.8
12	357 21.5	S 4 25.8
13	12 21.7	24.8
14	27 21.8	23.9
15	42 22.0	·· 22.9
16	57 22.2	21.9
17	72 22.3	20.9
18	87 22.5	S 4 19.9
19	102 22.6	19.0
20	117 22.8	18.0
21	132 23.0	·· 17.0
22	147 23.1	16.0
23	162 23.3	15.1
10 00	177 23.4	S 4 14.1
01	192 23.6	13.1
02	207 23.8	12.1
03	222 23.9	·· 11.1
04	237 24.1	10.2
05	252 24.2	09.2
06	267 24.4	S 4 08.2
07	282 24.6	07.2
08	297 24.7	06.2
09	312 24.9	·· 05.3
10	327 25.0	04.3
11	342 25.2	03.3
12	357 25.4	S 4 02.3
13	12 25.5	01.3
14	27 25.7	4 00.4
15	42 25.9	3 59.4
16	57 26.0	58.4
17	72 26.2	57.4
18	87 26.3	S 3 56.4
19	102 26.5	55.5
20	117 26.7	54.5
21	132 26.8	·· 53.5
22	147 27.0	52.5
23	162 27.2	51.5

d h	G.H.A. ° '	Dec. ° '
11 00	177 27.3	S 3 50.6
01	192 27.5	49.6
02	207 27.7	48.6
03	222 27.8	·· 47.6
04	237 28.0	46.6
05	252 28.2	45.7
06	267 28.3	S 3 44.7
07	282 28.5	43.7
08	297 28.6	42.7
09	312 28.8	·· 41.7
10	327 29.0	40.8
11	342 29.1	39.8
12	357 29.3	S 3 38.8
13	12 29.5	37.8
14	27 29.6	36.8
15	42 29.8	·· 35.8
16	57 30.0	34.9
17	72 30.1	33.9
18	87 30.3	S 3 32.9
19	102 30.5	31.9
20	117 30.6	30.9
21	132 30.8	·· 29.9
22	147 31.0	29.0
23	162 31.1	28.0
12 00	177 31.3	S 3 27.0
01	192 31.5	26.0
02	207 31.6	25.0
03	222 31.8	·· 24.1
04	237 32.0	23.1
05	252 32.1	22.1
06	267 32.3	S 3 21.1
07	282 32.5	20.1
08	297 32.6	19.1
09	312 32.8	·· 18.2
10	327 33.0	17.2
11	342 33.1	16.2
12	357 33.3	S 3 15.2
13	12 33.5	14.2
14	27 33.6	13.2
15	42 33.8	·· 12.3
16	57 34.0	11.3
17	72 34.2	10.3
18	87 34.3	S 3 09.3
19	102 34.5	08.3
20	117 34.7	07.3
21	132 34.8	·· 06.3
22	147 35.0	05.4
23	162 35.2	04.4
13 00	177 35.3	S 3 03.4
01	192 35.5	02.4
02	207 35.7	01.4
03	222 35.8	3 00.4
04	237 36.0	2 59.5
05	252 36.2	58.5
06	267 36.4	S 2 57.5
07	282 36.5	56.5
08	297 36.7	55.5
09	312 36.9	·· 54.5
10	327 37.0	53.5
11	342 37.2	52.6
12	357 37.4	S 2 51.6
13	12 37.6	50.6
14	27 37.7	49.6
15	42 37.9	·· 48.6
16	57 38.1	47.6
17	72 38.2	46.7
18	87 38.4	S 2 45.7
19	102 38.6	44.7
20	117 38.8	43.7
21	132 38.9	·· 42.7
22	147 39.1	41.7
23	162 39.3	40.7

d h	G.H.A.	Dec.
14 00	177 39.4	S 2 39.8
01	192 39.6	38.8
02	207 39.8	37.8
03	222 40.0	·· 36.8
04	237 40.1	35.8
05	252 40.3	34.8
06	267 40.5	S 2 33.8
07	282 40.6	32.9
08	297 40.8	31.9
09	312 41.0	·· 30.9
10	327 41.2	29.9
11	342 41.3	28.9
12	357 41.5	S 2 27.9
13	12 41.7	26.9
14	27 41.9	26.0
15	42 42.0	·· 25.0
16	57 42.2	24.0
17	72 42.4	23.0
18	97 42.6	S 2 22.0
19	102 42.7	21.0
20	117 42.9	20.0
21	132 43.1	·· 19.0
22	147 43.3	18.1
23	162 43.4	17.1
15 00	177 43.6	S 2 16.1
01	192 43.8	15.1
02	207 44.0	14.1
03	222 44.1	·· 13.1
04	237 44.3	12.1
05	252 44.5	11.1
06	267 44.7	S 2 10.2
07	282 44.8	09.2
08	297 45.0	08.2
09	312 45.2	·· 07.2
10	327 45.4	06.2
11	342 45.5	05.2
12	357 45.7	S 2 04.2
13	12 45.9	03.3
14	27 46.1	02.3
15	42 46.2	·· 01.3
16	57 46.4	2 00.3
17	72 46.6	1 59.3
18	87 46.8	S 1 58.3
19	102 46.9	57.3
20	117 47.1	56.3
21	132 47.3	·· 55.4
22	147 47.5	54.4
23	162 47.6	53.4
16 00	177 47.8	S 1 52.4
01	192 48.0	51.4
02	207 48.2	50.4
03	222 48.4	·· 49.4
04	237 48.5	48.4
05	252 48.7	47.5
06	267 48.9	S 1 46.5
07	282 49.1	45.5
08	297 49.2	44.5
09	312 49.4	·· 43.5
10	327 49.6	42.5
11	342 49.8	41.5
12	357 50.0	S 1 40.5
13	12 50.1	39.5
14	27 50.3	38.6
15	42 50.5	·· 37.6
16	57 50.7	36.6
17	72 50.8	35.6
18	87 51.0	S 1 34.6
19	102 51.2	33.6
20	117 51.4	32.6
21	132 51.6	·· 31.6
22	147 51.7	30.7
23	162 51.9	29.7

d h	G.H.A.	Dec.
17 00	177 52.1	S 1 28.7
01	192 52.3	27.7
02	207 52.4	26.7
03	222 52.6	·· 25.7
04	237 52.8	24.7
05	252 53.0	23.7
06	267 53.2	S 1 22.7
07	282 53.3	21.8
08	297 53.5	20.8
09	312 53.7	·· 19.8
10	327 53.9	18.8
11	342 54.1	17.8
12	357 54.2	S 1 16.8
13	12 54.4	15.8
14	27 54.6	14.8
15	42 54.8	·· 13.9
16	57 55.0	12.9
17	72 55.1	11.9
18	87 55.3	S 1 10.9
19	102 55.5	09.9
20	117 55.7	08.9
21	132 55.9	·· 07.9
22	147 56.0	06.9
23	162 56.2	05.9
18 00	177 56.4	S 1 05.0
01	192 56.6	04.0
02	207 56.8	03.0
03	222 57.0	·· 02.0
04	237 57.1	01.0
05	252 57.3	1 00.0
06	267 57.5	S 0 59.0
07	282 57.7	58.0
08	297 57.9	57.0
09	312 58.0	·· 56.1
10	327 58.2	55.1
11	342 58.4	54.1
12	357 58.6	S 0 53.1
13	12 58.8	52.1
14	27 59.0	51.1
15	42 59.1	·· 50.1
16	57 59.3	49.1
17	72 59.5	48.1
18	87 59.7	S 0 47.2
19	102 59.9	46.2
20	118 00.0	45.2
21	133 00.2	·· 44.2
22	148 00.4	43.2
23	163 00.6	42.2
19 00	178 00.8	S 0 41.2
01	193 01.0	40.2
02	208 01.1	39.2
03	223 01.3	·· 38.3
04	238 01.5	37.3
05	253 01.7	36.3
06	268 01.9	S 0 35.3
07	283 02.1	34.3
08	298 02.2	33.3
09	313 02.4	·· 32.3
10	328 02.6	31.3
11	343 02.8	30.3
12	358 03.0	S 0 29.4
13	13 03.2	28.4
14	28 03.3	27.4
15	43 03.5	·· 26.4
16	58 03.7	25.4
17	73 03.9	24.4
18	88 04.1	S 0 23.4
19	103 04.3	22.4
20	118 04.4	21.5
21	133 04.6	·· 20.5
22	148 04.8	19.5
23	163 05.0	18.5

d h	G.H.A.	Dec.
20 00	178 05.2	S 0 17.5
01	193 05.4	16.5
02	208 05.5	15.5
03	223 05.7	·· 14.5
04	238 05.9	13.5
05	253 06.1	12.6
06	268 06.3	S 0 11.6
07	283 06.5	10.6
08	298 06.7	09.6
09	313 06.8	·· 08.6
10	328 07.0	07.6
11	343 07.2	06.6
12	358 07.4	S 0 05.6
13	13 07.6	04.6
14	28 07.8	03.7
15	43 08.0	·· 02.7
16	58 08.1	01.7
17	73 08.3	S 0 00.7
18	88 08.5	N 0 00.3
19	103 08.7	01.3
20	118 08.9	02.3
21	133 09.1	·· 03.3
22	148 09.3	04.2
23	163 09.4	05.2
21 00	178 09.6	N 0 06.2
01	193 09.8	07.2
02	208 10.0	08.2
03	223 10.2	·· 09.2
04	238 10.4	10.2
05	253 10.6	11.2
06	268 10.7	N 0 12.1
07	283 10.9	13.1
08	298 11.1	14.1
09	313 11.3	·· 15.1
10	328 11.5	16.1
11	343 11.7	17.1
12	358 11.9	N 0 18.1
13	13 12.0	19.1
14	28 12.2	20.0
15	43 12.4	·· 21.0
16	58 12.6	22.0
17	73 12.8	23.0
18	88 13.0	N 0 24.0
19	103 13.2	25.0
20	118 13.4	26.0
21	133 13.5	·· 27.0
22	148 13.7	27.9
23	163 13.9	28.9
22 00	178 14.1	N 0 29.9
01	193 14.3	30.9
02	208 14.5	31.9
03	223 14.7	·· 32.9
04	238 14.8	33.9
05	253 15.0	34.9
06	268 15.2	N 0 35.8
07	283 15.4	36.8
08	298 15.6	37.8
09	313 15.8	·· 38.8
10	328 16.0	39.8
11	343 16.2	40.8
12	358 16.3	N 0 41.8
13	13 16.5	42.8
14	28 16.7	43.7
15	43 16.9	·· 44.7
16	58 17.1	45.7
17	73 17.3	46.7
18	88 17.5	N 0 47.7
19	103 17.7	48.7
20	118 17.8	49.7
21	133 18.0	·· 50.6
22	148 18.2	51.6
23	163 18.4	52.6

	G.H.A.	Dec.
d h	° ′	° ′
23 00	178 18.6	N 0 53.6
01	193 18.8	54.6
02	208 19.0	55.6
03	223 19.2	·· 56.6
04	238 19.4	57.5
05	253 19.5	58.5
06	268 19.7	N 0 59.5
07	283 19.9	1 00.5
08	298 20.1	01.5
09	313 20.3	·· 02.5
10	328 20.5	03.5
11	343 20.7	04.4
12	358 20.9	N 1 05.4
13	13 21.0	06.4
14	28 21.2	07.4
15	43 21.4	·· 08.4
16	58 21.6	09.4
17	73 21.8	10.4
18	88 22.0	N 1 11.3
19	103 22.2	12.3
20	118 22.4	13.3
21	133 22.5	·· 14.3
22	148 22.7	15.3
23	163 22.9	16.3
24 00	178 23.1	N 1 17.2
01	193 23.3	18.2
02	208 23.5	19.2
03	223 23.7	·· 20.2
04	238 23.9	21.2
05	253 24.1	22.2
06	268 24.2	N 1 23.2
07	283 24.4	24.1
08	298 24.6	25.1
09	313 24.8	·· 26.1
10	328 25.0	27.1
11	343 25.2	28.1
12	358 25.4	N 1 29.1
13	13 25.6	30.0
14	28 25.7	31.0
15	43 25.9	·· 32.0
16	58 26.1	33.0
17	73 26.3	34.0
18	88 26.5	N 1 35.0
19	103 26.7	35.9
20	118 26.9	36.9
21	133 27.1	·· 37.9
22	148 27.2	38.9
23	163 27.4	39.9
25 00	178 27.6	N 1 40.9
01	193 27.8	41.8
02	208 28.0	42.8
03	223 28.2	·· 43.8
04	238 28.4	44.8
05	253 28.6	45.8
06	268 28.8	N 1 46.8
07	283 29.0	47.7
08	298 29.2	48.7
09	313 29.3	·· 49.7
10	328 29.5	50.7
11	343 29.7	51.7
12	358 29.9	N 1 52.6
13	13 30.1	53.6
14	28 30.3	54.6
15	43 30.5	·· 55.6
16	58 30.7	56.6
17	73 30.9	57.6
18	88 31.0	N 1 58.5
19	103 31.2	1 59.5
20	118 31.4	2 00.5
21	133 31.6	·· 01.5
22	148 31.8	02.5
23	163 32.0	03.5

	G.H.A.	Dec.
d h	° ′	° ′
26 00	178 32.2	N 2 04.4
01	193 32.4	05.4
02	208 32.6	06.4
03	223 32.8	·· 07.4
04	238 32.9	08.4
05	253 33.1	09.3
06	268 33.3	N 2 10.3
07	283 33.5	11.3
08	298 33.7	12.3
09	313 33.9	·· 13.3
10	328 34.1	14.2
11	343 34.3	15.2
12	358 34.5	N 2 16.2
13	13 34.7	17.2
14	28 34.8	18.2
15	43 35.0	·· 19.1
16	58 35.2	20.1
17	73 35.4	21.1
18	88 35.6	N 2 22.1
19	103 35.8	23.1
20	118 36.0	24.0
21	133 36.2	·· 25.0
22	148 36.4	26.0
23	163 36.5	27.0
27 00	178 36.7	N 2 28.0
01	193 36.9	28.9
02	208 37.1	29.9
03	223 37.3	·· 30.9
04	238 37.5	31.9
05	253 37.7	32.9
06	268 37.9	N 2 33.8
07	283 38.1	34.8
08	298 38.3	35.8
09	313 38.4	·· 36.8
10	328 38.6	37.8
11	343 38.8	38.7
12	358 39.0	N 2 39.7
13	13 39.2	40.7
14	28 39.4	41.7
15	43 39.6	·· 42.6
16	58 39.8	43.6
17	73 40.0	44.6
18	88 40.2	N 2 45.6
19	103 40.3	46.6
20	118 40.5	47.5
21	133 40.7	·· 48.5
22	148 40.9	49.5
23	163 41.1	50.5
28 00	178 41.3	N 2 51.4
01	193 41.5	52.4
02	208 41.7	53.4
03	223 41.9	·· 54.4
04	238 42.0	55.3
05	253 42.2	56.3
06	268 42.4	N 2 57.3
07	283 42.6	58.3
08	298 42.8	2 59.3
09	313 43.0	3 00.2
10	328 43.2	01.2
11	343 43.4	02.2
12	358 43.6	N 3 03.2
13	13 43.8	04.1
14	28 43.9	05.1
15	43 44.1	·· 06.1
16	58 44.3	07.1
17	73 44.5	08.0
18	88 44.7	N 3 09.0
19	103 44.9	10.0
20	118 45.1	11.0
21	133 45.3	·· 11.9
22	148 45.5	12.9
23	163 45.7	13.9

	G.H.A.	Dec.
d h	° ′	° ′
29 00	178 45.8	N 3 14.9
01	193 46.0	15.8
02	208 46.2	16.8
03	223 46.4	·· 17.8
04	238 46.6	18.8
05	253 46.8	19.7
06	268 47.0	N 3 20.7
07	283 47.2	21.7
08	298 47.4	22.6
09	313 47.5	·· 23.6
10	328 47.7	24.6
11	343 47.9	25.6
12	358 48.1	N 3 26.5
13	13 48.3	27.5
14	28 48.5	28.5
15	43 48.7	·· 29.5
16	58 48.9	30.4
17	73 49.1	31.4
18	88 49.3	N 3 32.4
19	103 49.4	33.4
20	118 49.6	34.3
21	133 49.8	·· 35.3
22	148 50.0	36.3
23	163 50.2	37.2
30 00	178 50.4	N 3 38.2
01	193 50.6	39.2
02	208 50.8	40.2
03	223 51.0	·· 41.1
04	238 51.1	42.1
05	253 51.3	43.1
06	268 51.5	N 3 44.0
07	283 51.7	45.0
08	298 51.9	46.0
09	313 52.1	47.0
10	328 52.3	47.9
11	343 52.5	48.9
12	358 52.7	N 3 49.9
13	13 52.8	50.8
14	28 53.0	51.8
15	43 53.2	·· 52.8
16	58 53.4	53.7
17	73 53.6	54.7
18	88 53.8	N 3 55.7
19	103 54.0	56.6
20	118 54.2	57.6
21	133 54.4	·· 58.6
22	148 54.5	3 59.6
23	163 54.7	4 00.5
31 00	178 54.9	N 4 01.5
01	193 55.1	02.5
02	208 55.3	03.4
03	223 55.5	·· 04.4
04	238 55.7	05.4
05	253 55.9	06.3
06	268 56.1	N 4 07.3
07	283 56.2	08.3
08	298 56.4	09.2
09	313 56.6	·· 10.2
10	328 56.8	11.2
11	343 57.0	12.1
12	358 57.2	N 4 13.1
13	13 57.4	14.1
14	28 57.6	15.0
15	43 57.8	·· 16.0
16	58 57.9	17.0
17	73 58.1	17.9
18	88 58.3	N 4 18.9
19	103 58.5	19.9
20	118 58.7	20.8
21	133 58.9	·· 21.8
22	148 59.1	22.8
23	163 59.3	23.7

Day 1

d h	G.H.A.	Dec.
1 00	178 59.4 N	4 24.7
01	193 59.6	25.7
02	208 59.8	26.6
03	224 00.0 ··	27.6
04	239 00.2	28.6
05	254 00.4	29.5
06	269 00.6 N	4 30.5
07	284 00.8	31.5
08	299 00.9	32.4
09	314 01.1 ··	33.4
10	329 01.3	34.3
11	344 01.5	35.3
12	359 01.7 N	4 36.3
13	14 01.9	37.2
14	29 02.1	38.2
15	44 02.3 ··	39.2
16	59 02.4	40.1
17	74 02.6	41.1
18	89 02.8 N	4 42.1
19	104 03.0	43.0
20	119 03.2	44.0
21	134 03.4 ··	44.9
22	149 03.6	45.9
23	164 03.8	46.9

Day 2

d h	G.H.A.	Dec.
2 00	179 03.9 N	4 47.8
01	194 04.1	48.8
02	209 04.3	49.8
03	224 04.5 ··	50.7
04	239 04.7	51.7
05	254 04.9	52.6
06	269 05.1 N	4 53.6
07	284 05.2	54.6
08	299 05.4	55.5
09	314 05.6 ··	56.5
10	329 05.8	57.4
11	344 06.0	58.4
12	359 06.2 N	4 59.4
13	14 06.4	5 00.3
14	29 06.6	01.3
15	44 06.7 ··	02.2
16	59 06.9	03.2
17	74 07.1	04.2
18	89 07.3 N	5 05.1
19	104 07.5	06.1
20	119 07.7	07.0
21	134 07.9 ··	08.0
22	149 08.0	09.0
23	164 08.2	09.9

Day 3

d h	G.H.A.	Dec.
3 00	179 08.4 N	5 10.9
01	194 08.6	11.8
02	209 08.8	12.8
03	224 09.0 ··	13.7
04	239 09.2	14.7
05	254 09.3	15.7
06	269 09.5 N	5 16.6
07	284 09.7	17.6
08	299 09.9	18.5
09	314 10.1 ··	19.5
10	329 10.3	20.4
11	344 10.4	21.4
12	359 10.6 N	5 22.4
13	14 10.8	23.3
14	29 11.0	24.3
15	44 11.2 ··	25.2
16	59 11.4	26.2
17	74 11.6	27.1
18	89 11.7 N	5 28.1
19	104 11.9	29.1
20	119 12.1	30.0
21	134 12.3 ··	31.0
22	149 12.5	31.9
23	164 12.7	32.9

Day 4

d h	G.H.A.	Dec.
4 00	179 12.8 N	5 33.8
01	194 13.0	34.8
02	209 13.2	35.7
03	224 13.4 ··	36.7
04	239 13.6	37.6
05	254 13.8	38.6
06	269 14.0 N	5 39.5
07	284 14.1	40.5
08	299 14.3	41.5
09	314 14.5 ··	42.4
10	329 14.7	43.4
11	344 14.9	44.3
12	359 15.1 N	5 45.3
13	14 15.2	46.2
14	29 15.4	47.2
15	44 15.6 ··	48.1
16	59 15.8	49.1
17	74 16.0	50.0
18	89 16.2 N	5 51.0
19	104 16.3	51.9
20	119 16.5	52.9
21	134 16.7 ··	53.8
22	149 16.9	54.8
23	164 17.1	55.7

Day 5

d h	G.H.A.	Dec.
5 00	179 17.2 N	5 56.7
01	194 17.4	57.6
02	209 17.6	58.6
03	224 17.8	5 59.5
04	239 18.0	6 00.5
05	254 18.2	01.4
06	269 18.3 N	6 02.4
07	284 18.5	03.3
08	299 18.7	04.3
09	314 18.9 ··	05.2
10	329 19.1	06.2
11	344 19.2	07.1
12	359 19.4 N	6 08.1
13	14 19.6	09.0
14	29 19.8	10.0
15	44 20.0 ··	10.9
16	59 20.2	11.9
17	74 20.3	12.8
18	89 20.5 N	6 13.8
19	104 20.7	14.7
20	119 20.9	15.7
21	134 21.1 ··	16.6
22	149 21.2	17.5
23	164 21.4	18.5

Day 6

d h	G.H.A.	Dec.
6 00	179 21.6 N	6 19.4
01	194 21.8	20.4
02	209 22.0	21.3
03	224 22.1 ··	22.3
04	239 22.3	23.2
05	254 22.5	24.2
06	269 22.7 N	6 25.1
07	284 22.9	26.1
08	299 23.0	27.0
09	314 23.2 ··	27.9
10	329 23.4	28.9
11	344 23.6	29.8
12	359 23.8 N	6 30.8
13	14 23.9	31.7
14	29 24.1	32.7
15	44 24.3 ··	33.6
16	59 24.5	34.5
17	74 24.6	35.5
18	89 24.8 N	6 36.4
19	104 25.0	37.4
20	119 25.2	38.3
21	134 25.4 ··	39.3
22	149 25.5	40.2
23	164 25.7	41.1

Day 7

d h	G.H.A.	Dec.
7 00	179 25.9 N	6 42.1
01	194 26.1	43.0
02	209 26.3	44.0
03	224 26.4 ··	44.9
04	239 26.6	45.8
05	254 26.8	46.8
06	269 27.0 N	6 47.7
07	284 27.1	48.7
08	299 27.3	49.6
09	314 27.5 ··	50.5
10	329 27.7	51.5
11	344 27.9	52.4
12	359 28.0 N	6 53.4
13	14 28.2	54.3
14	29 28.4	55.2
15	44 28.6 ··	56.2
16	59 28.7	57.1
17	74 28.9	58.1
18	89 29.1 N	6 59.0
19	104 29.3	6 59.9
20	119 29.4	7 00.9
21	134 29.6 ··	01.8
22	149 29.8	02.7
23	164 30.0	03.7

Day 8

d h	G.H.A.	Dec.
8 00	179 30.1 N	7 04.6
01	194 30.3	05.6
02	209 30.5	06.5
03	224 30.7 ··	07.4
04	239 30.8	08.4
05	254 31.0	09.3
06	269 31.2 N	7 10.2
07	284 31.4	11.2
08	299 31.5	12.1
09	314 31.7 ··	13.0
10	329 31.9	14.0
11	344 32.1	14.9
12	359 32.2 N	7 15.8
13	14 32.4	16.8
14	29 32.6	17.7
15	44 32.8 ··	18.6
16	59 32.9	19.6
17	74 33.1	20.5
18	89 33.3 N	7 21.4
19	104 33.5	22.4
20	119 33.6	23.3
21	134 33.8 ··	24.2
22	149 34.0	25.2
23	164 34.1	26.1

Day 9

d h	G.H.A.	Dec.
9 00	179 34.3 N	7 27.0
01	194 34.5	28.0
02	209 34.7	28.9
03	224 34.8 ··	29.8
04	239 35.0	30.8
05	254 35.2	31.7
06	269 35.4 N	7 32.6
07	284 35.5	33.5
08	299 35.7	34.5
09	314 35.9 ··	35.4
10	329 36.0	36.3
11	344 36.2	37.3
12	359 36.4 N	7 38.2
13	14 36.6	39.1
14	29 36.7	40.0
15	44 36.9 ··	41.0
16	59 37.1	41.9
17	74 37.2	42.8
18	89 37.4 N	7 43.8
19	104 37.6	44.7
20	119 37.8	45.6
21	134 37.9 ··	46.5
22	149 38.1	47.5
23	164 38.3	48.4

Day 10

d h	G.H.A.	Dec.
10 00	179 38.4	N 7 49.3
01	194 38.6	50.2
02	209 38.8	51.2
03	224 38.9 ··	52.1
04	239 39.1	53.0
05	254 39.3	53.9
06	269 39.5	N 7 54.9
07	284 39.6	55.8
08	299 39.8	56.7
09	314 40.0 ··	57.6
10	329 40.1	58.6
11	344 40.3	7 59.5
12	359 40.5	N 8 00.4
13	14 40.6	01.3
14	29 40.8	02.3
15	44 41.0 ··	03.2
16	59 41.1	04.1
17	74 41.3	05.0
18	89 41.5	N 8 06.0
19	104 41.6	06.9
20	119 41.8	07.8
21	134 42.0 ··	08.7
22	149 42.1	09.6
23	164 42.3	10.6

Day 11

d h	G.H.A.	Dec.
11 00	179 42.5	N 8 11.5
01	194 42.6	12.4
02	209 42.8	13.3
03	224 43.0 ··	14.2
04	239 43.1	15.2
05	254 43.3	16.1
06	269 43.5	N 8 17.0
07	284 43.6	17.9
08	299 43.8	18.8
09	314 44.0 ··	19.8
10	329 44.1	20.7
11	344 44.3	21.6
12	359 44.5	N 8 22.5
13	14 44.6	23.4
14	29 44.8	24.3
15	44 45.0 ··	25.3
16	59 45.1	26.2
17	74 45.3	27.1
18	89 45.5	N 8 28.0
19	104 45.6	28.9
20	119 45.8	29.8
21	134 46.0 ··	30.8
22	149 46.1	31.7
23	164 46.3	32.6

Day 12

d h	G.H.A.	Dec.
12 00	179 46.4	N 8 33.5
01	194 46.6	34.4
02	209 46.8	35.3
03	224 46.9 ··	36.2
04	239 47.1	37.2
05	254 47.3	38.1
06	269 47.4	N 8 39.0
07	284 47.6	39.9
08	299 47.7	40.8
09	314 47.9 ··	41.7
10	329 48.1	42.6
11	344 48.2	43.5
12	359 48.4	N 8 44.5
13	14 48.6	45.4
14	29 48.7	46.3
15	44 48.9 ··	47.2
16	59 49.0	48.1
17	74 49.2	49.0
18	89 49.4	N 8 49.9
19	104 49.5	50.8
20	119 49.7	51.7
21	134 49.8 ··	52.7
22	149 50.0	53.6
23	164 50.2	54.5

Day 13

d h	G.H.A.	Dec.
13 00	179 50.3	N 8 55.4
01	194 50.5	56.3
02	209 50.6	57.2
03	224 50.8 ··	58.1
04	239 51.0	59.0
05	254 51.1	8 59.9
06	269 51.3	N 9 00.8
07	284 51.4	01.7
08	299 51.6	02.6
09	314 51.8 ··	03.5
10	329 51.9	04.5
11	344 52.1	05.4
12	359 52.2	N 9 06.3
13	14 52.4	07.2
14	29 52.6	08.1
15	44 52.7 ··	09.0
16	59 52.9	09.9
17	74 53.0	10.8
18	89 53.2	N 9 11.7
19	104 53.3	12.6
20	119 53.5	13.5
21	134 53.7 ··	14.4
22	149 53.8	15.3
23	164 54.0	16.2

Day 14

d h	G.H.A.	Dec.
14 00	179 54.1	N 9 17.1
01	194 54.3	18.0
02	209 54.4	18.9
03	224 54.6 ··	19.8
04	239 54.8	20.7
05	254 54.9	21.6
06	269 55.1	N 9 22.5
07	284 55.2	23.4
08	299 55.4	24.3
09	314 55.5 ··	25.2
10	329 55.7	26.1
11	344 55.8	27.0
12	359 56.0	N 9 27.9
13	14 56.2	28.8
14	29 56.3	29.7
15	44 56.5 ··	30.6
16	59 56.6	31.5
17	74 56.8	32.4
18	89 56.9	N 9 33.3
19	104 57.1	34.2
20	119 57.2	35.1
21	134 57.4 ··	36.0
22	149 57.5	36.9
23	164 57.7	37.8

Day 15

d h	G.H.A.	Dec.
15 00	179 57.8	N 9 38.7
01	194 58.0	39.6
02	209 58.1	40.5
03	224 58.3 ··	41.4
04	239 58.5	42.3
05	254 58.6	43.2
06	269 58.8	N 9 44.1
07	284 58.9	45.0
08	299 59.1	45.8
09	314 59.2 ··	46.7
10	329 59.4	47.6
11	344 59.5	48.5
12	359 59.7	N 9 49.4
13	14 59.8	50.3
14	30 00.0	51.2
15	45 00.1 ··	52.1
16	60 00.3	53.0
17	75 00.4	53.9
18	90 00.6	N 9 54.8
19	105 00.7	55.7
20	120 00.9	56.5
21	135 01.0 ··	57.4
22	150 01.2	58.3
23	165 01.3	59.2

Day 16

d h	G.H.A.	Dec.
16 00	180 01.5	N10 00.1
01	195 01.6	01.0
02	210 01.8	01.9
03	225 01.9 ··	02.8
04	240 02.1	03.7
05	255 02.2	04.5
06	270 02.4	N10 05.4
07	285 02.5	06.3
08	300 02.7	07.2
09	315 02.8 ··	08.1
10	330 03.0	09.0
11	345 03.1	09.9
12	0 03.2	N10 10.8
13	15 03.4	11.6
14	30 03.5	12.5
15	45 03.7 ··	13.4
16	60 03.8	14.3
17	75 04.0	15.2
18	90 04.1	N10 16.1
19	105 04.3	16.9
20	120 04.4	17.8
21	135 04.6 ··	18.7
22	150 04.7	19.6
23	165 04.9	20.5

Day 17

d h	G.H.A.	Dec.
17 00	180 05.0	N10 21.4
01	195 05.1	22.2
02	210 05.3	23.1
03	225 05.4 ··	24.0
04	240 05.6	24.9
05	255 05.7	25.8
06	270 05.9	N10 26.6
07	285 06.0	27.5
08	300 06.2	28.4
09	315 06.3 ··	29.3
10	330 06.4	30.2
11	345 06.6	31.0
12	0 06.7	N10 31.9
13	15 06.9	32.8
14	30 07.0	33.7
15	45 07.2 ··	34.6
16	60 07.3	35.4
17	75 07.4	36.3
18	90 07.6	N10 37.2
19	105 07.7	38.1
20	120 07.9	38.9
21	135 08.0 ··	39.8
22	150 08.2	40.7
23	165 08.3	41.6

Day 18

d h	G.H.A.	Dec.
18 00	180 08.4	N10 42.4
01	195 08.5	43.3
02	210 08.7	44.2
03	225 08.9 ··	45.1
04	240 09.0	45.9
05	255 09.1	46.8
06	270 09.3	N10 47.7
07	285 09.4	48.6
08	300 09.6	49.4
09	315 09.7 ··	50.3
10	330 09.8	51.2
11	345 10.0	52.0
12	0 10.1	N10 52.9
13	15 10.3	53.8
14	30 10.4	54.7
15	45 10.5 ··	55.5
16	60 10.7	56.4
17	75 10.8	57.3
18	90 10.9	N10 58.1
19	105 11.1	59.0
20	120 11.2	10 59.9
21	135 11.4	11 00.7
22	150 11.5	01.6
23	165 11.6	02.5

d h	G.H.A.	Dec.		d h	G.H.A.	Dec.		d h	G.H.A.	Dec.
19 00	180 11.8	N11 03.3		22 00	180 21.2	N12 04.9		25 00	180 29.6	N13 04.8
01	195 11.9	04.2		01	195 21.3	05.8		01	195 29.7	05.6
02	210 12.0	05.1		02	210 21.4	06.6		02	210 29.8	06.4
03	225 12.2 ··	05.9		03	225 21.5 ··	07.5		03	225 29.9 ··	07.2
04	240 12.3	06.8		04	240 21.7	08.3		04	240 30.0	08.0
05	255 12.5	07.7		05	255 21.8	09.2		05	255 30.2	08.8
06	270 12.6	N11 08.5		06	270 21.9	N12 10.0		06	270 30.3	N13 09.7
07	285 12.7	09.4		07	285 22.0	10.8		07	285 30.4	10.5
08	300 12.9	10.3		08	300 22.2	11.7		08	300 30.5	11.3
09	315 13.0 ··	11.1		09	315 22.3 ··	12.5		09	315 30.6 ··	12.1
10	330 13.1	12.0		10	330 22.4	13.4		10	330 30.7	12.9
11	345 13.3	12.9		11	345 22.5	14.2		11	345 30.8	13.7
12	0 13.4	N11 13.7		12	0 22.6	N12 15.0		12	0 30.9	N13 14.5
13	15 13.5	14.6		13	15 22.8	15.9		13	15 31.0	15.4
14	30 13.7	15.5		14	30 22.9	16.7		14	30 31.1	16.2
15	45 13.8 ··	16.3		15	45 23.0 ··	17.6		15	45 31.2 ··	17.0
16	60 13.9	17.2		16	60 23.1	18.4		16	60 31.3	17.8
17	75 14.1	18.0		17	75 23.3	19.2		17	75 31.4	18.6
18	90 14.2	N11 18.9		18	90 23.4	N12 20.1		18	90 31.6	N13 19.4
19	105 14.3	19.8		19	105 23.5	20.9		19	105 31.7	20.2
20	120 14.5	20.6		20	120 23.6	21.7		20	120 31.8	21.0
21	135 14.6 ··	21.5		21	135 23.7 ··	22.6		21	135 31.9 ··	21.8
22	150 14.7	22.4		22	150 23.9	23.4		22	150 32.0	22.7
23	165 14.9	23.2		23	165 24.0	24.3		23	165 32.1	23.5
20 00	180 15.0	N11 24.1		23 00	180 24.1	N12 25.1		26 00	180 32.2	N13 24.3
01	195 15.1	24.9		01	195 24.2	25.9		01	195 32.3	25.1
02	210 15.3	25.8		02	210 24.3	26.8		02	210 32.4	25.9
03	225 15.4 ··	26.6		03	225 24.5 ··	27.6		03	225 32.5 ··	26.7
04	240 15.5	27.5		04	240 24.6	28.4		04	240 32.6	27.5
05	255 15.7	28.4		05	255 24.7	29.3		05	255 32.7	28.3
06	270 15.8	N11 29.2		06	270 24.8	N12 30.1		06	270 32.8	N13 29.1
07	285 15.9	30.1		07	285 24.9	30.9		07	285 32.9	29.9
08	300 16.1	30.9		08	300 25.0	31.8		08	300 33.0	30.7
09	315 16.2 ··	31.8		09	315 25.2 ··	32.6		09	315 33.1 ··	31.5
10	330 16.3	32.7		10	330 25.3	33.4		10	330 33.2	32.3
11	345 16.5	33.5		11	345 25.4	34.3		11	345 33.3	33.1
12	0 16.6	N11 34.4		12	0 25.5	N12 35.1		12	0 33.4	N13 33.9
13	15 16.7	35.2		13	15 25.6	35.9		13	15 33.5	34.8
14	30 16.9	36.1		14	30 25.7	36.7		14	30 33.6	35.6
15	45 17.0 ··	36.9		15	45 25.9 ··	37.6		15	45 33.7 ··	36.4
16	60 17.1	37.8		16	60 26.0	38.4		16	60 33.8	37.2
17	75 17.2	38.6		17	75 26.1	39.2		17	75 33.9	38.0
18	90 17.4	N11 39.5		18	90 26.2	N12 40.1		18	90 34.0	N13 38.8
19	105 17.5	40.3		19	105 26.3	40.9		19	105 34.1	39.6
20	120 17.6	41.2		20	120 26.4	41.7		20	120 34.2	40.4
21	135 17.8 ··	42.1		21	135 26.6 ··	42.5		21	135 34.3 ··	41.2
22	150 17.9	42.9		22	150 26.7	43.4		22	150 34.4	42.0
23	165 18.0	43.8		23	165 26.8	44.2		23	165 34.5	42.8
21 00	180 18.1	N11 44.6		24 00	180 26.9	N12 45.0		27 00	180 34.6	N13 43.6
01	195 18.3	45.5		01	195 27.0	45.9		01	195 34.7	44.4
02	210 18.4	46.3		02	210 27.1	46.7		02	210 34.8	45.2
03	225 18.5 ··	47.2		03	225 27.2 ··	47.5		03	225 34.9 ··	46.0
04	240 18.7	48.0		04	240 27.4	48.3		04	240 35.0	46.8
05	255 18.8	48.9		05	255 27.5	49.2		05	255 35.1	47.6
06	270 18.9	N11 49.7		06	270 27.6	N12 50.0		06	270 35.2	N13 48.4
07	285 19.0	50.6		07	285 27.7	50.8		07	285 35.3	49.1
08	300 19.2	51.4		08	300 27.8	51.6		08	300 35.4	49.9
09	315 19.3 ··	52.3		09	315 27.9 ··	52.5		09	315 35.5 ··	50.7
10	330 19.4	53.1		10	330 28.0	53.3		10	330 35.6	51.5
11	345 19.5	54.0		11	345 28.2	54.1		11	345 35.7	52.3
12	0 19.7	N11 54.8		12	0 28.3	N12 54.9		12	0 35.8	N13 53.1
13	15 19.8	55.6		13	15 28.4	55.7		13	15 35.9	53.9
14	30 19.9	56.5		14	30 28.5	56.6		14	30 36.0	54.7
15	45 20.0 ··	57.3		15	45 28.6 ··	57.4		15	45 36.1 ··	55.5
16	60 20.2	58.2		16	60 28.7	58.2		16	60 36.2	56.3
17	75 20.3	59.0		17	75 28.8	59.0		17	75 36.3	57.1
18	90 20.4	N11 59.9		18	90 28.9	N12 59.8		18	90 36.4	N13 57.9
19	105 20.6	12 00.7		19	105 29.1	13 00.7		19	105 36.5	58.7
20	120 20.7	01.6		20	120 29.2	01.5		20	120 36.6	13 59.5
21	135 20.8 ··	02.4		21	135 29.3 ··	02.3		21	135 36.7	14 00.3
22	150 20.9	03.3		22	150 29.4	03.1		22	150 36.8	01.1
23	165 21.0	04.1		23	165 29.5	03.9		23	165 36.9	01.8

d h	G.H.A. ° '	Dec. ° '
28 00	180 37.0	N14 02.6
01	195 37.1	03.4
02	210 37.2	04.2
03	225 37.3	·· 05.0
04	240 37.4	05.8
05	255 37.5	06.6
06	270 37.6	N14 07.4
07	285 37.7	08.1
08	300 37.7	08.9
09	315 37.8	·· 09.7
10	330 37.9	10.5
11	345 38.0	11.3
12	0 38.1	N14 12.1
13	15 38.2	12.9
14	30 38.3	13.6
15	45 38.4	·· 14.4
16	60 38.5	15.2
17	75 38.6	16.0
18	90 38.7	N14 16.8
19	105 38.8	17.6
20	120 38.9	18.3
21	135 38.9	·· 19.1
22	150 39.0	19.9
23	165 39.1	20.7
29 00	180 39.2	N14 21.5
01	195 39.3	22.2
02	210 39.4	23.0
03	225 39.5	·· 23.8
04	240 39.6	24.6
05	255 39.7	25.4
06	270 39.7	N14 26.1
07	285 39.8	26.9
08	300 39.9	27.7
09	315 40.0	·· 28.5
10	330 40.1	29.2
11	345 40.2	30.0
12	0 40.3	N14 30.8
13	15 40.4	31.6
14	30 40.4	32.3
15	45 40.5	·· 33.1
16	60 40.6	33.9
17	75 40.7	34.7
18	90 40.8	N14 35.4
19	105 40.9	36.2
20	120 41.0	37.0
21	135 41.1	·· 37.8
22	150 41.1	38.5
23	165 41.2	39.3
30 00	180 41.3	N14 40.1
01	195 41.4	40.8
02	210 41.5	41.6
03	225 41.6	·· 42.4
04	240 41.6	43.1
05	255 41.7	43.9
06	270 41.8	N14 44.7
07	285 41.9	45.4
08	300 42.0	46.2
09	315 42.1	·· 47.0
10	330 42.1	47.7
11	345 42.2	48.5
12	0 42.3	N14 49.3
13	15 42.4	50.0
14	30 42.5	50.8
15	45 42.5	·· 51.6
16	60 42.6	52.3
17	75 42.7	53.1
18	90 42.8	N14 53.9
19	105 42.9	54.6
20	120 43.0	55.4
21	135 43.0	·· 56.1
22	150 43.1	56.9
23	165 43.2	57.7

MAY

d h	G.H.A. ° '	Dec. ° '
1 00	180 43.3	N14 58.4
01	195 43.3	59.2
02	210 43.4	14 59.9
03	225 43.5	15 00.7
04	240 43.6	01.5
05	255 43.7	02.2
06	270 43.7	N15 03.0
07	285 43.8	03.7
08	300 43.9	04.5
09	315 44.0	·· 05.2
10	330 44.0	06.0
11	345 44.1	06.7
12	0 44.2	N15 07.5
13	15 44.3	08.3
14	30 44.4	09.0
15	45 44.4	·· 09.8
16	60 44.5	10.5
17	75 44.6	11.3
18	90 44.7	N15 12.0
19	105 44.7	12.8
20	120 44.8	13.5
21	135 44.9	·· 14.3
22	150 45.0	15.0
23	165 45.0	15.8
2 00	180 45.1	N15 16.5
01	195 45.2	17.3
02	210 45.2	18.0
03	225 45.3	·· 18.8
04	240 45.4	19.5
05	255 45.5	20.3
06	270 45.5	N15 21.0
07	285 45.6	21.8
08	300 45.7	22.5
09	315 45.8	·· 23.3
10	330 45.8	24.0
11	345 45.9	24.7
12	0 46.0	N15 25.5
13	15 46.0	26.2
14	30 46.1	27.0
15	45 46.2	·· 27.7
16	60 46.2	28.5
17	75 46.3	29.2
18	90 46.4	N15 29.9
19	105 46.5	30.7
20	120 46.5	31.4
21	135 46.6	·· 32.2
22	150 46.7	32.9
23	165 46.7	33.6
3 00	180 46.8	N15 34.4
01	195 46.9	35.1
02	210 46.9	35.9
03	225 47.0	·· 36.6
04	240 47.1	37.3
05	255 47.1	38.1
06	270 47.2	N15 38.8
07	285 47.3	39.5
08	300 47.3	40.3
09	315 47.4	·· 41.0
10	330 47.5	41.7
11	345 47.5	42.5
12	0 47.6	N15 43.2
13	15 47.7	43.9
14	30 47.7	44.7
15	45 47.8	·· 45.4
16	60 47.9	46.1
17	75 47.9	46.9
18	90 48.0	N15 47.6
19	105 48.0	48.3
20	120 48.1	49.1
21	135 48.2	·· 49.8
22	150 48.2	50.5
23	165 48.3	51.3

d h	G.H.A. ° '	Dec. ° '
4 00	180 48.4	N15 52.0
01	195 48.4	52.7
02	210 48.5	53.4
03	225 48.5	·· 54.2
04	240 48.6	54.9
05	255 48.7	55.6
06	270 48.7	N15 56.3
07	285 48.8	57.1
08	300 48.8	57.8
09	315 48.9	·· 58.5
10	330 49.0	15 59.2
11	345 49.0	16 00.0
12	0 49.1	N16 00.7
13	15 49.1	01.4
14	30 49.2	02.1
15	45 49.3	·· 02.8
16	60 49.3	03.6
17	75 49.4	04.3
18	90 49.4	N16 05.0
19	105 49.5	05.7
20	120 49.6	06.4
21	135 49.6	·· 07.2
22	150 49.7	07.9
23	165 49.7	08.6
5 00	180 49.8	N16 09.3
01	195 49.8	10.0
02	210 49.9	10.7
03	225 49.9	·· 11.5
04	240 50.0	12.2
05	255 50.1	12.9
06	270 50.1	N16 13.6
07	285 50.2	14.3
08	300 50.2	15.0
09	315 50.3	·· 15.7
10	330 50.3	16.5
11	345 50.4	17.2
12	0 50.4	N16 17.9
13	15 50.5	18.6
14	30 50.5	19.3
15	45 50.6	·· 20.0
16	60 50.6	20.7
17	75 50.7	21.4
18	90 50.8	N16 22.1
19	105 50.8	22.8
20	120 50.9	23.6
21	135 50.9	·· 24.3
22	150 51.0	25.0
23	165 51.0	25.7
6 00	180 51.1	N16 26.4
01	195 51.1	27.1
02	210 51.2	27.8
03	225 51.2	·· 28.5
04	240 51.3	29.2
05	255 51.3	29.9
06	270 51.4	N16 30.6
07	285 51.4	31.3
08	300 51.5	32.0
09	315 51.5	·· 32.7
10	330 51.5	33.4
11	345 51.6	34.1
12	0 51.6	N16 34.8
13	15 51.7	35.5
14	30 51.7	36.2
15	45 51.8	·· 36.9
16	60 51.8	37.6
17	75 51.9	38.3
18	90 51.9	N16 39.0
19	105 52.0	39.7
20	120 52.0	40.4
21	135 52.1	·· 41.1
22	150 52.1	41.8
23	165 52.1	42.5

d	h	G.H.A. ° ′	Dec. ° ′
7	00	180 52.2	N16 43.2
	01	195 52.2	43.9
	02	210 52.3	44.6
	03	225 52.3	·· 45.3
	04	240 52.4	45.9
	05	255 52.4	46.6
	06	270 52.5	N16 47.3
	07	285 52.5	48.0
	08	300 52.5	48.7
	09	315 52.6	·· 49.4
	10	330 52.6	50.1
	11	345 52.7	50.8
	12	0 52.7	N16 51.5
	13	15 52.7	52.2
	14	30 52.8	52.8
	15	45 52.8	·· 53.5
	16	60 52.9	54.2
	17	75 52.9	54.9
	18	90 52.9	N16 55.6
	19	105 53.0	56.3
	20	120 53.0	57.0
	21	135 53.1	·· 57.6
	22	150 53.1	58.3
	23	165 53.1	59.0
8	00	180 53.2	N16 59.7
	01	195 53.2	17 00.4
	02	210 53.3	01.1
	03	225 53.3	·· 01.7
	04	240 53.3	02.4
	05	255 53.4	03.1
	06	270 53.4	N17 03.8
	07	285 53.4	04.5
	08	300 53.5	05.1
	09	315 53.5	·· 05.8
	10	330 53.6	06.5
	11	345 53.6	07.2
	12	0 53.6	N17 07.8
	13	15 53.7	08.5
	14	30 53.7	09.2
	15	45 53.7	·· 09.9
	16	60 53.8	10.5
	17	75 53.8	11.2
	18	90 53.8	N17 11.9
	19	105 53.9	12.6
	20	120 53.9	13.2
	21	135 53.9	·· 13.9
	22	150 54.0	14.6
	23	165 54.0	15.2
9	00	180 54.0	N17 15.9
	01	195 54.1	16.6
	02	210 54.1	17.3
	03	225 54.1	·· 17.9
	04	240 54.2	18.6
	05	255 54.2	19.3
	06	270 54.2	N17 19.9
	07	285 54.2	20.6
	08	300 54.3	21.3
	09	315 54.3	·· 21.9
	10	330 54.3	22.6
	11	345 54.4	23.3
	12	0 54.4	N17 23.9
	13	15 54.4	24.6
	14	30 54.5	25.3
	15	45 54.5	·· 25.9
	16	60 54.5	26.6
	17	75 54.5	27.2
	18	90 54.6	N17 27.9
	19	105 54.6	28.6
	20	120 54.6	29.2
	21	135 54.6	·· 29.9
	22	150 54.7	30.5
	23	165 54.7	31.2

d	h	G.H.A. ° ′	Dec. ° ′
10	00	180 54.7	N17 31.9
	01	195 54.8	32.5
	02	210 54.8	33.2
	03	225 54.8	·· 33.8
	04	240 54.8	34.5
	05	255 54.9	35.1
	06	270 54.9	N17 35.8
	07	285 54.9	36.5
	08	300 54.9	37.1
	09	315 55.0	·· 37.8
	10	330 55.0	38.4
	11	345 55.0	39.1
	12	0 55.0	N17 39.7
	13	15 55.0	40.4
	14	30 55.1	41.0
	15	45 55.1	·· 41.7
	16	60 55.1	42.3
	17	75 55.1	43.0
	18	90 55.2	N17 43.6
	19	105 55.2	44.3
	20	120 55.2	44.9
	21	135 55.2	·· 45.6
	22	150 55.2	46.2
	23	165 55.3	46.9
11	00	180 55.3	N17 47.5
	01	195 55.3	48.2
	02	210 55.3	48.8
	03	225 55.3	·· 49.4
	04	240 55.4	50.1
	05	255 55.4	50.7
	06	270 55.4	N17 51.4
	07	285 55.4	52.0
	08	300 55.4	52.7
	09	315 55.4	·· 53.3
	10	330 55.5	53.9
	11	345 55.5	54.6
	12	0 55.5	N17 55.2
	13	15 55.5	55.9
	14	30 55.5	56.5
	15	45 55.5	·· 57.1
	16	60 55.6	57.8
	17	75 55.6	58.4
	18	90 55.6	N17 59.1
	19	105 55.6	17 59.7
	20	120 55.6	18 00.3
	21	135 55.6	·· 01.0
	22	150 55.7	01.6
	23	165 55.7	02.2
12	00	180 55.7	N18 02.9
	01	195 55.7	03.5
	02	210 55.7	04.1
	03	225 55.7	·· 04.8
	04	240 55.7	05.4
	05	255 55.7	06.0
	06	270 55.8	N18 06.7
	07	285 55.8	07.3
	08	300 55.8	07.9
	09	315 55.8	·· 08.5
	10	330 55.8	09.2
	11	345 55.8	09.8
	12	0 55.8	N18 10.4
	13	15 55.8	11.1
	14	30 55.8	11.7
	15	45 55.9	·· 12.3
	16	60 55.9	12.9
	17	75 55.9	13.6
	18	90 55.9	N18 14.2
	19	105 55.9	14.8
	20	120 55.9	15.4
	21	135 55.9	·· 16.0
	22	150 55.9	16.7
	23	165 55.9	17.3

d	h	G.H.A. ° ′	Dec. ° ′
13	00	180 55.9	N18 17.9
	01	195 55.9	18.5
	02	210 56.0	19.2
	03	225 56.0	·· 19.8
	04	240 56.0	20.4
	05	255 56.0	21.0
	06	270 56.0	N18 21.6
	07	285 56.0	22.2
	08	300 56.0	22.9
	09	315 56.0	·· 23.5
	10	330 56.0	24.1
	11	345 56.0	24.7
	12	0 56.0	N18 25.3
	13	15 56.0	25.9
	14	30 56.0	26.6
	15	45 56.0	·· 27.2
	16	60 56.0	27.8
	17	75 56.0	28.4
	18	90 56.0	N18 29.0
	19	105 56.0	29.6
	20	120 56.0	30.2
	21	135 56.0	·· 30.8
	22	150 56.0	31.4
	23	165 56.0	32.0
14	00	180 56.0	N18 32.7
	01	195 56.0	33.3
	02	210 56.0	33.9
	03	225 56.0	·· 34.5
	04	240 56.0	35.1
	05	255 56.0	35.7
	06	270 56.0	N18 36.3
	07	285 56.0	36.9
	08	300 56.0	37.5
	09	315 56.0	·· 38.1
	10	330 56.0	38.7
	11	345 56.0	39.3
	12	0 56.0	N18 39.9
	13	15 56.0	40.5
	14	30 56.0	41.1
	15	45 56.0	·· 41.7
	16	60 56.0	42.3
	17	75 56.0	42.9
	18	90 56.0	N18 43.5
	19	105 56.0	44.1
	20	120 56.0	44.7
	21	135 56.0	·· 45.3
	22	150 56.0	45.9
	23	165 56.0	46.5
15	00	180 56.0	N18 47.1
	01	195 56.0	47.7
	02	210 56.0	48.3
	03	225 56.0	·· 48.9
	04	240 56.0	49.5
	05	255 56.0	50.1
	06	270 56.0	N18 50.6
	07	285 56.0	51.2
	08	300 56.0	51.8
	09	315 56.0	·· 52.4
	10	330 55.9	53.0
	11	345 55.9	53.6
	12	0 55.9	N18 54.2
	13	15 55.9	54.8
	14	30 55.9	55.4
	15	45 55.9	·· 55.9
	16	60 55.9	56.5
	17	75 55.9	57.1
	18	90 55.9	N18 57.7
	19	105 55.9	58.3
	20	120 55.9	58.9
	21	135 55.9	18 59.5
	22	150 55.8	19 00.0
	23	165 55.8	00.6

d	h	G.H.A. °	′	Dec. °	′
16	00	180	55.8	N19	01.2
	01	195	55.8		01.8
	02	210	55.8		02.4
	03	225	55.8	· ·	02.9
	04	240	55.8		03.5
	05	255	55.8		04.1
	06	270	55.8	N19	04.7
	07	285	55.7		05.3
	08	300	55.7		05.8
	09	315	55.7	· ·	06.4
	10	330	55.7		07.0
	11	345	55.7		07.6
	12	0	55.7	N19	08.1
	13	15	55.7		08.7
	14	30	55.7		09.3
	15	45	55.6	· ·	09.9
	16	60	55.6		10.4
	17	75	55.6		11.0
	18	90	55.6	N19	11.6
	19	105	55.6		12.1
	20	120	55.6		12.7
	21	135	55.5	· ·	13.3
	22	150	55.5		13.9
	23	165	55.5		14.4
17	00	180	55.5	N19	15.0
	01	195	55.5		15.6
	02	210	55.5		16.1
	03	225	55.4	· ·	16.7
	04	240	55.4		17.3
	05	255	55.4		17.8
	06	270	55.4	N19	18.4
	07	285	55.4		19.0
	08	300	55.4		19.5
	09	315	55.3	· ·	20.1
	10	330	55.3		20.6
	11	345	55.3		21.2
	12	0	55.3	N19	21.8
	13	15	55.3		22.3
	14	30	55.2		22.9
	15	45	55.2	· ·	23.4
	16	60	55.2		24.0
	17	75	55.2		24.6
	18	90	55.2	N19	25.1
	19	105	55.1		25.7
	20	120	55.1		26.2
	21	135	55.1	· ·	26.8
	22	150	55.1		27.3
	23	165	55.1		27.9
18	00	180	55.0	N19	28.5
	01	195	55.0		29.0
	02	210	55.0		29.6
	03	225	55.0	· ·	30.1
	04	240	54.9		30.7
	05	255	54.9		31.2
	06	270	54.9	N19	31.8
	07	285	54.9		32.3
	08	300	54.8		32.9
	09	315	54.8	· ·	33.4
	10	330	54.8		34.0
	11	345	54.8		34.5
	12	0	54.7	N19	35.1
	13	15	54.7		35.6
	14	30	54.7		36.2
	15	45	54.7	· ·	36.7
	16	60	54.6		37.2
	17	75	54.6		37.8
	18	90	54.6	N19	38.3
	19	105	54.6		38.9
	20	120	54.5		39.4
	21	135	54.5	· ·	40.0
	22	150	54.5		40.5
	23	165	54.5		41.1

d	h	G.H.A. °	′	Dec. °	′
19	00	180	54.4	N19	41.6
	01	195	54.4		42.1
	02	210	54.4		42.7
	03	225	54.3	· ·	43.2
	04	240	54.3		43.7
	05	255	54.3		44.3
	06	270	54.3	N19	44.8
	07	285	54.2		45.4
	08	300	54.2		45.9
	09	315	54.2	· ·	46.4
	10	330	54.1		47.0
	11	345	54.1		47.5
	12	0	54.1	N19	48.0
	13	15	54.0		48.6
	14	30	54.0		49.1
	15	45	54.0	· ·	49.6
	16	60	53.9		50.2
	17	75	53.9		50.7
	18	90	53.9	N19	51.2
	19	105	53.8		51.8
	20	120	53.8		52.3
	21	135	53.8	· ·	52.8
	22	150	53.7		53.3
	23	165	53.7		53.9
20	00	180	53.7	N19	54.4
	01	195	53.6		54.9
	02	210	53.6		55.4
	03	225	53.6	· ·	56.0
	04	240	53.5		56.5
	05	255	53.5		57.0
	06	270	53.5	N19	57.5
	07	285	53.4		58.1
	08	300	53.4		58.6
	09	315	53.4	· ·	59.1
	10	330	53.3	19	59.6
	11	345	53.3	20	00.1
	12	0	53.3	N20	00.7
	13	15	53.2		01.2
	14	30	53.2		01.7
	15	45	53.1	· ·	02.2
	16	60	53.1		02.7
	17	75	53.1		03.3
	18	90	53.0	N20	03.8
	19	105	53.0		04.3
	20	120	53.0		04.8
	21	135	52.9	· ·	05.3
	22	150	52.9		05.8
	23	165	52.8		06.3
21	00	180	52.8	N20	06.9
	01	195	52.8		07.4
	02	210	52.7		07.9
	03	225	52.7	· ·	08.4
	04	240	52.6		08.9
	05	255	52.6		09.4
	06	270	52.6	N20	09.9
	07	285	52.5		10.4
	08	300	52.5		10.9
	09	315	52.4	· ·	11.4
	10	330	52.4		11.9
	11	345	52.4		12.5
	12	0	52.3	N20	13.0
	13	15	52.3		13.5
	14	30	52.2		14.0
	15	45	52.2	· ·	14.5
	16	60	52.1		15.0
	17	75	52.1		15.5
	18	90	52.1	N20	16.0
	19	105	52.0		16.5
	20	120	52.0		17.0
	21	135	51.9	· ·	17.5
	22	150	51.9		18.0
	23	165	51.8		18.5

d	h	G.H.A. °	′	Dec. °	′
22	00	180	51.8	N20	19.0
	01	195	51.7		19.5
	02	210	51.7		20.0
	03	225	51.7	· ·	20.5
	04	240	51.6		21.0
	05	255	51.6		21.5
	06	270	51.5	N20	21.9
	07	285	51.5		22.4
	08	300	51.4		22.9
	09	315	51.4	· ·	23.4
	10	330	51.3		23.9
	11	345	51.3		24.4
	12	0	51.2	N20	24.9
	13	15	51.2		25.4
	14	30	51.1		25.9
	15	45	51.1	· ·	26.4
	16	60	51.0		26.9
	17	75	51.0		27.3
	18	90	50.9	N20	27.8
	19	105	50.9		28.3
	20	120	50.9		28.8
	21	135	50.8	· ·	29.3
	22	150	50.8		29.8
	23	165	50.7		30.3
23	00	180	50.7	N20	30.7
	01	195	50.6		31.2
	02	210	50.6		31.7
	03	225	50.5	· ·	32.2
	04	240	50.4		32.7
	05	255	50.4		33.1
	06	270	50.3	N20	33.6
	07	285	50.3		34.1
	08	300	50.2		34.6
	09	315	50.2	· ·	35.1
	10	330	50.1		35.5
	11	345	50.1		36.0
	12	0	50.0	N20	36.5
	13	15	50.0		37.0
	14	30	49.9		37.4
	15	45	49.9	· ·	37.9
	16	60	49.8		38.4
	17	75	49.8		38.9
	18	90	49.7	N20	39.3
	19	105	49.7		39.8
	20	120	49.6		40.3
	21	135	49.6	· ·	40.8
	22	150	49.5		41.2
	23	165	49.4		41.7
24	00	180	49.4	N20	42.2
	01	195	49.3		42.6
	02	210	49.3		43.1
	03	225	49.2	· ·	43.6
	04	240	49.2		44.0
	05	255	49.1		44.5
	06	270	49.0	N20	45.0
	07	285	49.0		45.4
	08	300	48.9		45.9
	09	315	48.9	· ·	46.4
	10	330	48.8		46.8
	11	345	48.8		47.3
	12	0	48.7	N20	47.7
	13	15	48.6		48.2
	14	30	48.6		48.7
	15	45	48.5	· ·	49.1
	16	60	48.5		49.6
	17	75	48.4		50.0
	18	90	48.4	N20	50.5
	19	105	48.3		50.9
	20	120	48.2		51.4
	21	135	48.2	· ·	51.9
	22	150	48.1		52.3
	23	165	48.1		52.8

JUNE

d h	GHA	Dec.
25 00	180 48.0	N20 53.2
01	195 47.9	53.7
02	210 47.9	54.1
03	225 47.8 ··	54.6
04	240 47.8	55.0
05	255 47.7	55.5
06	270 47.6	N20 55.9
07	285 47.6	56.4
08	300 47.5	56.8
09	315 47.4 ··	57.3
10	330 47.4	57.7
11	345 47.3	58.2
12	0 47.3	N20 58.6
13	15 47.2	59.1
14	30 47.1	20 59.5
15	45 47.1	21 00.0
16	60 47.0	00.4
17	75 46.9	00.8
18	90 46.9	N21 01.3
19	105 46.8	01.7
20	120 46.7	02.2
21	135 46.7 ··	02.6
22	150 46.6	03.0
23	165 46.6	03.5
26 00	180 46.5	N21 03.9
01	195 46.4	04.4
02	210 46.4	04.8
03	225 46.3 ··	05.2
04	240 46.2	05.7
05	255 46.2	06.1
06	270 46.1	N21 06.5
07	285 46.0	07.0
08	300 46.0	07.4
09	315 45.9 ··	07.8
10	330 45.8	08.3
11	345 45.8	08.7
12	0 45.7	N21 09.1
13	15 45.6	09.6
14	30 45.6	10.0
15	45 45.5 ··	10.4
16	60 45.4	10.9
17	75 45.3	11.3
18	90 45.3	N21 11.7
19	105 45.2	12.1
20	120 45.1	12.6
21	135 45.1 ··	13.0
22	150 45.0	13.4
23	165 44.9	13.8
27 00	180 44.9	N21 14.3
01	195 44.8	14.7
02	210 44.7	15.1
03	225 44.6 ··	15.5
04	240 44.6	16.0
05	255 44.5	16.4
06	270 44.4	N21 16.8
07	285 44.4	17.2
08	300 44.3	17.6
09	315 44.2 ··	18.1
10	330 44.1	18.5
11	345 44.1	18.9
12	0 44.0	N21 19.3
13	15 43.9	19.7
14	30 43.9	20.1
15	45 43.8 ··	20.5
16	60 43.7	21.0
17	75 43.6	21.4
18	90 43.6	N21 21.8
19	105 43.5	22.2
20	120 43.4	22.6
21	135 43.3 ··	23.0
22	150 43.3	23.4
23	165 43.2	23.8

d h	GHA	Dec.
28 00	180 43.1	N21 24.2
01	195 43.0	24.7
02	210 43.0	25.1
03	225 42.9 ··	25.5
04	240 42.8	25.9
05	255 42.7	26.3
06	270 42.7	N21 26.7
07	285 42.6	27.1
08	300 42.5	27.5
09	315 42.4 ··	27.9
10	330 42.4	28.3
11	345 42.3	28.7
12	0 42.2	N21 29.1
13	15 42.1	29.5
14	30 42.1	29.9
15	45 42.0 ··	30.3
16	60 41.9	30.7
17	75 41.8	31.1
18	90 41.7	N21 31.5
19	105 41.7	31.9
20	120 41.6	32.3
21	135 41.5 ··	32.7
22	150 41.4	33.1
23	165 41.3	33.5
29 00	180 41.3	N21 33.8
01	195 41.2	34.2
02	210 41.1	34.6
03	225 41.0 ··	35.0
04	240 40.9	35.4
05	255 40.9	35.8
06	270 40.8	N21 36.2
07	285 40.7	36.6
08	300 40.6	37.0
09	315 40.5 ··	37.4
10	330 40.5	37.7
11	345 40.4	38.1
12	0 40.3	N21 38.5
13	15 40.2	38.9
14	30 40.1	39.3
15	45 40.1 ··	39.7
16	60 40.0	40.0
17	75 39.9	40.4
18	90 39.8	N21 40.8
19	105 39.7	41.2
20	120 39.6	41.6
21	135 39.6 ··	41.9
22	150 39.5	42.3
23	165 39.4	42.7
30 00	180 39.3	N21 43.1
01	195 39.2	43.5
02	210 39.1	43.8
03	225 39.1 ··	44.2
04	240 39.0	44.6
05	255 38.9	45.0
06	270 38.8	N21 45.3
07	285 38.7	45.7
08	300 38.6	46.1
09	315 38.5 ··	46.4
10	330 38.5	46.8
11	345 38.4	47.2
12	0 38.3	N21 47.5
13	15 38.2	47.9
14	30 38.1	48.3
15	45 38.0 ··	48.7
16	60 37.9	49.0
17	75 37.9	49.4
18	90 37.8	N21 49.8
19	105 37.7	50.1
20	120 37.6	50.5
21	135 37.5 ··	50.8
22	150 37.4	51.2
23	165 37.3	51.6

d h	GHA	Dec.
31 00	180 37.2	N21 51.9
01	195 37.2	52.3
02	210 37.1	52.6
03	225 37.0 ··	53.0
04	240 36.9	53.4
05	255 36.8	53.7
06	270 36.7	N21 54.1
07	285 36.6	54.4
08	300 36.5	54.8
09	315 36.4 ··	55.1
10	330 36.3	55.5
11	345 36.3	55.9
12	0 36.2	N21 56.2
13	15 36.1	56.6
14	30 36.0	56.9
15	45 35.9 ··	57.3
16	60 35.8	57.6
17	75 35.7	58.0
18	90 35.6	N21 58.3
19	105 35.5	58.7
20	120 35.4	59.0
21	135 35.3 ··	59.4
22	150 35.3	21 59.7
23	165 35.2	22 00.1
1 00	180 35.1	N22 00.4
01	195 35.0	00.7
02	210 34.9	01.1
03	225 34.8 ··	01.4
04	240 34.7	01.8
05	255 34.6	02.1
06	270 34.5	N22 02.5
07	285 34.4	02.8
08	300 34.3	03.1
09	315 34.2 ··	03.5
10	330 34.1	03.8
11	345 34.0	04.2
12	0 33.9	N22 04.5
13	15 33.9	04.8
14	30 33.8	05.2
15	45 33.7 ··	05.5
16	60 33.6	05.8
17	75 33.5	06.2
18	90 33.4	N22 06.5
19	105 33.3	06.8
20	120 33.2	07.2
21	135 33.1 ··	07.5
22	150 33.0	07.8
23	165 32.9	08.2
2 00	180 32.8	N22 08.5
01	195 32.7	08.8
02	210 32.6	09.1
03	225 32.5 ··	09.4
04	240 32.4	09.8
05	255 32.3	10.1
06	270 32.2	N22 10.4
07	285 32.1	10.8
08	300 32.0	11.1
09	315 31.9 ··	11.4
10	330 31.8	11.7
11	345 31.7	12.1
12	0 31.6	N22 12.4
13	15 31.5	12.7
14	30 31.4	13.0
15	45 31.3 ··	13.3
16	60 31.2	13.7
17	75 31.1	14.0
18	90 31.0	N22 14.3
19	105 30.9	14.6
20	120 30.8	14.9
21	135 30.7 ··	15.2
22	150 30.6	15.6
23	165 30.5	15.9

d h	G.H.A.	Dec.		d h	G.H.A.	Dec.		d h	G.H.A.	Dec.
3 00	180 30.4	N22 16.2		6 00	180 22.8	N22 36.9		9 00	180 14.4	N22 54.1
01	195 30.3	16.5		01	195 22.7	37.2		01	195 14.3	54.4
02	210 30.2	16.8		02	210 22.6	37.5		02	210 14.2	54.6
03	225 30.1	·· 17.1		03	225 22.5	·· 37.7		03	225 14.1	·· 54.8
04	240 30.0	17.4		04	240 22.4	38.0		04	240 13.9	55.0
05	255 29.9	17.7		05	255 22.2	38.3		05	255 13.8	55.2
06	270 29.8	N22 18.1		06	270 22.1	N22 38.5		06	270 13.7	N22 55.4
07	285 29.7	18.4		07	285 22.0	38.8		07	285 13.6	55.6
08	300 29.6	18.7		08	300 21.9	39.0		08	300 13.5	55.8
09	315 29.5	·· 19.0		09	315 21.8	·· 39.3		09	315 13.3	·· 56.0
10	330 29.4	19.3		10	330 21.7	39.6		10	330 13.2	56.2
11	345 29.3	19.6		11	345 21.6	39.8		11	345 13.1	56.4
12	0 29.2	N22 19.9		12	0 21.5	N22 40.1		12	0 13.0	N22 56.7
13	15 29.1	20.2		13	15 21.3	40.3		13	15 12.9	56.9
14	30 29.0	20.5		14	30 21.2	40.6		14	30 12.7	57.1
15	45 28.9	·· 20.8		15	45 21.1	·· 40.8		15	45 12.6	·· 57.3
16	60 28.8	21.1		16	60 21.0	41.1		16	60 12.5	57.5
17	75 28.7	21.4		17	75 20.9	41.3		17	75 12.4	57.7
18	90 28.6	N22 21.7		18	90 20.8	N22 41.6		18	90 12.2	N22 57.9
19	105 28.5	22.0		19	105 20.7	41.8		19	105 12.1	58.1
20	120 28.4	22.3		20	120 20.6	42.1		20	120 12.0	58.3
21	135 28.3	·· 22.6		21	135 20.4	·· 42.3		21	135 11.9	·· 58.5
22	150 28.2	22.9		22	150 20.3	42.6		22	150 11.7	58.7
23	165 28.1	23.2		23	165 20.2	42.8		23	165 11.6	58.9
4 00	180 28.0	N22 23.5		7 00	180 20.1	N22 43.1		10 00	180 11.5	N22 59.1
01	195 27.9	23.8		01	195 20.0	43.3		01	195 11.4	59.3
02	210 27.8	24.1		02	210 19.9	43.6		02	210 11.2	59.5
03	225 27.7	·· 24.4		03	225 19.7	·· 43.8		03	225 11.1	·· 59.7
04	240 27.6	24.7		04	240 19.6	44.1		04	240 11.0	22 59.8
05	255 27.5	25.0		05	255 19.5	44.3		05	255 10.9	23 00.0
06	270 27.4	N22 25.3		06	270 19.4	N22 44.5		06	270 10.8	N23 00.2
07	285 27.3	25.6		07	285 19.3	44.8		07	285 10.6	00.4
08	300 27.1	25.9		08	300 19.2	45.0		08	300 10.5	00.6
09	315 27.0	·· 26.1		09	315 19.1	·· 45.3		09	315 10.4	·· 00.8
10	330 26.9	26.4		10	330 18.9	45.5		10	330 10.3	01.0
11	345 26.8	26.7		11	345 18.8	45.8		11	345 10.1	01.2
12	0 26.7	N22 27.0		12	0 18.7	N22 46.0		12	0 10.0	N23 01.4
13	15 26.6	27.3		13	15 18.6	46.2		13	15 09.9	01.6
14	30 26.5	27.6		14	30 18.5	46.5		14	30 09.8	01.8
15	45 26.4	·· 27.9		15	45 18.4	·· 46.7		15	45 09.6	·· 01.9
16	60 26.3	28.2		16	60 18.2	46.9		16	60 09.5	02.1
17	75 26.2	28.4		17	75 18.1	47.2		17	75 09.4	02.3
18	90 26.1	N22 28.7		18	90 18.0	N22 47.4		18	90 09.3	N23 02.5
19	105 26.0	29.0		19	105 17.9	47.6		19	105 09.1	02.7
20	120 25.9	29.3		20	120 17.8	47.9		20	120 09.0	02.9
21	135 25.8	·· 29.6		21	135 17.7	·· 48.1		21	135 08.9	·· 03.0
22	150 25.7	29.9		22	150 17.5	48.3		22	150 08.8	03.2
23	165 25.5	30.1		23	165 17.4	48.6		23	165 08.6	03.4
5 00	180 25.4	N22 30.4		8 00	180 17.3	N22 48.8		11 00	180 08.5	N23 03.6
01	195 25.3	30.7		01	195 17.2	49.0		01	195 08.4	03.8
02	210 25.2	31.0		02	210 17.1	49.3		02	210 08.2	03.9
03	225 25.1	·· 31.3		03	225 16.9	·· 49.5		03	225 08.1	·· 04.1
04	240 25.0	31.5		04	240 16.8	49.7		04	240 08.0	04.3
05	255 24.9	31.8		05	255 16.7	50.0		05	255 07.9	04.5
06	270 24.8	N22 32.1		06	270 16.6	N22 50.2		06	270 07.7	N23 04.7
07	285 24.7	32.4		07	285 16.5	50.4		07	285 07.6	04.8
08	300 24.6	32.6		08	300 16.4	50.6		08	300 07.5	05.0
09	315 24.5	·· 32.9		09	315 16.2	·· 50.9		09	315 07.4	·· 05.2
10	330 24.4	33.2		10	330 16.1	51.1		10	330 07.2	05.4
11	345 24.2	33.5		11	345 16.0	51.3		11	345 07.1	05.5
12	0 24.1	N22 33.7		12	0 15.9	N22 51.5		12	0 07.0	N23 05.7
13	15 24.0	34.0		13	15 15.8	51.7		13	15 06.8	05.9
14	30 23.9	34.3		14	30 15.6	52.0		14	30 06.7	06.0
15	45 23.8	·· 34.5		15	45 15.5	·· 52.2		15	45 06.6	·· 06.2
16	60 23.7	34.8		16	60 15.4	52.4		16	60 06.5	06.4
17	75 23.6	35.1		17	75 15.3	52.6		17	75 06.3	06.5
18	90 23.5	N22 35.4		18	90 15.2	N22 52.8		18	90 06.2	N23 06.7
19	105 23.4	35.6		19	105 15.0	53.1		19	105 06.1	06.9
20	120 23.3	35.9		20	120 14.9	53.3		20	120 06.0	07.0
21	135 23.1	·· 36.2		21	135 14.8	·· 53.5		21	135 05.8	·· 07.2
22	150 23.0	36.4		22	150 14.7	53.7		22	150 05.7	07.4
23	165 22.9	36.7		23	165 14.6	53.9		23	165 05.6	07.5

d h	G.H.A.	Dec.	d h	G.H.A.	Dec.	d h	G.H.A.	Dec.
12 00	180 05.4	N23 07.7	15 00	179 56.0	N23 17.6	18 00	179 46.2	N23 23.8
01	195 05.3	07.9	01	194 55.8	17.7	01	194 46.1	23.9
02	210 05.2	08.0	02	209 55.7	17.8	02	209 45.9	23.9
03	225 05.1 ··	08.2	03	224 55.6 ··	17.9	03	224 45.8 ··	24.0
04	240 04.9	08.4	04	239 55.4	18.0	04	239 45.7	24.0
05	255 04.8	08.5	05	254 55.3	18.2	05	254 45.5	24.1
06	270 04.7	N23 08.7	06	269 55.2	N23 18.3	06	269 45.4	N23 24.2
07	285 04.5	08.8	07	284 55.0	18.4	07	284 45.3	24.2
08	300 04.4	09.0	08	299 54.9	18.5	08	299 45.1	24.3
09	315 04.3 ··	09.1	09	314 54.8 ··	18.6	09	314 45.0 ··	24.3
10	330 04.1	09.3	10	329 54.6	18.7	10	329 44.8	24.4
11	345 04.0	09.5	11	344 54.5	18.8	11	344 44.7	24.4
12	0 03.9	N23 09.6	12	359 54.4	N23 18.9	12	359 44.6	N23 24.5
13	15 03.8	09.8	13	14 54.2	19.0	13	14 44.4	24.5
14	30 03.6	09.9	14	29 54.1	19.1	14	29 44.3	24.6
15	45 03.5 ··	10.1	15	44 54.0 ··	19.2	15	44 44.2 ··	24.6
16	60 03.4	10.2	16	59 53.8	19.3	16	59 44.0	24.7
17	75 03.2	10.4	17	74 53.7	19.4	17	74 43.9	24.7
18	90 03.1	N23 10.5	18	89 53.6	N23 19.5	18	89 43.7	N23 24.8
19	105 03.0	10.7	19	104 53.4	19.6	19	104 43.6	24.8
20	120 02.9	10.8	20	119 53.3	19.7	20	119 43.5	24.9
21	135 02.7 ··	11.0	21	134 53.1 ··	19.8	21	134 43.3 ··	24.9
22	150 02.6	11.1	22	149 53.0	19.9	22	149 43.2	25.0
23	165 02.5	11.3	23	164 52.9	20.0	23	164 43.1	25.0
13 00	180 02.3	N23 11.4	16 00	179 52.7	N23 20.1	19 00	179 42.9	N23 25.1
01	195 02.2	11.6	01	194 52.6	20.2	01	194 42.8	25.1
02	210 02.1	11.7	02	209 52.5	20.3	02	209 42.6	25.1
03	225 01.9 ··	11.8	03	224 52.3 ··	20.4	03	224 42.5 ··	25.2
04	240 01.8	12.0	04	239 52.2	20.5	04	239 42.4	25.2
05	255 01.7	12.1	05	254 52.1	20.5	05	254 42.2	25.3
06	270 01.5	N23 12.3	06	269 51.9	N23 20.6	06	269 42.1	N23 25.3
07	285 01.4	12.4	07	284 51.8	20.7	07	284 42.0	25.3
08	300 01.3	12.6	08	299 51.7	20.8	08	299 41.8	25.4
09	315 01.1 ··	12.7	09	314 51.5 ··	20.9	09	314 41.7 ··	25.4
10	330 01.0	12.8	10	329 51.4	21.0	10	329 41.6	25.4
11	345 00.9	13.0	11	344 51.3	21.1	11	344 41.4	25.5
12	0 00.8	N23 13.1	12	359 51.1	N23 21.2	12	359 41.3	N23 25.5
13	15 00.6	13.3	13	14 51.0	21.3	13	14 41.1	25.6
14	30 00.5	13.4	14	29 50.8	21.3	14	29 41.0	25.6
15	45 00.4 ··	13.5	15	44 50.7 ··	21.4	15	44 40.9 ··	25.6
16	60 00.2	13.7	16	59 50.6	21.5	16	59 40.7	25.7
17	75 00.1	13.8	17	74 50.4	21.6	17	74 40.6	25.7
18	90 00.0	N23 13.9	18	89 50.3	N23 21.7	18	89 40.5	N23 25.7
19	104 59.8	14.1	19	104 50.2	21.8	19	104 40.3	25.7
20	119 59.7	14.2	20	119 50.0	21.8	20	119 40.2	25.8
21	134 59.6 ··	14.3	21	134 49.9 ··	21.9	21	134 40.0 ··	25.8
22	149 59.4	14.5	22	149 49.8	22.0	22	149 39.9	25.8
23	164 59.3	14.6	23	164 49.6	22.1	23	164 39.8	25.9
14 00	179 59.2	N23 14.7	17 00	179 49.5	N23 22.2	20 00	179 39.6	N23 25.9
01	194 59.0	14.8	01	194 49.3	22.2	01	194 39.5	25.9
02	209 58.9	15.0	02	209 49.2	22.3	02	209 39.4	25.9
03	224 58.8 ··	15.1	03	224 49.1 ··	22.4	03	224 39.2 ··	26.0
04	239 58.6	15.2	04	239 48.9	22.5	04	239 39.1	26.0
05	254 58.5	15.4	05	254 48.8	22.5	05	254 38.9	26.0
06	269 58.4	N23 15.5	06	269 48.7	N23 22.6	06	269 38.8	N23 26.0
07	284 58.2	15.6	07	284 48.5	22.7	07	284 38.7	26.0
08	299 58.1	15.7	08	299 48.4	22.8	08	299 38.5	26.1
09	314 58.0 ··	15.8	09	314 48.3 ··	22.8	09	314 38.4 ··	26.1
10	329 57.8	16.0	10	329 48.1	22.9	10	329 38.3	26.1
11	344 57.7	16.1	11	344 48.0	23.0	11	344 38.1	26.1
12	359 57.6	N23 16.2	12	359 47.8	N23 23.0	12	359 38.0	N23 26.1
13	14 57.4	16.3	13	14 47.7	23.1	13	14 37.9	26.2
14	29 57.3	16.5	14	29 47.6	23.2	14	29 37.7	26.2
15	44 57.2 ··	16.6	15	44 47.4 ··	23.2	15	44 37.6 ··	26.2
16	59 57.0	16.7	16	59 47.3	23.3	16	59 37.4	26.2
17	74 56.9	16.8	17	74 47.2	23.4	17	74 37.3	26.2
18	89 56.8	N23 16.9	18	89 47.0	N23 23.4	18	89 37.2	N23 26.2
19	104 56.6	17.0	19	104 46.9	23.5	19	104 37.0	26.2
20	119 56.5	17.2	20	119 46.8	23.6	20	119 36.9	26.3
21	134 56.4 ··	17.3	21	134 46.6 ··	23.6	21	134 36.8 ··	26.3
22	149 56.2	17.4	22	149 46.5	23.7	22	149 36.6	26.3
23	164 56.1	17.5	23	164 46.3	23.7	23	164 36.5	26.3

d h	G.H.A.	Dec.		d h	G.H.A.	Dec.		d h	G.H.A.	Dec.
21 00	179 36.3	N23 26.3		24 00	179 26.6	N23 25.1		27 00	179 17.1	N23 20.2
01	194 36.2	26.3		01	194 26.4	25.0		01	194 16.9	20.1
02	209 36.1	26.3		02	209 26.3	25.0		02	209 16.8	20.0
03	224 35.9 ··	26.3		03	224 26.2 ··	25.0		03	224 16.7 ··	19.9
04	239 35.8	26.3		04	239 26.0	24.9		04	239 16.5	19.8
05	254 35.7	26.3		05	254 25.9	24.9		05	254 16.4	19.7
06	269 35.5	N23 26.3		06	269 25.8	N23 24.8		06	269 16.3	N23 19.6
07	284 35.4	26.3		07	284 25.6	24.8		07	284 16.2	19.5
08	299 35.3	26.4		08	299 25.5	24.7		08	299 16.0	19.4
09	314 35.1 ··	26.4		09	314 25.4 ··	24.7		09	314 15.9 ··	19.3
10	329 35.0	26.4		10	329 25.2	24.6		10	329 15.8	19.2
11	344 34.8	26.4		11	344 25.1	24.6		11	344 15.6	19.1
12	359 34.7	N23 26.4		12	359 25.0	N23 24.5		12	359 15.5	N23 19.0
13	14 34.6	26.4		13	14 24.8	24.5		13	14 15.4	18.9
14	29 34.4	26.4		14	29 24.7	24.4		14	29 15.2	18.8
15	44 34.3 ··	26.4		15	44 24.6 ··	24.4		15	44 15.1 ··	18.7
16	59 34.2	26.4		16	59 24.4	24.3		16	59 15.0	18.6
17	74 34.0	26.3		17	74 24.3	24.3		17	74 14.9	18.5
18	89 33.9	N23 26.3		18	89 24.2	N23 24.2		18	89 14.7	N23 18.3
19	104 33.8	26.3		19	104 24.0	24.1		19	104 14.6	18.2
20	119 33.6	26.3		20	119 23.9	24.1		20	119 14.5	18.1
21	134 33.5 ··	26.3		21	134 23.8 ··	24.0		21	134 14.3 ··	18.0
22	149 33.3	26.3		22	149 23.6	24.0		22	149 14.2	17.9
23	164 33.2	26.3		23	164 23.5	23.9		23	164 14.1	17.8
22 00	179 33.1	N23 26.3		25 00	179 23.4	N23 23.9		28 00	179 14.0	N23 17.7
01	194 32.9	26.3		01	194 23.2	23.8		01	194 13.8	17.6
02	209 32.8	26.3		02	209 23.1	23.7		02	209 13.7	17.5
03	224 32.7 ··	26.3		03	224 23.0 ··	23.7		03	224 13.6 ··	17.4
04	239 32.5	26.3		04	239 22.8	23.6		04	239 13.5	17.2
05	254 32.4	26.3		05	254 22.7	23.5		05	254 13.3	17.1
06	269 32.3	N23 26.2		06	269 22.6	N23 23.5		06	269 13.2	N23 17.0
07	284 32.1	26.2		07	284 22.4	23.4		07	284 13.1	16.9
08	299 32.0	26.2		08	299 22.3	23.3		08	299 12.9	16.8
09	314 31.8 ··	26.2		09	314 22.2 ··	23.3		09	314 12.8 ··	16.7
10	329 31.7	26.2		10	329 22.0	23.2		10	329 12.7	16.5
11	344 31.6	26.2		11	344 21.9	23.1		11	344 12.6	16.4
12	359 31.4	N23 26.2		12	359 21.8	N23 23.1		12	359 12.4	N23 16.3
13	14 31.3	26.1		13	14 21.6	23.0		13	14 12.3	16.2
14	29 31.2	26.1		14	29 21.5	22.9		14	29 12.2	16.1
15	44 31.0 ··	26.1		15	44 21.4 ··	22.9		15	44 12.1 ··	15.9
16	59 30.9	26.1		16	59 21.3	22.8		16	59 11.9	15.8
17	74 30.8	26.1		17	74 21.1	22.7		17	74 11.8	15.7
18	89 30.6	N23 26.0		18	89 21.0	N23 22.7		18	89 11.7	N23 15.6
19	104 30.5	26.0		19	104 20.9	22.6		19	104 11.5	15.5
20	119 30.4	26.0		20	119 20.7	22.5		20	119 11.4	15.3
21	134 30.2 ··	26.0		21	134 20.6 ··	22.4		21	134 11.3 ··	15.2
22	149 30.1	26.0		22	149 20.5	22.4		22	149 11.2	15.1
23	164 29.9	25.9		23	164 20.3	22.3		23	164 11.0	14.9
23 00	179 29.8	N23 25.9		26 00	179 20.2	N23 22.2		29 00	179 10.9	N23 14.8
01	194 29.7	25.9		01	194 20.1	22.1		01	194 10.8	14.7
02	209 29.5	25.8		02	209 19.9	22.1		02	209 10.7	14.6
03	224 29.4 ··	25.8		03	224 19.8 ··	22.0		03	224 10.5 ··	14.4
04	239 29.3	25.8		04	239 19.7	21.9		04	239 10.4	14.3
05	254 29.1	25.8		05	254 19.5	21.8		05	254 10.3	14.2
06	269 29.0	N23 25.7		06	269 19.4	N23 21.7		06	269 10.2	N23 14.0
07	284 28.9	25.7		07	284 19.3	21.7		07	284 10.0	13.9
08	299 28.7	25.7		08	299 19.1	21.6		08	299 09.9	13.8
09	314 28.6 ··	25.6		09	314 19.0 ··	21.5		09	314 09.8 ··	13.6
10	329 28.5	25.6		10	329 18.9	21.4		10	329 09.7	13.5
11	344 28.3	25.6		11	344 18.8	21.3		11	344 09.5	13.4
12	359 28.2	N23 25.5		12	359 18.6	N23 21.2		12	359 09.4	N23 13.2
13	14 28.1	25.5		13	14 18.5	21.1		13	14 09.3	13.1
14	29 27.9	25.5		14	29 18.4	21.1		14	29 09.2	13.0
15	44 27.8 ··	25.4		15	44 18.2 ··	21.0		15	44 09.0 ··	12.8
16	59 27.7	25.4		16	59 18.1	20.9		16	59 08.9	12.7
17	74 27.5	25.4		17	74 18.0	20.8		17	74 08.8	12.5
18	89 27.4	N23 25.3		18	89 17.8	N23 20.7		18	89 08.7	N23 12.4
19	104 27.3	25.3		19	104 17.7	20.6		19	104 08.5	12.3
20	119 27.1	25.2		20	119 17.6	20.5		20	119 08.4	12.1
21	134 27.0 ··	25.2		21	134 17.4 ··	20.4		21	134 08.3 ··	12.0
22	149 26.8	25.2		22	149 17.3	20.3		22	149 08.2	11.8
23	164 26.7	25.1		23	164 17.2	20.2		23	164 08.0	11.7

d h	G.H.A.	Dec.		d h	G.H.A.	Dec.		d h	G.H.A.	Dec.
30 00	179 07.9	N23 11.5		3 00	178 59.3	N22 59.3		6 00	178 51.3	N22 43.4
01	194 07.8	11.4		01	193 59.2	59.1		01	193 51.2	43.1
02	209 07.7	11.2		02	208 59.1	58.9		02	208 51.1	42.9
03	224 07.5 ··	11.1		03	223 59.0 ··	58.7		03	223 51.0 ··	42.6
04	239 07.4	11.0		04	238 58.9	58.5		04	238 50.9	42.4
05	254 07.3	10.8		05	253 58.7	58.3		05	253 50.8	42.1
06	269 07.2	N23 10.7		06	268 58.6	N22 58.1		06	268 50.7	N22 41.9
07	284 07.1	10.5		07	283 58.5	57.9		07	283 50.6	41.6
08	299 06.9	10.4		08	298 58.4	57.7		08	298 50.5	41.4
09	314 06.8 ··	10.2		09	313 58.3 ··	57.5		09	313 50.4 ··	41.1
10	329 06.7	10.1		10	328 58.2	57.3		10	328 50.3	40.9
11	344 06.6	09.9		11	343 58.0	57.1		11	343 50.2	40.6
12	359 06.4	N23 09.7		12	358 57.9	N22 56.9		12	358 50.1	N22 40.4
13	14 06.3	09.6		13	13 57.8	56.7		13	13 50.0	40.1
14	29 06.2	09.4		14	28 57.7	56.4		14	28 49.9	39.9
15	44 06.1 ··	09.3		15	43 57.6 ··	56.2		15	43 49.8 ··	39.6
16	59 05.9	09.1		16	58 57.5	56.0		16	58 49.7	39.4
17	74 05.8	09.0		17	73 57.4	55.8		17	73 49.6	39.1
18	89 05.7	N23 08.8		18	88 57.2	N22 55.6		18	88 49.5	N22 38.8
19	104 05.6	08.7		19	103 57.1	55.4		19	103 49.4	38.6
20	119 05.5	08.5		20	118 57.0	55.2		20	118 49.2	38.3
21	134 05.3 ··	08.3		21	133 56.9 ··	55.0		21	133 49.1 ··	38.1
22	149 05.2	08.2		22	148 56.8	54.8		22	148 49.0	37.8
23	164 05.1	08.0		23	163 56.7	54.6		23	163 48.9	37.5
1 00	179 05.0	N23 07.8		4 00	178 56.6	N22 54.4		7 00	178 48.8	N22 37.3
01	194 04.9	07.7		01	193 56.4	54.1		01	193 48.7	37.0
02	209 04.7	07.5		02	208 56.3	53.9		02	208 48.6	36.8
03	224 04.6 ··	07.4		03	223 56.2 ··	53.7		03	223 48.5 ··	36.5
04	239 04.5	07.2		04	238 56.1	53.5		04	238 48.4	36.2
05	254 04.4	07.0		05	253 56.0	53.3		05	253 48.3	36.0
06	269 04.3	N23 06.9		06	268 55.9	N22 53.1		06	268 48.2	N22 35.7
07	284 04.1	06.7		07	283 55.8	52.9		07	283 48.1	35.4
08	299 04.0	06.5		08	298 55.7	52.6		08	298 48.0	35.2
09	314 03.9 ··	06.4		09	313 55.6 ··	52.4		09	313 47.9 ··	34.9
10	329 03.8	06.2		10	328 55.4	52.2		10	328 47.8	34.6
11	344 03.7	06.0		11	343 55.3	52.0		11	343 47.7	34.4
12	359 03.5	N23 05.9		12	358 55.2	N22 51.8		12	358 47.6	N22 34.1
13	14 03.4	05.7		13	13 55.1	51.5		13	13 47.5	33.8
14	29 03.3	05.5		14	28 55.0	51.3		14	28 47.4	33.5
15	44 03.2 ··	05.3		15	43 54.9 ··	51.1		15	43 47.3 ··	33.3
16	59 03.1	05.2		16	58 54.8	50.9		16	58 47.2	33.0
17	74 02.9	05.0		17	73 54.7	50.7		17	73 47.1	32.7
18	89 02.8	N23 04.8		18	88 54.6	N22 50.4		18	88 47.0	N22 32.5
19	104 02.7	04.6		19	103 54.4	50.2		19	103 46.9	32.2
20	119 02.6	04.5		20	118 54.3	50.0		20	118 46.8	31.9
21	134 02.5 ··	04.3		21	133 54.2 ··	49.8		21	133 46.7 ··	31.6
22	149 02.3	04.1		22	148 54.1	49.5		22	148 46.6	31.4
23	164 02.2	03.9		23	163 54.0	49.3		23	163 46.5	31.1
2 00	179 02.1	N23 03.8		5 00	178 53.9	N22 49.1		8 00	178 46.4	N22 30.8
01	194 02.0	03.6		01	193 53.8	48.8		01	193 46.3	30.5
02	209 01.9	03.4		02	208 53.7	48.6		02	208 46.2	30.2
03	224 01.7 ··	03.2		03	223 53.6 ··	48.4		03	223 46.1 ··	30.0
04	239 01.6	03.0		04	238 53.5	48.1		04	238 46.1	29.7
05	254 01.5	02.8		05	253 53.4	47.9		05	253 46.0	29.4
06	269 01.4	N23 02.7		06	268 53.2	N22 47.7		06	268 45.9	N22 29.1
07	284 01.3	02.5		07	283 53.1	47.4		07	283 45.8	28.8
08	299 01.2	02.3		08	298 53.0	47.2		08	298 45.7	28.6
09	314 01.0 ··	02.1		09	313 52.9 ··	47.0		09	313 45.6 ··	28.3
10	329 00.9	01.9		10	328 52.8	46.7		10	328 45.5	28.0
11	344 00.8	01.7		11	343 52.7	46.5		11	343 45.4	27.7
12	359 00.7	N23 01.5		12	358 52.6	N22 46.3		12	358 45.3	N22 27.4
13	14 00.6	01.4		13	13 52.5	46.0		13	13 45.2	27.1
14	29 00.5	01.2		14	28 52.4	45.8		14	28 45.1	26.8
15	44 00.3 ··	01.0		15	43 52.3 ··	45.6		15	43 45.0 ··	26.5
16	59 00.2	00.8		16	58 52.2	45.3		16	58 44.9	26.3
17	74 00.1	00.6		17	73 52.1	45.1		17	73 44.8	26.0
18	89 00.0	N23 00.4		18	88 52.0	N22 44.8		18	88 44.7	N22 25.7
19	103 59.9	00.2		19	103 51.9	44.6		19	103 44.6	25.4
20	118 59.8	23 00.0		20	118 51.7	44.4		20	118 44.5	25.1
21	133 59.7	22 59.8		21	133 51.6 ··	44.1		21	133 44.4 ··	24.8
22	148 59.5	59.6		22	148 51.5	43.9		22	148 44.3	24.5
23	163 59.4	59.4		23	163 51.4	43.6		23	163 44.2	24.2

d	h	G.H.A. °	′	Dec. °	′
9	00	178	44.1	N22	23.9
	01	193	44.0		23.6
	02	208	43.9		23.3
	03	223	43.8	··	23.0
	04	238	43.8		22.7
	05	253	43.7		22.4
	06	268	43.6	N22	22.1
	07	283	43.5		21.8
	08	298	43.4		21.5
	09	313	43.3	··	21.2
	10	328	43.2		20.9
	11	343	43.1		20.6
	12	358	43.0	N22	20.3
	13	13	42.9		20.0
	14	28	42.8		19.7
	15	43	42.7	··	19.4
	16	58	42.6		19.1
	17	73	42.6		18.8
	18	88	42.5	N22	18.5
	19	103	42.4		18.2
	20	118	42.3		17.9
	21	133	42.2	··	17.6
	22	148	42.1		17.3
	23	163	42.0		17.0
10	00	178	41.9	N22	16.7
	01	193	41.8		16.4
	02	208	41.7		16.0
	03	223	41.6	··	15.7
	04	238	41.6		15.4
	05	253	41.5		15.1
	06	268	41.4	N22	14.8
	07	283	41.3		14.5
	08	298	41.2		14.2
	09	313	41.1	··	13.8
	10	328	41.0		13.5
	11	343	40.9		13.2
	12	358	40.8	N22	12.9
	13	13	40.8		12.6
	14	28	40.7		12.2
	15	43	40.6	··	11.9
	16	58	40.5		11.6
	17	73	40.4		11.3
	18	88	40.3	N22	11.0
	19	103	40.2		10.6
	20	118	40.1		10.3
	21	133	40.1	··	10.0
	22	148	40.0		09.7
	23	163	39.9		09.3
11	00	178	39.8	N22	09.0
	01	193	39.7		08.7
	02	208	39.6		08.4
	03	223	39.5	··	08.0
	04	238	39.5		07.7
	05	253	39.4		07.4
	06	268	39.3	N22	07.0
	07	283	39.2		06.7
	08	298	39.1		06.4
	09	313	39.0	··	06.1
	10	328	38.9		05.7
	11	343	38.9		05.4
	12	358	38.8	N22	05.0
	13	13	38.7		04.7
	14	28	38.6		04.4
	15	43	38.5	··	04.0
	16	58	38.4		03.7
	17	73	38.4		03.4
	18	88	38.3	N22	03.0
	19	103	38.2		02.7
	20	118	38.1		02.4
	21	133	38.0	··	02.0
	22	148	37.9		01.7
	23	163	37.9		01.3

d	h	G.H.A. °	′	Dec. °	′
12	00	178	37.8	N22	01.0
	01	193	37.7		00.6
	02	208	37.6		00.3
	03	223	37.5	22	00.0
	04	238	37.5	21	59.6
	05	253	37.4		59.3
	06	268	37.3	N21	58.9
	07	283	37.2		58.6
	08	298	37.1		58.2
	09	313	37.0	··	57.9
	10	328	37.0		57.5
	11	343	36.9		57.2
	12	358	36.8	N21	56.8
	13	13	36.7		56.5
	14	28	36.7		56.1
	15	43	36.6	··	55.8
	16	58	36.5		55.4
	17	73	36.4		55.1
	18	88	36.3	N21	54.7
	19	103	36.3		54.4
	20	118	36.2		54.0
	21	133	36.1	··	53.7
	22	148	36.0		53.3
	23	163	35.9		52.9
13	00	178	35.9	N21	52.6
	01	193	35.8		52.2
	02	208	35.7		51.9
	03	223	35.6	··	51.5
	04	238	35.6		51.1
	05	253	35.5		50.8
	06	268	35.4	N21	50.4
	07	283	35.3		50.1
	08	298	35.3		49.7
	09	313	35.2	··	49.3
	10	328	35.1		49.0
	11	343	35.0		48.6
	12	358	35.0	N21	48.2
	13	13	34.9		47.9
	14	28	34.8		47.5
	15	43	34.7	··	47.1
	16	58	34.7		46.8
	17	73	34.6		46.4
	18	88	34.5	N21	46.0
	19	103	34.4		45.7
	20	118	34.4		45.3
	21	133	34.3	··	44.9
	22	148	34.2		44.6
	23	163	34.1		44.2
14	00	178	34.1	N21	43.8
	01	193	34.0		43.4
	02	208	33.9		43.1
	03	223	33.9	··	42.7
	04	238	33.8		42.3
	05	253	33.7		41.9
	06	268	33.6	N21	41.6
	07	283	33.6		41.2
	08	298	33.5		40.8
	09	313	33.4	··	40.4
	10	328	33.4		40.0
	11	343	33.3		39.7
	12	358	33.2	N21	39.3
	13	13	33.2		38.9
	14	28	33.1		38.5
	15	43	33.0	··	38.1
	16	58	32.9		37.7
	17	73	32.9		37.4
	18	88	32.8	N21	37.0
	19	103	32.7		36.6
	20	118	32.7		36.2
	21	133	32.6	··	35.8
	22	148	32.5		35.4
	23	163	32.5		35.0

d	h	G.H.A. °	′	Dec. °	′
15	00	178	32.4	N21	34.7
	01	193	32.3		34.3
	02	208	32.3		33.9
	03	223	32.2	··	33.5
	04	238	32.1		33.1
	05	253	32.1		32.7
	06	268	32.0	N21	32.3
	07	283	31.9		31.9
	08	298	31.9		31.5
	09	313	31.8	··	31.1
	10	328	31.7		30.7
	11	343	31.7		30.3
	12	358	31.6	N21	29.9
	13	13	31.5		29.6
	14	28	31.5		29.2
	15	43	31.4	··	28.8
	16	58	31.4		28.4
	17	73	31.3		28.0
	18	88	31.2	N21	27.6
	19	103	31.2		27.2
	20	118	31.1		26.8
	21	133	31.0	··	26.4
	22	148	31.0		26.0
	23	163	30.9		25.5
16	00	178	30.8	N21	25.1
	01	193	30.8		24.7
	02	208	30.7		24.3
	03	223	30.7	··	23.9
	04	238	30.6		23.5
	05	253	30.5		23.1
	06	268	30.5	N21	22.7
	07	283	30.4		22.3
	08	298	30.4		21.9
	09	313	30.3	··	21.5
	10	328	30.2		21.1
	11	343	30.2		20.7
	12	358	30.1	N21	20.3
	13	13	30.1		19.8
	14	28	30.0		19.4
	15	43	29.9	··	19.0
	16	58	29.9		18.6
	17	73	29.8		18.2
	18	88	29.8	N21	17.8
	19	103	29.7		17.4
	20	118	29.7		16.9
	21	133	29.6	··	16.5
	22	148	29.5		16.1
	23	163	29.5		15.7
17	00	178	29.4	N21	15.3
	01	193	29.4		14.8
	02	208	29.3		14.4
	03	223	29.3	··	14.0
	04	238	29.2		13.6
	05	253	29.1		13.2
	06	268	29.1	N21	12.7
	07	283	29.0		12.3
	08	298	29.0		11.9
	09	313	28.9	··	11.5
	10	328	28.9		11.0
	11	343	28.8		10.6
	12	358	28.8	N21	10.2
	13	13	28.7		09.8
	14	28	28.7		09.3
	15	43	28.6	··	08.9
	16	58	28.6		08.5
	17	73	28.5		08.1
	18	88	28.4	N21	07.6
	19	103	28.4		07.2
	20	118	28.3		06.8
	21	133	28.3	··	06.3
	22	148	28.2		05.9
	23	163	28.2		05.5

d h	G.H.A.	Dec.
18 00	178 28.1	N21 05.0
01	193 28.1	04.6
02	208 28.0	04.2
03	223 28.0 ··	03.7
04	238 27.9	03.3
05	253 27.9	02.8
06	268 27.8	N21 02.4
07	283 27.8	02.0
08	298 27.7	01.5
09	313 27.7 ··	01.1
10	328 27.6	00.7
11	343 27.6	21 00.2
12	358 27.5	N20 59.8
13	13 27.5	59.3
14	28 27.5	58.9
15	43 27.4 ··	58.4
16	58 27.4	58.0
17	73 27.3	57.6
18	88 27.3	N20 57.1
19	103 27.2	56.7
20	118 27.2	56.2
21	133 27.1 ··	55.8
22	148 27.1	55.3
23	163 27.0	54.9
19 00	178 27.0	N20 54.4
01	193 26.9	54.0
02	208 26.9	53.5
03	223 26.9 ··	53.1
04	238 26.8	52.6
05	253 26.8	52.2
06	268 26.7	N20 51.7
07	283 26.7	51.3
08	298 26.6	50.8
09	313 26.6 ··	50.4
10	328 26.6	49.9
11	343 26.5	49.5
12	358 26.5	N20 49.0
13	13 26.4	48.5
14	28 26.4	48.1
15	43 26.3 ··	47.6
16	58 26.3	47.2
17	73 26.3	46.7
18	88 26.2	N20 46.3
19	103 26.2	45.8
20	118 26.1	45.3
21	133 26.1 ··	44.9
22	148 26.1	44.4
23	163 26.0	44.0
20 00	178 26.0	N20 43.5
01	193 25.9	43.0
02	208 25.9	42.6
03	223 25.9 ··	42.1
04	238 25.8	41.6
05	253 25.8	41.2
06	268 25.8	N20 40.7
07	283 25.7	40.2
08	298 25.7	39.8
09	313 25.6 ··	39.3
10	328 25.6	38.8
11	343 25.6	38.4
12	358 25.5	N20 37.9
13	13 25.5	37.4
14	28 25.5	36.9
15	43 25.4 ··	36.5
16	58 25.4	36.0
17	73 25.4	35.5
18	88 25.3	N20 35.0
19	103 25.3	34.6
20	118 25.3	34.1
21	133 25.2 ··	33.6
22	148 25.2	33.1
23	163 25.1	32.7

d h	G.H.A.	Dec.
21 00	178 25.1	N20 32.2
01	193 25.1	31.7
02	208 25.1	31.2
03	223 25.0 ··	30.8
04	238 25.0	30.3
05	253 25.0	29.8
06	268 24.9	N20 29.3
07	283 24.9	28.8
08	298 24.9	28.4
09	313 24.8 ··	27.9
10	328 24.8	27.4
11	343 24.8	26.9
12	358 24.7	N20 26.4
13	13 24.7	25.9
14	28 24.7	25.4
15	43 24.6 ··	25.0
16	58 24.6	24.5
17	73 24.6	24.0
18	88 24.6	N20 23.5
19	103 24.5	23.0
20	118 24.5	22.5
21	133 24.5 ··	22.0
22	148 24.5	21.5
23	163 24.4	21.0
22 00	178 24.4	N20 20.6
01	193 24.4	20.1
02	208 24.3	19.6
03	223 24.3 ··	19.1
04	238 24.3	18.6
05	253 24.3	18.1
06	268 24.2	N20 17.6
07	283 24.2	17.1
08	298 24.2	16.6
09	313 24.2 ··	16.1
10	328 24.1	15.6
11	343 24.1	15.1
12	358 24.1	N20 14.6
13	13 24.1	14.1
14	28 24.0	13.6
15	43 24.0 ··	13.1
16	58 24.0	12.6
17	73 24.0	12.1
18	88 24.0	N20 11.6
19	103 23.9	11.1
20	118 23.9	10.6
21	133 23.9 ··	10.1
22	148 23.9	09.6
23	163 23.8	09.1
23 00	178 23.8	N20 08.6
01	193 23.8	08.1
02	208 23.8	07.6
03	223 23.8 ··	07.1
04	238 23.7	06.5
05	253 23.7	06.0
06	268 23.7	N20 05.5
07	283 23.7	05.0
08	298 23.7	04.5
09	313 23.7 ··	04.0
10	328 23.6	03.5
11	343 23.6	03.0
12	358 23.6	N20 02.5
13	13 23.6	01.9
14	28 23.6	01.4
15	43 23.5 ··	00.9
16	58 23.5	20 00.4
17	73 23.5	19 59.9
18	88 23.5	N19 59.4
19	103 23.5	58.9
20	118 23.5	58.3
21	133 23.5 ··	57.8
22	148 23.4	57.3
23	163 23.4	56.8

d h	G.H.A.	Dec.
24 00	178 23.4	N19 56.3
01	193 23.4	55.7
02	208 23.4	55.2
03	223 23.4 ··	54.7
04	238 23.4	54.2
05	253 23.3	53.7
06	268 23.3	N19 53.1
07	283 23.3	52.6
08	298 23.3	52.1
09	313 23.3 ··	51.6
10	328 23.3	51.0
11	343 23.3	50.5
12	358 23.3	N19 50.0
13	13 23.2	49.4
14	28 23.2	48.9
15	43 23.2 ··	48.4
16	58 23.2	47.9
17	73 23.2	47.3
18	88 23.2	N19 46.8
19	103 23.2	46.3
20	118 23.2	45.7
21	133 23.2 ··	45.2
22	148 23.2	44.7
23	163 23.1	44.1
25 00	178 23.1	N19 43.6
01	193 23.1	43.1
02	208 23.1	42.5
03	223 23.1 ··	42.0
04	238 23.1	41.5
05	253 23.1	40.9
06	268 23.1	N19 40.4
07	283 23.1	39.9
08	298 23.1	39.3
09	313 23.1 ··	38.8
10	328 23.1	38.2
11	343 23.1	37.7
12	358 23.1	N19 37.2
13	13 23.1	36.6
14	28 23.1	36.1
15	43 23.0 ··	35.5
16	58 23.0	35.0
17	73 23.0	34.5
18	88 23.0	N19 33.9
19	103 23.0	33.4
20	118 23.0	32.8
21	133 23.0 ··	32.3
22	148 23.0	31.7
23	163 23.0	31.2
26 00	178 23.0	N19 30.6
01	193 23.0	30.1
02	208 23.0	29.5
03	223 23.0 ··	29.0
04	238 23.0	28.4
05	253 23.0	27.9
06	268 23.0	N19 27.3
07	283 23.0	26.8
08	298 23.0	26.2
09	313 23.0 ··	25.7
10	328 23.0	25.1
11	343 23.0	24.6
12	358 23.0	N19 24.0
13	13 23.0	23.5
14	28 23.0	22.9
15	43 23.0 ··	22.4
16	58 23.0	21.8
17	73 23.0	21.2
18	88 23.0	N19 20.7
19	103 23.0	20.1
20	118 23.0	19.6
21	133 23.0 ··	19.0
22	148 23.0	18.5
23	163 23.0	17.9

	G.H.A.	Dec.		G.H.A.	Dec.		G.H.A.	Dec.
d h	° ′	° ′	d h	° ′	° ′	d h	° ′	° ′
27 00	178 23.1	N19 17.3	30 00	178 24.1	N18 35.5	2 00	178 26.4	N17 51.0
01	193 23.1	16.8	01	193 24.1	34.9	01	193 26.5	50.4
02	208 23.1	16.2	02	208 24.1	34.3	02	208 26.5	49.7
03	223 23.1	·· 15.6	03	223 24.1	·· 33.7	03	223 26.6	·· 49.1
04	238 23.1	15.1	04	238 24.2	33.1	04	238 26.6	48.4
05	253 23.1	14.5	05	253 24.2	32.5	05	253 26.6	47.8
06	268 23.1	N19 14.0	06	268 24.2	N18 31.9	06	268 26.7	N17 47.2
07	283 23.1	13.4	07	283 24.2	31.3	07	283 26.7	46.5
08	298 23.1	12.8	08	298 24.3	30.7	08	298 26.8	45.9
09	313 23.1	·· 12.3	09	313 24.3	·· 30.1	09	313 26.8	·· 45.2
10	328 23.1	11.7	10	328 24.3	29.5	10	328 26.9	44.6
11	343 23.1	11.1	11	343 24.3	28.9	11	343 26.9	44.0
12	358 23.1	N19 10.6	12	358 24.4	N18 28.3	12	358 27.0	N17 43.3
13	13 23.1	10.0	13	13 24.4	27.7	13	13 27.0	42.7
14	28 23.1	09.4	14	28 24.4	27.1	14	28 27.0	42.0
15	43 23.2	·· 08.9	15	43 24.4	·· 26.5	15	43 27.1	·· 41.4
16	58 23.2	08.3	16	58 24.5	25.9	16	58 27.1	40.7
17	73 23.2	07.7	17	73 24.5	25.3	17	73 27.2	40.1
18	88 23.2	N19 07.2	18	88 24.5	N18 24.7	18	88 27.2	N17 39.4
19	103 23.2	06.6	19	103 24.6	24.0	19	103 27.3	38.8
20	118 23.2	06.0	20	118 24.6	23.4	20	118 27.3	38.1
21	133 23.2	·· 05.4	21	133 24.6	·· 22.8	21	133 27.4	·· 37.5
22	148 23.2	04.9	22	148 24.6	22.2	22	148 27.4	36.9
23	163 23.2	04.3	23	163 24.7	21.6	23	163 27.5	36.2
28 00	178 23.2	N19 03.7	31 00	178 24.7	N18 21.0	3 00	178 27.5	N17 35.6
01	193 23.2	03.1	01	193 24.7	20.4	01	193 27.6	34.9
02	208 23.3	02.6	02	208 24.8	19.8	02	208 27.6	34.3
03	223 23.3	·· 02.0	03	223 24.8	·· 19.1	03	223 27.7	·· 33.6
04	230 23.3	01.4	04	238 24.8	18.5	04	238 27.7	33.0
05	253 23.3	00.8	05	253 24.9	17.9	05	253 27.8	32.3
06	268 23.3	N19 00.3	06	268 24.9	N18 17.3	06	268 27.8	N17 31.7
07	283 23.3	18 59.7	07	283 24.9	16.7	07	283 27.9	31.0
08	298 23.3	59.1	08	298 24.9	16.1	08	298 27.9	30.3
09	313 23.3	·· 58.5	09	313 25.0	·· 15.5	09	313 28.0	·· 29.7
10	328 23.4	57.9	10	328 25.0	14.8	10	328 28.0	29.0
11	343 23.4	57.4	11	343 25.0	14.2	11	343 28.1	28.4
12	358 23.4	N18 56.8	12	358 25.1	N18 13.6	12	358 28.1	N17 27.7
13	13 23.4	56.2	13	13 25.1	13.0	13	13 28.2	27.1
14	28 23.4	55.6	14	28 25.1	12.4	14	28 28.2	26.4
15	43 23.4	·· 55.0	15	43 25.2	·· 11.7	15	43 28.3	·· 25.8
16	58 23.4	54.5	16	58 25.2	11.1	16	58 28.3	25.1
17	73 23.5	53.9	17	73 25.2	10.5	17	73 28.4	24.4
18	88 23.5	N18 53.3	18	88 25.3	N18 09.9	18	88 28.4	N17 23.8
19	103 23.5	52.7	19	103 25.3	09.3	19	103 28.5	23.1
20	118 23.5	52.1	20	118 25.3	08.6	20	118 28.5	22.5
21	133 23.5	·· 51.5	21	133 25.4	·· 08.0	21	133 28.6	·· 21.8
22	148 23.5	51.0	22	148 25.4	07.4	22	148 28.6	21.2
23	163 23.6	50.4	23	163 25.5	06.8	23	163 28.7	20.5
29 00	178 23.6	N18 49.8	1 00	178 25.5	N18 06.1	4 00	178 28.8	N17 19.8
01	193 23.6	49.2	01	193 25.5	05.5	01	193 28.8	19.2
02	208 23.6	48.6	02	208 25.6	04.9	02	208 28.9	18.5
03	223 23.6	·· 48.0	03	223 25.6	·· 04.3	03	223 28.9	·· 17.8
04	238 23.6	47.4	04	238 25.6	03.6	04	238 29.0	17.2
05	253 23.7	46.8	05	253 25.7	03.0	05	253 29.0	16.5
06	268 23.7	N18 46.3	06	268 25.7	N18 02.4	06	268 29.1	N17 15.9
07	283 23.7	45.7	07	283 25.7	01.8	07	283 29.1	15.2
08	298 23.7	45.1	08	298 25.8	01.1	08	298 29.2	14.5
09	313 23.7	·· 44.5	09	313 25.8	18 00.5	09	313 29.3	·· 13.9
10	328 23.8	43.9	10	328 25.9	17 59.9	10	328 29.3	13.2
11	343 23.8	43.3	11	343 25.9	59.2	11	343 29.4	12.5
12	358 23.8	N18 42.7	12	358 25.9	N17 58.6	12	358 29.4	N17 11.9
13	13 23.8	42.1	13	13 26.0	58.0	13	13 29.5	11.2
14	28 23.8	41.5	14	28 26.0	57.3	14	28 29.5	10.5
15	43 23.9	··· 40.9	15	43 26.1	·· 56.7	15	43 29.6	·· 09.9
16	58 23.9	40.3	16	58 26.1	56.1	16	58 29.7	09.2
17	73 23.9	39.7	17	73 26.1	55.4	17	73 29.7	08.5
18	88 23.9	N18 39.1	18	88 26.2	N17 54.8	18	88 29.8	N17 07.9
19	103 23.9	38.5	19	103 26.2	54.2	19	103 29.8	07.2
20	118 24.0	37.9	20	118 26.3	53.5	20	118 29.9	06.5
21	133 24.0	·· 37.3	21	133 26.3	·· 52.9	21	133 30.0	·· 05.8
22	148 24.0	36.7	22	148 26.3	52.3	22	148 30.0	05.2
23	163 24.0	36.1	23	163 26.4	51.6	23	163 30.1	04.5

AUGUST

197

5, 6, 7

d h	G.H.A.	Dec.
5 00	178 30.1	N17 03.8
01	193 30.2	03.2
02	208 30.3	02.5
03	223 30.3 ··	01.8
04	238 30.4	01.1
05	253 30.4	17 00.5
06	268 30.5	N16 59.8
07	283 30.6	59.1
08	298 30.6	58.4
09	313 30.7 ··	57.8
10	328 30.8	57.1
11	343 30.8	56.4
12	358 30.9	N16 55.7
13	13 30.9	55.0
14	28 31.0	54.4
15	43 31.1 ··	53.7
16	58 31.1	53.0
17	73 31.2	52.3
18	88 31.3	N16 51.6
19	103 31.3	51.0
20	118 31.4	50.3
21	133 31.5 ··	49.6
22	148 31.5	48.9
23	163 31.6	48.2
6 00	178 31.7	N16 47.5
01	193 31.7	46.9
02	208 31.8	46.2
03	223 31.9 ··	45.5
04	238 31.9	44.8
05	253 32.0	44.1
06	268 32.1	N16 43.4
07	283 32.1	42.7
08	298 32.2	42.1
09	313 32.3 ··	41.4
10	328 32.3	40.7
11	343 32.4	40.0
12	358 32.5	N16 39.3
13	13 32.6	38.6
14	28 32.6	37.9
15	43 32.7 ··	37.2
16	58 32.8	36.5
17	73 32.8	35.8
18	88 32.9	N16 35.2
19	103 33.0	34.5
20	118 33.1	33.8
21	133 33.1 ··	33.1
22	148 33.2	32.4
23	163 33.3	31.7
7 00	178 33.3	N16 31.0
01	193 33.4	30.3
02	208 33.5	29.6
03	223 33.6 ··	28.9
04	238 33.6	28.2
05	253 33.7	27.5
06	268 33.8	N16 26.8
07	283 33.9	26.1
08	298 33.9	25.4
09	313 34.0 ··	24.7
10	328 34.1	24.0
11	343 34.2	23.3
12	358 34.2	N16 22.6
13	13 34.3	21.9
14	28 34.4	21.2
15	43 34.5 ··	20.5
16	58 34.5	19.8
17	73 34.6	19.1
18	88 34.7	N16 18.4
19	103 34.8	17.7
20	118 34.8	17.0
21	133 34.9 ··	16.3
22	148 35.0	15.6
23	163 35.1	14.9

8, 9, 10

d h	G.H.A.	Dec.
8 00	178 35.2	N16 14.2
01	193 35.2	13.5
02	208 35.3	12.8
03	223 35.4 ··	12.1
04	238 35.5	11.3
05	253 35.6	10.6
06	268 35.6	N16 09.9
07	283 35.7	09.2
08	298 35.8	08.5
09	313 35.9 ··	07.8
10	328 36.0	07.1
11	343 36.0	06.4
12	358 36.1	N16 05.7
13	13 36.2	05.0
14	28 36.3	04.2
15	43 36.4 ··	03.5
16	58 36.5	02.8
17	73 36.5	02.1
18	88 36.6	N16 01.4
19	103 36.7	00.7
20	118 36.8	16 00.0
21	133 36.9	15 59.2
22	148 37.0	58.5
23	163 37.0	57.8
9 00	178 37.1	N15 57.1
01	193 37.2	56.4
02	208 37.3	55.7
03	223 37.4 ··	54.9
04	238 37.5	54.2
05	253 37.5	53.5
06	268 37.6	N15 52.8
07	283 37.7	52.1
08	298 37.8	51.3
09	313 37.9 ··	50.6
10	328 38.0	49.9
11	343 38.1	49.2
12	358 38.2	N15 48.5
13	13 38.2	47.7
14	28 38.3	47.0
15	43 38.4 ··	46.3
16	58 38.5	45.6
17	73 38.6	44.8
18	88 38.7	N15 44.1
19	103 38.8	43.4
20	118 38.9	42.7
21	133 39.0 ··	41.9
22	148 39.0	41.2
23	163 39.1	40.5
10 00	178 39.2	N15 39.8
01	193 39.3	39.0
02	208 39.4	38.3
03	223 39.5 ··	37.6
04	238 39.6	36.8
05	253 39.7	36.1
06	268 39.8	N15 35.4
07	283 39.9	34.7
08	298 40.0	33.9
09	313 40.1 ··	33.2
10	328 40.1	32.5
11	343 40.2	31.7
12	358 40.3	N15 31.0
13	13 40.4	30.3
14	28 40.5	29.5
15	43 40.6 ··	28.8
16	58 40.7	28.1
17	73 40.8	27.3
18	88 40.9	N15 26.6
19	103 41.0	25.9
20	118 41.1	25.1
21	133 41.2 ··	24.4
22	148 41.3	23.6
23	163 41.4	22.9

11, 12, 13

d h	G.H.A.	Dec.
11 00	178 41.5	N15 22.2
01	193 41.6	21.4
02	208 41.7	20.7
03	223 41.8 ··	20.0
04	238 41.9	19.2
05	253 42.0	18.5
06	268 42.1	N15 17.7
07	283 42.2	17.0
08	298 42.3	16.3
09	313 42.4 ··	15.5
10	328 42.5	14.8
11	343 42.6	14.0
12	358 42.7	N15 13.3
13	13 42.7	12.5
14	28 42.8	11.8
15	43 42.9 ··	11.1
16	58 43.1	10.3
17	73 43.2	09.6
18	88 43.3	N15 08.8
19	103 43.4	08.1
20	118 43.5	07.3
21	133 43.6 ··	06.6
22	148 43.7	05.8
23	163 43.8	05.1
12 00	178 43.9	N15 04.3
01	193 44.0	03.6
02	208 44.1	02.8
03	223 44.2 ··	02.1
04	238 44.3	01.3
05	253 44.4	15 00.6
06	268 44.5	N14 59.8
07	283 44.6	59.1
08	298 44.7	58.3
09	313 44.8 ··	57.6
10	328 44.9	56.8
11	343 45.0	56.1
12	358 45.1	N14 55.3
13	13 45.2	54.6
14	28 45.3	53.8
15	43 45.4 ··	53.1
16	58 45.5	52.3
17	73 45.6	51.6
18	88 45.7	N14 50.8
19	103 45.9	50.0
20	118 46.0	49.3
21	133 46.1 ··	48.5
22	148 46.2	47.8
23	163 46.3	47.0
13 00	178 46.4	N14 46.3
01	193 46.5	45.5
02	208 46.6	44.7
03	223 46.7 ··	44.0
04	238 46.8	43.2
05	253 46.9	42.5
06	268 47.0	N14 41.7
07	283 47.2	40.9
08	298 47.3	40.2
09	313 47.4 ··	39.4
10	328 47.5	38.7
11	343 47.6	37.9
12	358 47.7	N14 37.1
13	13 47.8	36.4
14	28 47.9	35.6
15	43 48.0 ··	34.8
16	58 48.1	34.1
17	73 48.3	33.3
18	88 48.4	N14 32.5
19	103 48.5	31.8
20	118 48.6	31.0
21	133 48.7 ··	30.3
22	148 48.8	29.5
23	163 48.9	28.7

		G.H.A.	Dec.			G.H.A.	Dec.			G.H.A.	Dec.
d h		° ′	° ′	d h		° ′	° ′	d h		° ′	° ′
14	00	178 49.1	N14 28.0	17	00	178 57.9	N13 31.7	20	00	179 07.8	N12 33.4
	01	193 49.2	27.2		01	193 58.0	30.9		01	194 08.0	32.6
	02	208 49.3	26.4		02	208 58.1	30.1		02	209 08.1	31.8
	03	223 49.4	·· 25.6		03	223 58.2	·· 29.3		03	224 08.3	·· 31.0
	04	238 49.5	24.9		04	238 58.4	28.5		04	239 08.4	30.2
	05	253 49.6	24.1		05	253 58.5	27.7		05	254 08.6	29.3
	06	268 49.7	N14 23.3		06	268 58.6	N13 26.9		06	269 08.7	N12 28.5
	07	283 49.9	22.6		07	283 58.8	26.1		07	284 08.9	27.7
	08	298 50.0	21.8		08	298 58.9	25.3		08	299 09.0	26.9
	09	313 50.1	·· 21.0		09	313 59.0	·· 24.5		09	314 09.2	·· 26.0
	10	328 50.2	20.3		10	328 59.2	23.7		10	329 09.3	25.2
	11	343 50.3	19.5		11	343 59.3	22.9		11	344 09.5	24.4
	12	358 50.4	N14 18.7		12	358 59.4	N13 22.1		12	359 09.6	N12 23.6
	13	13 50.6	17.9		13	13 59.6	21.3		13	14 09.8	22.7
	14	28 50.7	17.2		14	28 59.7	20.5		14	29 09.9	21.9
	15	43 50.8	·· 16.4		15	43 59.8	·· 19.7		15	44 10.0	·· 21.1
	16	58 50.9	15.6		16	59 00.0	18.9		16	59 10.2	20.3
	17	73 51.0	14.8		17	74 00.1	18.1		17	74 10.3	19.4
	18	00 51.1	N14 14.1		18	07 00.2	N13 17.3		18	07 10.5	N12 10.6
	19	103 51.3	13.3		19	104 00.4	16.5		19	104 10.6	17.8
	20	118 51.4	12.5		20	119 00.5	15.7		20	119 10.8	16.9
	21	133 51.5	·· 11.7		21	134 00.6	·· 14.9		21	134 10.9	·· 16.1
	22	148 51.6	11.0		22	149 00.8	14.1		22	149 11.1	15.3
	23	163 51.7	10.2		23	164 00.9	13.3		23	164 11.3	14.5
15	00	178 51.9	N14 09.4	18	00	179 01.1	N13 12.5	21	00	179 11.4	N12 13.6
	01	193 52.0	08.6		01	194 01.2	11.7		01	194 11.6	12.8
	02	208 52.1	07.9		02	209 01.3	10.9		02	209 11.7	12.0
	03	223 52.2	·· 07.1		03	224 01.5	·· 10.0		03	224 11.9	·· 11.1
	04	238 52.3	06.3		04	239 01.6	09.2		04	239 12.0	10.3
	05	253 52.5	05.5		05	254 01.7	08.4		05	254 12.2	09.5
	06	268 52.6	N14 04.7		06	269 01.9	N13 07.6		06	269 12.3	N12 08.7
	07	283 52.7	04.0		07	284 02.0	06.8		07	284 12.5	07.8
	08	298 52.8	03.2		08	299 02.1	06.0		08	299 12.6	07.0
	09	313 52.9	·· 02.4		09	314 02.3	·· 05.2		09	314 12.8	·· 06.2
	10	328 53.1	01.6		10	329 02.4	04.4		10	329 12.9	05.3
	11	343 53.2	00.8		11	344 02.6	03.6		11	344 13.1	04.5
	12	358 53.3	N14 00.1		12	359 02.7	N13 02.8		12	359 13.2	N12 03.7
	13	13 53.4	13 59.3		13	14 02.8	02.0		13	14 13.4	02.8
	14	28 53.5	58.5		14	29 03.0	01.2		14	29 13.5	02.0
	15	43 53.7	·· 57.7		15	44 03.1	13 00.4		15	44 13.7	·· 01.2
	16	58 53.8	56.9		16	59 03.3	12 59.5		16	59 13.9	12 00.3
	17	73 53.9	56.1		17	74 03.4	58.7		17	74 14.0	11 59.5
	18	88 54.0	N13 55.4		18	89 03.5	N12 57.9		18	89 14.2	N11 58.6
	19	103 54.2	54.6		19	104 03.7	57.1		19	104 14.3	57.8
	20	118 54.3	53.8		20	119 03.8	56.3		20	119 14.5	57.0
	21	133 54.4	·· 53.0		21	134 04.0	·· 55.5		21	134 14.6	·· 56.1
	22	148 54.5	52.2		22	149 04.1	54.7		22	149 14.8	55.3
	23	163 54.7	51.4		23	164 04.2	53.9		23	164 14.9	54.5
16	00	178 54.8	N13 50.6	19	00	179 04.4	N12 53.1	22	00	179 15.1	N11 53.6
	01	193 54.9	49.9		01	194 04.5	52.2		01	194 15.3	52.8
	02	208 55.0	49.1		02	209 04.7	51.4		02	209 15.4	52.0
	03	223 55.2	·· 48.3		03	224 04.8	·· 50.6		03	224 15.6	·· 51.1
	04	238 55.3	47.5		04	239 04.9	49.8		04	239 15.7	50.3
	05	253 55.4	46.7		05	254 05.1	49.0		05	254 15.9	49.4
	06	268 55.5	N13 45.9		06	269 05.2	N12 48.2		06	269 16.0	N11 48.6
	07	283 55.7	45.1		07	284 05.4	47.4		07	284 16.2	47.8
	08	298 55.8	44.3		08	299 05.5	46.5		08	299 16.4	46.9
	09	313 55.9	·· 43.6		09	314 05.7	·· 45.7		09	314 16.5	·· 46.1
	10	328 56.0	42.8		10	329 05.8	44.9		10	329 16.7	45.2
	11	343 56.2	42.0		11	344 05.9	44.1		11	344 16.8	44.4
	12	358 56.3	N13 41.2		12	359 06.1	N12 43.3		12	359 17.0	N11 43.6
	13	13 56.4	40.4		13	14 06.2	42.5		13	14 17.1	42.7
	14	28 56.6	39.6		14	29 06.4	41.6		14	29 17.3	41.9
	15	43 56.7	·· 38.8		15	44 06.5	·· 40.8		15	44 17.5	·· 41.0
	16	58 56.8	38.0		16	59 06.7	40.0		16	59 17.6	40.2
	17	73 56.9	37.2		17	74 06.8	39.2		17	74 17.8	39.3
	18	88 57.1	N13 36.4		18	89 07.0	N12 38.4		18	89 17.9	N11 38.5
	19	103 57.2	35.6		19	104 07.1	37.5		19	104 18.1	37.7
	20	118 57.3	34.8		20	119 07.2	36.7		20	119 18.3	36.8
	21	133 57.5	·· 34.0		21	134 07.4	·· 35.9		21	134 18.4	·· 36.0
	22	148 57.6	33.3		22	149 07.5	35.1		22	149 18.6	35.1
	23	163 57.7	32.5		23	164 07.7	34.3		23	164 18.7	34.3

d h	G.H.A.	Dec.
23 00	179 18.9	N11 33.4
01	194 19.1	32.6
02	209 19.2	31.7
03	224 19.4	·· 30.9
04	239 19.6	30.1
05	254 19.7	29.2
06	269 19.9	N11 28.4
07	284 20.0	27.5
08	299 20.2	26.7
09	314 20.4	·· 25.8
10	329 20.5	25.0
11	344 20.7	24.1
12	359 20.9	N11 23.3
13	14 21.0	22.4
14	29 21.2	21.6
15	44 21.3	·· 20.7
16	59 21.5	19.9
17	74 21.7	19.0
18	89 21.8	N11 18.2
19	104 22.0	17.3
20	119 22.2	16.5
21	134 22.3	·· 15.6
22	149 22.5	14.8
23	164 22.7	13.9
24 00	179 22.8	N11 13.1
01	194 23.0	12.2
02	209 23.2	11.4
03	224 23.3	·· 10.5
04	239 23.5	09.7
05	254 23.7	08.8
06	269 23.8	N11 07.9
07	284 24.0	07.1
08	299 24.2	06.2
09	314 24.3	·· 05.4
10	329 24.5	04.5
11	344 24.7	03.7
12	359 24.8	N11 02.8
13	14 25.0	02.0
14	29 25.2	01.1
15	44 25.3	11 00.2
16	59 25.5	10 59.4
17	74 25.7	58.5
18	89 25.8	N10 57.7
19	104 26.0	56.8
20	119 26.2	56.0
21	134 26.4	·· 55.1
22	149 26.5	54.2
23	164 26.7	53.4
25 00	179 26.9	N10 52.5
01	194 27.0	51.7
02	209 27.2	50.8
03	224 27.4	·· 49.9
04	239 27.5	49.1
05	254 27.7	48.2
06	269 27.9	N10 47.3
07	284 28.1	46.5
08	299 28.2	45.6
09	314 28.4	·· 44.8
10	329 28.6	43.9
11	344 28.8	43.0
12	359 28.9	N10 42.2
13	14 29.1	41.3
14	29 29.3	40.4
15	44 29.4	·· 39.6
16	59 29.6	38.7
17	74 29.8	37.9
18	89 30.0	N10 37.0
19	104 30.1	36.1
20	119 30.3	35.3
21	134 30.5	·· 34.4
22	149 30.7	33.5
23	164 30.8	32.7

d h	G.H.A.	Dec.
26 00	179 31.0	N10 31.8
01	194 31.2	30.9
02	209 31.4	30.1
03	224 31.5	·· 29.2
04	239 31.7	28.3
05	254 31.9	27.5
06	269 32.1	N10 26.6
07	284 32.2	25.7
08	299 32.4	24.8
09	314 32.6	·· 24.0
10	329 32.8	23.1
11	344 32.9	22.2
12	359 33.1	N10 21.4
13	14 33.3	20.5
14	29 33.5	19.6
15	44 33.6	·· 18.8
16	59 33.8	17.9
17	74 34.0	17.0
18	89 34.2	N10 16.1
19	104 34.4	15.3
20	119 34.5	14.4
21	134 34.7	·· 13.5
22	149 34.9	12.6
23	164 35.1	11.8
27 00	179 35.2	N10 10.9
01	194 35.4	10.0
02	209 35.6	09.2
03	224 35.8	·· 08.3
04	239 36.0	07.4
05	254 36.1	06.5
06	269 36.3	N10 05.7
07	284 36.5	04.8
08	299 36.7	03.9
09	314 36.9	·· 03.0
10	329 37.0	02.2
11	344 37.2	01.3
12	359 37.4	N10 00.4
13	14 37.6	9 59.5
14	29 37.8	58.6
15	44 37.9	·· 57.8
16	59 38.1	56.9
17	74 38.3	56.0
18	89 38.5	N 9 55.1
19	104 38.7	54.2
20	119 38.9	53.4
21	134 39.0	·· 52.5
22	149 39.2	51.6
23	164 39.4	50.7
28 00	179 39.6	N 9 49.9
01	194 39.8	49.0
02	209 40.0	48.1
03	224 40.1	·· 47.2
04	239 40.3	46.3
05	254 40.5	45.4
06	269 40.7	N 9 44.6
07	284 40.9	43.7
08	299 41.1	42.8
09	314 41.2	·· 41.9
10	329 41.4	41.0
11	344 41.6	40.1
12	359 41.8	N 9 39.3
13	14 42.0	38.4
14	29 42.2	37.5
15	44 42.3	·· 36.6
16	59 42.5	35.7
17	74 42.7	34.8
18	89 42.9	N 9 34.0
19	104 43.1	33.1
20	119 43.3	32.2
21	134 43.5	·· 31.3
22	149 43.6	30.4
23	164 43.8	29.5

d h	G.H.A.	Dec.
29 00	179 44.0	N 9 28.6
01	194 44.2	27.8
02	209 44.4	26.9
03	224 44.6	·· 26.0
04	239 44.8	25.1
05	254 45.0	24.2
06	269 45.1	N 9 23.3
07	284 45.3	22.4
08	299 45.5	21.5
09	314 45.7	·· 20.7
10	329 45.9	19.8
11	344 46.1	18.9
12	359 46.3	N 9 18.0
13	14 46.5	17.1
14	29 46.6	16.2
15	44 46.8	·· 15.3
16	59 47.0	14.4
17	74 47.2	13.5
18	89 47.4	N 9 12.6
19	104 47.6	11.7
20	119 47.8	10.9
21	134 48.0	·· 10.0
22	149 48.2	09.1
23	164 48.4	08.2
30 00	179 48.5	N 9 07.3
01	194 48.7	06.4
02	209 48.9	05.5
03	224 49.1	·· 04.6
04	239 49.3	03.7
05	254 49.5	02.8
06	269 49.7	N 9 01.9
07	284 49.9	01.0
08	299 50.1	9 00.1
09	314 50.3	8 59.2
10	329 50.5	58.3
11	344 50.6	57.4
12	359 50.8	N 8 56.6
13	14 51.0	55.7
14	29 51.2	54.8
15	44 51.4	·· 53.9
16	59 51.6	53.0
17	74 51.8	52.1
18	89 52.0	N 8 51.2
19	104 52.2	50.3
20	119 52.4	49.4
21	134 52.6	·· 48.5
22	149 52.8	47.6
23	164 53.0	46.7
31 00	179 53.1	N 8 45.8
01	194 53.3	44.9
02	209 53.5	44.0
03	224 53.7	·· 43.1
04	239 53.9	42.2
05	254 54.1	41.3
06	269 54.3	N 8 40.4
07	284 54.5	39.5
08	299 54.7	38.6
09	314 54.9	·· 37.7
10	329 55.1	36.8
11	344 55.3	35.9
12	359 55.5	N 8 35.0
13	14 55.7	34.1
14	29 55.9	33.2
15	44 56.1	·· 32.3
16	59 56.3	31.4
17	74 56.5	30.5
18	89 56.7	N 8 29.6
19	104 56.9	28.7
20	119 57.0	27.8
21	134 57.2	·· 26.9
22	149 57.4	25.9
23	164 57.6	25.0

SEPTEMBER

d h	G.H.A.	Dec.
1 00	179 57.8	N 8 24.1
01	194 58.0	23.2
02	209 58.2	22.3
03	224 58.4 ··	21.4
04	239 58.6	20.5
05	254 58.8	19.6
06	269 59.0	N 8 18.7
07	284 59.2	17.8
08	299 59.4	16.9
09	314 59.6 ··	16.0
10	329 59.8	15.1
11	345 00.0	14.2
12	0 00.2	N 8 13.3
13	15 00.4	12.4
14	30 00.6	11.4
15	45 00.8 ··	10.5
16	60 01.0	09.6
17	75 01.2	08.7
18	90 01.4	N 8 07.8
19	105 01.6	06.9
20	120 01.8	06.0
21	135 02.0 ··	05.1
22	150 02.2	04.2
23	165 02.4	03.3
2 00	180 02.6	N 8 02.4
01	195 02.8	01.4
02	210 03.0	8 00.5
03	225 03.2	7 59.6
04	240 03.4	58.7
05	255 03.6	57.8
06	270 03.8	N 7 56.9
07	285 04.0	56.0
08	300 04.2	55.1
09	315 04.4 ··	54.2
10	330 04.6	53.2
11	345 04.8	52.3
12	0 05.0	N 7 51.4
13	15 05.2	50.5
14	30 05.4	49.6
15	45 05.6 ··	48.7
16	60 05.8	47.8
17	75 06.0	46.9
18	90 06.2	N 7 45.9
19	105 06.4	45.0
20	120 06.6	44.1
21	135 06.8 ··	43.2
22	150 07.0	42.3
23	165 07.2	41.4
3 00	180 07.4	N 7 40.5
01	195 07.6	39.5
02	210 07.8	38.6
03	225 08.0 ··	37.7
04	240 08.2	36.8
05	255 08.4	35.9
06	270 08.6	N 7 35.0
07	285 08.8	34.0
08	300 09.0	33.1
09	315 09.3 ··	32.2
10	330 09.5	31.3
11	345 09.7	30.4
12	0 09.9	N 7 29.4
13	15 10.1	28.5
14	30 10.3	27.6
15	45 10.5 ··	26.7
16	60 10.7	25.8
17	75 10.9	24.9
18	90 11.1	N 7 23.9
19	105 11.3	23.0
20	120 11.5	22.1
21	135 11.7 ··	21.2
22	150 11.9	20.3
23	165 12.1	19.3

d h	G.H.A.	Dec.
4 00	180 12.3	N 7 18.4
01	195 12.5	17.5
02	210 12.7	16.6
03	225 12.9 ··	15.7
04	240 13.1	14.7
05	255 13.3	13.8
06	270 13.6	N 7 12.9
07	285 13.8	12.0
08	300 14.0	11.0
09	315 14.2 ··	10.1
10	330 14.4	09.2
11	345 14.6	08.3
12	0 14.8	N 7 07.4
13	15 15.0	06.4
14	30 15.2	05.5
15	45 15.4 ··	04.6
16	60 15.6	03.7
17	75 15.8	02.7
18	90 16.0	N 7 01.8
19	105 16.2	00.9
20	120 16.4	7 00.0
21	135 16.7	6 59.0
22	150 16.9	58.1
23	165 17.1	57.2
5 00	180 17.3	N 6 56.3
01	195 17.5	55.3
02	210 17.7	54.4
03	225 17.9 ··	53.5
04	240 18.1	52.6
05	255 18.3	51.6
06	270 18.5	N 6 50.7
07	285 18.7	49.8
08	300 18.9	48.9
09	315 19.1 ··	47.9
10	330 19.4	47.0
11	345 19.6	46.1
12	0 19.8	N 6 45.1
13	15 20.0	44.2
14	30 20.2	43.3
15	45 20.4 ··	42.4
16	60 20.6	41.4
17	75 20.8	40.5
18	90 21.0	N 6 39.6
19	105 21.2	38.6
20	120 21.4	37.7
21	135 21.7 ··	36.8
22	150 21.9	35.9
23	165 22.1	34.9
6 00	180 22.3	N 6 34.0
01	195 22.5	33.1
02	210 22.7	32.1
03	225 22.9 ··	31.2
04	240 23.1	30.3
05	255 23.3	29.3
06	270 23.5	N 6 28.4
07	285 23.8	27.5
08	300 24.0	26.6
09	315 24.2 ··	25.6
10	330 24.4	24.7
11	345 24.6	23.8
12	0 24.8	N 6 22.8
13	15 25.0	21.9
14	30 25.2	21.0
15	45 25.4 ··	20.0
16	60 25.7	19.1
17	75 25.9	18.2
18	90 26.1	N 6 17.2
19	105 26.3	16.3
20	120 26.5	15.4
21	135 26.7 ··	14.4
22	150 26.9	13.5
23	165 27.1	12.6

d h	G.H.A.	Dec.
7 00	180 27.3	N 6 11.6
01	195 27.6	10.7
02	210 27.8	09.8
03	225 28.0 ··	08.8
04	240 28.2	07.9
05	255 28.4	07.0
06	270 28.6	N 6 06.0
07	285 28.8	05.1
08	300 29.0	04.1
09	315 29.3 ··	03.2
10	330 29.5	02.3
11	345 29.7	01.3
12	0 29.9	N 6 00.4
13	15 30.1	5 59.5
14	30 30.3	58.5
15	45 30.5 ··	57.6
16	60 30.7	56.7
17	75 31.0	55.7
18	90 31.2	N 5 54.8
19	105 31.4	53.8
20	120 31.6	52.9
21	135 31.8 ··	52.0
22	150 32.0	51.0
23	165 32.2	50.1
8 00	180 32.5	N 5 49.2
01	195 32.7	48.2
02	210 32.9	47.3
03	225 33.1 ··	46.3
04	240 33.3	45.4
05	255 33.5	44.5
06	270 33.7	N 5 43.5
07	285 34.0	42.6
08	300 34.2	41.6
09	315 34.4 ··	40.7
10	330 34.6	39.8
11	345 34.8	38.8
12	0 35.0	N 5 37.9
13	15 35.2	36.9
14	30 35.5	36.0
15	45 35.7 ··	35.1
16	60 35.9	34.1
17	75 36.1	33.2
18	90 36.3	N 5 32.2
19	105 36.5	31.3
20	120 36.7	30.3
21	135 37.0 ··	29.4
22	150 37.2	28.5
23	165 37.4	27.5
9 00	180 37.6	N 5 26.6
01	195 37.8	25.6
02	210 38.0	24.7
03	225 38.2 ··	23.8
04	240 38.5	22.8
05	255 38.7	21.9
06	270 38.9	N 5 20.9
07	285 39.1	20.0
08	300 39.3	19.0
09	315 39.5 ··	18.1
10	330 39.8	17.1
11	345 40.0	16.2
12	0 40.2	N 5 15.3
13	15 40.4	14.3
14	30 40.6	13.4
15	45 40.8 ··	12.4
16	60 41.1	11.5
17	75 41.3	10.5
18	90 41.5	N 5 09.6
19	105 41.7	08.6
20	120 41.9	07.7
21	135 42.1 ··	06.8
22	150 42.4	05.8
23	165 42.6	04.9

d h	G.H.A				Dec	
10 00	180	42.8	N	5	03.9	
01	195	43.0			03.0	
02	210	43.2			02.0	
03	225	43.4	··		01.1	
04	240	43.7		5	00.1	
05	255	43.9		4	59.2	
06	270	44.1	N	4	58.2	
07	285	44.3			57.3	
08	300	44.5			56.3	
09	315	44.7	··		55.4	
10	330	45.0			54.4	
11	345	45.2			53.5	
12	0	45.4	N	4	52.5	
13	15	45.6			51.6	
14	30	45.8			50.7	
15	45	46.0	··		49.7	
16	60	46.3			48.8	
17	75	46.5			47.8	
18	90	46.7	N	4	46.9	
19	105	46.9			45.9	
20	120	47.1			45.0	
21	135	47.3	··		44.0	
22	150	47.6			43.1	
23	165	47.8			42.1	
11 00	180	48.0	N	4	41.2	
01	195	48.2			40.2	
02	210	48.4			39.3	
03	225	48.7	··		38.3	
04	240	48.9			37.4	
05	255	49.1			36.4	
06	270	49.3	N	4	35.5	
07	285	49.5			34.5	
08	300	49.7			33.6	
09	315	50.0	··		32.6	
10	330	50.2			31.7	
11	345	50.4			30.7	
12	0	50.6	N	4	29.8	
13	15	50.8			28.8	
14	30	51.1			27.9	
15	45	51.3	··		26.9	
16	60	51.5			25.9	
17	75	51.7			25.0	
18	90	51.9	N	4	24.0	
19	105	52.2			23.1	
20	120	52.4			22.1	
21	135	52.6	··		21.2	
22	150	52.8			20.2	
23	165	53.0			19.3	
12 00	180	53.2	N	4	18.3	
01	195	53.5			17.4	
02	210	53.7			16.4	
03	225	53.9	··		15.5	
04	240	54.1			14.5	
05	255	54.3			13.6	
06	270	54.6	N	4	12.6	
07	285	54.8			11.7	
08	300	55.0			10.7	
09	315	55.2	··		09.7	
10	330	55.4			08.8	
11	345	55.7			07.8	
12	0	55.9	N	4	06.9	
13	15	56.1			05.9	
14	30	56.3			05.0	
15	45	56.5	··		04.0	
16	60	56.8			03.1	
17	75	57.0			02.1	
18	90	57.2	N	4	01.2	
19	105	57.4		4	00.2	
20	120	57.6		3	59.2	
21	135	57.9	··		58.3	
22	150	58.1			57.3	
23	165	58.3			56.4	

d h	G.H.A.				Dec.	
13 00	180	58.5	N	3	55.4	
01	195	58.7			54.5	
02	210	59.0			53.5	
03	225	59.2	··		52.6	
04	240	59.4			51.6	
05	255	59.6			50.6	
06	270	59.8	N	3	49.7	
07	286	00.1			48.7	
08	301	00.3			47.8	
09	316	00.5	··		46.8	
10	331	00.7			45.9	
11	346	00.9			44.9	
12	1	01.2	N	3	43.9	
13	16	01.4			43.0	
14	31	01.6			42.0	
15	46	01.8	··		41.1	
16	61	02.0			40.1	
17	76	02.3			39.2	
18	91	02.5	N	3	38.2	
19	106	02.7			37.2	
20	121	02.9			36.3	
21	136	03.1	··		35.3	
22	151	03.4			34.4	
23	166	03.6			33.4	
14 00	181	03.8	N	3	32.4	
01	196	04.0			31.5	
02	211	04.2			30.5	
03	226	04.5	··		29.6	
04	241	04.7			28.6	
05	256	04.9			27.6	
06	271	05.1	N	3	26.7	
07	286	05.3			25.7	
08	301	05.6			24.8	
09	316	05.8	··		23.8	
10	331	06.0			22.8	
11	346	06.2			21.9	
12	1	06.5	N	3	20.9	
13	16	06.7			20.0	
14	31	06.9			19.0	
15	46	07.1	··		18.0	
16	61	07.3			17.1	
17	76	07.6			16.1	
18	91	07.8	N	3	15.2	
19	106	08.0			14.2	
20	121	08.2			13.2	
21	136	08.4	··		12.3	
22	151	08.7			11.3	
23	166	08.9			10.4	
15 00	181	09.1	N	3	09.4	
01	196	09.3			08.4	
02	211	09.5			07.5	
03	226	09.8	··		06.5	
04	241	10.0			05.6	
05	256	10.2			04.6	
06	271	10.4	N	3	03.6	
07	286	10.7			02.7	
08	301	10.9			01.7	
09	316	11.1		3	00.7	
10	331	11.3		2	59.8	
11	346	11.5			58.8	
12	1	11.8	N	2	57.9	
13	16	12.0			56.9	
14	31	12.2			55.9	
15	46	12.4	··		55.0	
16	61	12.6			54.0	
17	76	12.9			53.0	
18	91	13.1	N	2	52.1	
19	106	13.3			51.1	
20	121	13.5			50.2	
21	136	13.8	··		49.2	
22	151	14.0			48.2	
23	166	14.2			47.3	

d h	G.H.A.				Dec.	
16 00	181	14.4	N	2	46.3	
01	196	14.6			45.3	
02	211	14.9			44.4	
03	226	15.1	··		43.4	
04	241	15.3			42.4	
05	256	15.5			41.5	
06	271	15.7	N	2	40.5	
07	286	16.0			39.5	
08	301	16.2			38.6	
09	316	16.4	··		37.6	
10	331	16.6			36.7	
11	346	16.9			35.7	
12	1	17.1	N	2	34.7	
13	16	17.3			33.8	
14	31	17.5			32.8	
15	46	17.7	··		31.8	
16	61	18.0			30.9	
17	76	18.2			29.9	
18	91	18.4	N	2	28.9	
19	106	18.6			28.0	
20	121	18.9			27.0	
21	136	19.1	··		26.0	
22	151	19.3			25.1	
23	166	19.5			24.1	
17 00	181	19.7	N	2	23.1	
01	196	20.0			22.2	
02	211	20.2			21.2	
03	226	20.4	··		20.2	
04	241	20.6			19.3	
05	256	20.8			18.3	
06	271	21.1	N	2	17.3	
07	286	21.3			16.4	
08	301	21.5			15.4	
09	316	21.7	··		14.4	
10	331	22.0			13.5	
11	346	22.2			12.5	
12	1	22.4	N	2	11.5	
13	16	22.6			10.6	
14	31	22.8			09.6	
15	46	23.1	··		08.6	
16	61	23.3			07.7	
17	76	23.5			06.7	
18	91	23.7	N	2	05.7	
19	106	24.0			04.8	
20	121	24.2			03.8	
21	136	24.4	··		02.8	
22	151	24.6			01.9	
23	166	24.8		2	00.9	
18 00	181	25.1	N	1	59.9	
01	196	25.3			59.0	
02	211	25.5			58.0	
03	226	25.7	··		57.0	
04	241	25.9			56.1	
05	256	26.2			55.1	
06	271	26.4	N	1	54.1	
07	286	26.6			53.2	
08	301	26.8			52.2	
09	316	27.1	··		51.2	
10	331	27.3			50.3	
11	346	27.5			49.3	
12	1	27.7	N	1	48.3	
13	16	27.9			47.4	
14	31	28.2			46.4	
15	46	28.4	··		45.4	
16	61	28.6			44.4	
17	76	28.8			43.5	
18	91	29.1	N	1	42.5	
19	106	29.3			41.5	
20	121	29.5			40.6	
21	136	29.7	··		39.6	
22	151	29.9			38.6	
23	166	30.2			37.7	

		G.H.A.	Dec.
d 19	h 00	181 30.4	N 1 36.7
	01	196 30.6	35.7
	02	211 30.8	34.8
	03	226 31.0	·· 33.8
	04	241 31.3	32.8
	05	256 31.5	31.8
	06	271 31.7	N 1 30.9
	07	286 31.9	29.9
	08	301 32.2	28.9
	09	316 32.4	·· 28.0
	10	331 32.6	27.0
	11	346 32.8	26.0
	12	1 33.0	N 1 25.1
	13	16 33.3	24.1
	14	31 33.5	23.1
	15	46 33.7	·· 22.1
	16	61 33.9	21.2
	17	76 34.2	20.2
	18	91 34.4	N 1 19.2
	19	106 34.6	18.3
	20	121 34.8	17.3
	21	136 35.0	·· 16.3
	22	151 35.3	15.4
	23	166 35.5	14.4
20	00	181 35.7	N 1 13.4
	01	196 35.9	12.4
	02	211 36.1	11.5
	03	226 36.4	·· 10.5
	04	241 36.6	09.5
	05	256 36.8	08.6
	06	271 37.0	N 1 07.6
	07	286 37.2	06.6
	08	301 37.5	05.6
	09	316 37.7	·· 04.7
	10	331 37.9	03.7
	11	346 38.1	02.7
	12	1 38.4	N 1 01.8
	13	16 38.6	1 00.8
	14	31 38.8	0 59.8
	15	46 39.0	·· 58.8
	16	61 39.2	57.9
	17	76 39.5	56.9
	18	91 39.7	N 0 55.9
	19	106 39.9	55.0
	20	121 40.1	54.0
	21	136 40.3	·· 53.0
	22	151 40.6	52.0
	23	166 40.8	51.1
21	00	181 41.0	N 0 50.1
	01	196 41.2	49.1
	02	211 41.4	48.2
	03	226 41.7	·· 47.2
	04	241 41.9	46.2
	05	256 42.1	45.2
	06	271 42.3	N 0 44.3
	07	286 42.6	43.3
	08	301 42.8	42.3
	09	316 43.0	·· 41.4
	10	331 43.2	40.4
	11	346 43.4	39.4
	12	1 43.7	N 0 38.4
	13	16 43.9	37.5
	14	31 44.1	36.5
	15	46 44.3	·· 35.5
	16	61 44.5	34.5
	17	76 44.8	33.6
	18	91 45.0	N 0 32.6
	19	106 45.2	31.6
	20	121 45.4	30.7
	21	136 45.6	·· 29.7
	22	151 45.9	28.7
	23	166 46.1	27.7

		G.H.A.	Dec.
d 22	h 00	181 46.3	N 0 26.8
	01	196 46.5	25.8
	02	211 46.7	24.8
	03	226 47.0	·· 23.8
	04	241 47.2	22.9
	05	256 47.4	21.9
	06	271 47.6	N 0 20.9
	07	286 47.8	20.0
	08	301 48.1	19.0
	09	316 48.3	·· 18.0
	10	331 48.5	17.0
	11	346 48.7	16.1
	12	1 48.9	N 0 15.1
	13	16 49.2	14.1
	14	31 49.4	13.1
	15	46 49.6	·· 12.2
	16	61 49.8	11.2
	17	76 50.0	10.2
	18	91 50.3	N 0 09.3
	19	106 50.5	08.3
	20	121 50.7	07.3
	21	136 50.9	·· 06.3
	22	151 51.1	05.4
	23	166 51.4	04.4
23	00	181 51.6	N 0 03.4
	01	196 51.8	02.4
	02	211 52.0	01.5
	03	226 52.2	N 0 00.5
	04	241 52.4	S 0 00.5
	05	256 52.7	01.5
	06	271 52.9	S 0 02.4
	07	286 53.1	03.4
	08	301 53.3	04.4
	09	316 53.5	·· 05.3
	10	331 53.8	06.3
	11	346 54.0	07.3
	12	1 54.2	S 0 08.3
	13	16 54.4	09.2
	14	31 54.6	10.2
	15	46 54.9	·· 11.2
	16	61 55.1	12.2
	17	76 55.3	13.1
	18	91 55.5	S 0 14.1
	19	106 55.7	15.1
	20	121 55.9	16.1
	21	136 56.2	·· 17.0
	22	151 56.4	18.0
	23	166 56.6	19.0
24	00	181 56.8	S 0 20.0
	01	196 57.0	20.9
	02	211 57.3	21.9
	03	226 57.5	·· 22.9
	04	241 57.7	23.8
	05	256 57.9	24.8
	06	271 58.1	S 0 25.8
	07	286 58.3	26.8
	08	301 58.6	27.7
	09	316 58.8	·· 28.7
	10	331 59.0	29.7
	11	346 59.2	30.7
	12	1 59.4	S 0 31.6
	13	16 59.7	32.6
	14	31 59.9	33.6
	15	47 00.1	·· 34.6
	16	62 00.3	35.5
	17	77 00.5	36.5
	18	92 00.7	S 0 37.5
	19	107 01.0	38.5
	20	122 01.2	39.4
	21	137 01.4	·· 40.4
	22	152 01.6	41.4
	23	167 01.8	42.4

		G.H.A.	Dec.
d 25	h 00	182 02.0	S 0 43.3
	01	197 02.3	44.3
	02	212 02.5	45.3
	03	227 02.7	·· 46.3
	04	242 02.9	47.2
	05	257 03.1	48.2
	06	272 03.3	S 0 49.2
	07	287 03.6	50.1
	08	302 03.8	51.1
	09	317 04.0	·· 52.1
	10	332 04.2	53.1
	11	347 04.4	54.0
	12	2 04.6	S 0 55.0
	13	17 04.9	56.0
	14	32 05.1	57.0
	15	47 05.3	·· 57.9
	16	62 05.5	58.9
	17	77 05.7	0 59.9
	18	92 05.9	S 1 00.9
	19	107 06.2	01.8
	20	122 06.4	02.8
	21	137 06.6	·· 03.8
	22	152 06.8	04.8
	23	167 07.0	05.7
26	00	182 07.2	S 1 06.7
	01	197 07.4	07.7
	02	212 07.7	08.7
	03	227 07.9	·· 09.6
	04	242 08.1	10.6
	05	257 08.3	11.6
	06	272 08.5	S 1 12.5
	07	287 08.7	13.5
	08	302 08.9	14.5
	09	317 09.2	·· 15.5
	10	332 09.4	16.4
	11	347 09.6	17.4
	12	2 09.8	S 1 18.4
	13	17 10.0	19.4
	14	32 10.2	20.3
	15	47 10.4	·· 21.3
	16	62 10.7	22.3
	17	77 10.9	23.3
	18	92 11.1	S 1 24.2
	19	107 11.3	25.2
	20	122 11.5	26.2
	21	137 11.7	·· 27.2
	22	152 11.9	28.1
	23	167 12.2	29.1
27	00	182 12.4	S 1 30.1
	01	197 12.6	31.1
	02	212 12.8	32.0
	03	227 13.0	·· 33.0
	04	242 13.2	34.0
	05	257 13.4	34.9
	06	272 13.7	S 1 35.9
	07	287 13.9	36.9
	08	302 14.1	37.9
	09	317 14.3	·· 38.8
	10	332 14.5	39.8
	11	347 14.7	40.8
	12	2 14.9	S 1 41.8
	13	17 15.1	42.7
	14	32 15.4	43.7
	15	47 15.6	·· 44.7
	16	62 15.8	45.7
	17	77 16.0	46.6
	18	92 16.2	S 1 47.6
	19	107 16.4	48.6
	20	122 16.6	49.5
	21	137 16.8	·· 50.5
	22	152 17.0	51.5
	23	167 17.3	52.5

d h	G.H.A.	Dec.
28 00	182 17.5	S 1 53.4
01	197 17.7	54.4
02	212 17.9	55.4
03	227 18.1	·· 56.4
04	242 18.3	57.3
05	257 18.5	58.3
06	272 18.7	S 1 59.3
07	287 18.9	2 00.3
08	302 19.2	01.2
09	317 19.4	·· 02.2
10	332 19.6	03.2
11	347 19.8	04.1
12	2 20.0	S 2 05.1
13	17 20.2	06.1
14	32 20.4	07.1
15	47 20.6	·· 08.0
16	62 20.8	09.0
17	77 21.0	10.0
18	92 21.3	S 2 11.0
19	107 21.5	11.9
20	122 21.7	12.9
21	137 21.9	·· 13.9
22	152 22.1	14.8
23	167 22.3	15.8
29 00	182 22.5	S 2 16.8
01	197 22.7	17.8
02	212 22.9	18.7
03	227 23.1	·· 19.7
04	242 23.3	20.7
05	257 23.6	21.6
06	272 23.8	S 2 22.6
07	287 24.0	23.6
08	302 24.2	24.6
09	317 24.4	·· 25.5
10	332 24.6	26.5
11	347 24.8	27.5
12	2 25.0	S 2 28.5
13	17 25.2	29.4
14	32 25.4	30.4
15	47 25.6	·· 31.4
16	62 25.8	32.3
17	77 26.0	33.3
18	92 26.3	S 2 34.3
19	107 26.5	35.3
20	122 26.7	36.2
21	137 26.9	·· 37.2
22	152 27.1	38.2
23	167 27.3	39.1
30 00	182 27.5	S 2 40.1
01	197 27.7	41.1
02	212 27.9	42.1
03	227 28.1	·· 43.0
04	242 28.3	44.0
05	257 28.5	45.0
06	272 28.7	S 2 45.9
07	287 28.9	46.9
08	302 29.1	47.9
09	317 29.3	·· 48.8
10	332 29.6	49.8
11	347 29.8	50.8
12	2 30.0	S 2 51.8
13	17 30.2	52.7
14	32 30.4	53.7
15	47 30.6	·· 54.7
16	62 30.8	55.6
17	77 31.0	56.6
18	92 31.2	S 2 57.6
19	107 31.4	58.6
20	122 31.6	2 59.5
21	137 31.8	3 00.5
22	152 32.0	01.5
23	167 32.2	02.4

d h	G.H.A.	Dec.
1 00	182 32.4	S 3 03.4
01	197 32.6	04.4
02	212 32.8	05.3
03	227 33.0	·· 06.3
04	242 33.2	07.3
05	257 33.4	08.3
06	272 33.6	S 3 09.2
07	287 33.8	10.2
08	302 34.0	11.2
09	317 34.2	·· 12.1
10	332 34.4	13.1
11	347 34.6	14.1
12	2 34.8	S 3 15.0
13	17 35.0	16.0
14	32 35.2	17.0
15	47 35.4	·· 18.0
16	62 35.6	18.9
17	77 35.8	19.9
18	92 36.1	S 3 20.9
19	107 36.3	21.8
20	122 36.5	22.8
21	137 36.7	·· 23.8
22	152 36.9	24.7
23	167 37.1	25.7
2 00	182 37.3	S 3 26.7
01	197 37.5	27.6
02	212 37.7	28.6
03	227 37.9	·· 29.6
04	242 38.1	30.5
05	257 38.3	31.5
06	272 38.5	S 3 32.5
07	287 38.7	33.4
08	302 38.8	34.4
09	317 39.0	·· 35.4
10	332 39.2	36.4
11	347 39.4	37.3
12	2 39.6	S 3 38.3
13	17 39.8	39.3
14	32 40.0	40.2
15	47 40.2	·· 41.2
16	62 40.4	42.2
17	77 40.6	43.1
18	92 40.8	S 3 44.1
19	107 41.0	45.1
20	122 41.2	46.0
21	137 41.4	·· 47.0
22	152 41.6	48.0
23	167 41.8	48.9
3 00	182 42.0	S 3 49.9
01	197 42.2	50.9
02	212 42.4	51.8
03	227 42.6	·· 52.8
04	242 42.8	53.8
05	257 43.0	54.7
06	272 43.2	S 3 55.7
07	287 43.4	56.7
08	302 43.6	57.6
09	317 43.8	58.6
10	332 44.0	3 59.6
11	347 44.2	4 00.5
12	2 44.4	S 4 01.5
13	17 44.6	02.5
14	32 44.8	03.4
15	47 44.9	·· 04.4
16	62 45.1	05.4
17	77 45.3	06.3
18	92 45.5	S 4 07.3
19	107 45.7	08.3
20	122 45.9	09.2
21	137 46.1	·· 10.2
22	152 46.3	11.1
23	167 46.5	12.1

d h	G.H.A.	Dec.
4 00	182 46.7	S 4 13.1
01	197 46.9	14.0
02	212 47.1	15.0
03	227 47.3	·· 16.0
04	242 47.4	16.9
05	257 47.6	17.9
06	272 47.8	S 4 18.9
07	287 48.0	19.8
08	302 48.2	20.8
09	317 48.4	·· 21.8
10	332 48.6	22.7
11	347 48.8	23.7
12	2 49.0	S 4 24.7
13	17 49.2	25.6
14	32 49.4	26.6
15	47 49.6	·· 27.5
16	62 49.7	28.5
17	77 49.9	29.5
18	92 50.1	S 4 30.4
19	107 50.3	31.4
20	122 50.5	32.4
21	137 50.7	·· 33.3
22	152 50.9	34.3
23	167 51.1	35.2
5 00	182 51.3	S 4 36.2
01	197 51.5	37.2
02	212 51.6	38.1
03	227 51.8	·· 39.1
04	242 52.0	40.1
05	257 52.2	41.0
06	272 52.4	S 4 42.0
07	287 52.6	42.9
08	302 52.8	43.9
09	317 53.0	·· 44.9
10	332 53.1	45.8
11	347 53.3	46.8
12	2 53.5	S 4 47.8
13	17 53.7	48.7
14	32 53.9	49.7
15	47 54.1	·· 50.6
16	62 54.3	51.6
17	77 54.4	52.6
18	92 54.6	S 4 53.5
19	107 54.8	54.5
20	122 55.0	55.4
21	137 55.2	·· 56.4
22	152 55.4	57.4
23	167 55.6	58.3
6 00	182 55.7	S 4 59.3
01	197 55.9	5 00.2
02	212 56.1	01.2
03	227 56.3	·· 02.2
04	242 56.5	03.1
05	257 56.7	04.1
06	272 56.8	S 5 05.0
07	287 57.0	06.0
08	302 57.2	07.0
09	317 57.4	·· 07.9
10	332 57.6	08.9
11	347 57.8	09.8
12	2 57.9	S 5 10.8
13	17 58.1	11.8
14	32 58.3	12.7
15	47 58.5	·· 13.7
16	62 58.7	14.6
17	77 58.9	15.6
18	92 59.0	S 5 16.6
19	107 59.2	17.5
20	122 59.4	18.5
21	137 59.6	·· 19.4
22	152 59.8	20.4
23	167 59.9	21.3

d h	G.H.A.	Dec.
7 00	183 00.1	S 5 22.3
01	198 00.3	23.3
02	213 00.5	24.2
03	228 00.7	·· 25.2
04	243 00.8	26.1
05	258 01.0	27.1
06	273 01.2	S 5 28.1
07	288 01.4	29.0
08	303 01.6	30.0
09	318 01.7	·· 30.9
10	333 01.9	31.9
11	348 02.1	32.8
12	3 02.3	S 5 33.8
13	18 02.4	34.7
14	33 02.6	35.7
15	48 02.8	·· 36.7
16	63 03.0	37.6
17	78 03.2	38.6
18	93 03.3	S 5 39.5
19	108 03.5	40.5
20	123 03.7	41.4
21	138 03.9	·· 42.4
22	153 04.0	43.3
23	168 04.2	44.3
8 00	183 04.4	S 5 45.3
01	198 04.6	46.2
02	213 04.7	47.2
03	228 04.9	·· 48.1
04	243 05.1	49.1
05	258 05.3	50.0
06	273 05.4	S 5 51.0
07	288 05.6	51.9
08	303 05.8	52.9
09	318 06.0	·· 53.9
10	333 06.1	54.8
11	348 06.3	55.8
12	3 06.5	S 5 56.7
13	18 06.7	57.7
14	33 06.8	58.6
15	48 07.0	5 59.6
16	63 07.2	6 00.5
17	78 07.4	01.5
18	93 07.5	S 6 02.4
19	108 07.7	03.4
20	123 07.9	04.3
21	138 08.0	·· 05.3
22	153 08.2	06.2
23	168 08.4	07.2
9 00	183 08.6	S 6 08.1
01	198 08.7	09.1
02	213 08.9	10.0
03	228 09.1	·· 11.0
04	243 09.2	12.0
05	258 09.4	12.9
06	273 09.6	S 6 13.9
07	288 09.8	14.8
08	303 09.9	15.8
09	318 10.1	·· 16.7
10	333 10.3	17.7
11	348 10.4	18.6
12	3 10.6	S 6 19.6
13	18 10.8	20.5
14	33 10.9	21.5
15	48 11.1	·· 22.4
16	63 11.3	23.4
17	78 11.5	24.3
18	93 11.6	S 6 25.3
19	108 11.8	26.2
20	123 12.0	27.2
21	138 12.1	·· 28.1
22	153 12.3	29.1
23	168 12.5	30.0

d h	G.H.A.	Dec.
10 00	183 12.6	S 6 31.0
01	198 12.8	31.9
02	213 13.0	32.9
03	228 13.1	·· 33.8
04	243 13.3	34.7
05	258 13.5	35.7
06	273 13.6	S 6 36.6
07	288 13.8	37.6
08	303 14.0	38.5
09	318 14.1	·· 39.5
10	333 14.3	40.4
11	348 14.5	41.4
12	3 14.6	S 6 42.3
13	18 14.8	43.3
14	33 15.0	44.2
15	48 15.1	·· 45.2
16	63 15.3	46.1
17	78 15.5	47.1
18	93 15.6	S 6 48.0
19	108 15.8	49.0
20	123 15.9	49.9
21	138 16.1	·· 50.8
22	153 16.3	51.8
23	168 16.4	52.7
11 00	183 16.6	S 6 53.7
01	198 16.7	54.6
02	213 16.9	55.6
03	228 17.1	·· 56.5
04	243 17.2	57.4
05	258 17.4	58.4
06	273 17.5	S 6 59.3
07	288 17.7	7 00.3
08	303 17.9	01.2
09	318 18.0	·· 02.2
10	333 18.2	03.1
11	348 18.3	04.1
12	3 18.5	S 7 05.0
13	18 18.7	05.9
14	33 18.8	06.9
15	48 19.0	·· 07.8
16	63 19.1	08.8
17	78 19.3	09.7
18	93 19.4	S 7 10.7
19	108 19.6	11.6
20	123 19.8	12.5
21	138 19.9	·· 13.5
22	153 20.1	14.4
23	168 20.2	15.4
12 00	183 20.4	S 7 16.3
01	198 20.5	17.2
02	213 20.7	18.2
03	228 20.9	·· 19.1
04	243 21.0	20.1
05	258 21.2	21.0
06	273 21.3	S 7 21.9
07	288 21.5	22.9
08	303 21.6	23.8
09	318 21.8	·· 24.8
10	333 21.9	25.7
11	348 22.1	26.6
12	3 22.3	S 7 27.6
13	18 22.4	28.5
14	33 22.6	29.5
15	48 22.7	·· 30.4
16	63 22.9	31.3
17	78 23.0	32.3
18	93 23.2	S 7 33.2
19	108 23.3	34.1
20	123 23.5	35.1
21	138 23.6	·· 36.0
22	153 23.8	37.0
23	168 23.9	37.9

d h	G.H.A.	Dec.
13 00	183 24.1	S 7 38.8
01	198 24.2	39.8
02	213 24.4	40.7
03	228 24.5	·· 41.6
04	243 24.7	42.6
05	258 24.8	43.5
06	273 25.0	S 7 44.4
07	288 25.1	45.4
08	303 25.3	46.3
09	318 25.4	·· 47.3
10	333 25.6	48.2
11	348 25.7	49.1
12	3 25.9	S 7 50.1
13	18 26.0	51.0
14	33 26.2	51.9
15	48 26.3	·· 52.9
16	63 26.5	53.8
17	78 26.6	54.8
18	93 26.8	S 7 55.7
19	108 26.9	56.6
20	123 27.1	57.5
21	138 27.2	·· 58.5
22	153 27.4	7 59.4
23	168 27.5	8 00.3
14 00	183 27.7	S 8 01.3
01	198 27.8	02.2
02	213 27.9	03.1
03	228 28.1	·· 04.0
04	243 28.2	05.0
05	258 28.4	05.9
06	273 28.5	S 8 06.8
07	288 28.7	07.8
08	303 28.8	08.7
09	318 29.0	·· 09.6
10	333 29.1	10.6
11	348 29.2	11.5
12	3 29.4	S 8 12.4
13	18 29.5	13.4
14	33 29.7	14.3
15	48 29.8	·· 15.2
16	63 30.0	16.1
17	78 30.1	17.1
18	93 30.2	S 8 18.0
19	108 30.4	18.9
20	123 30.5	19.9
21	138 30.7	·· 20.8
22	153 30.8	21.7
23	168 31.0	22.6
15 00	183 31.1	S 8 23.6
01	198 31.2	24.5
02	213 31.4	25.4
03	228 31.5	·· 26.3
04	243 31.7	27.3
05	258 31.8	28.2
06	273 31.9	S 8 29.1
07	288 32.1	30.0
08	303 32.2	31.0
09	318 32.3	·· 31.9
10	333 32.5	32.8
11	348 32.6	33.7
12	3 32.8	S 8 34.7
13	18 32.9	35.6
14	33 33.0	36.5
15	48 33.2	·· 37.4
16	63 33.3	38.4
17	78 33.4	39.3
18	93 33.6	S 8 40.2
19	108 33.7	41.1
20	123 33.9	42.1
21	138 34.0	·· 43.0
22	153 34.1	43.9
23	168 34.3	44.8

		G.H.A.	Dec.
d	h	° '	° '
16	00	183 34.4	S 8 45.8
	01	198 34.5	46.7
	02	213 34.7	47.6
	03	228 34.8	·· 48.5
	04	243 34.9	49.4
	05	258 35.1	50.4
	06	273 35.2	S 8 51.3
	07	288 35.3	52.2
	08	303 35.5	53.1
	09	318 35.6	·· 54.0
	10	333 35.7	55.0
	11	348 35.9	55.9
	12	3 36.0	S 8 56.8
	13	18 36.1	57.7
	14	33 36.2	58.6
	15	48 36.4	8 59.6
	16	63 36.5	9 00.5
	17	78 36.6	01.4
	18	93 36.8	S 9 02.3
	19	108 36.9	03.2
	20	123 37.0	04.1
	21	138 37.2	·· 05.1
	22	153 37.3	06.0
	23	168 37.4	06.9
17	00	183 37.5	S 9 07.8
	01	198 37.7	08.7
	02	213 37.8	09.6
	03	228 37.9	·· 10.6
	04	243 38.1	11.5
	05	258 38.2	12.4
	06	273 38.3	S 9 13.3
	07	288 38.4	14.2
	08	303 38.6	15.1
	09	318 38.7	·· 16.1
	10	333 38.8	17.0
	11	348 38.9	17.9
	12	3 39.1	S 9 18.8
	13	18 39.2	19.7
	14	33 39.3	20.6
	15	48 39.4	·· 21.5
	16	63 39.6	22.4
	17	78 39.7	23.4
	18	93 39.8	S 9 24.3
	19	108 39.9	25.2
	20	123 40.1	26.1
	21	138 40.2	·· 27.0
	22	153 40.3	27.9
	23	168 40.4	28.8
18	00	183 40.6	S 9 29.7
	01	198 40.7	30.7
	02	213 40.8	31.6
	03	228 40.9	·· 32.5
	04	243 41.0	33.4
	05	258 41.2	34.3
	06	273 41.3	S 9 35.2
	07	288 41.4	36.1
	08	303 41.5	37.0
	09	318 41.7	·· 37.9
	10	333 41.8	38.8
	11	348 41.9	39.7
	12	3 42.0	S 9 40.7
	13	18 42.1	41.6
	14	33 42.2	42.5
	15	48 42.4	·· 43.4
	16	63 42.5	44.3
	17	78 42.6	45.2
	18	93 42.7	S 9 46.1
	19	108 42.8	47.0
	20	123 43.0	47.9
	21	138 43.1	·· 48.8
	22	153 43.2	49.7
	23	168 43.3	50.6

		G.H.A.	Dec.
d	h	° '	° '
19	00	183 43.4	S 9 51.5
	01	198 43.5	52.4
	02	213 43.7	53.3
	03	228 43.8	·· 54.2
	04	243 43.9	55.2
	05	258 44.0	56.1
	06	273 44.1	S 9 57.0
	07	288 44.2	57.9
	08	303 44.3	58.8
	09	318 44.5	9 59.7
	10	333 44.6	10 00.6
	11	348 44.7	01.5
	12	3 44.8	S10 02.4
	13	18 44.9	03.3
	14	33 45.0	04.2
	15	48 45.1	·· 05.1
	16	63 45.2	06.0
	17	78 45.4	06.9
	18	93 45.5	S10 07.8
	19	108 45.6	08.7
	20	123 45.7	09.6
	21	138 45.8	·· 10.5
	22	153 45.9	11.4
	23	168 46.0	12.3
20	00	183 46.1	S10 13.2
	01	198 46.2	14.1
	02	213 46.4	15.0
	03	228 46.5	·· 15.9
	04	243 46.6	16.8
	05	258 46.7	17.7
	06	273 46.8	S10 18.6
	07	288 46.9	19.5
	08	303 47.0	20.4
	09	318 47.1	·· 21.3
	10	333 47.2	22.2
	11	348 47.3	23.1
	12	3 47.4	S10 23.9
	13	18 47.5	24.8
	14	33 47.6	25.7
	15	48 47.8	·· 26.6
	16	63 47.9	27.5
	17	78 48.0	28.4
	18	93 48.1	S10 29.3
	19	108 48.2	30.2
	20	123 48.3	31.1
	21	138 48.4	·· 32.0
	22	153 48.5	32.9
	23	168 48.6	33.8
21	00	183 48.7	S10 34.7
	01	198 48.8	35.6
	02	213 48.9	36.5
	03	228 49.0	·· 37.4
	04	243 49.1	38.2
	05	258 49.2	39.1
	06	273 49.3	S10 40.0
	07	288 49.4	40.9
	08	303 49.5	41.8
	09	318 49.6	·· 42.7
	10	333 49.7	43.6
	11	348 49.8	44.5
	12	3 49.9	S10 45.4
	13	18 50.0	46.3
	14	33 50.1	47.1
	15	48 50.2	·· 48.0
	16	63 50.3	48.9
	17	78 50.4	49.8
	18	93 50.5	S10 50.7
	19	108 50.6	51.6
	20	123 50.7	52.5
	21	138 50.8	·· 53.4
	22	153 50.9	54.2
	23	168 51.0	55.1

		G.H.A.	Dec.
d	h	° '	° '
22	00	183 51.1	S10 56.0
	01	198 51.2	56.9
	02	213 51.3	57.8
	03	228 51.4	·· 58.7
	04	243 51.5	10 59.6
	05	258 51.6	11 00.4
	06	273 51.7	S11 01.3
	07	288 51.8	02.2
	08	303 51.9	03.1
	09	318 52.0	·· 04.0
	10	333 52.0	04.9
	11	348 52.1	05.7
	12	3 52.2	S11 06.6
	13	18 52.3	07.5
	14	33 52.4	08.4
	15	48 52.5	·· 09.3
	16	63 52.6	10.2
	17	78 52.7	11.0
	18	93 52.8	S11 11.9
	19	108 52.9	12.8
	20	123 53.0	13.7
	21	138 53.1	·· 14.6
	22	153 53.2	15.4
	23	168 53.2	16.3
23	00	183 53.3	S11 17.2
	01	198 53.4	18.1
	02	213 53.5	19.0
	03	228 53.6	·· 19.8
	04	243 53.7	20.7
	05	258 53.8	21.6
	06	273 53.9	S11 22.5
	07	288 54.0	23.3
	08	303 54.0	24.2
	09	318 54.1	·· 25.1
	10	333 54.2	26.0
	11	348 54.3	26.8
	12	3 54.4	S11 27.7
	13	18 54.5	28.6
	14	33 54.6	29.5
	15	48 54.7	·· 30.3
	16	63 54.7	31.2
	17	78 54.8	32.1
	18	93 54.9	S11 33.0
	19	108 55.0	33.8
	20	123 55.1	34.7
	21	138 55.2	·· 35.6
	22	153 55.2	36.5
	23	168 55.3	37.3
24	00	183 55.4	S11 38.2
	01	198 55.5	39.1
	02	213 55.6	39.9
	03	228 55.7	·· 40.8
	04	243 55.7	41.7
	05	258 55.8	42.6
	06	273 55.9	S11 43.4
	07	288 56.0	44.3
	08	303 56.1	45.2
	09	318 56.1	·· 46.0
	10	333 56.2	46.9
	11	348 56.3	47.8
	12	3 56.4	S11 48.6
	13	18 56.5	49.5
	14	33 56.5	50.4
	15	48 56.6	·· 51.2
	16	63 56.7	52.1
	17	78 56.8	53.0
	18	93 56.9	S11 53.8
	19	108 56.9	54.7
	20	123 57.0	55.6
	21	138 57.1	·· 56.4
	22	153 57.2	57.3
	23	168 57.2	58.2

NOVEMBER

25 – 27

d h	G.H.A.	Dec.
25 00	183 57.3	S11 59.0
01	198 57.4	11 59.9
02	213 57.5	12 00.8
03	228 57.5	·· 01.6
04	243 57.6	02.5
05	258 57.7	03.3
06	273 57.8	S12 04.2
07	288 57.8	05.1
08	303 57.9	05.9
09	318 58.0	·· 06.8
10	333 58.1	07.7
11	348 58.1	08.5
12	3 58.2	S12 09.4
13	18 58.3	10.2
14	33 58.3	11.1
15	48 58.4	·· 12.0
16	63 58.5	12.8
17	78 58.6	13.7
18	93 58.6	S12 14.5
19	108 58.7	15.4
20	123 58.8	16.2
21	138 58.8	·· 17.1
22	153 58.9	18.0
23	168 59.0	18.8
26 00	183 59.0	S12 19.7
01	198 59.1	20.5
02	213 59.2	21.4
03	228 59.2	·· 22.2
04	243 59.3	23.1
05	258 59.4	24.0
06	273 59.4	S12 24.8
07	288 59.5	25.7
08	303 59.6	26.5
09	318 59.6	·· 27.4
10	333 59.7	28.2
11	348 59.8	29.1
12	3 59.8	S12 29.9
13	18 59.9	30.8
14	34 00.0	31.6
15	49 00.0	·· 32.5
16	64 00.1	33.3
17	79 00.2	34.2
18	94 00.2	S12 35.0
19	109 00.3	35.9
20	124 00.3	36.7
21	139 00.4	·· 37.6
22	154 00.5	38.4
23	169 00.5	39.3
27 00	184 00.6	S12 40.1
01	199 00.7	41.0
02	214 00.7	41.8
03	229 00.8	·· 42.7
04	244 00.8	43.5
05	259 00.9	44.4
06	274 00.9	S12 45.2
07	289 01.0	46.1
08	304 01.1	46.9
09	319 01.1	·· 47.8
10	334 01.2	48.6
11	349 01.2	49.4
12	4 01.3	S12 50.3
13	19 01.4	51.1
14	34 01.4	52.0
15	49 01.5	·· 52.8
16	64 01.5	53.7
17	79 01.6	54.5
18	94 01.6	S12 55.3
19	109 01.7	56.2
20	124 01.7	57.0
21	139 01.8	·· 57.9
22	154 01.8	58.7
23	169 01.9	59.6

28 – 30

d h	G.H.A.	Dec.
28 00	184 02.0	S13 00.4
01	199 02.0	01.2
02	214 02.1	02.1
03	229 02.1	·· 02.9
04	244 02.2	03.8
05	259 02.2	04.6
06	274 02.3	S13 05.4
07	289 02.3	06.3
08	304 02.4	07.1
09	319 02.4	·· 07.9
10	334 02.5	08.8
11	349 02.5	09.6
12	4 02.6	S13 10.5
13	19 02.6	11.3
14	34 02.7	12.1
15	49 02.7	·· 13.0
16	64 02.8	13.8
17	79 02.8	14.6
18	94 02.9	S13 15.5
19	109 02.9	16.3
20	124 02.9	17.1
21	139 03.0	·· 18.0
22	154 03.0	18.8
23	169 03.1	19.6
29 00	184 03.1	S13 20.5
01	199 03.2	21.3
02	214 03.2	22.1
03	229 03.3	·· 22.9
04	244 03.3	23.8
05	259 03.4	24.6
06	274 03.4	S13 25.4
07	289 03.4	26.3
08	304 03.5	27.1
09	319 03.5	·· 27.9
10	334 03.6	28.8
11	349 03.6	29.6
12	4 03.6	S13 30.4
13	19 03.7	31.2
14	34 03.7	32.1
15	49 03.8	·· 32.9
16	64 03.8	33.7
17	79 03.8	34.5
18	94 03.9	S13 35.4
19	109 03.9	36.2
20	124 04.0	37.0
21	139 04.0	·· 37.8
22	154 04.0	38.7
23	169 04.1	39.5
30 00	184 04.1	S13 40.3
01	199 04.1	41.1
02	214 04.2	41.9
03	229 04.2	·· 42.8
04	244 04.3	43.6
05	259 04.3	44.4
06	274 04.3	S13 45.2
07	289 04.4	46.1
08	304 04.4	46.9
09	319 04.4	·· 47.7
10	334 04.5	48.5
11	349 04.5	49.3
12	4 04.5	S13 50.1
13	19 04.6	51.0
14	34 04.6	51.8
15	49 04.6	·· 52.6
16	64 04.7	53.4
17	79 04.7	54.2
18	94 04.7	S13 55.1
19	109 04.7	55.9
20	124 04.8	56.7
21	139 04.8	·· 57.5
22	154 04.8	58.3
23	169 04.9	59.1

31 – 2

d h	G.H.A.	Dec.
31 00	184 04.9	S13 59.9
01	199 04.9	14 00.8
02	214 05.0	01.6
03	229 05.0	·· 02.4
04	244 05.0	03.2
05	259 05.0	04.0
06	274 05.1	S14 04.8
07	289 05.1	05.6
08	304 05.1	06.4
09	319 05.1	·· 07.2
10	334 05.2	08.1
11	349 05.2	08.9
12	4 05.2	S14 09.7
13	19 05.2	10.5
14	34 05.3	11.3
15	49 05.3	·· 12.1
16	64 05.3	12.9
17	79 05.3	13.7
18	94 05.4	S14 14.5
19	109 05.4	15.3
20	124 05.4	16.1
21	139 05.4	·· 16.9
22	154 05.4	17.7
23	169 05.5	18.5
1 00	184 05.5	S14 19.3
01	199 05.5	20.2
02	214 05.5	21.0
03	229 05.5	·· 21.8
04	244 05.6	22.6
05	259 05.6	23.4
06	274 05.6	S14 24.2
07	289 05.6	25.0
08	304 05.6	25.8
09	319 05.6	·· 26.6
10	334 05.7	27.4
11	349 05.7	28.2
12	4 05.7	S14 29.0
13	19 05.7	29.8
14	34 05.7	30.6
15	49 05.7	·· 31.4
16	64 05.8	32.2
17	79 05.8	33.0
18	94 05.8	S14 33.8
19	109 05.8	34.6
20	124 05.8	35.4
21	139 05.8	·· 36.1
22	154 05.8	36.9
23	169 05.8	37.7
2 00	184 05.9	S14 38.5
01	199 05.9	39.3
02	214 05.9	40.1
03	229 05.9	·· 40.9
04	244 05.9	41.7
05	259 05.9	42.5
06	274 05.9	S14 43.3
07	289 05.9	44.1
08	304 05.9	44.9
09	319 05.9	·· 45.7
10	334 06.0	46.5
11	349 06.0	47.2
12	4 06.0	S14 48.0
13	19 06.0	48.8
14	34 06.0	49.6
15	49 06.0	·· 50.4
16	64 06.0	51.2
17	79 06.0	52.0
18	94 06.0	S14 52.8
19	109 06.0	53.5
20	124 06.0	54.3
21	139 06.0	·· 55.1
22	154 06.0	55.9
23	169 06.0	56.7

SUN

d h	G.H.A.	Dec.
3 00	184 06.0	S14 57.5
01	199 06.0	58.3
02	214 06.0	59.0
03	229 06.0	14 59.8
04	244 06.0	15 00.6
05	259 06.0	01.4
06	274 06.0	S15 02.2
07	289 06.0	03.0
08	304 06.0	03.7
09	319 06.0	·· 04.5
10	334 06.0	05.3
11	349 06.0	06.1
12	4 06.0	S15 06.9
13	19 06.0	07.6
14	34 06.0	08.4
15	49 06.0	·· 09.2
16	64 06.0	10.0
17	79 06.0	10.7
18	94 06.0	S15 11.5
19	109 06.0	12.3
20	124 06.0	13.1
21	139 06.0	·· 13.9
22	154 06.0	14.6
23	169 06.0	15.4
4 00	184 06.0	S15 16.2
01	199 06.0	17.0
02	214 06.0	17.7
03	229 06.0	·· 18.5
04	244 06.0	19.3
05	259 06.0	20.0
06	274 06.0	S15 20.8
07	289 05.9	21.6
08	304 05.9	22.4
09	319 05.9	·· 23.1
10	334 05.9	23.9
11	349 05.9	24.7
12	4 05.9	S15 25.4
13	19 05.9	26.2
14	34 05.9	27.0
15	49 05.9	·· 27.7
16	64 05.9	28.5
17	79 05.8	29.3
18	94 05.8	S15 30.0
19	109 05.8	30.8
20	124 05.8	31.6
21	139 05.8	·· 32.3
22	154 05.8	33.1
23	169 05.8	33.9
5 00	184 05.7	S15 34.6
01	199 05.7	35.4
02	214 05.7	36.2
03	229 05.7	·· 36.9
04	244 05.7	37.7
05	259 05.7	38.4
06	274 05.7	S15 39.2
07	289 05.6	40.0
08	304 05.6	40.7
09	319 05.6	·· 41.5
10	334 05.6	42.2
11	349 05.6	43.0
12	4 05.5	S15 43.8
13	19 05.5	44.5
14	34 05.5	45.3
15	49 05.5	·· 46.0
16	64 05.5	46.8
17	79 05.4	47.5
18	94 05.4	S15 48.3
19	109 05.4	49.1
20	124 05.4	49.8
21	139 05.4	·· 50.6
22	154 05.3	51.3
23	169 05.3	52.1

d h	G.H.A.	Dec.
6 00	184 05.3	S15 52.8
01	199 05.3	53.6
02	214 05.2	54.3
03	229 05.2	·· 55.1
04	244 05.2	55.8
05	259 05.2	56.6
06	274 05.1	S15 57.3
07	289 05.1	58.1
08	304 05.1	58.8
09	319 05.1	15 59.6
10	334 05.0	16 00.3
11	349 05.0	01.1
12	4 05.0	S16 01.8
13	19 05.0	02.6
14	34 04.9	03.3
15	49 04.9	·· 04.1
16	64 04.9	04.8
17	79 04.8	05.6
18	94 04.8	S16 06.3
19	109 04.8	07.0
20	124 04.7	07.8
21	139 04.7	·· 08.5
22	154 04.7	09.3
23	169 04.7	10.0
7 00	184 04.6	S16 10.8
01	199 04.6	11.5
02	214 04.6	12.2
03	229 04.5	·· 13.0
04	244 04.5	13.7
05	259 04.5	14.5
06	274 04.4	S16 15.2
07	289 04.4	15.9
08	304 04.3	16.7
09	319 04.3	·· 17.4
10	334 04.3	18.2
11	349 04.2	18.9
12	4 04.2	S16 19.6
13	19 04.2	20.4
14	34 04.1	21.1
15	49 04.1	·· 21.8
16	64 04.1	22.6
17	79 04.0	23.3
18	94 04.0	S16 24.0
19	109 03.9	24.8
20	124 03.9	25.5
21	139 03.9	·· 26.2
22	154 03.8	27.0
23	169 03.8	27.7
8 00	184 03.7	S16 28.4
01	199 03.7	29.2
02	214 03.7	29.9
03	229 03.6	·· 30.6
04	244 03.6	31.3
05	259 03.5	32.1
06	274 03.5	S16 32.8
07	289 03.4	33.5
08	304 03.4	34.2
09	319 03.3	·· 35.0
10	334 03.3	35.7
11	349 03.3	36.4
12	4 03.2	S16 37.1
13	19 03.2	37.9
14	34 03.1	38.6
15	49 03.1	·· 39.3
16	64 03.0	40.0
17	79 03.0	40.8
18	94 02.9	S16 41.5
19	109 02.9	42.2
20	124 02.8	42.9
21	139 02.8	·· 43.6
22	154 02.7	44.4
23	169 02.7	45.1

d h	G.H.A.	Dec.
9 00	184 02.6	S16 45.8
01	199 02.6	46.5
02	214 02.5	47.2
03	229 02.5	·· 48.0
04	244 02.4	48.7
05	259 02.4	49.4
06	274 02.3	S16 50.1
07	289 02.3	50.8
08	304 02.2	51.5
09	319 02.2	·· 52.3
10	334 02.1	53.0
11	349 02.1	53.7
12	4 02.0	S16 54.4
13	19 02.0	55.1
14	34 01.9	55.8
15	49 01.8	·· 56.5
16	64 01.8	57.2
17	79 01.7	58.0
18	94 01.7	S16 58.7
19	109 01.6	16 59.4
20	124 01.6	17 00.1
21	139 01.5	·· 00.8
22	154 01.4	01.5
23	169 01.4	02.2
10 00	184 01.3	S17 02.9
01	199 01.3	03.6
02	214 01.2	04.3
03	229 01.1	·· 05.0
04	244 01.1	05.7
05	259 01.0	06.4
06	274 01.0	S17 07.1
07	289 00.9	07.8
08	304 00.8	08.5
09	319 00.8	·· 09.2
10	334 00.7	09.9
11	349 00.7	10.6
12	4 00.6	S17 11.4
13	19 00.5	12.1
14	34 00.5	12.8
15	49 00.4	·· 13.4
16	64 00.3	14.1
17	79 00.3	14.8
18	94 00.2	S17 15.5
19	109 00.1	16.2
20	124 00.1	16.9
21	139 00.0	·· 17.6
22	153 59.9	18.3
23	168 59.9	19.0
11 00	183 59.8	S17 19.7
01	198 59.7	20.4
02	213 59.7	21.1
03	228 59.6	·· 21.8
04	243 59.5	22.5
05	258 59.5	23.2
06	273 59.4	S17 23.9
07	288 59.3	24.6
08	303 59.3	25.3
09	318 59.2	·· 25.9
10	333 59.1	26.6
11	348 59.0	27.3
12	3 59.0	S17 28.0
13	18 58.9	28.7
14	33 58.8	29.4
15	48 58.7	·· 30.1
16	63 58.7	30.8
17	78 58.6	31.4
18	93 58.5	S17 32.1
19	108 58.5	32.8
20	123 58.4	33.5
21	138 58.3	·· 34.2
22	153 58.2	34.9
23	168 58.1	35.6

		G.H.A.	Dec.
d h		° ′	° ′
12 00		183 58.1	S17 36.2
01		198 58.0	36.9
02		213 57.9	37.6
03		228 57.8 ··	38.3
04		243 57.8	39.0
05		258 57.7	39.6
06		273 57.6	S17 40.3
07		288 57.5	41.0
08		303 57.4	41.7
09		318 57.4 ··	42.3
10		333 57.3	43.0
11		348 57.2	43.7
12		3 57.1	S17 44.4
13		18 57.0	45.1
14		33 57.0	45.7
15		48 56.9 ··	46.4
16		63 56.8	47.1
17		78 56.7	47.7
18		93 56.6	S17 48.4
19		108 56.6	49.1
20		123 56.5	49.8
21		138 56.4 ··	50.4
22		153 56.3	51.1
23		168 56.2	51.8
13 00		183 56.1	S17 52.4
01		198 56.0	53.1
02		213 56.0	53.8
03		228 55.9 ··	54.4
04		243 55.8	55.1
05		258 55.7	55.8
06		273 55.6	S17 56.4
07		288 55.5	57.1
08		303 55.4	57.8
09		318 55.3 ··	58.4
10		333 55.3	59.1
11		348 55.2	17 59.8
12		3 55.1	S18 00.4
13		18 55.0	01.1
14		33 54.9	01.8
15		48 54.8 ··	02.4
16		63 54.7	03.1
17		78 54.6	03.7
18		93 54.5	S18 04.4
19		108 54.4	05.1
20		123 54.3	05.7
21		138 54.3 ··	06.4
22		153 54.2	07.0
23		168 54.1	07.7
14 00		183 54.0	S18 08.3
01		198 53.9	09.0
02		213 53.8	09.7
03		228 53.7 ··	10.3
04		243 53.6	11.0
05		258 53.5	11.6
06		273 53.4	S18 12.3
07		288 53.3	12.9
08		303 53.2	13.6
09		318 53.1 ··	14.2
10		333 53.0	14.9
11		348 52.9	15.5
12		3 52.8	S18 16.2
13		18 52.7	16.8
14		33 52.6	17.5
15		48 52.5 ··	18.1
16		63 52.4	18.8
17		78 52.3	19.4
18		93 52.2	S18 20.1
19		108 52.1	20.7
20		123 52.0	21.3
21		138 51.9 ··	22.0
22		153 51.8	22.6
23		168 51.7	23.3

		G.H.A.	Dec.
d h		° ′	° ′
15 00		183 51.6	S18 23.9
01		198 51.5	24.6
02		213 51.4	25.2
03		228 51.3 ··	25.8
04		243 51.2	26.5
05		258 51.1	27.1
06		273 51.0	S18 27.8
07		288 50.9	28.4
08		303 50.8	29.0
09		318 50.7 ··	29.7
10		333 50.6	30.3
11		348 50.5	31.0
12		3 50.4	S18 31.6
13		18 50.2	32.2
14		33 50.1	32.9
15		48 50.0 ··	33.5
16		63 49.9	34.1
17		78 49.8	34.8
18		93 49.7	S18 35.4
19		108 49.6	36.0
20		123 49.5	36.7
21		138 49.4 ··	37.3
22		153 49.3	37.9
23		168 49.2	38.6
16 00		183 49.0	S18 39.2
01		198 48.9	39.8
02		213 48.8	40.4
03		228 48.7 ··	41.1
04		243 48.6	41.7
05		258 48.5	42.3
06		273 48.4	S18 42.9
07		288 48.3	43.6
08		303 48.1	44.2
09		318 48.0 ··	44.8
10		333 47.9	45.4
11		348 47.8	46.1
12		3 47.7	S18 46.7
13		18 47.6	47.3
14		33 47.5	47.9
15		48 47.3 ··	48.6
16		63 47.2	49.2
17		78 47.1	49.8
18		93 47.0	S18 50.4
19		108 46.9	51.0
20		123 46.7	51.6
21		138 46.6 ··	52.3
22		153 46.5	52.9
23		168 46.4	53.5
17 00		183 46.3	S18 54.1
01		198 46.1	54.7
02		213 46.0	55.3
03		228 45.9 ··	56.0
04		243 45.8	56.6
05		258 45.7	57.2
06		273 45.5	S18 57.8
07		288 45.4	58.4
08		303 45.3	59.0
09		318 45.2	18 59.6
10		333 45.1	19 00.2
11		348 44.9	00.8
12		3 44.8	S19 01.4
13		18 44.7	02.1
14		33 44.6	02.7
15		48 44.4 ··	03.3
16		63 44.3	03.9
17		78 44.2	04.5
18		93 44.1	S19 05.1
19		108 43.9	05.7
20		123 43.8	06.3
21		138 43.7 ··	06.9
22		153 43.5	07.5
23		168 43.4	08.1

		G.H.A.	Dec.
d h		° ′	° ′
18 00		183 43.3	S19 08.7
01		198 43.2	09.3
02		213 43.0	09.9
03		228 42.9 ··	10.5
04		243 42.8	11.1
05		258 42.6	11.7
06		273 42.5	S19 12.3
07		288 42.4	12.9
08		303 42.3	13.5
09		318 42.1 ··	14.1
10		333 42.0	14.7
11		348 41.9	15.3
12		3 41.7	S19 15.9
13		18 41.6	16.5
14		33 41.5	17.1
15		48 41.3 ··	17.6
16		63 41.2	18.2
17		78 41.1	18.8
18		93 40.9	S19 19.4
19		108 40.8	20.0
20		123 40.7	20.6
21		138 40.5 ··	21.2
22		153 40.4	21.8
23		168 40.2	22.4
19 00		183 40.1	S19 23.0
01		198 40.0	23.5
02		213 39.8	24.1
03		228 39.7 ··	24.7
04		243 39.6	25.3
05		258 39.4	25.9
06		273 39.3	S19 26.5
07		288 39.1	27.0
08		303 39.0	27.6
09		318 38.9 ··	28.2
10		333 38.7	28.8
11		348 38.6	29.4
12		3 38.4	S19 29.9
13		18 38.3	30.5
14		33 38.2	31.1
15		48 38.0 ··	31.7
16		63 37.9	32.3
17		78 37.7	32.8
18		93 37.6	S19 33.4
19		108 37.4	34.0
20		123 37.3	34.6
21		138 37.2 ··	35.1
22		153 37.0	35.7
23		168 36.9	36.3
20 00		183 36.7	S19 36.9
01		198 36.6	37.4
02		213 36.4	38.0
03		228 36.3 ··	38.6
04		243 36.1	39.1
05		258 36.0	39.7
06		273 35.8	S19 40.3
07		288 35.7	40.8
08		303 35.6	41.4
09		318 35.4 ··	42.0
10		333 35.3	42.5
11		348 35.1	43.1
12		3 35.0	S19 43.7
13		18 34.8	44.2
14		33 34.7	44.8
15		48 34.5 ··	45.4
16		63 34.4	45.9
17		78 34.2	46.5
18		93 34.1	S19 47.0
19		108 33.9	47.6
20		123 33.8	48.2
21		138 33.6 ··	48.7
22		153 33.4	49.3
23		168 33.3	49.8

21

d h	G.H.A.	Dec.
00	183 33.1	S19 50.4
01	198 33.0	51.0
02	213 32.8	51.5
03	228 32.7	·· 52.1
04	243 32.5	52.6
05	258 32.4	53.2
06	273 32.2	S19 53.7
07	288 32.1	54.3
08	303 31.9	54.8
09	318 31.7	·· 55.4
10	333 31.6	55.9
11	348 31.4	56.5
12	3 31.3	S19 57.0
13	18 31.1	57.6
14	33 31.0	58.1
15	48 30.8	·· 58.7
16	63 30.6	59.2
17	78 30.5	19 59.8
18	93 30.3	S20 00.3
19	108 30.2	00.9
20	123 30.0	01.4
21	138 29.8	·· 02.0
22	153 29.7	02.5
23	168 29.5	03.0

22

d h	G.H.A.	Dec.
00	183 29.4	S20 03.6
01	198 29.2	04.1
02	213 29.0	04.7
03	228 28.9	·· 05.2
04	243 28.7	05.7
05	258 28.5	06.3
06	273 28.4	S20 06.8
07	288 28.2	07.4
08	303 28.1	07.9
09	318 27.9	·· 08.4
10	333 27.7	09.0
11	348 27.6	09.5
12	3 27.4	S20 10.0
13	18 27.2	10.6
14	33 27.1	11.1
15	48 26.9	·· 11.6
16	63 26.7	12.2
17	78 26.6	12.7
18	93 26.4	S20 13.2
19	108 26.2	13.8
20	123 26.1	14.3
21	138 25.9	·· 14.8
22	153 25.7	15.3
23	168 25.6	15.9

23

d h	G.H.A.	Dec.
00	183 25.4	S20 16.4
01	198 25.2	16.9
02	213 25.0	17.4
03	228 24.9	·· 18.0
04	243 24.7	18.5
05	258 24.5	19.0
06	273 24.4	S20 19.5
07	288 24.2	20.1
08	303 24.0	20.6
09	318 23.8	·· 21.1
10	333 23.7	21.6
11	348 23.5	22.1
12	3 23.3	S20 22.7
13	18 23.1	23.2
14	33 23.0	23.7
15	48 22.8	·· 24.2
16	63 22.6	24.7
17	78 22.4	25.2
18	93 22.3	S20 25.8
19	108 22.1	26.3
20	123 21.9	26.8
21	138 21.7	·· 27.3
22	153 21.6	27.8
23	168 21.4	28.3

24

d h	G.H.A.	Dec.
00	183 21.2	S20 28.8
01	198 21.0	29.3
02	213 20.9	29.9
03	228 20.7	·· 30.4
04	243 20.5	30.9
05	258 20.3	31.4
06	273 20.1	S20 31.9
07	288 20.0	32.4
08	303 19.8	32.9
09	318 19.6	·· 33.4
10	333 19.4	33.9
11	348 19.2	34.4
12	3 19.1	S20 34.9
13	18 18.9	35.4
14	33 18.7	35.9
15	48 18.5	·· 36.4
16	63 18.3	36.9
17	78 18.1	37.4
18	93 18.0	S20 37.9
19	108 17.8	38.4
20	123 17.6	38.9
21	138 17.4	·· 39.4
22	153 17.2	39.9
23	168 17.0	40.4

25

d h	G.H.A.	Dec.
00	183 16.8	S20 40.9
01	198 16.7	41.4
02	213 16.5	41.9
03	228 16.3	·· 42.4
04	243 16.1	42.9
05	258 15.9	43.4
06	273 15.7	S20 43.9
07	288 15.5	44.3
08	303 15.3	44.8
09	318 15.2	·· 45.3
10	333 15.0	45.8
11	348 14.8	46.3
12	3 14.6	S20 46.8
13	18 14.4	47.3
14	33 14.2	47.8
15	48 14.0	·· 48.2
16	63 13.8	48.7
17	78 13.6	49.2
18	93 13.4	S20 49.7
19	108 13.3	50.2
20	123 13.1	50.7
21	138 12.9	·· 51.1
22	153 12.7	51.6
23	168 12.5	52.1

26

d h	G.H.A.	Dec.
00	183 12.3	S20 52.6
01	198 12.1	53.1
02	213 11.9	53.5
03	228 11.7	·· 54.0
04	243 11.5	54.5
05	258 11.3	55.0
06	273 11.1	S20 55.4
07	288 10.9	55.9
08	303 10.7	56.4
09	318 10.5	·· 56.9
10	333 10.3	57.3
11	348 10.1	57.8
12	3 09.9	S20 58.3
13	18 09.7	58.7
14	33 09.6	59.2
15	48 09.4	20 59.7
16	63 09.2	21 00.1
17	78 09.0	00.6
18	93 08.8	S21 01.1
19	108 08.6	01.5
20	123 08.4	02.0
21	138 08.2	·· 02.5
22	153 08.0	02.9
23	168 07.8	03.4

27

d h	G.H.A.	Dec.
00	183 07.6	S21 03.9
01	198 07.4	04.3
02	213 07.2	04.8
03	228 06.9	·· 05.2
04	243 06.7	05.7
05	258 06.5	06.2
06	273 06.3	S21 06.6
07	288 06.1	07.1
08	303 05.9	07.5
09	318 05.7	·· 08.0
10	333 05.5	08.5
11	348 05.3	08.9
12	3 05.1	S21 09.4
13	18 04.9	09.8
14	33 04.7	10.3
15	48 04.5	·· 10.7
16	63 04.3	11.2
17	78 04.1	11.6
18	93 03.9	S21 12.1
19	108 03.7	12.5
20	123 03.5	13.0
21	138 03.3	·· 13.4
22	153 03.1	13.9
23	168 02.8	14.3

28

d h	G.H.A.	Dec.
00	183 02.6	S21 14.8
01	198 02.4	15.2
02	213 02.2	15.6
03	228 02.0	·· 16.1
04	243 01.8	16.5
05	258 01.6	17.0
06	273 01.4	S21 17.4
07	288 01.2	17.9
08	303 01.0	18.3
09	318 00.7	·· 18.7
10	333 00.5	19.2
11	348 00.3	19.6
12	3 00.1	S21 20.1
13	17 59.9	20.5
14	32 59.7	20.9
15	47 59.5	·· 21.4
16	62 59.3	21.8
17	77 59.0	22.2
18	92 58.8	S21 22.7
19	107 58.6	23.1
20	122 58.4	23.5
21	137 58.2	·· 24.0
22	152 58.0	24.4
23	167 57.7	24.8

29

d h	G.H.A.	Dec.
00	182 57.5	S21 25.2
01	197 57.3	25.7
02	212 57.1	26.1
03	227 56.9	·· 26.5
04	242 56.7	27.0
05	257 56.4	27.4
06	272 56.2	S21 27.8
07	287 56.0	28.2
08	302 55.8	28.7
09	317 55.6	·· 29.1
10	332 55.4	29.5
11	347 55.1	29.9
12	2 54.9	S21 30.3
13	17 54.7	30.8
14	32 54.5	31.2
15	47 54.3	·· 31.6
16	62 54.0	32.0
17	77 53.8	32.4
18	92 53.6	S21 32.9
19	107 53.4	33.3
20	122 53.1	33.7
21	137 52.9	34.1
22	152 52.7	34.5
23	167 52.5	34.9

d	h	G.H.A. ° '	Dec. ° '
30	00	182 52.3	S21 35.3
	01	197 52.0	35.7
	02	212 51.8	36.2
	03	227 51.6 ··	36.6
	04	242 51.4	37.0
	05	257 51.1	37.4
	06	272 50.9	S21 37.8
	07	287 50.7	38.2
	08	302 50.5	38.6
	09	317 50.2 ··	39.0
	10	332 50.0	39.4
	11	347 49.8	39.8
	12	2 49.6	S21 40.2
	13	17 49.3	40.6
	14	32 49.1	41.0
	15	47 48.9 ··	41.4
	16	62 48.6	41.8
	17	77 48.4	42.2
	18	92 48.2	S21 42.6
	19	107 48.0	43.0
	20	122 47.7	43.4
	21	137 47.5 ··	43.8
	22	152 47.3	44.2
	23	167 47.0	44.6
1	00	182 46.8	S21 45.0
	01	197 46.6	45.4
	02	212 46.3	45.8
	03	227 46.1 ··	46.2
	04	242 45.9	46.6
	05	257 45.6	47.0
	06	272 45.4	S21 47.4
	07	287 45.2	47.8
	08	302 45.0	48.1
	09	317 44.7 ··	48.5
	10	332 44.5	48.9
	11	347 44.3	49.3
	12	2 44.0	S21 49.7
	13	17 43.8	50.1
	14	32 43.5	50.5
	15	47 43.3 ··	50.8
	16	62 43.1	51.2
	17	77 42.8	51.6
	18	92 42.6	S21 52.0
	19	107 42.4	52.4
	20	122 42.1	52.8
	21	137 41.9 ··	53.1
	22	152 41.7	53.5
	23	167 41.4	53.9
2	00	182 41.2	S21 54.3
	01	197 41.0	54.6
	02	212 40.7	55.0
	03	227 40.5 ··	55.4
	04	242 40.2	55.8
	05	257 40.0	56.1
	06	272 39.8	S21 56.5
	07	287 39.5	56.9
	08	302 39.3	57.3
	09	317 39.0 ··	57.6
	10	332 38.8	58.0
	11	347 38.6	58.4
	12	2 38.3	S21 58.7
	13	17 38.1	59.1
	14	32 37.8	59.5
	15	47 37.6	21 59.8
	16	62 37.4	22 00.2
	17	77 37.1	00.6
	18	92 36.9	S22 00.9
	19	107 36.6	01.3
	20	122 36.4	01.7
	21	137 36.1 ··	02.0
	22	152 35.9	02.4
	23	167 35.7	02.7

d	h	G.H.A. ° '	Dec. ° '
3	00	182 35.4	S22 03.1
	01	197 35.2	03.5
	02	212 34.9	03.8
	03	227 34.7 ··	04.2
	04	242 34.4	04.5
	05	257 34.2	04.9
	06	272 33.9	S22 05.2
	07	287 33.7	05.6
	08	302 33.4	06.0
	09	317 33.2 ··	06.3
	10	332 33.0	06.7
	11	347 32.7	07.0
	12	2 32.5	S22 07.4
	13	17 32.2	07.7
	14	32 32.0	08.1
	15	47 31.7 ··	08.4
	16	62 31.5	08.8
	17	77 31.2	09.1
	18	92 31.0	S22 09.5
	19	107 30.7	09.8
	20	122 30.5	10.1
	21	137 30.2 ··	10.5
	22	152 30.0	10.8
	23	167 29.7	11.2
4	00	182 29.5	S22 11.5
	01	197 29.2	11.9
	02	212 29.0	12.2
	03	227 28.7 ··	12.5
	04	242 28.5	12.9
	05	257 28.2	13.2
	06	272 28.0	S22 13.6
	07	287 27.7	13.9
	08	302 27.5	14.2
	09	317 27.2 ··	14.6
	10	332 27.0	14.9
	11	347 26.7	15.2
	12	2 26.5	S22 15.6
	13	17 26.2	15.9
	14	32 25.9	16.2
	15	47 25.7 ··	16.6
	16	62 25.4	16.9
	17	77 25.2	17.2
	18	92 24.9	S22 17.5
	19	107 24.7	17.9
	20	122 24.4	18.2
	21	137 24.2 ··	18.5
	22	152 23.9	18.9
	23	167 23.6	19.2
5	00	182 23.4	S22 19.5
	01	197 23.1	19.8
	02	212 22.9	20.1
	03	227 22.6 ··	20.5
	04	242 22.4	20.8
	05	257 22.1	21.1
	06	272 21.8	S22 21.4
	07	287 21.6	21.8
	08	302 21.3	22.1
	09	317 21.1 ··	22.4
	10	332 20.8	22.7
	11	347 20.6	23.0
	12	2 20.3	S22 23.3
	13	17 20.0	23.6
	14	32 19.8	24.0
	15	47 19.5 ··	24.3
	16	62 19.3	24.6
	17	77 19.0	24.9
	18	92 18.7	S22 25.2
	19	107 18.5	25.5
	20	122 18.2	25.8
	21	137 17.9 ··	26.1
	22	152 17.7	26.4
	23	167 17.4	26.7

d	h	G.H.A. ° '	Dec. ° '
6	00	182 17.2	S22 27.1
	01	197 16.9	27.4
	02	212 16.6	27.7
	03	227 16.4 ··	28.0
	04	242 16.1	28.3
	05	257 15.8	28.6
	06	272 15.6	S22 28.9
	07	287 15.3	29.2
	08	302 15.1	29.5
	09	317 14.8 ··	29.8
	10	332 14.5	30.1
	11	347 14.3	30.4
	12	2 14.0	S22 30.7
	13	17 13.7	31.0
	14	32 13.5	31.3
	15	47 13.2 ··	31.6
	16	62 12.9	31.8
	17	77 12.7	32.1
	18	92 12.4	S22 32.4
	19	107 12.1	32.7
	20	122 11.9	33.0
	21	137 11.6 ··	33.3
	22	152 11.3	33.6
	23	167 11.1	33.9
7	00	182 10.8	S22 34.2
	01	197 10.5	34.5
	02	212 10.3	34.7
	03	227 10.0 ··	35.0
	04	242 09.7	35.3
	05	257 09.5	35.6
	06	272 09.2	S22 35.9
	07	287 08.9	36.2
	08	302 08.6	36.4
	09	317 08.4 ··	36.7
	10	332 08.1	37.0
	11	347 07.8	37.3
	12	2 07.6	S22 37.6
	13	17 07.3	37.8
	14	32 07.0	38.1
	15	47 06.8 ··	38.4
	16	62 06.5	38.7
	17	77 06.2	38.9
	18	92 05.9	S22 39.2
	19	107 05.7	39.5
	20	122 05.4	39.8
	21	137 05.1 ··	40.0
	22	152 04.8	40.3
	23	167 04.6	40.6
8	00	182 04.3	S22 40.8
	01	197 04.0	41.1
	02	212 03.8	41.4
	03	227 03.5 ··	41.6
	04	242 03.2	41.9
	05	257 02.9	42.2
	06	272 02.7	S22 42.4
	07	287 02.4	42.7
	08	302 02.1	43.0
	09	317 01.8 ··	43.2
	10	332 01.6	43.5
	11	347 01.3	43.8
	12	2 01.0	S22 44.0
	13	17 00.7	44.3
	14	32 00.5	44.5
	15	47 00.2 ··	44.8
	16	61 59.9	45.0
	17	76 59.6	45.3
	18	91 59.3	S22 45.6
	19	106 59.1	45.8
	20	121 58.8	46.1
	21	136 58.5 ··	46.3
	22	151 58.2	46.6
	23	166 58.0	46.8

d h	G.H.A. ° '	Dec. ° '		d h	G.H.A. ° '	Dec. ° '		d h	G.H.A. ° '	Dec. ° '
9 00	181 57.7	S22 47.1		**12** 00	181 37.2	S23 03.1		**15** 00	181 15.8	S23 14.9
01	196 57.4	47.3		01	196 36.9	03.2		01	196 15.5	15.1
02	211 57.1	47.6		02	211 36.6	03.4		02	211 15.2	15.2
03	226 56.8	·· 47.8		03	226 36.3	·· 03.6		03	226 14.9	·· 15.3
04	241 56.6	48.1		04	241 36.0	03.8		04	241 14.6	15.5
05	256 56.3	48.3		05	256 35.7	04.0		05	256 14.3	15.6
06	271 56.0	S22 48.6		06	271 35.4	S23 04.2		06	271 14.0	S23 15.7
07	286 55.7	48.8		07	286 35.1	04.4		07	286 13.7	15.9
08	301 55.4	49.0		08	301 34.8	04.6		08	301 13.4	16.0
09	316 55.2	·· 49.3		09	316 34.5	·· 04.8		09	316 13.1	·· 16.1
10	331 54.9	49.5		10	331 34.2	05.0		10	331 12.8	16.3
11	346 54.6	49.8		11	346 34.0	05.1		11	346 12.5	16.4
12	1 54.3	S22 50.0		12	1 33.7	S23 05.3		12	1 12.2	S23 16.5
13	16 54.0	50.3		13	16 33.4	05.5		13	16 11.9	16.6
14	31 53.8	50.5		14	31 33.1	05.7		14	31 11.6	16.8
15	46 53.5	·· 50.7		15	46 32.8	·· 05.9		15	46 11.3	·· 16.9
16	61 53.2	51.0		16	61 32.5	06.1		16	61 11.0	17.0
17	76 52.9	51.2		17	76 32.2	06.2		17	76 10.7	17.1
18	91 52.6	S22 51.4		18	91 31.9	S23 06.4		18	91 10.4	S23 17.3
19	106 52.4	51.7		19	106 31.6	06.6		19	106 10.1	17.4
20	121 52.1	51.9		20	121 31.3	06.8		20	121 09.8	17.5
21	136 51.8	·· 52.2		21	136 31.0	·· 06.9		21	136 09.5	·· 17.6
22	151 51.5	52.4		22	151 30.7	07.1		22	151 09.2	17.7
23	166 51.2	52.6		23	166 30.4	07.3		23	166 08.9	17.9
10 00	181 50.9	S22 52.9		**13** 00	181 30.1	S23 07.5		**16** 00	181 08.6	S23 18.0
01	196 50.7	53.1		01	196 29.8	07.6		01	196 08.3	18.1
02	211 50.4	53.3		02	211 29.5	07.8		02	211 08.0	18.2
03	226 50.1	·· 53.5		03	226 29.3	·· 08.0		03	226 07.7	·· 18.3
04	241 49.8	53.8		04	241 29.0	08.2		04	241 07.4	18.4
05	256 49.5	54.0		05	256 28.7	08.3		05	256 07.1	18.5
06	271 49.2	S22 54.2		06	271 28.4	S23 08.5		06	271 06.8	S23 18.7
07	286 49.0	54.5		07	286 28.1	08.7		07	286 06.5	18.8
08	301 48.7	54.7		08	301 27.8	08.8		08	301 06.2	18.9
09	316 48.4	·· 54.9		09	316 27.5	·· 09.0		09	316 05.9	·· 19.0
10	331 48.1	55.1		10	331 27.2	09.2		10	331 05.6	19.1
11	346 47.8	55.3		11	346 26.9	09.3		11	346 05.3	19.2
12	1 47.5	S22 55.6		12	1 26.6	S23 09.5		12	1 05.0	S23 19.3
13	16 47.3	55.8		13	16 26.3	09.7		13	16 04.6	19.4
14	31 47.0	56.0		14	31 26.0	09.8		14	31 04.3	19.5
15	46 46.7	·· 56.2		15	46 25.7	·· 10.0		15	46 04.0	·· 19.6
16	61 46.4	56.5		16	61 25.4	10.2		16	61 03.7	19.7
17	76 46.1	56.7		17	76 25.1	10.3		17	76 03.4	19.8
18	91 45.8	S22 56.9		18	91 24.8	S23 10.5		18	91 03.1	S23 19.9
19	106 45.5	57.1		19	106 24.5	10.6		19	106 02.8	20.0
20	121 45.3	57.3		20	121 24.2	10.8		20	121 02.5	20.1
21	136 45.0	·· 57.5		21	136 23.9	·· 11.0		21	136 02.2	·· 20.2
22	151 44.7	57.8		22	151 23.6	11.1		22	151 01.9	20.3
23	166 44.4	58.0		23	166 23.3	11.3		23	166 01.6	20.4
11 00	181 44.1	S22 58.2		**14** 00	181 23.0	S23 11.4		**17** 00	181 01.3	S23 20.5
01	196 43.8	58.4		01	196 22.7	11.6		01	196 01.0	20.6
02	211 43.5	58.6		02	211 22.4	11.7		02	211 00.7	20.7
03	226 43.2	·· 58.8		03	226 22.1	·· 11.9		03	226 00.4	·· 20.8
04	241 43.0	59.0		04	241 21.8	12.0		04	241 00.1	20.9
05	256 42.7	59.2		05	256 21.5	12.2		05	255 59.8	21.0
06	271 42.4	S22 59.4		06	271 21.2	S23 12.4		06	270 59.5	S23 21.1
07	286 42.1	59.6		07	286 20.9	12.5		07	285 59.2	21.2
08	301 41.8	22 59.9		08	301 20.6	12.7		08	300 58.9	21.3
09	316 41.5	23 00.1		09	316 20.3	·· 12.8		09	315 58.5	·· 21.4
10	331 41.2	00.3		10	331 20.0	12.9		10	330 58.2	21.5
11	346 40.9	00.5		11	346 19.7	13.1		11	345 57.9	21.6
12	1 40.6	S23 00.7		12	1 19.4	S23 13.2		12	0 57.6	S23 21.6
13	16 40.4	00.9		13	16 19.1	13.4		13	15 57.3	21.7
14	31 40.1	01.1		14	31 18.8	13.5		14	30 57.0	21.8
15	46 39.8	·· 01.3		15	46 18.5	·· 13.7		15	45 56.7	·· 21.9
16	61 39.5	01.5		16	61 18.2	13.8		16	60 56.4	22.0
17	76 39.2	01.7		17	76 17.9	14.0		17	75 56.1	22.1
18	91 38.9	S23 01.9		18	91 17.6	S23 14.1		18	90 55.8	S23 22.2
19	106 38.6	02.1		19	106 17.3	14.2		19	105 55.5	22.2
20	121 38.3	02.3		20	121 17.0	14.4		20	120 55.2	22.3
21	136 38.0	·· 02.5		21	136 16.7	·· 14.5		21	135 54.9	·· 22.4
22	151 37.7	02.7		22	151 16.4	14.7		22	150 54.6	22.5
23	166 37.5	02.9		23	166 16.1	14.8		23	165 54.3	22.6

d h	G.H.A.	Dec.
18 00	180 53.9	S23 22.6
01	195 53.6	22.7
02	210 53.3	22.8
03	225 53.0 ··	22.9
04	240 52.7	22.9
05	255 52.4	23.0
06	270 52.1	S23 23.1
07	285 51.8	23.2
08	300 51.5	23.2
09	315 51.2 ··	23.3
10	330 50.9	23.4
11	345 50.6	23.5
12	0 50.3	S23 23.5
13	15 50.0	23.6
14	30 49.6	23.7
15	45 49.3 ··	23.7
16	60 49.0	23.8
17	75 48.7	23.8
18	90 48.4	S23 23.9
19	105 48.1	24.0
20	120 47.8	24.0
21	135 47.5	24.1
22	150 47.2	24.2
23	165 46.9	24.2
19 00	180 46.6	S23 24.3
01	195 46.3	24.3
02	210 45.9	24.4
03	225 45.6 ··	24.4
04	240 45.3	24.5
05	255 45.0	24.6
06	270 44.7	S23 24.6
07	285 44.4	24.7
08	300 44.1	24.7
09	315 43.8 ··	24.8
10	330 43.5	24.8
11	345 43.2	24.9
12	0 42.9	S23 24.9
13	15 42.5	25.0
14	30 42.2	25.0
15	45 41.9 ··	25.1
16	60 41.6	25.1
17	75 41.3	25.2
18	90 41.0	S23 25.2
19	105 40.7	25.2
20	120 40.4	25.3
21	135 40.1 ··	25.3
22	150 39.8	25.4
23	165 39.5	25.4
20 00	180 39.1	S23 25.4
01	195 38.8	25.5
02	210 38.5	25.5
03	225 38.2 ··	25.6
04	240 37.9	25.6
05	255 37.6	25.6
06	270 37.3	S23 25.7
07	285 37.0	25.7
08	300 36.7	25.7
09	315 36.4 ··	25.8
10	330 36.1	25.8
11	345 35.7	25.8
12	0 35.4	S23 25.9
13	15 35.1	25.9
14	30 34.8	25.9
15	45 34.5 ··	25.9
16	60 34.2	26.0
17	75 33.9	26.0
18	90 33.6	S23 26.0
19	105 33.3	26.0
20	120 33.0	26.1
21	135 32.6 ··	26.1
22	150 32.3	26.1
23	165 32.0	26.1

d h	G.H.A.	Dec.
21 00	180 31.7	S23 26.1
01	195 31.4	26.2
02	210 31.1	26.2
03	225 30.8 ··	26.2
04	240 30.5	26.2
05	255 30.2	26.2
06	270 29.9	S23 26.2
07	285 29.5	26.3
08	300 29.2	26.3
09	315 28.9 ··	26.3
10	330 28.6	26.3
11	345 28.3	26.3
12	0 28.0	S23 26.3
13	15 27.7	26.3
14	30 27.4	26.3
15	45 27.1 ··	26.3
16	60 26.8	26.3
17	75 26.4	26.3
18	90 26.1	S23 26.4
19	105 25.8	26.4
20	120 25.5	26.4
21	135 25.2 ··	26.4
22	150 24.9	26.4
23	165 24.6	26.4
22 00	180 24.3	S23 26.4
01	195 24.0	26.4
02	210 23.6	26.4
03	225 23.3 ··	26.4
04	240 23.0	26.4
05	255 22.7	26.4
06	270 22.4	S23 26.3
07	285 22.1	26.3
08	300 21.8	26.3
09	315 21.5 ··	26.3
10	330 21.2	26.3
11	345 20.9	26.3
12	0 20.5	S23 26.3
13	15 20.2	26.3
14	30 19.9	26.3
15	45 19.6 ··	26.3
16	60 19.3	26.3
17	75 19.0	26.2
18	90 18.7	S23 26.2
19	105 18.4	26.2
20	120 18.1	26.2
21	135 17.8 ··	26.2
22	150 17.4	26.2
23	165 17.1	26.1
23 00	180 16.8	S23 26.1
01	195 16.5	26.1
02	210 16.2	26.1
03	225 15.9 ··	26.1
04	240 15.6	26.0
05	255 15.3	26.0
06	270 15.0	S23 26.0
07	285 14.6	26.0
08	300 14.3	25.9
09	315 14.0 ··	25.9
10	330 13.7	25.9
11	345 13.4	25.9
12	0 13.1	S23 25.8
13	15 12.8	25.8
14	30 12.5	25.8
15	45 12.2 ··	25.7
16	60 11.9	25.7
17	75 11.5	25.7
18	90 11.2	S23 25.6
19	105 10.9	25.6
20	120 10.6	25.6
21	135 10.3 ··	25.5
22	150 10.0	25.5
23	165 09.7	25.4

d h	G.H.A.	Dec.
24 00	180 09.4	S23 25.4
01	195 09.1	25.4
02	210 08.8	25.3
03	225 08.4 ··	25.3
04	240 08.1	25.2
05	255 07.8	25.2
06	270 07.5	S23 25.2
07	285 07.2	25.1
08	300 06.9	25.1
09	315 06.6 ··	25.0
10	330 06.3	25.0
11	345 06.0	24.9
12	0 05.7	S23 24.9
13	15 05.3	24.8
14	30 05.0	24.8
15	45 04.7 ··	24.7
16	60 04.4	24.7
17	75 04.1	24.6
18	90 03.8	S23 24.6
19	105 03.5	24.5
20	120 03.2	24.4
21	135 02.9 ··	24.4
22	150 02.6	24.3
23	165 02.2	24.3
25 00	180 01.9	S23 24.2
01	195 01.6	24.2
02	210 01.3	24.1
03	225 01.0 ··	24.0
04	240 00.7	24.0
05	255 00.4	23.9
06	270 00.1	S23 23.8
07	284 59.8	23.8
08	299 59.5	23.7
09	314 59.2 ··	23.7
10	329 58.8	23.6
11	344 58.5	23.5
12	359 58.2	S23 23.4
13	14 57.9	23.4
14	29 57.6	23.3
15	44 57.3 ··	23.2
16	59 57.0	23.2
17	74 56.7	23.1
18	89 56.4	S23 23.0
19	104 56.1	22.9
20	119 55.8	22.9
21	134 55.4 ··	22.8
22	149 55.1	22.7
23	164 54.8	22.6
26 00	179 54.5	S23 22.6
01	194 54.2	22.5
02	209 53.9	22.4
03	224 53.6 ··	22.3
04	239 53.3	22.2
05	254 53.0	22.2
06	269 52.7	S23 22.1
07	284 52.4	22.0
08	299 52.1	21.9
09	314 51.7 ··	21.8
10	329 51.4	21.7
11	344 51.1	21.6
12	359 50.8	S23 21.6
13	14 50.5	21.5
14	29 50.2	21.4
15	44 49.9 ··	21.3
16	59 49.6	21.2
17	74 49.3	21.1
18	89 49.0	S23 21.0
19	104 48.7	20.9
20	119 48.4	20.8
21	134 48.1 ··	20.7
22	149 47.7	20.6
23	164 47.4	20.5

d h	G.H.A. ° '	Dec. ° '
27 00	179 47.1	S23 20.4
01	194 46.8	20.3
02	209 46.5	20.2
03	224 46.2 ··	20.1
04	239 45.9	20.0
05	254 45.6	19.9
06	269 45.3	S23 19.8
07	284 45.0	19.7
08	299 44.7	19.6
09	314 44.4 ··	19.5
10	329 44.1	19.4
11	344 43.8	19.3
12	359 43.5	S23 19.2
13	14 43.1	19.1
14	29 42.8	19.0
15	44 42.5 ··	18.9
16	59 42.2	18.8
17	74 41.9	18.6
18	89 41.6	S23 18.5
19	104 41.3	18.4
20	119 41.0	18.3
21	134 40.7 ··	18.2
22	149 40.4	18.1
23	164 40.1	18.0
28 00	179 39.8	S23 17.8
01	194 39.5	17.7
02	209 39.2	17.6
03	224 38.9 ··	17.5
04	239 38.6	17.4
05	254 38.3	17.2
06	269 37.9	S23 17.1
07	284 37.6	17.0
08	299 37.3	16.9
09	314 37.0 ··	16.7
10	329 36.7	16.6
11	344 36.4	16.5
12	359 36.1	S23 16.4
13	14 35.8	16.2
14	29 35.5	16.1
15	44 35.2 ··	16.0
16	59 34.9	15.8
17	74 34.6	15.7
18	89 34.3	S23 15.6
19	104 34.0	15.4
20	119 33.7	15.3
21	134 33.4 ··	15.2
22	149 33.1	15.0
23	164 32.8	14.9
29 00	179 32.5	S23 14.8
01	194 32.2	14.6
02	209 31.9	14.5
03	224 31.6 ··	14.4
04	239 31.3	14.2
05	254 30.9	14.1
06	269 30.6	S23 13.9
07	284 30.3	13.8
08	299 30.0	13.6
09	314 29.7 ··	13.5
10	329 29.4	13.4
11	344 29.1	13.2
12	359 28.8	S23 13.1
13	14 28.5	12.9
14	29 28.2	12.8
15	44 27.9 ··	12.6
16	59 27.6	12.5
17	74 27.3	12.3
18	89 27.0	S23 12.2
19	104 26 7	12.0
20	119 26.4	11.9
21	134 26.1 ··	11.7
22	149 25.8	11.6
23	164 25.5	11.4

d h	G.H.A. ° '	Dec. ° '
30 00	179 25.2	S23 11.2
01	194 24.9	11.1
02	209 24.6	10.9
03	224 24.3 ··	10.8
04	239 24.0	10.6
05	254 23.7	10.4
06	269 23.4	S23 10.3
07	284 23.1	10.1
08	299 22.8	10.0
09	314 22.5 ··	09.8
10	329 22.2	09.6
11	344 21.9	09.5
12	359 21.6	S23 09.3
13	14 21.3	09.1
14	29 21.0	09.0
15	44 20.7 ··	08.8
16	59 20.4	08.6
17	74 20.1	08.5
18	89 19.8	S23 08.3
19	104 19.5	08.1
20	119 19.2	07.9
21	134 18.9 ··	07.8
22	149 18.6	07.6
23	164 18.3	07.4
31 00	179 18.0	S23 07.2
01	194 17.7	07.1
02	209 17.4	06.9
03	224 17.1 ··	06.7
04	239 16.8	06.5
05	254 16.5	06.4
06	269 16.2	S23 06.2
07	284 15.9	06.0
08	299 15.6	05.8
09	314 15.3 ··	05.6
10	329 15.0	05.4
11	344 14.7	05.3
12	359 14.4	S23 05.1
13	14 14.1	04.9
14	29 13.8	04.7
15	44 13.5 ··	04.5
16	59 13.2	04.3
17	74 12.9	04.1
18	89 12.6	S23 03.9
19	104 12.3	03.8
20	119 12.0	03.6
21	134 11.7 ··	03.4
22	149 11.4	03.2
23	164 11.1	03.0

Table 7

DAILY LONG-TERM ALMANAC OF THE SUN FOR THE YEARS 1978, 1982, 1986, 1990, 1994, 1998, 2002, ETC.*

This table is a compilation of the columns for Greenwich Hour Angle and declination of the sun from the 1978 *Nautical Almanac*. For use beyond 1978, a correction may be necessary, depending upon the degree of precision desired. The amount of change in *4 years* in the GHA and declination (DEC) is given for each 3-day interval in Table 9.

Since the correction factors are changing slowly with the passage of time, this table is *not* a perpetual almanac, but an almanac designed for use in practical navigation for a period of about 30 years.

*See note at end of table on use of 4-year correction factors.

JANUARY

d h	G.H.A. ° '	Dec. ° '
1 00	179 10.8	S23 02.8
01	194 10.5	02.6
02	209 10.2	02.4
03	224 09.9	·· 02.2
04	239 09.6	02.0
05	254 09.4	01.8
06	269 09.1	S23 01.6
07	284 08.8	01.4
08	299 08.5	01.2
09	314 08.2	·· 01.0
10	329 07.9	00.8
11	344 07.6	00.6
12	359 07.3	S23 00.4
13	14 07.0	00.2
14	29 06.7	23 00.0
15	44 06.4	22 59.8
16	59 06.1	59.6
17	74 05.8	59.3
18	89 05.5	S22 59.1
19	104 05.2	58.9
20	119 04.9	58.7
21	134 04.6	·· 58.5
22	149 04.3	58.3
23	164 04.0	58.1
2 00	179 03.8	S22 57.9
01	194 03.5	57.6
02	209 03.2	57.4
03	224 02.9	·· 57.2
04	239 02.6	57.0
05	254 02.3	56.8
06	269 02.0	S22 56.6
07	284 01.7	56.3
08	299 01.4	56.1
09	314 01.1	·· 55.9
10	329 00.8	55.7
11	344 00.5	55.5
12	359 00.2	S22 55.2
13	13 59.9	55.0
14	28 59.7	54.8
15	43 59.4	·· 54.6
16	58 59.1	54.3
17	73 58.8	54.1
18	88 58.5	S22 53.9
19	103 58.2	53.6
20	118 57.9	53.4
21	133 57.6	·· 53.2
22	148 57.3	52.9
23	163 57.0	52.7
3 00	178 56.7	S22 52.5
01	193 56.5	52.2
02	208 56.2	52.0
03	223 55.9	·· 51.8
04	238 55.6	51.5
05	253 55.3	51.3
06	268 55.0	S22 51.1
07	283 54.7	50.8
08	298 54.4	50.6
09	313 54.1	·· 50.3
10	328 53.9	50.1
11	343 53.6	49.9
12	358 53.3	S22 49.6
13	13 53.0	49.4
14	28 52.7	49.1
15	43 52.4	·· 48.9
16	58 52.1	48.6
17	73 51.8	48.4
18	88 51.6	S22 48.2
19	103 51.3	47.9
20	118 51.0	47.7
21	133 50.7	·· 47.4
22	148 50.4	47.2
23	163 50.1	46.9

d h	G.H.A. ° '	Dec. ° '
4 00	178 49.8	S22 46.7
01	193 49.5	46.4
02	208 49.3	46.1
03	223 49.0	·· 45.9
04	238 48.7	45.6
05	253 48.4	45.4
06	268 48.1	S22 45.1
07	283 47.8	44.9
08	298 47.5	44.6
09	313 47.3	·· 44.3
10	328 47.0	44.1
11	343 46.7	43.8
12	358 46.4	S22 43.6
13	13 46.1	43.3
14	28 45.8	43.0
15	43 45.6	·· 42.8
16	58 45.3	42.5
17	73 45.0	42.2
18	88 44.7	S22 42.0
19	103 44.4	41.7
20	118 44.1	41.4
21	133 43.9	·· 41.2
22	148 43.6	40.9
23	163 43.3	40.6
5 00	178 43.0	S22 40.4
01	193 42.7	40.1
02	208 42.4	39.8
03	223 42.2	·· 39.6
04	238 41.9	39.3
05	253 41.6	39.0
06	268 41.3	S22 38.7
07	283 41.0	38.5
08	298 40.8	38.2
09	313 40.5	·· 37.9
10	328 40.2	37.6
11	343 39.9	37.3
12	358 39.6	S22 37.1
13	13 39.3	36.8
14	28 39.1	36.5
15	43 38.8	·· 36.2
16	58 38.5	35.9
17	73 38.2	35.6
18	88 38.0	S22 35.4
19	103 37.7	35.1
20	118 37.4	34.8
21	133 37.1	·· 34.5
22	148 36.8	34.2
23	163 36.6	33.9
6 00	178 36.3	S22 33.6
01	193 36.0	33.3
02	208 35.7	33.1
03	223 35.4	·· 32.8
04	238 35.2	32.5
05	253 34.9	32.2
06	268 34.6	S22 31.9
07	283 34.3	31.6
08	298 34.1	31.3
09	313 33.8	·· 31.0
10	328 33.5	30.7
11	343 33.2	30.4
12	358 33.0	S22 30.1
13	13 32.7	29.8
14	28 32.4	29.5
15	43 32.1	·· 29.2
16	58 31.9	28.9
17	73 31.6	28.6
18	88 31.3	S22 28.3
19	103 31.0	28.0
20	118 30.8	27.7
21	133 30.5	·· 27.4
22	148 30.2	27.1
23	163 29.9	26.8

d h	G.H.A. ° '	Dec. ° '
7 00	178 29.7	S22 26.5
01	193 29.4	26.1
02	208 29.1	25.8
03	223 28.8	·· 25.5
04	238 28.6	25.2
05	253 28.3	24.9
06	268 28.0	S22 24.6
07	283 27.8	24.3
08	298 27.5	24.0
09	313 27.2	·· 23.7
10	328 26.9	23.3
11	343 26.7	23.0
12	358 26.4	S22 22.7
13	13 26.1	22.4
14	28 25.9	22.1
15	43 25.6	·· 21.7
16	58 25.3	21.4
17	73 25.1	21.1
18	88 24.8	S22 20.8
19	103 24.5	20.5
20	118 24.2	20.1
21	133 24.0	·· 19.8
22	148 23.7	19.5
23	163 23.4	19.2
8 00	178 23.2	S22 18.8
01	193 22.9	18.5
02	208 22.6	18.2
03	223 22.4	·· 17.9
04	238 22.1	17.5
05	253 21.8	17.2
06	268 21.6	S22 16.9
07	283 21.3	16.5
08	298 21.0	16.2
09	313 20.8	·· 15.9
10	328 20.5	15.5
11	343 20.2	15.2
12	358 20.0	S22 14.9
13	13 19.7	14.5
14	28 19.4	14.2
15	43 19.2	·· 13.9
16	58 18.9	13.5
17	73 18.6	13.2
18	88 18.4	S22 12.8
19	103 18.1	12.5
20	118 17.9	12.2
21	133 17.6	·· 11.8
22	148 17.3	11.5
23	163 17.1	11.1
9 00	178 16.8	S22 10.8
01	193 16.5	10.4
02	208 16.3	10.1
03	223 16.0	·· 09.7
04	238 15.8	09.4
05	253 15.5	09.0
06	268 15.2	S22 08.7
07	283 15.0	08.3
08	298 14.7	08.0
09	313 14.4	·· 07.6
10	328 14.2	07.3
11	343 13.9	06.9
12	358 13.7	S22 06.6
13	13 13.4	06.2
14	28 13.1	05.9
15	43 12.9	·· 05.5
16	58 12.6	05.2
17	73 12.4	04.8
18	88 12.1	S22 04.4
19	103 11.8	04.1
20	118 11.6	03.7
21	133 11.3	·· 03.4
22	148 11.1	03.0
23	163 10.8	02.6

d h	G.H.A.	Dec.		d h	G.H.A.	Dec.		d h	G.H.A.	Dec.
10 00	178 10.6	S22 02.3		13 00	177 52.7	S21 34.2		16 00	177 36.3	S21 02.4
01	193 10.3	01.9		01	192 52.5	33.8		01	192 36.1	02.0
02	208 10.0	01.6		02	207 52.2	33.4		02	207 35.9	01.5
03	223 09.8 ··	01.2		03	222 52.0 ··	33.0		03	222 35.6 ··	01.0
04	238 09.5	00.8		04	237 51.8	32.6		04	237 35.4	00.6
05	253 09.3	00.5		05	252 51.5	32.2		05	252 35.2	21 00.1
06	268 09.0	S22 00.1		06	267 51.3	S21 31.7		06	267 35.0	S20 59.6
07	283 08.8	21 59.7		07	282 51.0	31.3		07	282 34.8	59.2
08	298 08.5	59.4		08	297 50.8	30.9		08	297 34.6	58.7
09	313 08.3 ··	59.0		09	312 50.6 ··	30.5		09	312 34.3 ··	58.2
10	328 08.0	58.6		10	327 50.3	30.0		10	327 34.1	57.7
11	343 07.7	58.2		11	342 50.1	29.6		11	342 33.9	57.3
12	358 07.5	S21 57.9		12	357 49.9	S21 29.2		12	357 33.7	S20 56.8
13	13 07.2	57.5		13	12 49.6	28.8		13	12 33.5	56.3
14	28 07.0	57.1		14	27 49.4	28.3		14	27 33.3	55.8
15	43 06.7 ··	56.8		15	42 49.2 ··	27.9		15	42 33.1 ··	55.4
16	58 06.5	56.4		16	57 48.9	27.5		16	57 32.9	54.9
17	73 06.2	56.0		17	72 48.7	27.1		17	72 32.6	54.4
18	88 06.0	S21 55.6		18	87 48.5	S21 26.6		18	87 32.4	S20 53.9
19	103 05.7	55.3		19	102 48.2	26.2		19	102 32.2	53.4
20	118 05.5	54.9		20	117 48.0	25.8		20	117 32.0	53.0
21	133 05.2 ··	54.5		21	132 47.8 ··	25.3		21	132 31.8 ··	52.5
22	148 05.0	54.1		22	147 47.5	24.9		22	147 31.6	52.0
23	163 04.7	53.7		23	162 47.3	24.5		23	162 31.4	51.5
11 00	178 04.5	S21 53.4		14 00	177 47.1	S21 24.1		17 00	177 31.2	S20 51.0
01	193 04.2	53.0		01	192 46.8	23.6		01	192 31.0	50.5
02	208 04.0	52.6		02	207 46.6	23.2		02	207 30.7	50.1
03	223 03.7 ··	52.2		03	222 46.4 ··	22.7		03	222 30.5 ··	49.6
04	238 03.5	51.8		04	237 46.1	22.3		04	237 30.3	49.1
05	253 03.2	51.4		05	252 45.9	21.9		05	252 30.1	48.6
06	268 03.0	S21 51.1		06	267 45.7	S21 21.4		06	267 29.9	S20 48.1
07	283 02.7	50.7		07	282 45.5	21.0		07	282 29.7	47.6
08	298 02.5	50.3		08	297 45.2	20.6		08	297 29.5	47.1
09	313 02.2 ··	49.9		09	312 45.0 ··	20.1		09	312 29.3 ··	46.6
10	328 02.0	49.5		10	327 44.8	19.7		10	327 29.1	46.2
11	343 01.7	49.1		11	342 44.5	19.2		11	342 28.9	45.7
12	358 01.5	S21 48.7		12	357 44.3	S21 18.8		12	357 28.7	S20 45.2
13	13 01.2	48.3		13	12 44.1	18.4		13	12 28.5	44.7
14	28 01.0	48.0		14	27 43.9	17.9		14	27 28.3	44.2
15	43 00.7 ··	47.6		15	42 43.6 ··	17.5		15	42 28.0 ··	43.7
16	58 00.5	47.2		16	57 43.4	17.0		16	57 27.8	43.2
17	73 00.2	46.8		17	72 43.2	16.6		17	72 27.6	42.7
18	88 00.0	S21 46.4		18	87 42.9	S21 16.1		18	87 27.4	S20 42.2
19	102 59.7	46.0		19	102 42.7	15.7		19	102 27.2	41.7
20	117 59.5	45.6		20	117 42.5	15.2		20	117 27.0	41.2
21	132 59.2 ··	45.2		21	132 42.3 ··	14.8		21	132 26.8 ··	40.7
22	147 59.0	44.8		22	147 42.0	14.3		22	147 26.6	40.2
23	162 58.8	44.4		23	162 41.8	13.9		23	162 26.4	39.7
12 00	177 58.5	S21 44.0		15 00	177 41.6	S21 13.4		18 00	177 26.2	S20 39.2
01	192 58.3	43.6		01	192 41.4	13.0		01	192 26.0	38.7
02	207 58.0	43.2		02	207 41.1	12.5		02	207 25.8	38.2
03	222 57.8 ··	42.8		03	222 40.9 ··	12.1		03	222 25.6 ··	37.7
04	237 57.5	42.4		04	237 40.7	11.6		04	237 25.4	37.2
05	252 57.3	42.0		05	252 40.5	11.2		05	252 25.2	36.7
06	267 57.0	S21 41.6		06	267 40.3	S21 10.7		06	267 25.0	S20 36.2
07	282 56.8	41.2		07	282 40.0	10.3		07	282 24.8	35.7
08	297 56.6	40.8		08	297 39.8	09.8		08	297 24.6	35.2
09	312 56.3 ··	40.4		09	312 39.6 ··	09.4		09	312 24.4 ··	34.7
10	327 56.1	40.0		10	327 39.4	08.9		10	327 24.2	34.2
11	342 55.8	39.6		11	342 39.1	08.5		11	342 24.0	33.7
12	357 55.6	S21 39.2		12	357 38.9	S21 08.0		12	357 23.8	S20 33.2
13	12 55.3	38.8		13	12 38.7	07.5		13	12 23.6	32.7
14	27 55.1	38.4		14	27 38.5	07.1		14	27 23.4	32.2
15	42 54.9 ··	38.0		15	42 38.3 ··	06.6		15	42 23.2 ··	31.6
16	57 54.6	37.5		16	57 38.0	06.2		16	57 23.0	31.1
17	72 54.4	37.1		17	72 37.8	05.7		17	72 22.8	30.6
18	87 54.1	S21 36.7		18	87 37.6	S21 05.2		18	87 22.6	S20 30.1
19	102 53.9	36.3		19	102 37.4	04.8		19	102 22.4	29.6
20	117 53.7	35.9		20	117 37.2	04.3		20	117 22.2	29.1
21	132 53.4 ··	35.5		21	132 36.9 ··	03.8		21	132 22.0 ··	28.6
22	147 53.2	35.1		22	147 36.7	03.4		22	147 21.8	28.1
23	162 52.9	34.7		23	162 36.5	02.9		23	162 21.6	27.5

		G.H.A.		Dec.	
19	d h	° '		° '	
	00	177	21.4	S20	27.0
	01	192	21.2		26.5
	02	207	21.1		26.0
	03	222	20.9	··	25.5
	04	237	20.7		25.0
	05	252	20.5		24.4
	06	267	20.3	S20	23.9
	07	282	20.1		23.4
	08	297	19.9		22.9
	09	312	19.7	··	22.4
	10	327	19.5		21.8
	11	342	19.3		21.3
	12	357	19.1	S20	20.8
	13	12	18.9		20.3
	14	27	18.7		19.7
	15	42	18.6	··	19.2
	16	57	18.4		18.7
	17	72	18.2		18.2
	18	87	18.0	S20	17.6
	19	102	17.8		17.1
	20	117	17.6		16.6
	21	132	17.4	··	16.0
	22	147	17.2		15.5
	23	162	17.0		15.0
20	00	177	16.9	S20	14.4
	01	192	16.7		13.9
	02	207	16.5		13.4
	03	222	16.3	··	12.8
	04	237	16.1		12.3
	05	252	15.9		11.8
	06	267	15.7	S20	11.2
	07	282	15.6		10.7
	08	297	15.4		10.2
	09	312	15.2	··	09.6
	10	327	15.0		09.1
	11	342	14.8		08.5
	12	357	14.6	S20	08.0
	13	12	14.5		07.5
	14	27	14.3		06.9
	15	42	14.1	··	06.4
	16	57	13.9		05.8
	17	72	13.7		05.3
	18	87	13.6	S20	04.8
	19	102	13.4		04.2
	20	117	13.2		03.7
	21	132	13.0	··	03.1
	22	147	12.8		02.6
	23	162	12.6		02.0
21	00	177	12.5	S20	01.5
	01	192	12.3		00.9
	02	207	12.1	20	00.4
	03	222	11.9	19	59.8
	04	237	11.8		59.3
	05	252	11.6		58.7
	06	267	11.4	S19	58.2
	07	282	11.2		57.6
	08	297	11.0		57.1
	09	312	10.9	··	56.5
	10	327	10.7		56.0
	11	342	10.5		55.4
	12	357	10.3	S19	54.9
	13	12	10.2		54.3
	14	27	10.0		53.7
	15	42	09.8	··	53.2
	16	57	09.6		52.6
	17	72	09.5		52.1
	18	87	09.3	S19	51.5
	19	102	09.1		50.9
	20	117	09.0		50.4
	21	132	08.8	··	49.8
	22	147	08.6		49.3
	23	162	08.4		48.7

		G.H.A.		Dec.	
22	d h	° '		° '	
	00	177	08.3	S19	48.1
	01	192	08.1		47.6
	02	207	07.9		47.0
	03	222	07.8	··	46.4
	04	237	07.6		45.9
	05	252	07.4		45.3
	06	267	07.2	S19	44.7
	07	282	07.1		44.2
	08	297	06.9		43.6
	09	312	06.7	··	43.0
	10	327	06.6		42.5
	11	342	06.4		41.9
	12	357	06.2	S19	41.3
	13	12	06.1		40.8
	14	27	05.9		40.2
	15	42	05.7	··	39.6
	16	57	05.6		39.0
	17	72	05.4		38.5
	18	87	05.2	S19	37.9
	19	102	05.1		37.3
	20	117	04.9		36.7
	21	132	04.8	··	36.2
	22	147	04.6		35.6
	23	162	04.4		35.0
23	00	177	04.3	S19	34.4
	01	192	04.1		33.8
	02	207	03.9		33.3
	03	222	03.8	··	32.7
	04	237	03.6		32.1
	05	252	03.4		31.5
	06	267	03.3	S19	30.9
	07	282	03.1		30.4
	08	297	03.0		29.8
	09	312	02.8	··	29.2
	10	327	02.6		28.6
	11	342	02.5		28.0
	12	357	02.3	S19	27.4
	13	12	02.2		26.9
	14	27	02.0		26.3
	15	42	01.9	··	25.7
	16	57	01.7		25.1
	17	72	01.5		24.5
	18	87	01.4	S19	23.9
	19	102	01.2		23.3
	20	117	01.1		22.7
	21	132	00.9	··	22.1
	22	147	00.8		21.5
	23	162	00.6		20.9
24	00	177	00.4	S19	20.4
	01	192	00.3		19.8
	02	207	00.1		19.2
	03	222	00.0	··	18.6
	04	236	59.8		18.0
	05	251	59.7		17.4
	06	266	59.5	S19	16.8
	07	281	59.4		16.2
	08	296	59.2		15.6
	09	311	59.1	··	15.0
	10	326	58.9		14.4
	11	341	58.8		13.8
	12	356	58.6	S19	13.2
	13	11	58.5		12.6
	14	26	58.3		12.0
	15	41	58.2	··	11.4
	16	56	58.0		10.8
	17	71	57.9		10.2
	18	86	57.7	S19	09.6
	19	101	57.6		09.0
	20	116	57.4		08.4
	21	131	57.3	··	07.8
	22	146	57.1		07.1
	23	161	57.0		06.5

		G.H.A.		Dec.	
25	d h	° '		° '	
	00	176	56.8	S19	05.9
	01	191	56.7		05.3
	02	206	56.5		04.7
	03	221	56.4	··	04.1
	04	236	56.2		03.5
	05	251	56.1		02.9
	06	266	56.0	S19	02.3
	07	281	55.8		01.7
	08	296	55.7		01.0
	09	311	55.5	19	00.4
	10	326	55.4	18	59.8
	11	341	55.2		59.2
	12	356	55.1	S18	58.6
	13	11	55.0		58.0
	14	26	54.8		57.4
	15	41	54.7	··	56.7
	16	56	54.5		56.1
	17	71	54.4		55.5
	18	86	54.2	S18	54.9
	19	101	54.1		54.3
	20	116	54.0		53.6
	21	131	53.8	··	53.0
	22	146	53.7		52.4
	23	161	53.5		51.8
26	00	176	53.4	S18	51.2
	01	191	53.3		50.5
	02	206	53.1		49.9
	03	221	53.0	··	49.3
	04	236	52.9		48.7
	05	251	52.7		48.0
	06	266	52.6	S18	47.4
	07	281	52.4		46.8
	08	296	52.3		46.1
	09	311	52.2	··	45.5
	10	326	52.0		44.9
	11	341	51.9		44.3
	12	356	51.8	S18	43.6
	13	11	51.6		43.0
	14	26	51.5		42.4
	15	41	51.4	··	41.7
	16	56	51.2		41.1
	17	71	51.1		40.5
	18	86	51.0	S18	39.8
	19	101	50.8		39.2
	20	116	50.7		38.6
	21	131	50.6	··	37.9
	22	146	50.4		37.3
	23	161	50.3		36.7
27	00	176	50.2	S18	36.0
	01	191	50.1		35.4
	02	206	49.9		34.8
	03	221	49.8	··	34.1
	04	236	49.7		33.5
	05	251	49.5		32.8
	06	266	49.4	S18	32.2
	07	281	49.3		31.6
	08	296	49.2		30.9
	09	311	49.0	··	30.3
	10	326	48.9		29.6
	11	341	48.8		29.0
	12	356	48.7	S18	28.3
	13	11	48.5		27.7
	14	26	48.4		27.1
	15	41	48.3	··	26.4
	16	56	48.2		25.8
	17	71	48.0		25.1
	18	86	47.9	S18	24.5
	19	101	47.8		23.8
	20	116	47.7		23.2
	21	131	47.5	··	22.5
	22	146	47.4		21.9
	23	161	47.3		21.2

d h	G.H.A.	Dec.
28 00	176 47.2	S18 20.6
01	191 47.0	19.9
02	206 46.9	19.3
03	221 46.8 ··	18.6
04	236 46.7	18.0
05	251 46.6	17.3
06	266 46.4	S18 16.7
07	281 46.3	16.0
08	296 46.2	15.3
09	311 46.1 ··	14.7
10	326 46.0	14.0
11	341 45.8	13.4
12	356 45.7	S18 12.7
13	11 45.6	12.1
14	26 45.5	11.4
15	41 45.4 ··	10.7
16	56 45.3	10.1
17	71 45.1	09.4
18	86 45.0	S18 08.8
19	101 44.9	08.1
20	116 44.8	07.4
21	131 44.7 ··	06.0
22	146 44.6	06.1
23	161 44.5	05.4
29 00	176 44.3	S18 04.8
01	191 44.2	04.1
02	206 44.1	03.4
03	221 44.0 ··	02.8
04	236 43.9	02.1
05	251 43.8	01.4
06	266 43.7	S18 00.8
07	281 43.6	18 00.1
08	296 43.4	17 59.4
09	311 43.3 ··	58.8
10	326 43.2	58.1
11	341 43.1	57.4
12	356 43.0	S17 56.8
13	11 42.9	56.1
14	26 42.8	55.4
15	41 42.7 ··	54.7
16	56 42.6	54.1
17	71 42.5	53.4
18	86 42.4	S17 52.7
19	101 42.2	52.0
20	116 42.1	51.4
21	131 42.0 ··	50.7
22	146 41.9	50.0
23	161 41.8	49.3
30 00	176 41.7	S17 48.7
01	191 41.6	48.0
02	206 41.5	47.3
03	221 41.4 ··	46.6
04	236 41.3	45.9
05	251 41.2	45.3
06	266 41.1	S17 44.6
07	281 41.0	43.9
08	296 40.9	43.2
09	311 40.8 ··	42.5
10	326 40.7	41.9
11	341 40.6	41.2
12	356 40.5	S17 40.5
13	11 40.4	39.8
14	26 40.3	39.1
15	41 40.2 ··	38.4
16	56 40.1	37.7
17	71 40.0	37.1
18	86 39.9	S17 36.4
19	101 39.8	35.7
20	116 39.7	35.0
21	131 39.6 ··	34.3
22	146 39.5	33.6
23	161 39.4	32.9

FEBRUARY

d h	G.H.A.	Dec.
31 00	176 39.3	S17 32.2
01	191 39.2	31.5
02	206 39.1	30.8
03	221 39.0 ··	30.2
04	236 38.9	29.5
05	251 38.8	28.8
06	266 38.7	S17 28.1
07	281 38.6	27.4
08	296 38.5	26.7
09	311 38.4 ··	26.0
10	326 38.3	25.3
11	341 38.2	24.6
12	356 38.2	S17 23.9
13	11 38.1	23.2
14	26 38.0	22.5
15	41 37.9 ··	21.8
16	56 37.8	21.1
17	71 37.7	20.4
18	86 37.6	S17 19.7
19	101 37.5	19.0
20	116 37.4	18.3
21	131 37.3 ··	17.6
22	146 37.2	16.9
23	161 37.2	16.2
1 00	176 37.1	S17 15.5
01	191 37.0	14.8
02	206 36.9	14.1
03	221 36.8 ··	13.4
04	236 36.7	12.7
05	251 36.6	12.0
06	266 36.5	S17 11.3
07	281 36.5	10.5
08	296 36.4	09.8
09	311 36.3 ··	09.1
10	326 36.2	08.4
11	341 36.1	07.7
12	356 36.0	S17 07.0
13	11 35.9	06.3
14	26 35.9	05.6
15	41 35.8 ··	04.9
16	56 35.7	04.2
17	71 35.6	03.4
18	86 35.5	S17 02.7
19	101 35.4	02.0
20	116 35.4	01.3
21	131 35.3	17 00.6
22	146 35.2	16 59.9
23	161 35.1	59.2
2 00	176 35.0	S16 58.4
01	191 35.0	57.7
02	206 34.9	57.0
03	221 34.8 ··	56.3
04	236 34.7	55.6
05	251 34.6	54.8
06	266 34.6	S16 54.1
07	281 34.5	53.4
08	296 34.4	52.7
09	311 34.3 ··	52.0
10	326 34.3	51.2
11	341 34.2	50.5
12	356 34.1	S16 49.8
13	11 34.0	49.1
14	26 34.0	48.4
15	41 33.9 ··	47.6
16	56 33.8	46.9
17	71 33.7	46.2
18	86 33.7	S16 45.5
'19	101 33.6	44.7
20	116 33.5	44.0
21	131 33.4 ··	43.3
22	146 33.4	42.5
23	161 33.3	41.8

d h	G.H.A.	Dec.
3 00	176 33.2	S16 41.1
01	191 33.1	40.4
02	206 33.1	39.6
03	221 33.0 ··	38.9
04	236 32.9	38.2
05	251 32.9	37.4
06	266 32.8	S16 36.7
07	281 32.7	36.0
08	296 32.7	35.2
09	311 32.6 ··	34.5
10	326 32.5	33.8
11	341 32.5	33.0
12	356 32.4	S16 32.3
13	11 32.3	31.6
14	26 32.2	30.8
15	41 32.2 ··	30.1
16	56 32.1	29.4
17	71 32.0	28.6
18	86 32.0	S16 27.9
19	101 31.9	27.2
20	116 31.9	26.4
21	131 31.8 ··	25.7
22	146 31.7	24.9
23	161 31.7	24.2
4 00	176 31.6	S16 23.5
01	191 31.5	22.7
02	206 31.5	22.0
03	221 31.4 ··	21.2
04	236 31.3	20.5
05	251 31.3	19.7
06	266 31.2	S16 19.0
07	281 31.2	18.3
08	296 31.1	17.5
09	311 31.0 ··	16.8
10	326 31.0	16.0
11	341 30.9	15.3
12	356 30.9	S16 14.5
13	11 30.8	13.8
14	26 30.7	13.0
15	41 30.7 ··	12.3
16	56 30.6	11.5
17	71 30.6	10.8
18	86 30.5	S16 10.0
19	101 30.5	09.3
20	116 30.4	08.5
21	131 30.3 ··	07.8
22	146 30.3	07.0
23	161 30.2	06.3
5 00	176 30.2	S16 05.5
01	191 30.1	04.8
02	206 30.1	04.0
03	221 30.0 ··	03.3
04	236 30.0	02.5
05	251 29.9	01.8
06	266 29.8	S16 01.0
07	281 29.8	16 00.3
08	296 29.7	15 59.5
09	311 29.7 ··	58.7
10	326 29.6	58.0
11	341 29.6	57.2
12	356 29.5	S15 56.5
13	11 29.5	55.7
14	26 29.4	54.9
15	41 29.4 ··	54.2
16	56 29.3	53.4
17	71 29.3	52.7
18	86 29.2	S15 51.9
19	101 29.2	51.1
20	116 29.1	50.4
21	131 29.1 ··	49.6
22	146 29.0	48.9
23	161 29.0	48.1

d	h	G.H.A. ° ′	Dec. ° ′
6	00	176 28.9	S15 47.3
	01	191 28.9	46.6
	02	206 28.9	45.8
	03	221 28.8	·· 45.0
	04	236 28.8	44.3
	05	251 28.7	43.5
	06	266 28.7	S15 42.7
	07	281 28.6	42.0
	08	296 28.6	41.2
	09	311 28.5	·· 40.4
	10	326 28.5	39.7
	11	341 28.5	38.9
	12	356 28.4	S15 38.1
	13	11 28.4	37.4
	14	26 28.3	36.6
	15	41 28.3	·· 35.8
	16	56 28.2	35.0
	17	71 28.2	34.3
	18	86 28.2	S15 33.5
	19	101 28.1	32.7
	20	116 28.1	32.0
	21	131 28.0	·· 31.2
	22	146 28.0	30.4
	23	161 28.0	29.6
7	00	176 27.9	S15 28.9
	01	191 27.9	28.1
	02	206 27.8	27.3
	03	221 27.8	·· 26.5
	04	236 27.8	25.8
	05	251 27.7	25.0
	06	266 27.7	S15 24.2
	07	281 27.7	23.4
	08	296 27.6	22.6
	09	311 27.6	·· 21.9
	10	326 27.6	21.1
	11	341 27.5	20.3
	12	356 27.5	S15 19.5
	13	11 27.5	18.7
	14	26 27.4	18.0
	15	41 27.4	·· 17.2
	16	56 27.4	16.4
	17	71 27.3	15.6
	18	86 27.3	S15 14.8
	19	101 27.3	14.1
	20	116 27.2	13.3
	21	131 27.2	·· 12.5
	22	146 27.2	11.7
	23	161 27.1	10.9
8	00	176 27.1	S15 10.1
	01	191 27.1	09.3
	02	206 27.0	08.6
	03	221 27.0	·· 07.8
	04	236 27.0	07.0
	05	251 27.0	06.2
	06	266 26.9	S15 05.4
	07	281 26.9	04.6
	08	296 26.9	03.8
	09	311 26.8	·· 03.0
	10	326 26.8	02.2
	11	341 26.8	01.5
	12	356 26.8	S15 00.7
	13	11 26.7	14 59.9
	14	26 26.7	59.1
	15	41 26.7	·· 58.3
	16	56 26.7	57.5
	17	71 26.6	56.7
	18	86 26.6	S14 55.9
	19	101 26.6	55.1
	20	116 26.6	54.3
	21	131 26.5	·· 53.5
	22	146 26.5	52.7
	23	161 26.5	51.9

d	h	G.H.A. ° ′	Dec. ° ′
9	00	176 26.5	S14 51.1
	01	191 26.4	50.3
	02	206 26.4	49.5
	03	221 26.4	·· 48.7
	04	236 26.4	47.9
	05	251 26.4	47.1
	06	266 26.3	S14 46.4
	07	281 26.3	45.6
	08	296 26.3	44.8
	09	311 26.3	·· 44.0
	10	326 26.3	43.2
	11	341 26.2	42.3
	12	356 26.2	S14 41.5
	13	11 26.2	40.7
	14	26 26.2	39.9
	15	41 26.2	·· 39.1
	16	56 26.2	38.3
	17	71 26.1	37.5
	18	86 26.1	S14 36.7
	19	101 26.1	35.9
	20	116 26.1	35.1
	21	131 26.1	·· 34.3
	22	146 26.1	33.5
	23	161 26.1	32.7
10	00	176 26.0	S14 31.9
	01	191 26.0	31.1
	02	206 26.0	30.3
	03	221 26.0	·· 29.5
	04	236 26.0	28.7
	05	251 26.0	27.9
	06	266 26.0	S14 27.0
	07	281 26.0	26.2
	08	296 25.9	25.4
	09	311 25.9	·· 24.6
	10	326 25.9	23.8
	11	341 25.9	23.0
	12	356 25.9	S14 22.2
	13	11 25.9	21.4
	14	26 25.9	20.6
	15	41 25.9	·· 19.7
	16	56 25.9	18.9
	17	71 25.9	18.1
	18	86 25.8	S14 17.3
	19	101 25.8	16.5
	20	116 25.8	15.7
	21	131 25.8	·· 14.9
	22	146 25.8	14.0
	23	161 25.8	13.2
11	00	176 25.8	S14 12.4
	01	191 25.8	11.6
	02	206 25.8	10.8
	03	221 25.8	·· 10.0
	04	236 25.8	09.1
	05	251 25.8	08.3
	06	266 25.8	S14 07.5
	07	281 25.8	06.7
	08	296 25.8	05.9
	09	311 25.8	·· 05.0
	10	326 25.8	04.2
	11	341 25.8	03.4
	12	356 25.8	S14 02.6
	13	11 25.8	01.8
	14	26 25.8	00.9
	15	41 25.8	14 00.1
	16	56 25.8	13 59.3
	17	71 25.8	58.5
	18	86 25.8	S13 57.6
	19	101 25.8	56.8
	20	116 25.8	56.0
	21	131 25.8	·· 55.2
	22	146 25.8	54.3
	23	161 25.8	53.5

d	h	G.H.A. ° ′	Dec. ° ′
12	00	176 25.8	S13 52.7
	01	191 25.8	51.9
	02	206 25.8	51.0
	03	221 25.8	·· 50.2
	04	236 25.8	49.4
	05	251 25.8	48.6
	06	266 25.8	S13 47.7
	07	281 25.8	46.9
	08	296 25.8	46.1
	09	311 25.8	·· 45.2
	10	326 25.8	44.4
	11	341 25.8	43.6
	12	356 25.8	S13 42.8
	13	11 25.8	41.9
	14	26 25.8	41.1
	15	41 25.8	·· 40.3
	16	56 25.9	39.4
	17	71 25.9	38.6
	18	86 25.9	S13 37.8
	19	101 25.9	36.9
	20	116 25.9	36.1
	21	131 25.9	·· 35.3
	22	146 25.9	34.4
	23	161 25.9	33.6
13	00	176 25.9	S13 32.8
	01	191 25.9	31.9
	02	206 25.9	31.1
	03	221 26.0	·· 30.2
	04	236 26.0	29.4
	05	251 26.0	28.6
	06	266 26.0	S13 27.7
	07	281 26.0	26.9
	08	296 26.0	26.1
	09	311 26.0	·· 25.2
	10	326 26.0	24.4
	11	341 26.1	23.5
	12	356 26.1	S13 22.7
	13	11 26.1	21.9
	14	26 26.1	21.0
	15	41 26.1	·· 20.2
	16	56 26.1	19.3
	17	71 26.2	18.5
	18	86 26.2	S13 17.6
	19	101 26.2	16.8
	20	116 26.2	16.0
	21	131 26.2	·· 15.1
	22	146 26.2	14.3
	23	161 26.3	13.4
14	00	176 26.3	S13 12.6
	01	191 26.3	11.7
	02	206 26.3	10.9
	03	221 26.3	·· 10.0
	04	236 26.3	09.2
	05	251 26.4	08.4
	06	266 26.4	S13 07.5
	07	281 26.4	06.7
	08	296 26.4	05.8
	09	311 26.5	·· 05.0
	10	326 26.5	04.1
	11	341 26.5	03.3
	12	356 26.5	S13 02.4
	13	11 26.5	01.6
	14	26 26.6	13 00.7
	15	41 26.6	12 59.9
	16	56 26.6	59.0
	17	71 26.6	58.2
	18	86 26.7	S12 57.3
	19	101 26.7	56.5
	20	116 26.7	55.6
	21	131 26.7	·· 54.8
	22	146 26.8	53.9
	23	161 26.8	53.1

	G.H.A.	Dec.		G.H.A.	Dec.		G.H.A.	Dec.
d h	° ′	° ′	**d h**	° ′	° ′	**d h**	° ′	° ′
15 00	176 26.8	S12 52.2	**18** 00	176 29.5	S11 49.8	**21** 00	176 33.9	S10 45.8
01	191 26.8	51.3	01	191 29.6	49.0	01	191 33.9	44.9
02	206 26.9	50.5	02	206 29.6	48.1	02	206 34.0	44.0
03	221 26.9 ··	49.6	03	221 29.7 ··	47.2	03	221 34.1 ··	43.1
04	236 26.9	48.8	04	236 29.7	46.3	04	236 34.1	42.2
05	251 26.9	47.9	05	251 29.8	45.4	05	251 34.2	41.3
06	266 27.0	S12 47.1	06	266 29.8	S11 44.6	06	266 34.3	S10 40.4
07	281 27.0	46.2	07	281 29.9	43.7	07	281 34.4	39.5
08	296 27.0	45.4	08	296 29.9	42.8	08	296 34.4	38.6
09	311 27.1 ··	44.5	09	311 30.0 ··	41.9	09	311 34.5 ··	37.7
10	326 27.1	43.6	10	326 30.0	41.0	10	326 34.6	36.8
11	341 27.1	42.8	11	341 30.1	40.2	11	341 34.7	35.9
12	356 27.1	S12 41.9	12	356 30.1	S11 39.3	12	356 34.7	S10 35.0
13	11 27.2	41.1	13	11 30.2	38.4	13	11 34.8	34.1
14	26 27.2	40.2	14	26 30.3	37.5	14	26 34.9	33.1
15	41 27.2 ··	39.4	15	41 30.3 ··	36.6	15	41 34.9 ··	32.2
16	56 27.3	38.5	16	56 30.4	35.7	16	56 35.0	31.3
17	71 27.3	37.6	17	71 30.4	34.9	17	71 35.1	30.4
18	86 27.3	S12 36.8	18	86 30.5	S11 34.0	18	86 35.2	S10 29.5
19	101 27.4	35.9	19	101 30.5	33.1	19	101 35.3	28.6
20	116 27.4	35.1	20	116 30.6	32.2	20	116 35.3	27.7
21	131 27.4 ··	34.2	21	131 30.6 ··	31.3	21	131 35.4 ··	26.8
22	146 27.5	33.3	22	146 30.7	30.4	22	146 35.5	25.9
23	161 27.5	32.5	23	161 30.7	29.5	23	161 35.6	25.0
16 00	176 27.5	S12 31.6	**19** 00	176 30.8	S11 28.7	**22** 00	176 35.6	S10 24.1
01	191 27.6	30.7	01	191 30.9	27.8	01	191 35.7	23.2
02	206 27.6	29.9	02	206 30.9	26.9	02	206 35.8	22.3
03	221 27.6 ··	29.0	03	221 31.0 ··	26.0	03	221 35.9 ··	21.4
04	236 27.7	28.2	04	236 31.0	25.1	04	236 35.9	20.5
05	251 27.7	27.3	05	251 31.1	24.2	05	251 36.0	19.5
06	266 27.7	S12 26.4	06	266 31.1	S11 23.3	06	266 36.1	S10 18.6
07	281 27.8	25.6	07	281 31.2	22.5	07	281 36.2	17.7
08	296 27.8	24.7	08	296 31.3	21.6	08	296 36.3	16.8
09	311 27.9 ··	23.8	09	311 31.3 ··	20.7	09	311 36.3 ··	15.9
10	326 27.9	23.0	10	326 31.4	19.8	10	326 36.4	15.0
11	341 27.9	22.1	11	341 31.4	18.9	11	341 36.5	14.1
12	356 28.0	S12 21.2	12	356 31.5	S11 18.0	12	356 36.6	S10 13.2
13	11 28.0	20.4	13	11 31.6	17.1	13	11 36.7	12.3
14	26 28.0	19.5	14	26 31.6	16.2	14	26 36.7	11.4
15	41 28.1 ··	18.6	15	41 31.7 ··	15.3	15	41 36.8 ··	10.4
16	56 28.1	17.8	16	56 31.7	14.4	16	56 36.9	09.5
17	71 28.2	16.9	17	71 31.8	13.6	17	71 37.0	08.6
18	86 28.2	S12 16.0	18	86 31.9	S11 12.7	18	86 37.1	S10 07.7
19	101 28.2	15.2	19	101 31.9	11.8	19	101 37.2	06.8
20	116 28.3	14.3	20	116 32.0	10.9	20	116 37.2	05.9
21	131 28.3 ··	13.4	21	131 32.1 ··	10.0	21	131 37.3 ··	05.0
22	146 28.4	12.6	22	146 32.1	09.1	22	146 37.4	04.1
23	161 28.4	11.7	23	161 32.2	08.2	23	161 37.5	03.1
17 00	176 28.4	S12 10.8	**20** 00	176 32.2	S11 07.3	**23** 00	176 37.6	S10 02.2
01	191 28.5	09.9	01	191 32.3	06.4	01	191 37.7	01.3
02	206 28.5	09.1	02	206 32.4	05.5	02	206 37.7	10 00.4
03	221 28.6 ··	08.2	03	221 32.4 ··	04.6	03	221 37.8	9 59.5
04	236 28.6	07.3	04	236 32.5	03.7	04	236 37.9	58.6
05	251 28.7	06.5	05	251 32.6	02.8	05	251 38.0	57.7
06	266 28.7	S12 05.6	06	266 32.6	S11 01.9	06	266 38.1	S 9 56.7
07	281 28.7	04.7	07	281 32.7	01.0	07	281 38.2	55.8
08	296 28.8	03.8	08	296 32.8	11 00.2	08	296 38.3	54.9
09	311 28.8 ··	03.0	09	311 32.8	10 59.3	09	311 38.3 ··	54.0
10	326 28.9	02.1	10	326 32.9	58.4	10	326 38.4	53.1
11	341 28.9	01.2	11	341 33.0	57.5	11	341 38.5	52.2
12	356 29.0	S12 00.3	12	356 33.0	S10 56.6	12	356 38.6	S 9 51.3
13	11 29.0	11 59.5	13	11 33.1	55.7	13	11 38.7	50.3
14	26 29.1	58.6	14	26 33.2	54.8	14	26 38.8	49.4
15	41 29.1 ··	57.7	15	41 33.2 ··	53.9	15	41 38.9 ··	48.5
16	56 29.2	56.8	16	56 33.3	53.0	16	56 39.0	47.6
17	71 29.2	56.0	17	71 33.4	52.1	17	71 39.0	46.7
18	86 29.2	S11 55.1	18	86 33.4	S10 51.2	18	86 39.1	S 9 45.7
19	101 29.3	54.2	19	101 33.5	50.3	19	101 39.2	44.8
20	116 29.3	53.3	20	116 33.6	49.4	20	116 39.3	43.9
21	131 29.4 ··	52.5	21	131 33.6 ··	48.5	21	131 39.4 ··	43.0
22	146 29.4	51.6	22	146 33.7	47.6	22	146 39.5	42.1
23	161 29.5	50.7	23	161 33.8	46.7	23	161 39.6	41.2

d h	G.H.A. ° '	Dec. ° '
24 00	176 39.7	S 9 40.2
01	191 39.8	39.3
02	206 39.8	38.4
03	221 39.9	·· 37.5
04	236 40.0	36.5
05	251 40.1	35.6
06	266 40.2	S 9 34.7
07	281 40.3	33.8
08	296 40.4	32.9
09	311 40.5	·· 31.9
10	326 40.6	31.0
11	341 40.7	30.1
12	356 40.8	S 9 29.2
13	11 40.9	28.2
14	26 41.0	27.3
15	41 41.0	·· 26.4
16	56 41.1	25.5
17	71 41.2	24.6
18	86 41.3	S 9 23.6
19	101 41.4	22.7
20	116 41.5	21.8
21	131 41.6	·· 20.9
22	146 41.7	19.9
23	161 41.8	19.0
25 00	176 41.9	S 9 18.1
01	191 42.0	17.2
02	206 42.1	16.2
03	221 42.2	·· 15.3
04	236 42.3	14.4
05	251 42.4	13.4
06	266 42.5	S 9 12.5
07	281 42.6	11.6
08	296 42.7	10.7
09	311 42.8	·· 09.7
10	326 42.9	08.8
11	341 43.0	07.9
12	356 43.1	S 9 07.0
13	11 43.2	06.0
14	26 43.3	05.1
15	41 43.4	·· 04.2
16	56 43.5	03.2
17	71 43.6	02.3
18	86 43.7	S 9 01.4
19	101 43.8	9 00.4
20	116 43.9	8 59.5
21	131 44.0	·· 58.6
22	146 44.1	57.7
23	161 44.2	56.7
26 00	176 44.3	S 8 55.8
01	191 44.4	54.9
02	206 44.5	53.9
03	221 44.6	·· 53.0
04	236 44.7	52.1
05	251 44.8	51.1
06	266 44.9	S 8 50.2
07	281 45.0	49.3
08	296 45.1	48.3
09	311 45.2	·· 47.4
10	326 45.3	46.5
11	341 45.4	45.5
12	356 45.5	S 8 44.6
13	11 45.6	43.7
14	26 45.8	42.7
15	41 45.9	·· 41.8
16	56 46.0	40.9
17	71 46.1	39.9
18	86 46.2	S 8 39.0
19	101 46.3	38.0
20	116 46.4	37.1
21	131 46.5	·· 36.2
22	146 46.6	35.2
23	161 46.7	34.3

d h	G.H.A. ° '	Dec. ° '
27 00	176 46.8	S 8 33.4
01	191 46.9	32.4
02	206 47.0	31.5
03	221 47.1	·· 30.6
04	236 47.3	29.6
05	251 47.4	28.7
06	266 47.5	S 8 27.7
07	281 47.6	26.8
08	296 47.7	25.9
09	311 47.8	·· 24.9
10	326 47.9	24.0
11	341 48.0	23.0
12	356 48.1	S 8 22.1
13	11 48.2	21.2
14	26 48.4	20.2
15	41 48.5	·· 19.3
16	56 48.6	18.3
17	71 48.7	17.4
18	86 48.8	S 8 16.5
19	101 48.9	15.5
20	116 49.0	14.6
21	131 49.1	·· 13.6
22	146 49.3	12.7
23	161 49.4	11.8
28 00	176 49.5	S 8 10.8
01	191 49.6	09.9
02	206 49.7	09.0
03	221 49.8	·· 08.0
04	236 49.9	07.0
05	251 50.1	06.1
06	266 50.2	S 8 05.2
07	281 50.3	04.2
08	296 50.4	03.3
09	311 50.5	·· 02.3
10	326 50.6	01.4
11	341 50.7	8 00.4
12	356 50.9	S 7 59.5
13	11 51.0	58.5
14	26 51.1	57.6
15	41 51.2	·· 56.7
16	56 51.3	55.7
17	71 51.5	54.8
18	86 51.6	S 7 53.8
19	101 51.7	52.9
20	116 51.8	51.9
21	131 51.9	·· 51.0
22	146 52.0	50.0
23	161 52.2	49.1
MARCH 1 00	176 52.3	S 7 48.1
01	191 52.4	47.2
02	206 52.5	46.2
03	221 52.6	·· 45.3
04	236 52.8	44.4
05	251 52.9	43.4
06	266 53.0	S 7 42.5
07	281 53.1	41.5
08	296 53.2	40.6
09	311 53.4	·· 39.6
10	326 53.5	38.7
11	341 53.6	37.7
12	356 53.7	S 7 36.8
13	11 53.8	35.8
14	26 54.0	34.9
15	41 54.1	·· 33.9
16	56 54.2	33.0
17	71 54.3	32.0
18	86 54.5	S 7 31.1
19	101 54.6	30.1
20	116 54.7	29.2
21	131 54.8	·· 28.2
22	146 54.9	27.3
23	161 55.1	26.3

d h	G.H.A. ° '	Dec. ° '
2 00	176 55.2	S 7 25.4
01	191 55.3	24.4
02	206 55.4	23.5
03	221 55.6	·· 22.5
04	236 55.7	21.6
05	251 55.8	20.6
06	266 55.9	S 7 19.6
07	281 56.1	18.7
08	296 56.2	17.7
09	311 56.3	·· 16.8
10	326 56.4	15.8
11	341 56.6	14.9
12	356 56.7	S 7 13.9
13	11 56.8	13.0
14	26 56.9	12.0
15	41 57.1	·· 11.1
16	56 57.2	10.1
17	71 57.3	09.2
18	86 57.5	S 7 08.2
19	101 57.6	07.2
20	116 57.7	06.3
21	131 57.8	·· 05.3
22	146 58.0	04.4
23	161 58.1	03.4
3 00	176 58.2	S 7 02.5
01	191 58.4	01.5
02	206 58.5	7 00.6
03	221 58.6	6 59.6
04	236 58.7	58.6
05	251 58.9	57.7
06	266 59.0	S 6 56.7
07	281 59.1	55.8
08	296 59.3	54.8
09	311 59.4	·· 53.9
10	326 59.5	52.9
11	341 59.7	51.9
12	356 59.8	S 6 51.0
13	11 59.9	50.0
14	27 00.1	49.1
15	42 00.2	·· 48.1
16	57 00.3	47.2
17	72 00.4	46.2
18	87 00.6	S 6 45.2
19	102 00.7	44.3
20	117 00.8	43.3
21	132 01.0	·· 42.4
22	147 01.1	41.4
23	162 01.2	40.4
4 00	177 01.4	S 6 39.5
01	192 01.5	38.5
02	207 01.6	37.6
03	222 01.8	·· 36.6
04	237 01.9	35.6
05	252 02.0	34.7
06	267 02.2	S 6 33.7
07	282 02.3	32.8
08	297 02.5	31.8
09	312 02.6	·· 30.8
10	327 02.7	29.9
11	342 02.9	28.9
12	357 03.0	S 6 27.9
13	12 03.1	27.0
14	27 03.3	26.0
15	42 03.4	·· 25.1
16	57 03.5	24.1
17	72 03.7	23.1
18	87 03.8	S 6 22.2
19	102 04.0	21.2
20	117 04.1	20.2
21	132 04.2	·· 19.3
22	147 04.4	18.3
23	162 04.5	17.4

d/h	G.H.A.	Dec.
5 00	177 04.6	S 6 16.4
01	192 04.8	15.4
02	207 04.9	14.5
03	222 05.1	·· 13.5
04	237 05.2	12.5
05	252 05.3	11.6
06	267 05.5	S 6 10.6
07	282 05.6	09.6
08	297 05.7	08.7
09	312 05.9	·· 07.7
10	327 06.0	06.8
11	342 06.2	05.8
12	357 06.3	S 6 04.8
13	12 06.4	03.9
14	27 06.6	02.9
15	42 06.7	·· 01.9
16	57 06.9	01.0
17	72 07.0	6 00.0
18	87 07.2	S 5 59.0
19	102 07.3	58.1
20	117 07.4	57.1
21	132 07.6	·· 56.1
22	147 07.7	55.2
23	162 07.9	54.2
6 00	177 08.0	S 5 53.2
01	192 08.1	52.3
02	207 08.3	51.3
03	222 08.4	·· 50.3
04	237 08.6	49.4
05	252 08.7	48.4
06	267 08.9	S 5 47.4
07	282 09.0	46.5
08	297 09.1	45.5
09	312 09.3	·· 44.5
10	327 09.4	43.5
11	342 09.6	42.6
12	357 09.7	S 5 41.6
13	12 09.9	40.6
14	27 10.0	39.7
15	42 10.2	·· 38.7
16	57 10.3	37.7
17	72 10.5	36.8
18	87 10.6	S 5 35.8
19	102 10.7	34.8
20	117 10.9	33.9
21	132 11.0	·· 32.9
22	147 11.2	31.9
23	162 11.3	30.9
7 00	177 11.5	S 5 30.0
01	192 11.6	29.0
02	207 11.8	28.0
03	222 11.9	·· 27.1
04	237 12.1	26.1
05	252 12.2	25.1
06	267 12.4	S 5 24.1
07	282 12.5	23.2
08	297 12.7	22.2
09	312 12.8	·· 21.2
10	327 12.9	20.3
11	342 13.1	19.3
12	357 13.2	S 5 18.3
13	12 13.4	17.3
14	27 13.5	16.4
15	42 13.7	·· 15.4
16	57 13.8	14.4
17	72 14.0	13.5
18	87 14.1	S 5 12.5
19	102 14.3	11.5
20	117 14.4	10.5
21	132 14.6	·· 09.6
22	147 14.7	08.6
23	162 14.9	07.6

d/h	G.H.A.	Dec.
8 00	177 15.0	S 5 06.6
01	192 15.2	05.7
02	207 15.3	04.7
03	222 15.5	·· 03.7
04	237 15.6	02.8
05	252 15.8	01.8
06	267 15.9	S 5 00.8
07	282 16.1	4 59.8
08	297 16.2	58.9
09	312 16.4	·· 57.9
10	327 16.6	56.9
11	342 16.7	55.9
12	357 16.9	S 4 55.0
13	12 17.0	54.0
14	27 17.2	53.0
15	42 17.3	·· 52.0
16	57 17.5	51.1
17	72 17.6	50.1
18	87 17.8	S 4 49.1
19	102 17.9	48.1
20	117 18.1	47.2
21	132 18.2	·· 46.2
22	147 18.4	45.2
23	162 18.5	44.2
9 00	177 18.7	S 4 43.3
01	192 18.9	42.3
02	207 19.0	41.3
03	222 19.2	·· 40.3
04	237 19.3	39.3
05	252 19.5	38.4
06	267 19.6	S 4 37.4
07	282 19.8	36.4
08	297 19.9	35.4
09	312 20.1	·· 34.5
10	327 20.3	33.5
11	342 20.4	32.5
12	357 20.6	S 4 31.5
13	12 20.7	30.6
14	27 20.9	29.6
15	42 21.0	·· 28.6
16	57 21.2	27.6
17	72 21.4	26.6
18	87 21.5	S 4 25.7
19	102 21.7	24.7
20	117 21.8	23.7
21	132 22.0	·· 22.7
22	147 22.1	21.7
23	162 22.3	20.8
10 00	177 22.5	S 4 19.8
01	192 22.6	18.8
02	207 22.8	17.8
03	222 22.9	·· 16.9
04	237 23.1	15.9
05	252 23.2	14.9
06	267 23.4	S 4 13.9
07	282 23.6	12.9
08	297 23.7	12.0
09	312 23.9	·· 11.0
10	327 24.0	10.0
11	342 24.2	09.0
12	357 24.4	S 4 08.0
13	12 24.5	07.1
14	27 24.7	06.1
15	42 24.8	·· 05.1
16	57 25.0	04.1
17	72 25.2	03.1
18	87 25.3	S 4 02.2
19	102 25.5	01.2
20	117 25.6	4 00.2
21	132 25.8	3 59.2
22	147 26.0	58.2
23	162 26.1	57.3

d/h	G.H.A.	Dec.
11 00	177 26.3	S 3 56.3
01	192 26.5	55.3
02	207 26.6	54.3
03	222 26.8	·· 53.3
04	237 26.9	52.4
05	252 27.1	51.4
06	267 27.3	S 3 50.4
07	282 27.4	49.4
08	297 27.6	48.4
09	312 27.8	·· 47.5
10	327 27.9	46.5
11	342 28.1	45.5
12	357 28.2	S 3 44.5
13	12 28.4	43.5
14	27 28.6	42.5
15	42 28.7	·· 41.6
16	57 28.9	40.6
17	72 29.1	39.6
18	87 29.2	S 3 38.6
19	102 29.4	37.6
20	117 29.6	36.7
21	132 29.7	·· 35.7
22	147 29.9	34.7
23	162 30.1	33.7
12 00	177 30.2	S 3 32.7
01	192 30.4	31.7
02	207 30.6	30.8
03	222 30.7	·· 29.8
04	237 30.9	28.8
05	252 31.0	27.8
06	267 31.2	S 3 26.8
07	282 31.4	25.8
08	297 31.5	24.9
09	312 31.7	·· 23.9
10	327 31.9	22.9
11	342 32.0	21.9
12	357 32.2	S 3 20.9
13	12 32.4	19.9
14	27 32.5	19.0
15	42 32.7	·· 18.0
16	57 32.9	17.0
17	72 33.0	16.0
18	87 33.2	S 3 15.0
19	102 33.4	14.0
20	117 33.6	13.1
21	132 33.7	·· 12.1
22	147 33.9	11.1
23	162 34.1	10.1
13 00	177 34.2	S 3 09.1
01	192 34.4	08.1
02	207 34.6	07.2
03	222 34.7	·· 06.2
04	237 34.9	05.2
05	252 35.1	04.2
06	267 35.2	S 3 03.2
07	282 35.4	02.2
08	297 35.6	01.2
09	312 35.7	3 00.3
10	327 35.9	2 59.3
11	342 36.1	58.3
12	357 36.3	S 2 57.3
13	12 36.4	56.3
14	27 36.6	55.3
15	42 36.8	·· 54.4
16	57 36.9	53.4
17	72 37.1	52.4
18	87 37.3	S 2 51.4
19	102 37.5	50.4
20	117 37.6	49.4
21	132 37.8	·· 48.4
22	147 38.0	47.5
23	162 38.1	46.5

		G.H.A.	Dec.				G.H.A.	Dec.				G.H.A.	Dec.
d h		° ′	° ′		d h		° ′	° ′		d h		° ′	° ′
14 00		177 38.3	S 2 45.5		**17** 00		177 51.0	S 1 34.4		**20** 00		178 04.1	S 0 23.3
01		192 38.5	44.5		01		192 51.1	33.4		01		193 04.3	22.3
02		207 38.7	43.5		02		207 51.3	32.4		02		208 04.5	21.3
03		222 38.8	·· 42.5		03		222 51.5	·· 31.5		03		223 04.7	·· 20.3
04		237 39.0	41.5		04		237 51.7	30.5		04		238 04.8	19.3
05		252 39.2	40.6		05		252 51.9	29.5		05		253 05.0	18.3
06		267 39.3	S 2 39.6		06		267 52.0	S 1 28.5		06		268 05.2	S 0 17.3
07		282 39.5	38.6		07		282 52.2	27.5		07		283 05.4	16.4
08		297 39.7	37.6		08		297 52.4	26.5		08		298 05.6	15.4
09		312 39.9	·· 36.6		09		312 52.6	·· 25.5		09		313 05.8	·· 14.4
10		327 40.0	35.6		10		327 52.8	24.5		10		328 06.0	13.4
11		342 40.2	34.6		11		342 52.9	23.6		11		343 06.1	12.4
12		357 40.4	S 2 33.7		12		357 53.1	S 1 22.6		12		358 06.3	S 0 11.4
13		12 40.5	32.7		13		12 53.3	21.6		13		13 06.5	10.4
14		27 40.7	31.7		14		27 53.5	20.6		14		28 06.7	09.4
15		42 40.9	·· 30.7		15		42 53.7	·· 19.6		15		43 06.9	·· 08.4
16		57 41.1	29.7		16		57 53.8	18.6		16		58 07.1	07.5
17		72 41.2	28.7		17		72 54.0	17.6		17		73 07.3	06.5
18		87 41.4	S 2 27.7		18		87 54.2	S 1 16.6		18		88 07.4	S 0 05.5
19		102 41.6	26.8		19		102 54.4	15.6		19		103 07.6	04.5
20		117 41.8	.25.8		20		117 54.6	14.7		20		118 07.8	03.5
21		132 41.9	·· 24.8		21		132 54.7	·· 13.7		21		133 08.0	·· 02.5
22		147 42.1	23.8		22		147 54.9	12.7		22		148 08.2	01.5
23		162 42.3	22.8		23		162 55.1	11.7		23		163 08.4	S 0 00.5
15 00		177 42.5	S 2 21.8		**18** 00		177 55.3	S 1 10.7		**21** 00		178 08.6	N 0 00.4
01		192 42.6	20.8		01		192 55.5	09.7		01		193 08.7	01.4
02		207 42.8	19.8		02		207 55.7	08.7		02		208 08.9	02.4
03		222 43.0	·· 18.9		03		222 55.8	·· 07.7		03		223 09.1	·· 03.4
04		237 43.2	17.9		04		237 56.0	06.8		04		238 09.3	04.4
05		252 43.3	16.9		05		252 56.2	05.8		05		253 09.5	05.4
06		267 43.5	S 2 15.9		06		267 56.4	S 1 04.8		06		268 09.7	N 0 06.4
07		282 43.7	14.9		07		282 56.6	03.8		07		283 09.9	07.4
08		297 43.9	13.9		08		297 56.8	02.8		08		298 10.1	08.3
09		312 44.0	·· 12.9		09		312 56.9	·· 01.8		09		313 10.2	·· 09.3
10		327 44.2	12.0		10		327 57.1	1 00.8		10		328 10.4	10.3
11		342 44.4	11.0		11		342 57.3	0 59.8		11		343 10.6	11.3
12		357 44.6	S 2 10.0		12		357 57.5	S 0 58.8		12		358 10.8	N 0 12.3
13		12 44.7	09.0		13		12 57.7	57.9		13		13 11.0	13.3
14		27 44.9	08.0		14		27 57.8	56.9		14		28 11.2	14.3
15		42 45.1	·· 07.0		15		42 58.0	·· 55.9		15		43 11.4	·· 15.2
16		57 45.3	06.0		16		57 58.2	54.9		16		58 11.6	16.2
17		72 45.4	05.0		17		72 58.4	53.9		17		73 11.7	17.2
18		87 45.6	S 2 04.1		18		87 58.6	S 0 52.9		18		88 11.9	N 0 18.2
19		102 45.8	03.1		19		102 58.8	51.9		19		103 12.1	19.2
20		117 46.0	02.1		20		117 58.9	50.9		20		118 12.3	20.2
21		132 46.1	·· 01.1		21		132 59.1	·· 50.0		21		133 12.5	·· 21.2
22		147 46.3	2 00.1		22		147 59.3	49.0		22		148 12.7	22.2
23		162 46.5	1 59.1		23		162 59.5	48.0		23		163 12.9	23.1
16 00		177 46.7	S 1 58.1		**19** 00		177 59.7	S 0 47.0		**22** 00		178 13.1	N 0 24.1
01		192 46.9	57.1		01		192 59.9	46.0		01		193 13.2	25.1
02		207 47.0	56.2		02		208 00.0	45.0		02		208 13.4	26.1
03		222 47.2	·· 55.2		03		223 00.2	·· 44.0		03		223 13.6	·· 27.1
04		237 47.4	54.2		04		238 00.4	43.0		04		238 13.8	28.1
05		252 47.6	53.2		05		253 00.6	42.0		05		253 14.0	29.1
06		267 47.7	S 1 52.2		06		268 00.8	S 0 41.1		06		268 14.2	N 0 30.0
07		282 47.9	51.2		07		283 01.0	40.1		07		283 14.4	31.0
08		297 48.1	50.2		08		298 01.1	39.1		08		298 14.6	32.0
09		312 48.3	·· 49.2		09		313 01.3	·· 38.1		09		313 14.7	·· 33.0
10		327 48.5	48.3		10		328 01.5	37.1		10		328 14.9	34.0
11		342 48.6	47.3		11		343 01.7	36.1		11		343 15.1	35.0
12		357 48.8	S 1 46.3		12		358 01.9	S 0 35.1		12		358 15.3	N 0 36.0
13		12 49.0	45.3		13		13 02.1	34.1		13		13 15.5	37.0
14		27 49.2	44.3		14		28 02.3	33.1		14		28 15.7	37.9
15		42 49.3	·· 43.3		15		43 02.4	·· 32.2		15		43 15.9	·· 38.9
16		57 49.5	42.3		16		58 02.6	31.2		16		58 16.1	39.9
17		72 49.7	41.3		17		73 02.8	30.2		17		73 16.3	40.9
18		87 49.9	S 1 40.4		18		88 03.0	S 0 29.2		18		88 16.4	N 0 41.9
19		102 50.1	39.4		19		103 03.2	28.2		19		103 16.6	42.9
20		117 50.2	38.4		20		118 03.4	27.2		20		118 16.8	43.9
21		132 50.4	·· 37.4		21		133 03.5	·· 26.2		21		133 17.0	·· 44.8
22		147 50.6	36.4		22		148 03.7	25.2		22		148 17.2	45.8
23		162 50.8	35.4		23		163 03.9	24.3		23		163 17.4	46.8

G.H.A. and Dec. tables

d h	G.H.A. °	′	Dec.	′
23 00	178	17.6	N 0	47.8
01	193	17.8		48.8
02	208	18.0		49.8
03	223	18.1 ··		50.8
04	238	18.3		51.7
05	253	18.5		52.7
06	268	18.7	N 0	53.7
07	283	18.9		54.7
08	298	19.1		55.7
09	313	19.3 ··		56.7
10	328	19.5		57.6
11	343	19.7		58.6
12	358	19.8	N 0	59.6
13	13	20.0	1	00.6
14	28	20.2		01.6
15	43	20.4 ··		02.6
16	58	20.6		03.6
17	73	20.8		04.5
18	88	21.0	N 1	05.5
19	103	21.2		06.5
20	118	21.4		07.5
21	133	21.6 ··		08.5
22	148	21.7		09.5
23	163	21.9		10.4
24 00	178	22.1	N 1	11.4
01	193	22.3		12.4
02	208	22.5		13.4
03	223	22.7 ··		14.4
04	238	22.9		15.4
05	253	23.1		16.4
06	268	23.3	N 1	17.3
07	283	23.5		18.3
08	298	23.6		19.3
09	313	23.8 ··		20.3
10	328	24.0		21.3
11	343	24.2		22.3
12	358	24.4	N 1	23.2
13	13	24.6		24.2
14	28	24.8		25.2
15	43	25.0 ··		26.2
16	58	25.2		27.2
17	73	25.4		28.2
18	88	25.6	N 1	29.1
19	103	25.8		30.1
20	118	25.9		31.1
21	133	26.1 ··		32.1
22	148	26.3		33.1
23	163	26.5		34.1
25 00	178	26.7	N 1	35.1
01	193	26.9		36.0
02	208	27.1		37.0
03	223	27.3 ··		38.0
04	238	27.5		39.0
05	253	27.6		40.0
06	268	27.8	N 1	40.9
07	283	28.0		41.9
08	298	28.2		42.9
09	313	28.4 ··		43.9
10	328	28.6		44.9
11	343	28.8		45.9
12	358	29.0	N 1	46.8
13	13	29.2		47.8
14	28	29.4		48.8
15	43	29.6 ··		49.8
16	58	29.8		50.8
17	73	29.9		51.7
18	88	30.1	N 1	52.7
19	103	30.3		53.7
20	118	30.5		54.7
21	133	30.7 ··		55.7
22	148	30.9		56.7
23	163	31.1		57.6

d h	G.H.A. °	′	Dec.	′
26 00	178	31.3	N 1	58.6
01	193	31.5	1	59.6
02	208	31.7	2	00.6
03	223	31.9 ··		01.6
04	238	32.0		02.5
05	253	32.2		03.5
06	268	32.4	N 2	04.5
07	283	32.6		05.5
08	298	32.8		06.5
09	313	33.0 ··		07.4
10	328	33.2		08.4
11	343	33.4		09.4
12	358	33.6	N 2	10.4
13	13	33.8		11.4
14	28	34.0		12.3
15	43	34.1 ··		13.3
16	58	34.3		14.3
17	73	34.5		15.3
18	88	34.7	N 2	16.3
19	103	34.9		17.2
20	118	35.1		18.2
21	133	35.3 ··		19.2
22	148	35.5		20.2
23	163	35.7		21.2
27 00	178	35.9	N 2	22.1
01	193	36.1		23.1
02	208	36.2		24.1
03	223	36.4 ··		25.1
04	238	36.6		26.1
05	253	36.8		27.0
06	268	37.0	N 2	28.0
07	283	37.2		29.0
08	298	37.4		30.0
09	313	37.6 ··		31.0
10	328	37.8		31.9
11	343	38.0		32.9
12	358	38.1	N 2	33.9
13	13	38.3		34.9
14	28	38.5		35.9
15	43	38.7 ··		36.8
16	58	38.9		37.8
17	73	39.1		38.8
18	88	39.3	N 2	39.8
19	103	39.5		40.7
20	118	39.7		41.7
21	133	39.9 ··		42.7
22	148	40.1		43.7
23	163	40.2		44.7
28 00	178	40.4	N 2	45.6
01	193	40.6		46.6
02	208	40.8		47.6
03	223	41.0 ··		48.6
04	238	41.2		49.5
05	253	41.4		50.5
06	268	41.6	N 2	51.5
07	283	41.8		52.5
08	298	42.0		53.4
09	313	42.1 ··		54.4
10	328	42.3		55.4
11	343	42.5		56.4
12	358	42.7	N 2	57.4
13	13	42.9		58.3
14	28	43.1	2	59.3
15	43	43.3	3	00.3
16	58	43.5		01.3
17	73	43.7		02.2
18	88	43.9	N 3	03.2
19	103	44.0		04.2
20	118	44.2		05.2
21	133	44.4 ··		06.1
22	148	44.6		07.1
23	163	44.8		08.1

d h	G.H.A. °	′	Dec.	′
29 00	178	45.0	N 3	09.1
01	193	45.2		10.0
02	208	45.4		11.0
03	223	45.6 ··		12.0
04	238	45.8		13.0
05	253	45.9		13.9
06	268	46.1	N 3	14.9
07	283	46.3		15.9
08	298	46.5		16.9
09	313	46.7 ··		17.8
10	328	46.9		18.8
11	343	47.1		19.8
12	358	47.3	N 3	20.8
13	13	47.5		21.7
14	28	47.6		22.7
15	43	47.8 ··		23.7
16	58	48.0		24.7
17	73	48.2		25.6
18	88	48.4	N 3	26.6
19	103	48.6		27.6
20	118	48.8		28.5
21	133	49.0 ··		29.5
22	148	49.2		30.5
23	163	49.4		31.5
30 00	178	49.5	N 3	32.4
01	193	49.7		33.4
02	208	49.9		34.4
03	223	50.1 ··		35.4
04	238	50.3		36.3
05	253	50.5		37.3
06	268	50.7	N 3	38.3
07	283	50.9		39.2
08	298	51.1		40.2
09	313	51.2 ··		41.2
10	328	51.4		42.2
11	343	51.6		43.1
12	358	51.8	N 3	44.1
13	13	52.0		45.1
14	28	52.2		46.0
15	43	52.4 ··		47.0
16	58	52.6		48.0
17	73	52.8		49.0
18	88	52.9	N 3	49.9
19	103	53.1		50.9
20	118	53.3		51.9
21	133	53.5 ··		52.8
22	148	53.7		53.8
23	163	53.9		54.8
31 00	178	54.1	N 3	55.7
01	193	54.3		56.7
02	208	54.4		57.7
03	223	54.6 ··		58.7
04	238	54.8	3	59.6
05	253	55.0	4	00.6
06	268	55.2	N 4	01.6
07	283	55.4		02.5
08	298	55.6		03.5
09	313	55.8 ··		04.5
10	328	56.0		05.4
11	343	56.1		06.4
12	358	56.3	N 4	07.4
13	13	56.5		08.3
14	28	56.7		09.3
15	43	56.9 ··		10.3
16	58	57.1		11.2
17	73	57.3		12.2
18	88	57.5	N 4	13.2
19	103	57.6		14.1
20	118	57.8		15.1
21	133	58.0 ··		16.0
22	148	58.2		17.0
23	163	58.4		18.0

APRIL

1

d h	G.H.A.	Dec.
00	178 58.6	N 4 19.0
01	193 58.8	19.9
02	208 59.0	20.9
03	223 59.1	·· 21.9
04	238 59.3	22.8
05	253 59.5	23.8
06	268 59.7	N 4 24.8
07	283 59.9	25.7
08	299 00.1	26.7
09	314 00.3	·· 27.7
10	329 00.5	28.6
11	344 00.6	29.6
12	359 00.8	N 4 30.6
13	14 01.0	31.5
14	29 01.2	32.5
15	44 01.4	·· 33.5
16	59 01.6	34.4
17	74 01.8	35.4
18	89 01.9	N 4 36.4
19	104 02.1	37.3
20	119 02.3	38.3
21	134 02.5	·· 39.3
22	149 02.7	40.2
23	164 02.9	41.2

2

d h	G.H.A.	Dec.
00	179 03.1	N 4 42.1
01	194 03.3	43.1
02	209 03.4	44.1
03	224 03.6	·· 45.0
04	239 03.8	46.0
05	254 04.0	47.0
06	269 04.2	N 4 47.9
07	284 04.4	48.9
08	299 04.6	49.8
09	314 04.7	·· 50.8
10	329 04.9	51.8
11	344 05.1	52.7
12	359 05.3	N 4 53.7
13	14 05.5	54.7
14	29 05.7	55.6
15	44 05.9	·· 56.6
16	59 06.0	57.5
17	74 06.2	58.5
18	89 06.4	N 4 59.5
19	104 06.6	5 00.4
20	119 06.8	01.4
21	134 07.0	·· 02.3
22	149 07.1	03.3
23	164 07.3	04.3

3

d h	G.H.A.	Dec.
00	179 07.5	N 5 05.2
01	194 07.7	06.2
02	209 07.9	07.1
03	224 08.1	·· 08.1
04	239 08.3	09.1
05	254 08.4	10.0
06	269 08.6	N 5 11.0
07	284 08.8	11.9
08	299 09.0	12.9
09	314 09.2	·· 13.8
10	329 09.4	14.8
11	344 09.5	15.8
12	359 09.7	N 5 16.7
13	14 09.9	17.7
14	29 10.1	18.6
15	44 10.3	·· 19.6
16	59 10.5	20.6
17	74 10.6	21.5
18	89 10.8	N 5 22.5
19	104 11.0	23.4
20	119 11.2	24.4
21	134 11.4	·· 25.3
22	149 11.6	26.3
23	164 11.7	27.2

4

d h	G.H.A.	Dec.
00	179 11.9	N 5 28.2
01	194 12.1	29.2
02	209 12.3	30.1
03	224 12.5	·· 31.1
04	239 12.7	32.0
05	254 12.8	33.0
06	269 13.0	N 5 33.9
07	284 13.2	34.9
08	299 13.4	35.8
09	314 13.6	·· 36.8
10	329 13.8	37.8
11	344 13.9	38.7
12	359 14.1	N 5 39.7
13	14 14.3	40.6
14	29 14.5	41.6
15	44 14.7	·· 42.5
16	59 14.8	43.5
17	74 15.0	44.4
18	89 15.2	N 5 45.4
19	104 15.4	46.3
20	119 15.6	47.3
21	134 15.8	·· 48.2
22	149 15.9	49.2
23	164 16.1	50.1

5

d h	G.H.A.	Dec.
00	179 16.3	N 5 51.1
01	194 16.5	52.0
02	209 16.7	53.0
03	224 16.8	·· 53.9
04	239 17.0	54.9
05	254 17.2	55.8
06	269 17.4	N 5 56.8
07	284 17.6	57.7
08	299 17.7	58.7
09	314 17.9	5 59.7
10	329 18.1	6 00.6
11	344 18.3	01.5
12	359 18.5	N 6 02.5
13	14 18.6	03.4
14	29 18.8	04.4
15	44 19.0	·· 05.3
16	59 19.2	06.3
17	74 19.4	07.2
18	89 19.5	N 6 08.2
19	104 19.7	09.1
20	119 19.9	10.1
21	134 20.1	·· 11.0
22	149 20.3	12.0
23	164 20.4	12.9

6

d h	G.H.A.	Dec.
00	179 20.6	N 6 13.9
01	194 20.8	14.8
02	209 21.0	15.8
03	224 21.2	·· 16.7
04	239 21.3	17.7
05	254 21.5	18.6
06	269 21.7	N 6 19.6
07	284 21.9	20.5
08	299 22.0	21.5
09	314 22.2	·· 22.4
10	329 22.4	23.3
11	344 22.6	24.3
12	359 22.8	N 6 25.2
13	14 22.9	26.2
14	29 23.1	27.1
15	44 23.3	·· 28.1
16	59 23.5	29.0
17	74 23.6	30.0
18	89 23.8	N 6 30.9
19	104 24.0	31.8
20	119 24.2	32.8
21	134 24.4	·· 33.7
22	149 24.5	34.7
23	164 24.7	35.6

7

d h	G.H.A.	Dec.
00	179 24.9	N 6 36.6
01	194 25.1	37.5
02	209 25.2	38.4
03	224 25.4	·· 39.4
04	239 25.6	40.3
05	254 25.8	- 41.3
06	269 25.9	N 6 42.2
07	284 26.1	43.2
08	299 26.3	44.1
09	314 26.5	·· 45.0
10	329 26.7	46.0
11	344 26.8	46.9
12	359 27.0	N 6 47.9
13	14 27.2	48.8
14	29 27.4	49.7
15	44 27.5	·· 50.7
16	59 27.7	51.6
17	74 27.9	52.6
18	89 28.1	N 6 53.5
19	104 28.2	54.4
20	119 28.4	55.4
21	134 28.6	·· 56.3
22	149 28.8	57.3
23	164 28.9	58.2

8

d h	G.H.A.	Dec.
00	179 29.1	N 6 59.1
01	194 29.3	7 00.1
02	209 29.5	·· 01.0
03	224 29.6	·· 01.9
04	239 29.8	02.9
05	254 30.0	03.8
06	269 30.2	N 7 04.8
07	284 30.3	05.7
08	299 30.5	06.6
09	314 30.7	·· 07.6
10	329 30.8	08.5
11	344 31.0	09.4
12	359 31.2	N 7 10.4
13	14 31.4	11.3
14	29 31.5	12.2
15	44 31.7	·· 13.2
16	59 31.9	14.1
17	74 32.1	15.0
18	89 32.2	N 7 16.0
19	104 32.4	16.9
20	119 32.6	17.8
21	134 32.7	·· 18.8
22	149 32.9	19.7
23	164 33.1	20.6

9

d h	G.H.A.	Dec.
00	179 33.3	N 7 21.6
01	194 33.4	22.5
02	209 33.6	23.4
03	224 33.8	·· 24.4
04	239 34.0	25.3
05	254 34.1	26.2
06	269 34.3	N 7 27.2
07	284 34.5	28.1
08	299 34.6	29.0
09	314 34.8	·· 30.0
10	329 35.0	30.9
11	344 35.2	31.8
12	359 35.3	N 7 32.8
13	14 35.5	33.7
14	29 35.7	34.6
15	44 35.8	·· 35.5
16	59 36.0	36.5
17	74 36.2	37.4
18	89 36.3	N 7 38.3
19	104 36.5	39.3
20	119 36.7	40.2
21	134 36.9	·· 41.1
22	149 37.0	42.0
23	164 37.2	43.0

d h	G.H.A. ° '	Dec. ° '
10 00	179 37.4	N 7 43.9
01	194 37.5	44.8
02	209 37.7	45.8
03	224 37.9 ··	46.7
04	239 38.0	47.6
05	254 38.2	48.5
06	269 38.4	N 7 49.5
07	284 38.5	50.4
08	299 38.7	51.3
09	314 38.9 ··	52.2
10	329 39.1	53.2
11	344 39.2	54.1
12	359 39.4	N 7 55.0
13	14 39.6	55.9
14	29 39.7	56.9
15	44 39.9 ··	57.8
16	59 40.1	58.7
17	74 40.2	7 59.6
18	89 40.4	N 8 00.6
19	104 40.6	01.5
20	119 40.7	02.4
21	134 40.9 ··	03.3
22	149 41.1	04.2
23	164 41.2	05.2
11 00	179 41.4	N 8 06.1
01	194 41.6	07.0
02	209 41.7	07.9
03	224 41.9 ··	08.9
04	239 42.1	09.8
05	254 42.2	10.7
06	269 42.4	N 8 11.6
07	284 42.6	12.5
08	299 42.7	13.5
09	314 42.9 ··	14.4
10	329 43.1	15.3
11	344 43.2	16.2
12	359 43.4	N 8 17.1
13	14 43.6	18.1
14	29 43.7	19.0
15	44 43.9 ··	19.9
16	59 44.1	20.8
17	74 44.2	21.7
18	89 44.4	N 8 22.6
19	104 44.5	23.6
20	119 44.7	24.5
21	134 44.9 ··	25.4
22	149 45.0	26.3
23	164 45.2	27.2
12 00	179 45.4	N 8 28.1
01	194 45.5	29.1
02	209 45.7	30.0
03	224 45.9 ··	30.9
04	239 46.0	31.8
05	254 46.2	32.7
06	269 46.3	N 8 33.6
07	284 46.5	34.6
08	299 46.7	35.5
09	314 46.8 ··	36.4
10	329 47.0	37.3
11	344 47.2	38.2
12	359 47.3	N 8 39.1
13	14 47.5	40.0
14	29 47.6	40.9
15	44 47.8 ··	41.9
16	59 48.0	42.8
17	74 48.1	43.7
18	89 48.3	N 8 44.6
19	104 48.5	45.5
20	119 48.6	46.4
21	134 48.8 ··	47.3
22	149 48.9	48.2
23	164 49.1	49.2

d h	G.H.A. ° '	Dec. ° '
13 00	179 49.3	N 8 50.1
01	194 49.4	51.0
02	209 49.6	51.9
03	224 49.7 ··	52.8
04	239 49.9	53.7
05	254 50.1	54.6
06	269 50.2	N 8 55.5
07	284 50.4	56.4
08	299 50.5	57.3
09	314 50.7 ··	58.2
10	329 50.9	8 59.1
11	344 51.0	9 00.1
12	359 51.2	N 9 01.0
13	14 51.3	01.9
14	29 51.5	02.8
15	44 51.7 ··	03.7
16	59 51.8	04.6
17	74 52.0	05.5
18	89 52.1	N 9 06.4
19	104 52.3	07.3
20	119 52.5	08.2
21	134 52.6 ··	09.1
22	149 52.8	10.0
23	164 52.9	10.9
14 00	179 53.1	N 9 11.8
01	194 53.2	12.7
02	209 53.4	13.6
03	224 53.6 ··	14.5
04	239 53.7	15.4
05	254 53.9	16.3
06	269 54.0	N 9 17.2
07	284 54.2	18.1
08	299 54.3	19.0
09	314 54.5 ··	20.0
10	329 54.7	20.9
11	344 54.8	21.8
12	359 55.0	N 9 22.7
13	14 55.1	23.6
14	29 55.3	24.5
15	44 55.4 ··	25.4
16	59 55.6	26.3
17	74 55.8	27.2
18	89 55.9	N 9 28.1
19	104 56.1	29.0
20	119 56.2	29.8
21	134 56.4 ··	30.7
22	149 56.5	31.6
23	164 56.7	32.5
15 00	179 56.8	N 9 33.4
01	194 57.0	34.3
02	209 57.1	35.2
03	224 57.3 ··	36.1
04	239 57.5	37.0
05	254 57.6	37.9
06	269 57.8	N 9 38.8
07	284 57.9	39.7
08	299 58.1	40.6
09	314 58.2 ··	41.5
10	329 58.4	42.4
11	344 58.5	43.3
12	359 58.7	N 9 44.2
13	14 58.8	45.1
14	29 59.0	46.0
15	44 59.1 ··	46.9
16	59 59.3	47.8
17	74 59.4	48.7
18	89 59.6	N 9 49.5
19	104 59.7	50.4
20	119 59.9	51.3
21	135 00.1 ··	52.2
22	150 00.2	53.1
23	165 00.4	54.0

d h	G.H.A. ° '	Dec. ° '
16 00	180 00.5	N 9 54.9
01	195 00.7	55.8
02	210 00.8	56.7
03	225 01.0 ··	57.6
04	240 01.1	58.5
05	255 01.3	9 59.3
06	270 01.4	N10 00.2
07	285 01.6	01.1
08	300 01.7	02.0
09	315 01.9 ··	02.9
10	330 02.0	03.8
11	345 02.2	04.7
12	0 02.3	N10 05.6
13	15 02.5	06.4
14	30 02.6	07.3
15	45 02.8 ··	08.2
16	60 02.9	09.1
17	75 03.1	10.0
18	90 03.2	N10 10.9
19	105 03.3	11.8
20	120 03.5	12.6
21	135 03.6 ··	13.5
22	150 03.8	14.4
23	165 03.9	15.3
17 00	180 04.1	N10 16.2
01	195 04.2	17.1
02	210 04.4	18.0
03	225 04.5 ··	18.8
04	240 04.7	19.7
05	255 04.8	20.6
06	270 05.0	N10 21.5
07	285 05.1	22.4
08	300 05.3	23.2
09	315 05.4 ··	24.1
10	330 05.6	25.0
11	345 05.7	25.9
12	0 05.8	N10 26.8
13	15 06.0	27.6
14	30 06.1	28.5
15	45 06.3 ··	29.4
16	60 06.4	30.3
17	75 06.6	31.2
18	90 06.7	N10 32.0
19	105 06.9	32.9
20	120 07.0	33.8
21	135 07.1 ··	34.7
22	150 07.3	35.6
23	165 07.4	36.4
18 00	180 07.6	N10 37.3
01	195 07.7	38.2
02	210 07.9	39.1
03	225 08.0 ··	39.9
04	240 08.2	40.8
05	255 08.3	41.7
06	270 08.4	N10 42.6
07	285 08.6	43.4
08	300 08.7	44.3
09	315 08.9 ··	45.2
10	330 09.0	46.1
11	345 09.2	46.9
12	0 09.3	N10 47.8
13	15 09.4	48.7
14	30 09.6	49.5
15	45 09.7 ··	50.4
16	60 09.9	51.3
17	75 10.0	52.2
18	90 10.1	N10 53.0
19	105 10.3	53.9
20	120 10.4	54.8
21	135 10.6 ··	55.6
22	150 10.7	56.5
23	165 10.8	57.4

d h	G.H.A.	Dec.
19 00	180 11.0	N10 58.2
01	195 11.1	10 59.1
02	210 11.3	11 00.0
03	225 11.4	·· 00.9
04	240 11.5	01.7
05	255 11.7	02.6
06	270 11.8	N11 03.5
07	285 12.0	04.3
08	300 12.1	05.2
09	315 12.2	·· 06.1
10	330 12.4	06.9
11	345 12.5	07.8
12	0 12.6	N11 08.7
13	15 12.8	09.5
14	30 12.9	10.4
15	45 13.1	·· 11.2
16	60 13.2	12.1
17	75 13.3	13.0
18	90 13.5	N11 13.8
19	105 13.6	14.7
20	120 13.7	15.6
21	135 13.9	·· 16.4
22	150 14.0	17.3
23	165 14.1	18.2
20 00	180 14.3	N11 19.0
01	195 14.4	19.9
02	210 14.6	20.7
03	225 14.7	·· 21.6
04	240 14.8	22.5
05	255 15.0	23.3
06	270 15.1	N11 24.2
07	285 15.2	25.0
08	300 15.4	25.9
09	315 15.5	·· 26.8
10	330 15.6	27.6
11	345 15.8	28.5
12	0 15.9	N11 29.3
13	15 16.0	30.2
14	30 16.2	31.0
15	45 16.3	·· 31.9
16	60 16.4	32.8
17	75 16.6	33.6
18	90 16.7	N11 34.5
19	105 16.8	35.3
20	120 17.0	36.2
21	135 17.1	·· 37.0
22	150 17.2	37.9
23	165 17.4	38.7
21 00	180 17.5	N11 39.6
01	195 17.6	40.4
02	210 17.7	41.3
03	225 17.9	·· 42.1
04	240 18.0	43.0
05	255 18.1	43.9
06	270 18.3	N11 44.7
07	285 18.4	45.6
08	300 18.5	46.4
09	315 18.7	·· 47.3
10	330 18.8	48.1
11	345 18.9	49.0
12	0 19.0	N11 49.8
13	15 19.2	50.7
14	30 19.3	51.5
15	45 19.4	·· 52.4
16	60 19.6	53.2
17	75 19.7	54.1
18	90 19.8	N11 54.9
19	105 19.9	55.7
20	120 20.1	56.6
21	135 20.2	·· 57.4
22	150 20.3	58.3
23	165 20.5	59.1

d h	G.H.A.	Dec.
22 00	180 20.6	N12 00.0
01	195 20.7	00.8
02	210 20.8	01.7
03	225 21.0	·· 02.5
04	240 21.1	03.4
05	255 21.2	04.2
06	270 21.3	N12 05.0
07	285 21.5	05.9
08	300 21.6	06.7
09	315 21.7	·· 07.6
10	330 21.8	08.4
11	345 22.0	09.3
12	0 22.1	N12 10.1
13	15 22.2	10.9
14	30 22.3	11.8
15	45 22.5	·· 12.6
16	60 22.6	13.5
17	75 22.7	14.3
18	90 22.8	N12 15.1
19	105 23.0	16.0
20	120 23.1	16.8
21	135 23.2	·· 17.7
22	150 23.3	18.5
23	165 23.4	19.3
23 00	180 23.6	N12 20.2
01	195 23.7	21.0
02	210 23.8	21.8
03	225 23.9	·· 22.7
04	240 24.1	23.5
05	255 24.2	24.3
06	270 24.3	N12 25.2
07	285 24.4	26.0
08	300 24.5	26.8
09	315 24.7	·· 27.7
10	330 24.8	28.5
11	345 24.9	29.4
12	0 25.0	N12 30.2
13	15 25.1	31.0
14	30 25.3	31.8
15	45 25.4	·· 32.7
16	60 25.5	33.5
17	75 25.6	34.3
18	90 25.7	N12 35.2
19	105 25.8	36.0
20	120 26.0	36.8
21	135 26.1	·· 37.7
22	150 26.2	38.5
23	165 26.3	39.3
24 00	180 26.4	N12 40.2
01	195 26.6	41.0
02	210 26.7	41.8
03	225 26.8	·· 42.6
04	240 26.9	43.5
05	255 27.0	44.3
06	270 27.1	N12 45.1
07	285 27.3	45.9
08	300 27.4	46.8
09	315 27.5	·· 47.6
10	330 27.6	48.4
11	345 27.7	49.2
12	0 27.8	N12 50.1
13	15 27.9	50.9
14	30 28.1	51.7
15	45 28.2	·· 52.5
16	60 28.3	53.4
17	75 28.4	54.2
18	90 28.5	N12 55.0
19	105 28.6	55.8
20	120 28.7	56.6
21	135 28.9	·· 57.5
22	150 29.0	58.3
23	165 29.1	59.1

d h	G.H.A.	Dec.
25 00	180 29.2	N12 59.9
01	195 29.3	13 00.7
02	210 29.4	01.6
03	225 29.5	·· 02.4
04	240 29.6	03.2
05	255 29.7	04.0
06	270 29.9	N13 04.8
07	285 30.0	05.7
08	300 30.1	06.5
09	315 30.2	·· 07.3
10	330 30.3	08.1
11	345 30.4	08.9
12	0 30.5	N13 09.7
13	15 30.6	10.6
14	30 30.7	11.4
15	45 30.8	·· 12.2
16	60 31.0	13.0
17	75 31.1	13.8
18	90 31.2	N13 14.6
19	105 31.3	15.4
20	120 31.4	16.2
21	135 31.5	·· 17.1
22	150 31.6	17.9
23	165 31.7	18.7
26 00	180 31.8	N13 19.5
01	195 31.9	20.3
02	210 32.0	21.1
03	225 32.1	·· 21.9
04	240 32.2	22.7
05	255 32.4	23.5
06	270 32.5	N13 24.3
07	285 32.6	25.2
08	300 32.7	26.0
09	315 32.8	·· 26.8
10	330 32.9	27.6
11	345 33.0	28.4
12	0 33.1	N13 29.2
13	15 33.2	30.0
14	30 33.3	30.8
15	45 33.4	·· 31.6
16	60 33.5	32.4
17	75 33.6	33.2
18	90 33.7	N13 34.0
19	105 33.8	34.8
20	120 33.9	35.6
21	135 34.0	·· 36.4
22	150 34.1	37.2
23	165 34.2	38.0
27 00	180 34.3	N13 38.8
01	195 34.4	39.6
02	210 34.5	40.4
03	225 34.6	·· 41.2
04	240 34.7	42.0
05	255 34.8	42.8
06	270 34.9	N13 43.6
07	285 35.0	44.4
08	300 35.1	45.2
09	315 35.2	·· 46.0
10	330 35.3	46.8
11	345 35.4	47.6
12	0 35.5	N13 48.4
13	15 35.6	49.2
14	30 35.7	50.0
15	45 35.8	·· 50.8
16	60 35.9	51.6
17	75 36.0	52.4
18	90 36.1	N13 53.2
19	105 36.2	54.0
20	120 36.3	54.8
21	135 36.4	·· 55.6
22	150 36.5	56.4
23	165 36.6	57.2

28 (d)

h	G.H.A.	Dec.
00	180 36.7	N13 58.0
01	195 36.8	58.7
02	210 36.9	13 59.5
03	225 37.0	14 00.3
04	240 37.1	01.1
05	255 37.2	01.9
06	270 37.3	N14 02.7
07	285 37.4	03.5
08	300 37.5	04.3
09	315 37.6	·· 05.1
10	330 37.6	05.9
11	345 37.7	06.6
12	0 37.8	N14 07.4
13	15 37.9	08.2
14	30 38.0	09.0
15	45 38.1	·· 09.8
16	60 38.2	10.6
17	75 38.3	11.4
18	90 38.4	N14 12.1
19	105 38.5	12.9
20	120 38.6	13.7
21	135 38.7	·· 14.5
22	150 38.8	15.3
23	165 38.9	16.1

29

h	G.H.A.	Dec.
00	180 38.9	N14 16.8
01	195 39.0	17.6
02	210 39.1	18.4
03	225 39.2	·· 19.2
04	240 39.3	20.0
05	255 39.4	20.8
06	270 39.5	N14 21.5
07	285 39.6	22.3
08	300 39.7	23.1
09	315 39.8	·· 23.9
10	330 39.8	24.6
11	345 39.9	25.4
12	0 40.0	N14 26.2
13	15 40.1	27.0
14	30 40.2	27.8
15	45 40.3	·· 28.5
16	60 40.4	29.3
17	75 40.5	30.1
18	90 40.5	N14 30.9
19	105 40.6	31.6
20	120 40.7	32.4
21	135 40.8	·· 33.2
22	150 40.9	34.0
23	165 41.0	34.7

30

h	G.H.A.	Dec.
00	180 41.1	N14 35.5
01	195 41.1	36.3
02	210 41.2	37.0
03	225 41.3	·· 37.8
04	240 41.4	38.6
05	255 41.5	39.4
06	270 41.6	N14 40.1
07	285 41.6	40.9
08	300 41.7	41.7
09	315 41.8	·· 42.4
10	330 41.9	43.2
11	345 42.0	44.0
12	0 42.1	N14 44.7
13	15 42.1	45.5
14	30 42.2	46.3
15	45 42.3	·· 47.0
16	60 42.4	47.8
17	75 42.5	48.6
18	90 42.5	N14 49.3
19	105 42.6	50.1
20	120 42.7	50.9
21	135 42.8	·· 51.6
22	150 42.9	52.4
23	165 42.9	53.2

MAY 1

h	G.H.A.	Dec.
00	180 43.0	N14 53.9
01	195 43.1	54.7
02	210 43.2	55.4
03	225 43.3	·· 56.2
04	240 43.3	57.0
05	255 43.4	57.7
06	270 43.5	N14 58.5
07	285 43.6	14 59.2
08	300 43.7	15 00.0
09	315 43.7	·· 00.8
10	330 43.8	01.5
11	345 43.9	02.3
12	0 44.0	N15 03.0
13	15 44.0	03.8
14	30 44.1	04.5
15	45 44.2	·· 05.3
16	60 44.3	06.1
17	75 44.3	06.8
18	90 44.4	N15 07.6
19	105 44.5	08.3
20	120 44.6	09.1
21	135 44.6	·· 09.8
22	150 44.7	10.6
23	165 44.8	11.3

2

h	G.H.A.	Dec.
00	180 44.9	N15 12.1
01	195 44.9	12.8
02	210 45.0	13.6
03	225 45.1	·· 14.3
04	240 45.2	15.1
05	255 45.2	15.8
06	270 45.3	N15 16.6
07	285 45.4	17.3
08	300 45.4	18.1
09	315 45.5	·· 18.8
10	330 45.6	19.6
11	345 45.7	20.3
12	0 45.7	N15 21.1
13	15 45.8	21.8
14	30 45.9	22.6
15	45 45.9	·· 23.3
16	60 46.0	24.1
17	75 46.1	24.8
18	90 46.2	N15 25.5
19	105 46.2	26.3
20	120 46.3	27.0
21	135 46.4	·· 27.8
22	150 46.4	28.5
23	165 46.5	29.3

3

h	G.H.A.	Dec.
00	180 46.6	N15 30.0
01	195 46.6	30.7
02	210 46.7	31.5
03	225 46.8	·· 32.2
04	240 46.8	33.0
05	255 46.9	33.7
06	270 47.0	N15 34.4
07	285 47.0	35.2
08	300 47.1	35.9
09	315 47.2	·· 36.7
10	330 47.2	37.4
11	345 47.3	38.1
12	0 47.4	N15 38.9
13	15 47.4	39.6
14	30 47.5	40.3
15	45 47.6	·· 41.1
16	60 47.6	41.8
17	75 47.7	42.5
18	90 47.7	N15 43.3
19	105 47.8	44.0
20	120 47.9	44.7
21	135 47.9	·· 45.5
22	150 48.0	46.2
23	165 48.1	46.9

4

h	G.H.A.	Dec.
00	180 48.1	N15 47.7
01	195 48.2	48.4
02	210 48.2	49.1
03	225 48.3	·· 49.9
04	240 48.4	50.6
05	255 48.4	51.3
06	270 48.5	N15 52.0
07	285 48.5	52.8
08	300 48.6	53.5
09	315 48.7	·· 54.2
10	330 48.7	55.0
11	345 48.8	55.7
12	0 48.8	N15 56.4
13	15 48.9	57.1
14	30 49.0	57.9
15	45 49.0	·· 58.6
16	60 49.1	15 59.3
17	75 49.1	16 00.0
18	90 49.2	N16 00.7
19	105 49.3	01.5
20	120 49.3	02.2
21	135 49.4	·· 02.9
22	150 49.4	03.6
23	165 49.5	04.3

5

h	G.H.A.	Dec.
00	180 49.5	N16 05.1
01	195 49.6	05.8
02	210 49.6	06.5
03	225 49.7	·· 07.2
04	240 49.8	07.9
05	255 49.8	08.7
06	270 49.9	N16 09.4
07	285 49.9	10.1
08	300 50.0	10.8
09	315 50.0	·· 11.5
10	330 50.1	12.2
11	345 50.1	13.0
12	0 50.2	N16 13.7
13	15 50.2	14.4
14	30 50.3	15.1
15	45 50.3	·· 15.8
16	60 50.4	16.5
17	75 50.5	17.2
18	90 50.5	N16 17.9
19	105 50.6	18.7
20	120 50.6	19.4
21	135 50.7	·· 20.1
22	150 50.7	20.8
23	165 50.8	21.5

6

h	G.H.A.	Dec.
00	180 50.8	N16 22.2
01	195 50.9	22.9
02	210 50.9	23.6
03	225 51.0	·· 24.3
04	240 51.0	25.0
05	255 51.1	25.7
06	270 51.1	N16 26.4
07	285 51.2	27.1
08	300 51.2	27.9
09	315 51.3	·· 28.6
10	330 51.3	29.3
11	345 51.4	30.0
12	0 51.4	N16 30.7
13	15 51.4	31.4
14	30 51.5	32.1
15	45 51.5	·· 32.8
16	60 51.6	33.5
17	75 51.6	34.2
18	90 51.7	N16 34.9
19	105 51.7	35.6
20	120 51.8	36.3
21	135 51.8	·· 37.0
22	150 51.9	37.7
23	165 51.9	38.4

d	h	G.H.A. ° ′	Dec. ° ′
7	00	180 51.9	N16 39.1
	01	195 52.0	39.8
	02	210 52.0	40.5
	03	225 52.1	·· 41.2
	04	240 52.1	41.8
	05	255 52.2	42.5
	06	270 52.2	N16 43.2
	07	285 52.2	43.9
	08	300 52.3	44.6
	09	315 52.3	·· 45.3
	10	330 52.4	46.0
	11	345 52.4	46.7
	12	0 52.5	N16 47.4
	13	15 52.5	48.1
	14	30 52.5	48.8
	15	45 52.6	·· 49.5
	16	60 52.6	50.2
	17	75 52.7	50.8
	18	90 52.7	N16 51.5
	19	105 52.7	52.2
	20	120 52.8	52.9
	21	135 52.8	·· 53.6
	22	150 52.9	54.3
	23	165 52.9	55.0
8	00	180 52.9	N16 55.7
	01	195 53.0	56.3
	02	210 53.0	57.0
	03	225 53.1	·· 57.7
	04	240 53.1	58.4
	05	255 53.1	59.1
	06	270 53.2	N16 59.8
	07	285 53.2	17 00.4
	08	300 53.2	01.1
	09	315 53.3	·· 01.8
	10	330 53.3	02.5
	11	345 53.3	03.2
	12	0 53.4	N17 03.8
	13	15 53.4	04.5
	14	30 53.4	05.2
	15	45 53.5	·· 05.9
	16	60 53.5	06.6
	17	75 53.6	07.2
	18	90 53.6	N17 07.9
	19	105 53.6	08.6
	20	120 53.7	09.3
	21	135 53.7	·· 09.9
	22	150 53.7	10.6
	23	165 53.8	11.3
9	00	180 53.8	N17 12.0
	01	195 53.8	12.6
	02	210 53.9	13.3
	03	225 53.9	·· 14.0
	04	240 53.9	14.6
	05	255 53.9	15.3
	06	270 54.0	N17 16.0
	07	285 54.0	16.7
	08	300 54.0	17.3
	09	315 54.1	·· 18.0
	10	330 54.1	18.7
	11	345 54.1	19.3
	12	0 54.2	N17 20.0
	13	15 54.2	20.7
	14	30 54.2	21.3
	15	45 54.2	·· 22.0
	16	60 54.3	22.7
	17	75 54.3	23.3
	18	90 54.3	N17 24.0
	19	105 54.4	24.7
	20	120 54.4	25.3
	21	135 54.4	·· 26.0
	22	150 54.4	26.6
	23	165 54.5	27.3

d	h	G.H.A. ° ′	Dec. ° ′
10	00	180 54.5	N17 28.0
	01	195 54.5	28.6
	02	210 54.5	29.3
	03	225 54.6	·· 30.0
	04	240 54.6	30.6
	05	255 54.6	31.3
	06	270 54.7	N17 31.9
	07	285 54.7	32.6
	08	300 54.7	33.2
	09	315 54.7	·· 33.9
	10	330 54.7	34.6
	11	345 54.8	35.2
	12	0 54.8	N17 35.9
	13	15 54.8	36.5
	14	30 54.8	37.2
	15	45 54.9	·· 37.8
	16	60 54.9	38.5
	17	75 54.9	39.1
	18	90 54.9	N17 39.8
	19	105 55.0	40.4
	20	120 55.0	41.1
	21	135 55.0	·· 41.7
	22	150 55.0	42.4
	23	165 55.0	43.0
11	00	180 55.1	N17 43.7
	01	195 55.1	44.3
	02	210 55.1	45.0
	03	225 55.1	·· 45.6
	04	240 55.1	46.3
	05	255 55.2	46.9
	06	270 55.2	N17 47.6
	07	285 55.2	48.2
	08	300 55.2	48.9
	09	315 55.2	·· 49.5
	10	330 55.3	50.2
	11	345 55.3	50.8
	12	0 55.3	N17 51.4
	13	15 55.3	52.1
	14	30 55.3	52.7
	15	45 55.3	·· 53.4
	16	60 55.4	54.0
	17	75 55.4	54.7
	18	90 55.4	N17 55.3
	19	105 55.4	55.9
	20	120 55.4	56.6
	21	135 55.4	·· 57.2
	22	150 55.5	57.8
	23	165 55.5	58.5
12	00	180 55.5	N17 59.1
	01	195 55.5	17 59.8
	02	210 55.5	18 00.4
	03	225 55.5	·· 01.0
	04	240 55.6	01.7
	05	255 55.6	02.3
	06	270 55.6	N18 02.9
	07	285 55.6	03.6
	08	300 55.6	04.2
	09	315 55.6	·· 04.8
	10	330 55.6	05.5
	11	345 55.6	06.1
	12	0 55.7	N18 06.7
	13	15 55.7	07.4
	14	30 55.7	08.0
	15	45 55.7	·· 08.6
	16	60 55.7	09.2
	17	75 55.7	09.9
	18	90 55.7	N18 10.5
	19	105 55.7	11.1
	20	120 55.7	11.8
	21	135 55.8	·· 12.4
	22	150 55.8	13.0
	23	165 55.8	13.6

d	h	G.H.A. ° ′	Dec. ° ′
13	00	180 55.8	N18 14.3
	01	195 55.8	14.9
	02	210 55.8	15.5
	03	225 55.8	·· 16.1
	04	240 55.8	16.7
	05	255 55.8	17.4
	06	270 55.8	N18 18.0
	07	285 55.8	18.6
	08	300 55.8	19.2
	09	315 55.9	·· 19.8
	10	330 55.9	20.5
	11	345 55.9	21.1
	12	0 55.9	N18 21.7
	13	15 55.9	22.3
	14	30 55.9	22.9
	15	45 55.9	·· 23.5
	16	60 55.9	24.2
	17	75 55.9	24.8
	18	90 55.9	N18 25.4
	19	105 55.9	26.0
	20	120 55.9	26.6
	21	135 55.9	·· 27.2
	22	150 55.9	27.8
	23	165 55.9	28.5
14	00	180 55.9	N18 29.1
	01	195 55.9	29.7
	02	210 55.9	30.3
	03	225 55.9	·· 30.9
	04	240 55.9	31.5
	05	255 55.9	32.1
	06	270 55.9	N18 32.7
	07	285 55.9	33.3
	08	300 55.9	33.9
	09	315 56.0	·· 34.5
	10	330 56.0	35.2
	11	345 56.0	35.8
	12	0 56.0	N18 36.4
	13	15 56.0	37.0
	14	30 56.0	37.6
	15	45 56.0	·· 38.2
	16	60 56.0	38.8
	17	75 56.0	39.4
	18	90 56.0	N18 40.0
	19	105 56.0	40.6
	20	120 55.9	41.2
	21	135 55.9	·· 41.8
	22	150 55.9	42.4
	23	165 55.9	43.0
15	00	180 55.9	N18 43.6
	01	195 55.9	44.2
	02	210 55.9	44.8
	03	225 55.9	·· 45.4
	04	240 55.9	46.0
	05	255 55.9	46.6
	06	270 55.9	N18 47.2
	07	285 55.9	47.7
	08	300 55.9	48.3
	09	315 55.9	·· 48.9
	10	330 55.9	49.5
	11	345 55.9	50.1
	12	0 55.9	N18 50.7
	13	15 55.9	51.3
	14	30 55.9	51.9
	15	45 55.9	·· 52.5
	16	60 55.9	53.1
	17	75 55.9	53.7
	18	90 55.9	N18 54.3
	19	105 55.9	54.8
	20	120 55.8	55.4
	21	135 55.8	·· 56.0
	22	150 55.8	56.6
	23	165 55.8	57.2

d h	G.H.A.	Dec.
16 00	180 55.8	N18 57.8
01	195 55.8	58.4
02	210 55.8	58.9
03	225 55.8	18 59.5
04	240 55.8	19 00.1
05	255 55.8	00.7
06	270 55.8	N19 01.3
07	285 55.8	01.8
08	300 55.7	02.4
09	315 55.7	03.0
10	330 55.7	03.6
11	345 55.7	04.2
12	0 55.7	N19 04.7
13	15 55.7	05.3
14	30 55.7	05.9
15	45 55.7	06.5
16	60 55.7	07.1
17	75 55.6	07.6
18	90 55.6	N19 08.2
19	105 55.6	08.8
20	120 55.6	09.4
21	135 55.6	09.9
22	150 55.6	10.5
23	165 55.6	11.1
17 00	180 55.6	N19 11.6
01	195 55.5	12.2
02	210 55.5	12.8
03	225 55.5	13.4
04	240 55.5	13.9
05	255 55.5	14.5
06	270 55.5	N19 15.1
07	285 55.4	15.6
08	300 55.4	16.2
09	315 55.4	16.8
10	330 55.4	17.3
11	345 55.4	17.9
12	0 55.4	N19 18.5
13	15 55.4	19.0
14	30 55.3	19.6
15	45 55.3	20.1
16	60 55.3	20.7
17	75 55.3	21.3
18	90 55.3	N19 21.8
19	105 55.2	22.4
20	120 55.2	23.0
21	135 55.2	23.5
22	150 55.2	24.1
23	165 55.2	24.6
18 00	180 55.2	N19 25.2
01	195 55.1	25.7
02	210 55.1	26.3
03	225 55.1	26.9
04	240 55.1	27.4
05	255 55.0	28.0
06	270 55.0	N19 28.5
07	285 55.0	29.1
08	300 55.0	29.6
09	315 55.0	30.2
10	330 54.9	30.7
11	345 54.9	31.3
12	0 54.9	N19 31.8
13	15 54.9	32.4
14	30 54.9	32.9
15	45 54.8	33.5
16	60 54.8	34.0
17	75 54.8	34.6
18	90 54.8	N19 35.1
19	105 54.7	35.7
20	120 54.7	36.2
21	135 54.7	36.8
22	150 54.7	37.3
23	165 54.6	37.9

d h	G.H.A.	Dec.
19 00	180 54.6	N19 38.4
01	195 54.6	38.9
02	210 54.6	39.5
03	225 54.5	40.0
04	240 54.5	40.6
05	255 54.5	41.1
06	270 54.5	N19 41.7
07	285 54.4	42.2
08	300 54.4	42.7
09	315 54.4	43.3
10	330 54.4	43.8
11	345 54.3	44.3
12	0 54.3	N19 44.9
13	15 54.3	45.4
14	30 54.2	46.0
15	45 54.2	46.5
16	60 54.2	47.0
17	75 54.2	47.6
18	90 54.1	N19 48.1
19	105 54.1	48.6
20	120 54.1	49.2
21	135 54.0	49.7
22	150 54.0	50.2
23	165 54.0	50.8
20 00	180 53.9	N19 51.3
01	195 53.9	51.8
02	210 53.9	52.3
03	225 53.8	52.9
04	240 53.8	53.4
05	255 53.8	53.9
06	270 53.8	N19 54.5
07	285 53.7	55.0
08	300 53.7	55.5
09	315 53.7	56.0
10	330 53.6	56.6
11	345 53.6	57.1
12	0 53.6	N19 57.6
13	15 53.5	58.1
14	30 53.5	58.6
15	45 53.5	59.2
16	60 53.4	19 59.7
17	75 53.4	20 00.2
18	90 53.4	N20 00.7
19	105 53.3	01.2
20	120 53.3	01.8
21	135 53.3	02.3
22	150 53.2	02.8
23	165 53.2	03.3
21 00	180 53.1	N20 03.8
01	195 53.1	04.3
02	210 53.1	04.9
03	225 53.0	05.4
04	240 53.0	05.9
05	255 53.0	06.4
06	270 52.9	N20 06.9
07	285 52.9	07.4
08	300 52.8	07.9
09	315 52.8	08.4
10	330 52.8	09.0
11	345 52.7	09.5
12	0 52.7	N20 10.0
13	15 52.6	10.5
14	30 52.6	11.0
15	45 52.6	11.5
16	60 52.5	12.0
17	75 52.5	12.5
18	90 52.4	N20 13.0
19	105 52.4	13.5
20	120 52.4	14.0
21	135 52.3	14.5
22	150 52.3	15.0
23	165 52.2	15.5

d h	G.H.A.	Dec.
22 00	180 52.2	N20 16.0
01	195 52.2	16.5
02	210 52.1	17.0
03	225 52.1	17.5
04	240 52.0	18.0
05	255 52.0	18.5
06	270 51.9	N20 19.0
07	285 51.9	19.5
08	300 51.9	20.0
09	315 51.8	20.5
10	330 51.8	21.0
11	345 51.7	21.5
12	0 51.7	N20 22.0
13	15 51.6	22.5
14	30 51.6	23.0
15	45 51.5	23.5
16	60 51.5	24.0
17	75 51.5	24.5
18	90 51.4	N20 25.0
19	105 51.4	25.4
20	120 51.3	25.9
21	135 51.3	26.4
22	150 51.2	26.9
23	165 51.2	27.4
23 00	180 51.1	N20 27.9
01	195 51.1	28.4
02	210 51.0	28.9
03	225 51.0	29.3
04	240 50.9	29.8
05	255 50.9	30.3
06	270 50.8	N20 30.8
07	285 50.8	31.3
08	300 50.7	31.8
09	315 50.7	32.2
10	330 50.6	32.7
11	345 50.6	33.2
12	0 50.5	N20 33.7
13	15 50.5	34.2
14	30 50.4	34.6
15	45 50.4	35.1
16	60 50.3	35.6
17	75 50.3	36.1
18	90 50.2	N20 36.5
19	105 50.2	37.0
20	120 50.1	37.5
21	135 50.1	38.0
22	150 50.0	38.4
23	165 50.0	38.9
24 00	180 49.9	N20 39.4
01	195 49.9	39.9
02	210 49.8	40.3
03	225 49.8	40.8
04	240 49.7	41.3
05	255 49.7	41.7
06	270 49.6	N20 42.2
07	285 49.5	42.7
08	300 49.5	43.1
09	315 49.4	43.6
10	330 49.4	44.1
11	345 49.3	44.5
12	0 49.3	N20 45.0
13	15 49.2	45.5
14	30 49.2	45.9
15	45 49.1	46.4
16	60 49.1	46.9
17	75 49.0	47.3
18	90 48.9	N20 47.8
19	105 48.9	48.2
20	120 48.8	48.7
21	135 48.8	49.2
22	150 48.7	49.6
23	165 48.7	50.1

d h	G.H.A.	Dec.
25 00	180 48.6	N20 50.5
01	195 48.5	51.0
02	210 48.5	51.5
03	225 48.4	·· 51.9
04	240 48.4	52.4
05	255 48.3	52.8
06	270 48.2	N20 53.3
07	285 48.2	53.7
08	300 48.1	54.2
09	315 48.1	·· 54.6
10	330 48.0	55.1
11	345 47.9	55.5
12	0 47.9	N20 56.0
13	15 47.8	56.4
14	30 47.8	56.9
15	45 47.7	·· 57.3
16	60 47.6	57.8
17	75 47.6	58.2
18	90 47.5	N20 58.7
19	105 47.4	59.1
20	120 47.4	20 59.6
21	135 47.3	21 00.0
22	150 47.3	00.4
23	165 47.2	00.9
26 00	180 47.1	N21 01.3
01	195 47.1	01.8
02	210 47.0	02.2
03	225 46.9	·· 02.7
04	240 46.9	03.1
05	255 46.8	03.5
06	270 46.7	N21 04.0
07	285 46.7	04.4
08	300 46.6	04.8
09	315 46.6	·· 05.3
10	330 46.5	05.7
11	345 46.4	06.2
12	0 46.4	N21 06.6
13	15 46.3	07.0
14	30 46.2	07.5
15	45 46.2	·· 07.9
16	60 46.1	08.3
17	75 46.0	08.8
18	90 46.0	N21 09.2
19	105 45.9	09.6
20	120 45.8	10.0
21	135 45.8	·· 10.5
22	150 45.7	10.9
23	165 45.6	11.3
27 00	180 45.5	N21 11.8
01	195 45.5	12.2
02	210 45.4	12.6
03	225 45.3	·· 13.0
04	240 45.3	13.5
05	255 45.2	13.9
06	270 45 1	N21 14.3
07	285 45.1	14.7
08	300 45.0	15.2
09	315 44.9	·· 15.6
10	330 44.9	16.0
11	345 44.8	16.4
12	0 44.7	N21 16.8
13	15 44.6	17.3
14	30 44.6	17.7
15	45 44.5	·· 18.1
16	60 44.4	18.5
17	75 44.4	18.9
18	90 44.3	N21 19.3
19	105 44.2	19.8
20	120 44.1	20.2
21	135 44.1	·· 20.6
22	150 44.0	21.0
23	165 43.9	21.4

d h	G.H.A.	Dec.
28 00	180 43.8	N21 21.8
01	195 43.8	22.2
02	210 43.7	22.6
03	225 43.6	·· 23.0
04	240 43.5	23.5
05	255 43.5	23.9
06	270 43.4	N21 24.3
07	285 43.3	24.7
08	300 43.2	25.1
09	315 43.2	·· 25.5
10	330 43.1	25.9
11	345 43.0	26.3
12	0 42.9	N21 26.7
13	15 42.9	27.1
14	30 42.8	27.5
15	45 42.7	·· 27.9
16	60 42.6	28.3
17	75 42.6	28.7
18	90 42.5	N21 29.1
19	105 42.4	29.5
20	120 42.3	29.9
21	135 42.2	·· 30.3
22	150 42.2	30.7
23	165 42.1	31.1
29 00	180 42.0	N21 31.5
01	195 41.9	31.9
02	210 41.9	32.3
03	225 41.8	·· 32.7
04	240 41.7	33.1
05	255 41.6	33.5
06	270 41.5	N21 33.9
07	285 41.5	34.3
08	300 41.4	34.7
09	315 41.3	·· 35.0
10	330 41.2	35.4
11	345 41.1	35.8
12	0 41.1	N21 36.2
13	15 41.0	36.6
14	30 40.9	37.0
15	45 40.8	·· 37.4
16	60 40.7	37.8
17	75 40.6	38.2
18	90 40.6	N21 38.5
19	105 40.5	38.9
20	120 40.4	39.3
21	135 40.3	·· 39.7
22	150 40.2	40.1
23	165 40.1	40.4
30 00	180 40.1	N21 40.8
01	195 40.0	41.2
02	210 39.9	41.6
03	225 39.8	·· 42.0
04	240 39.7	42.3
05	255 39.6	42.7
06	270 39.6	N21 43.1
07	285 39.5	43.5
08	300 39.4	43.9
09	315 39.3	·· 44.2
10	330 39.2	44.6
11	345 39.1	45.0
12	0 39.0	N21 45.3
13	15 39.0	45.7
14	30 38.9	46.1
15	45 38.8	·· 46.5
16	60 38.7	46.8
17	75 38.6	47.2
18	90 38.5	N21 47.6
19	105 38.4	47.9
20	120 38.4	48.3
21	135 38.3	·· 48.7
22	150 38.2	49.0
23	165 38.1	49.4

d h	G.H.A.	Dec.
31 00	180 38.0	N21 49.8
01	195 37.9	50.1
02	210 37.8	50.5
03	225 37.7	·· 50.9
04	240 37.6	51.2
05	255 37.6	51.6
06	270 37.5	N21 51.9
07	285 37.4	52.3
08	300 37.3	52.7
09	315 37.2	·· 53.0
10	330 37.1	53.4
11	345 37.0	53.7
12	0 36.9	N21 54.1
13	15 36.8	54.5
14	30 36.7	54.8
15	45 36.7	·· 55.2
16	60 36.6	55.5
17	75 36.5	55.9
18	90 36.4	N21 56.2
19	105 36.3	56.6
20	120 36.2	56.9
21	135 36.1	·· 57.3
22	150 36.0	57.6
23	165 35.9	58.0
JUNE		
1 00	180 35.8	N21 58.3
01	195 35.7	58.7
02	210 35.6	59.0
03	225 35.6	·· 59.4
04	240 35.5	21 59.7
05	255 35.4	22 00.1
06	270 35.3	N22 00.4
07	285 35.2	00.8
08	300 35.1	01.1
09	315 35.0	·· 01.4
10	330 34.9	01.8
11	345 34.8	02.1
12	0 34.7	N22 02.5
13	15 34.6	02.8
14	30 34.5	03.2
15	45 34.4	·· 03.5
16	60 34.3	03.8
17	75 34.2	04.2
18	90 34.1	N22 04.5
19	105 34.0	04.8
20	120 33.9	05.2
21	135 33.8	·· 05.5
22	150 33.7	05.9
23	165 33.6	06.2
2 00	180 33.6	N22 06.5
01	195 33.5	06.8
02	210 33.4	07.2
03	225 33.3	·· 07.5
04	240 33.2	07.8
05	255 33.1	08.2
06	270 33.0	N22 08.5
07	285 32.9	08.8
08	300 32.8	09.2
09	315 32.7	·· 09.5
10	330 32.6	09.8
11	345 32.5	10.1
12	0 32.4	N22 10.5
13	15 32.3	10.8
14	30 32.2	11.1
15	45 32.1	·· 11.4
16	60 32.0	11.8
17	75 31.9	12.1
18	90 31.8	N22 12.4
19	105 31.7	12.7
20	120 31.6	13.0
21	135 31.5	·· 13.4
22	150 31.4	13.7
23	165 31.3	14.0

d h	G.H.A. ° '	Dec. ° '
3 00	180 31.2	N22 14.3
01	195 31.1	14.6
02	210 31.0	14.9
03	225 30.9	·· 15.3
04	240 30.8	15.6
05	255 30.7	15.9
06	270 30.6	N22 16.2
07	285 30.5	16.5
08	300 30.4	16.8
09	315 30.3	·· 17.1
10	330 30.2	17.4
11	345 30.0	17.8
12	0 29.9	N22 18.1
13	15 29.8	18.4
14	30 29.7	18.7
15	45 29.6	·· 19.0
16	60 29.5	19.3
17	75 29.4	19.6
18	90 29.3	N22 19.9
19	105 29.2	20.2
20	120 29.1	20.5
21	135 29.0	·· 20.8
22	150 28.9	21.1
23	165 28.8	21.4
4 00	180 28.7	N22 21.7
01	195 28.6	22.0
02	210 28.5	22.3
03	225 28.4	·· 22.6
04	240 28.3	22.9
05	255 28.2	23.2
06	270 28.1	N22 23.5
07	285 28.0	23.8
08	300 27.8	24.1
09	315 27.7	·· 24.4
10	330 27.6	24.7
11	345 27.5	25.0
12	0 27.4	N22 25.3
13	15 27.3	25.6
14	30 27.2	25.9
15	45 27.1	·· 26.2
16	60 27.0	26.4
17	75 26.9	26.7
18	90 26.8	N22 27.0
19	105 26.7	27.3
20	120 26.6	27.6
21	135 26.5	·· 27.9
22	150 26.3	28.2
23	165 26.2	28.5
5 00	180 26.1	N22 28.7
01	195 26.0	29.0
02	210 25.9	29.3
03	225 25.8	·· 29.6
04	240 25.7	29.9
05	255 25.6	30.1
06	270 25.5	N22 30.4
07	285 25.4	30.7
08	300 25.2	31.0
09	315 25.1	·· 31.3
10	330 25.0	31.5
11	345 24.9	31.8
12	0 24.8	N22 32.1
13	15 24.7	32.4
14	30 24.6	32.6
15	45 24.5	·· 32.9
16	60 24.4	33.2
17	75 24.3	33.5
18	90 24.1	N22 33.7
19	105 24.0	34.0
20	120 23.9	34.3
21	135 23.8	·· 34.6
22	150 23.7	34.8
23	165 23.6	35.1

d h	G.H.A. ° '	Dec. ° '
6 00	180 23.5	N22 35.4
01	195 23.4	35.6
02	210 23.2	35.9
03	225 23.1	·· 36.2
04	240 23.0	36.4
05	255 22.9	36.7
06	270 22.8	N22 37.0
07	285 22.7	37.2
08	300 22.6	37.5
09	315 22.5	·· 37.7
10	330 22.3	38.0
11	345 22.2	38.3
12	0 22.1	N22 38.5
13	15 22.0	38.8
14	30 21.9	39.0
15	45 21.8	·· 39.3
16	60 21.7	39.6
17	75 21.5	39.8
18	90 21.4	N22 40.1
19	105 21.3	40.3
20	120 21.2	40.6
21	135 21.1	·· 40.8
22	150 21.0	41.1
23	165 20.8	41.3
7 00	180 20.7	N22 41.6
01	195 20.6	41.8
02	210 20.5	42.1
03	225 20.4	·· 42.3
04	240 20.3	42.6
05	255 20.2	42.8
06	270 20.0	N22 43.1
07	285 19.9	43.3
08	300 19.8	43.6
09	315 19.7	·· 43.8
10	330 19.6	44.1
11	345 19.5	44.3
12	0 19.3	N22 44.6
13	15 19.2	44.8
14	30 19.1	45.0
15	45 19.0	·· 45.3
16	60 18.9	45.5
17	75 18.7	45.8
18	90 18.6	N22 46.0
19	105 18.5	46.3
20	120 18.4	46.5
21	135 18.3	·· 46.7
22	150 18.2	46.9
23	165 18.0	47.2
8 00	180 17.9	N22 47.4
01	195 17.8	47.7
02	210 17.7	47.9
03	225 17.6	·· 48.1
04	240 17.4	48.4
05	255 17.3	48.6
06	270 17.2	N22 48.8
07	285 17.1	49.0
08	300 17.0	49.3
09	315 16.9	·· 49.5
10	330 16.7	49.7
11	345 16.6	50.0
12	0 16.5	N22 50.2
13	15 16.4	50.4
14	30 16.3	50.6
15	45 16.1	·· 50.9
16	60 16.0	51.1
17	75 15.9	51.3
18	90 15.8	N22 51.5
19	105 15.6	51.7
20	120 15.5	52.0
21	135 15.4	·· 52.2
22	150 15.3	52.4
23	165 15.2	52.6

d h	G.H.A. ° '	Dec. ° '
9 00	180 15.0	N22 52.8
01	195 14.9	53.1
02	210 14.8	53.3
03	225 14.7	·· 53.5
04	240 14.6	53.7
05	255 14.4	53.9
06	270 14.3	N22 54.1
07	285 14.2	54.4
08	300 14.1	54.6
09	315 13.9	·· 54.8
10	330 13.8	55.0
11	345 13.7	55.2
12	0 13.6	N22 55.4
13	15 13.5	55.6
14	30 13.3	55.8
15	45 13.2	·· 56.0
16	60 13.1	56.2
17	75 13.0	56.5
18	90 12.8	N22 56.7
19	105 12.7	56.9
20	120 12.6	57.1
21	135 12.5	·· 57.3
22	150 12.4	57.5
23	165 12.2	57.7
10 00	180 12.1	N22 57.9
01	195 12.0	58.1
02	210 11.9	58.3
03	225 11.7	·· 58.5
04	240 11.6	58.7
05	255 11.5	58.9
06	270 11.4	N22 59.1
07	285 11.2	59.3
08	300 11.1	59.5
09	315 11.0	·· 59.7
10	330 10.9	22 59.9
11	345 10.7	23 00.0
12	0 10.6	N23 00.2
13	15 10.5	00.4
14	30 10.4	00.6
15	45 10.2	·· 00.8
16	60 10.1	01.0
17	75 10.0	01.2
18	90 09.9	N23 01.4
19	105 09.7	01.6
20	120 09.6	01.8
21	135 09.5	·· 01.9
22	150 09.4	02.1
23	165 09.2	02.3
11 00	180 09.1	N23 02.5
01	195 09.0	02.7
02	210 08.9	02.9
03	225 08.7	·· 03.0
04	240 08.6	03.2
05	255 08.5	03.4
06	270 08.3	N23 03.6
07	285 08.2	03.8
08	300 08.1	04.0
09	315 08.0	·· 04.1
10	330 07.8	04.3
11	345 07.7	04.5
12	0 07.6	N23 04.7
13	15 07.5	04.8
14	30 07.3	05.0
15	45 07.2	·· 05.2
16	60 07.1	05.4
17	75 07.0	05.5
18	90 06.8	N23 05.7
19	105 06.7	05.9
20	120 06.6	06.0
21	135 06.4	·· 06.2
22	150 06.3	06.4
23	165 06.2	06.6

		G.H.A.	Dec.
d 12	h 00	180 06.1 N23	06.7
	01	195 05.9	06.9
	02	210 05.8	07.1
	03	225 05.7 ··	07.2
	04	240 05.5	07.4
	05	255 05.4	07.5
	06	270 05.3 N23	07.7
	07	285 05.2	07.9
	08	300 05.0	08.0
	09	315 04.9 ··	08.2
	10	330 04.8	08.4
	11	345 04.6	08.5
	12	0 04.5 N23	08.7
	13	15 04.4	08.8
	14	30 04.3	09.0
	15	45 04.1 ··	09.1
	16	60 04.0	09.3
	17	75 03.9	09.5
	18	90 03.7 N23	09.6
	19	105 03.6	09.8
	20	120 03.5	09.9
	21	135 03.4 ··	10.1
	22	150 03.2	10.2
	23	165 03.1	10.4
13	00	180 03.0 N23	10.5
	01	195 02.8	10.7
	02	210 02.7	10.8
	03	225 02.6 ··	11.0
	04	240 02.5	11.1
	05	255 02.3	11.3
	06	270 02.2 N23	11.4
	07	285 02.1	11.6
	08	300 01.9	11.7
	09	315 01.8 ··	11.9
	10	330 01.7	12.0
	11	345 01.5	12.1
	12	0 01.4 N23	12.3
	13	15 01.3	12.4
	14	30 01.1	12.6
	15	45 01.0 ··	12.7
	16	60 00.9	12.8
	17	75 00.8	13.0
	18	90 00.6 N23	13.1
	19	105 00.5	13.3
	20	120 00.4	13.4
	21	135 00.2 ··	13.5
	22	150 00.1	13.7
	23	165 00.0	13.8
14	00	179 59.8 N23	13.9
	01	194 59.7	14.1
	02	209 59.6	14.2
	03	224 59.4 ··	14.3
	04	239 59.3	14.5
	05	254 59.2	14.6
	06	269 59.1 N23	14.7
	07	284 58.9	14.8
	08	299 58.8	15.0
	09	314 58.7 ··	15.1
	10	329 58.5	15.2
	11	344 58.4	15.4
	12	359 58.3 N23	15.5
	13	14 58.1	15.6
	14	29 58.0	15.7
	15	44 57.9 ··	15.9
	16	59 57.7	16.0
	17	74 57.6	16.1
	18	89 57.5 N23	16.2
	19	104 57.3	16.3
	20	119 57.2	16.5
	21	134 57.1 ··	16.6
	22	149 56.9	16.7
	23	164 56.8	16.8

		G.H.A.	Dec.
d 15	h 00	179 56.7 N23	16.9
	01	194 56.5	17.0
	02	209 56.4	17.2
	03	224 56.3 ··	17.3
	04	239 56.2	17.4
	05	254 56.0	17.5
	06	269 55.9 N23	17.6
	07	284 55.8	17.7
	08	299 55.6	17.8
	09	314 55.5 ··	17.9
	10	329 55.4	18.1
	11	344 55.2	18.2
	12	359 55.1 N23	18.3
	13	14 55.0	18.4
	14	29 54.8	18.5
	15	44 54.7 ··	18.6
	16	59 54.6	18.7
	17	74 54.4	18.8
	18	89 54.3 N23	18.9
	19	104 54.2	19.0
	20	119 54.0	19.1
	21	134 53.9 ··	19.2
	22	149 53.8	19.3
	23	164 53.6	19.4
16	00	179 53.5 N23	19.5
	01	194 53.4	19.6
	02	209 53.2	19.7
	03	224 53.1 ··	19.8
	04	239 53.0	19.9
	05	254 52.8	20.0
	06	269 52.7 N23	20.1
	07	284 52.6	20.2
	08	299 52.4	20.3
	09	314 52.3 ··	20.4
	10	329 52.2	20.5
	11	344 52.0	20.6
	12	359 51.9 N23	20.6
	13	14 51.8	20.7
	14	29 51.6	20.8
	15	44 51.5 ··	20.9
	16	59 51.4	21.0
	17	74 51.2	21.1
	18	89 51.1 N23	21.2
	19	104 51.0	21.3
	20	119 50.8	21.4
	21	134 50.7 ··	21.4
	22	149 50.6	21.5
	23	164 50.4	21.6
17	00	179 50.3 N23	21.7
	01	194 50.2	21.8
	02	209 50.0	21.8
	03	224 49.9 ··	21.9
	04	239 49.8	22.0
	05	254 49.6	22.1
	06	269 49.5 N23	22.2
	07	284 49.3	22.2
	08	299 49.2	22.3
	09	314 49.1 ··	22.4
	10	329 48.9	22.5
	11	344 48.8	22.5
	12	359 48.7 N23	22.6
	13	14 48.5	22.7
	14	29 48.4	22.8
	15	44 48.3 ··	22.8
	16	59 48.1	22.9
	17	74 48.0	23.0
	18	89 47.9 N23	23.0
	19	104 47.7	23.1
	20	119 47.6	23.2
	21	134 47.5 ··	23.2
	22	149 47.3	23.3
	23	164 47.2	23.4

		G.H.A.	Dec.
d 18	h 00	179 47.1 N23	23.4
	01	194 46.9	23.5
	02	209 46.8	23.6
	03	224 46.7 ··	23.6
	04	239 46.5	23.7
	05	254 46.4	23.8
	06	269 46.3 N23	23.8
	07	284 46.1	23.9
	08	299 46.0	23.9
	09	314 45.9 ··	24.0
	10	329 45.7	24.1
	11	344 45.6	24.1
	12	359 45.4 N23	24.2
	13	14 45.3	24.2
	14	29 45.2	24.3
	15	44 45.0 ··	24.3
	16	59 44.9	24.4
	17	74 44.8	24.4
	18	89 44.6 N23	24.5
	19	104 44.5	24.5
	20	119 44.4	24.6
	21	134 44.2 ··	24.6
	22	149 44.1	24.7
	23	164 44.0	24.7
19	00	179 43.8 N23	24.8
	01	194 43.7	24.8
	02	209 43.6	24.9
	03	224 43.4 ··	24.9
	04	239 43.3	25.0
	05	254 43.2	25.0
	06	269 43.0 N23	25.1
	07	284 42.9	25.1
	08	299 42.7	25.1
	09	314 42.6 ··	25.2
	10	329 42.5	25.2
	11	344 42.3	25.3
	12	359 42.2 N23	25.3
	13	14 42.1	25.3
	14	29 41.9	25.4
	15	44 41.8 ··	25.4
	16	59 41.7	25.5
	17	74 41.5	25.5
	18	89 41.4 N23	25.5
	19	104 41.3	25.6
	20	119 41.1	25.6
	21	134 41.0 ··	25.6
	22	149 40.9	25.7
	23	164 40.7	25.7
20	00	179 40.6 N23	25.7
	01	194 40.5	25.8
	02	209 40.3	25.8
	03	224 40.2 ··	25.8
	04	239 40.0	25.8
	05	254 39.9	25.9
	06	269 39.8 N23	25.9
	07	284 39.6	25.9
	08	299 39.5	25.9
	09	314 39.4 ··	26.0
	10	329 39.2	26.0
	11	344 39.1	26.0
	12	359 39.0 N23	26.0
	13	14 38.8	26.1
	14	29 38.7	26.1
	15	44 38.6 ··	26.1
	16	59 38.4	26.1
	17	74 38.3	26.1
	18	89 38.2 N23	26.2
	19	104 38.0	26.2
	20	119 37.9	26.2
	21	134 37.7 ··	26.2
	22	149 37.6	26.2
	23	164 37.5	26.2

21, 22, 23

d h	G.H.A. ° ′	Dec. ° ′
21 00	179 37.3	N23 26.2
01	194 37.2	26.3
02	209 37.1	26.3
03	224 36.9 ··	26.3
04	239 36.8	26.3
05	254 36.7	26.3
06	269 36.5	N23 26.3
07	284 36.4	26.3
08	299 36.3	26.3
09	314 36.1 ··	26.3
10	329 36.0	26.3
11	344 35.9	26.3
12	359 35.7	N23 26.4
13	14 35.6	26.4
14	29 35.5	26.4
15	44 35.3 ··	26.4
16	59 35.2	26.4
17	74 35.0	26.4
18	89 34.9	N23 26.4
19	104 34.8	26.4
20	119 34.6	26.4
21	134 34.5 ··	26.4
22	149 34.4	26.4
23	164 34.2	26.4
22 00	179 34.1	N23 26.4
01	194 34.0	26.4
02	209 33.8	26.3
03	224 33.7 ··	26.3
04	239 33.6	26.3
05	254 33.4	26.3
06	269 33.3	N23 26.3
07	284 33.2	26.3
08	299 33.0	26.3
09	314 32.9 ··	26.3
10	329 32.8	26.3
11	344 32.6	26.3
12	359 32.5	N23 26.3
13	14 32.3	26.2
14	29 32.2	26.2
15	44 32.1 ··	26.2
16	59 31.9	26.2
17	74 31.8	26.2
18	89 31.7	N23 26.2
19	104 31.5	26.1
20	119 31.4	26.1
21	134 31.3 ··	26.1
22	149 31.1	26.1
23	164 31.0	26.1
23 00	179 30.9	N23 26.0
01	194 30.7	26.0
02	209 30.6	26.0
03	224 30.5 ··	26.0
04	239 30.3	26.0
05	254 30.2	25.9
06	269 30.1	N23 25.9
07	284 29.9	25.9
08	299 29.8	25.9
09	314 29.7 ··	25.8
10	329 29.5	25.8
11	344 29.4	25.8
12	359 29.3	N23 25.7
13	14 29.1	25.7
14	29 29.0	25.7
15	44 28.9 ··	25.6
16	59 28.7	25.6
17	74 28.6	25.6
18	89 28.4	N23 25.5
19	104 28.3	25.5
20	119 28.2	25.5
21	134 28.0 ··	25.4
22	149 27.9	25.4
23	164 27.8	25.4

24, 25, 26

d h	G.H.A. ° ′	Dec. ° ′
24 00	179 27.6	N23 25.3
01	194 27.5	25.3
02	209 27.4	25.3
03	224 27.2 ··	25.2
04	239 27.1	25.2
05	254 27.0	25.1
06	269 26.8	N23 25.1
07	284 26.7	25.0
08	299 26.6	25.0
09	314 26.4 ··	25.0
10	329 26.3	24.9
11	344 26.2	24.9
12	359 26.0	N23 24.8
13	14 25.9	24.8
14	29 25.8	24.7
15	44 25.6 ··	24.7
16	59 25.5	24.6
17	74 25.4	24.6
18	89 25.2	N23 24.5
19	104 25.1	24.5
20	119 25.0	24.4
21	134 24.8 ··	24.4
22	149 24.7	24.3
23	164 24.6	24.3
25 00	179 24.4	N23 24.2
01	194 24.3	24.1
02	209 24.2	24.1
03	224 24.0 ··	24.0
04	239 23.9	24.0
05	254 23.8	23.9
06	269 23.6	N23 23.8
07	284 23.5	23.8
08	299 23.4	23.7
09	314 23.2 ··	23.7
10	329 23.1	23.6
11	344 23.0	23.5
12	359 22.8	N23 23.5
13	14 22.7	23.4
14	29 22.6	23.3
15	44 22.4 ··	23.3
16	59 22.3	23.2
17	74 22.2	23.1
18	89 22.0	N23 23.1
19	104 21.9	23.0
20	119 21.8	22.9
21	134 21.7 ··	22.9
22	149 21.5	22.8
23	164 21.4	22.7
26 00	179 21.3	N23 22.7
01	194 21.1	22.6
02	209 21.0	22.5
03	224 20.9 ··	22.4
04	239 20.7	22.4
05	254 20.6	22.3
06	269 20.5	N23 22.2
07	284 20.3	22.1
08	299 20.2	22.0
09	314 20.1 ··	22.0
10	329 19.9	21.9
11	344 19.8	21.8
12	359 19.7	N23 21.7
13	14 19.5	21.6
14	29 19.4	21.6
15	44 19.3 ··	21.5
16	59 19.2	21.4
17	74 19.0	21.3
18	89 18.9	N23 21.2
19	104 18.8	21.1
20	119 18.6	21.1
21	134 18.5 ··	21.0
22	149 18.4	20.9
23	164 18.2	20.8

27, 28, 29

d h	G.H.A. ° ′	Dec. ° ′
27 00	179 18.1	N23 20.7
01	194 18.0	20.6
02	209 17.8	20.5
03	224 17.7 ··	20.4
04	239 17.6	20.3
05	254 17.4	20.2
06	269 17.3	N23 20.1
07	284 17.2	20.0
08	299 17.1	20.0
09	314 16.9 ··	19.9
10	329 16.8	19.8
11	344 16.7	19.7
12	359 16.5	N23 19.6
13	14 16.4	19.5
14	29 16.3	19.4
15	44 16.1 ··	19.3
16	59 16.0	19.2
17	74 15.9	19.1
18	89 15.8	N23 19.0
19	104 15.6	18.9
20	119 15.5	18.8
21	134 15.4 ··	18.6
22	149 15.2	18.5
23	164 15.1	18.4
28 00	179 15.0	N23 18.3
01	194 14.9	18.2
02	209 14.7	18.1
03	224 14.6 ··	18.0
04	239 14.5	17.9
05	254 14.3	17.8
06	269 14.2	N23 17.7
07	284 14.1	17.6
08	299 13.9	17.4
09	314 13.8 ··	17.3
10	329 13.7	17.2
11	344 13.6	17.1
12	359 13.4	N23 17.0
13	14 13.3	16.9
14	29 13.2	16.8
15	44 13.1 ··	16.6
16	59 12.9	16.5
17	74 12.8	16.4
18	89 12.7	N23 16.3
19	104 12.5	16.2
20	119 12.4	16.0
21	134 12.3 ··	15.9
22	149 12.2	15.8
23	164 12.0	15.7
29 00	179 11.9	N23 15.6
01	194 11.8	15.4
02	209 11.6	15.3
03	224 11.5 ··	15.2
04	239 11.4	15.0
05	254 11.3	14.9
06	269 11.1	N23 14.8
07	284 11.0	14.7
08	299 10.9	14.5
09	314 10.8 ··	14.4
10	329 10.6	14.3
11	344 10.5	14.1
12	359 10.4	N23 14.0
13	14 10.3	13.9
14	29 10.1	13.7
15	44 10.0 ··	13.6
16	59 09.9	13.5
17	74 09.7	13.3
18	89 09.6	N23 13.2
19	104 09.5	13.1
20	119 09.4	12.9
21	134 09.2 ··	12.8
22	149 09.1	12.6
23	164 09.0	12.5

d h	G.H.A.	Dec.
30 00	179 08.9	N23 12.4
01	194 08.7	12.2
02	209 08.6	12.1
03	224 08.5	·· 11.9
04	239 08.4	11.8
05	254 08.2	11.6
06	269 08.1	N23 11.5
07	284 08.0	11.4
08	299 07.9	11.2
09	314 07.7	·· 11.1
10	329 07.6	10.9
11	344 07.5	10.8
12	359 07.4	N23 10.6
13	14 07.2	10.5
14	29 07.1	10.3
15	44 07.0	·· 10.2
16	59 06.9	10.0
17	74 06.7	09.9
18	89 06.6	N23 09.7
19	104 06.5	09.6
20	119 06.4	09.4
21	134 06.3	·· 09.2
22	149 06.1	09.1
23	164 06.0	08.9
1 00	179 05.9	N23 08.8
01	194 05.8	08.6
02	209 05.6	08.5
03	224 05.5	·· 08.3
04	239 05.4	08.1
05	254 05.3	08.0
06	269 05.1	N23 07.8
07	284 05.0	07.6
08	299 04.9	07.5
09	314 04.8	·· 07.3
10	329 04.7	07.1
11	344 04.5	07.0
12	359 04.4	N23 06.8
13	14 04.3	06.7
14	29 04.2	06.5
15	44 04.0	·· 06.3
16	59 03.9	06.1
17	74 03.8	06.0
18	89 03.7	N23 05.8
19	104 03.6	05.6
20	119 03.4	05.5
21	134 03.3	·· 05.3
22	149 03.2	05.1
23	164 03.1	04.9
2 00	179 03.0	N23 04.8
01	194 02.8	04.6
02	209 02.7	04.4
03	224 02.6	·· 04.2
04	239 02.5	04.1
05	254 02.4	03.9
06	269 02.2	N23 03.7
07	284 02.1	03.5
08	299 02.0	03.3
09	314 01.9	·· 03.2
10	329 01.8	03.0
11	344 01.6	02.8
12	359 01.5	N23 02.6
13	14 01.4	02.4
14	29 01.3	02.2
15	44 01.2	·· 02.1
16	59 01.0	01.9
17	74 00.9	01.7
18	89 00.8	N23 01.5
19	104 00.7	01.3
20	119 00.6	01.1
21	134 00.4	·· 00.9
22	149 00.3	00.7
23	164 00.2	00.6

d h	G.H.A.	Dec.
3 00	179 00.1	N23 00.4
01	194 00.0	00.2
02	208 59.8	23 00.0
03	223 59.7	22 59.8
04	238 59.6	59.6
05	253 59.5	59.4
06	268 59.4	N22 59.2
07	283 59.3	59.0
08	298 59.1	58.8
09	313 59.0	·· 58.6
10	328 58.9	58.4
11	343 58.8	58.2
12	358 58.7	N22 58.0
13	13 58.6	57.8
14	28 58.4	57.6
15	43 58.3	·· 57.4
16	58 58.2	57.2
17	73 58.1	57.0
18	88 58.0	N22 56.8
19	103 57.9	56.6
20	118 57.8	56.4
21	133 57.6	·· 56.2
22	148 57.5	56.0
23	163 57.4	55.8
4 00	178 57.3	N22 55.6
01	193 57.2	55.4
02	208 57.1	55.1
03	223 56.9	·· 54.9
04	238 56.8	54.7
05	253 56.7	54.5
06	268 56.6	N22 54.3
07	283 56.5	54.1
08	298 56.4	53.9
09	313 56.3	·· 53.7
10	328 56.2	53.4
11	343 56.0	53.2
12	358 55.9	N22 53.0
13	13 55.8	52.8
14	28 55.7	52.6
15	43 55.6	·· 52.4
16	58 55.5	52.1
17	73 55.4	51.9
18	88 55.2	N22 51.7
19	103 55.1	51.5
20	118 55.0	51.2
21	133 54.9	·· 51.0
22	148 54.8	50.8
23	163 54.7	50.6
5 00	178 54.6	N22 50.4
01	193 54.5	50.1
02	208 54.4	49.9
03	223 54.2	·· 49.7
04	238 54.1	49.4
05	253 54.0	49.2
06	268 53.9	N22 49.0
07	283 53.8	48.8
08	298 53.7	48.5
09	313 53.6	·· 48.3
10	328 53.5	48.1
11	343 53.4	47.8
12	358 53.2	N22 47.6
13	13 53.1	47.4
14	28 53.0	47.1
15	43 52.9	·· 46.9
16	58 52.8	46.7
17	73 52.7	46.4
18	88 52.6	N22 46.2
19	103 52.5	45.9
20	118 52.4	45.7
21	133 52.3	·· 45.5
22	148 52.2	45.2
23	163 52.0	45.0

d h	G.H.A.	Dec.
6 00	178 51.9	N22 44.7
01	193 51.8	44.5
02	208 51.7	44.3
03	223 51.6	·· 44.0
04	238 51.5	43.8
05	253 51.4	43.5
06	268 51.3	N22 43.3
07	283 51.2	43.0
08	298 51.1	42.8
09	313 51.0	·· 42.5
10	328 50.9	42.3
11	343 50.8	42.1
12	358 50.6	N22 41.8
13	13 50.5	41.6
14	28 50.4	41.3
15	43 50.3	·· 41.0
16	58 50.2	40.8
17	73 50.1	40.5
18	88 50.0	N22 40.3
19	103 49.9	40.0
20	118 49.8	39.8
21	133 49.7	·· 39.5
22	148 49.6	39.3
23	163 49.5	39.0
7 00	178 49.4	N22 38.8
01	193 49.3	38.5
02	208 49.2	38.2
03	223 49.1	·· 38.0
04	238 49.0	37.7
05	253 48.9	37.5
06	268 48.8	N22 37.2
07	283 48.7	36.9
08	298 48.6	36.7
09	313 48.4	·· 36.4
10	328 48.3	36.1
11	343 48.2	35.9
12	358 48.1	N22 35.6
13	13 48.0	35.3
14	28 47.9	35.1
15	43 47.8	·· 34.8
16	58 47.7	34.5
17	73 47.6	34.3
18	88 47.5	N22 34.0
19	103 47.4	33.7
20	118 47.3	33.5
21	133 47.2	·· 33.2
22	148 47.1	32.9
23	163 47.0	32.6
8 00	178 46.9	N22 32.4
01	193 46.8	32.1
02	208 46.7	31.8
03	223 46.6	·· 31.5
04	238 46.5	31.3
05	253 46.4	31.0
06	268 46.3	N22 30.7
07	283 46.2	30.4
08	298 46.1	30.2
09	313 46.0	29.9
10	328 45.9	29.6
11	343 45.8	29.3
12	358 45.7	N22 29.0
13	13 45.6	28.8
14	28 45.5	28.5
15	43 45.4	·· 28.2
16	58 45.3	27.9
17	73 45.2	27.6
18	88 45.1	N22 27.3
19	103 45.0	27.0
20	118 44.9	26.8
21	133 44.8	·· 26.5
22	148 44.8	26.2
23	163 44.7	25.9

Day 9

d h	G.H.A.	Dec.
9 00	178 44.6	N22 25.6
01	193 44.5	25.3
02	208 44.4	25.0
03	223 44.3	·· 24.7
04	238 44.2	24.4
05	253 44.1	24.1
06	268 44.0	N22 23.8
07	283 43.9	23.6
08	298 43.8	23.3
09	313 43.7	·· 23.0
10	328 43.6	22.7
11	343 43.5	22.4
12	358 43.4	N22 22.1
13	13 43.3	21.8
14	28 43.2	21.5
15	43 43.1	·· 21.2
16	58 43.0	20.9
17	73 42.9	20.6
18	88 42.9	N22 20.3
19	103 42.8	20.0
20	118 42.7	19.7
21	133 42.6	·· 19.4
22	148 42.5	19.1
23	163 42.4	18.7
10 00	178 42.3	N22 18.4
01	193 42.2	18.1
02	208 42.1	17.8
03	223 42.0	·· 17.5
04	238 41.9	17.2
05	253 41.8	16.9
06	268 41.8	N22 16.6
07	283 41.7	16.3
08	298 41.6	16.0
09	313 41.5	·· 15.7
10	328 41.4	15.3
11	343 41.3	15.0
12	358 41.2	N22 14.7
13	13 41.1	14.4
14	28 41.0	14.1
15	43 40.9	·· 13.8
16	58 40.9	13.5
17	73 40.8	13.1
18	88 40.7	N22 12.8
19	103 40.6	12.5
20	118 40.5	12.2
21	133 40.4	·· 11.9
22	148 40.3	11.5
23	163 40.2	11.2
11 00	178 40.1	N22 10.9
01	193 40.1	10.6
02	208 40.0	10.3
03	223 39.9	·· 09.9
04	238 39.8	09.6
05	253 39.7	09.3
06	268 39.6	N22 09.0
07	283 39.5	08.6
08	298 39.5	08.3
09	313 39.4	·· 08.0
10	328 39.3	07.6
11	343 39.2	07.3
12	358 39.1	N22 07.0
13	13 39.0	06.7
14	28 38.9	06.3
15	43 38.9	·· 06.0
16	58 38.8	05.7
17	73 38.7	05.3
18	88 38.6	N22 05.0
19	103 38.5	04.7
20	118 38.4	04.3
21	133 38.4	·· 04.0
22	148 38.3	03.7
23	163 38.2	03.3

Day 12

d h	G.H.A.	Dec.
12 00	178 38.1	N22 03.0
01	193 38.0	02.6
02	208 38.0	02.3
03	223 37.9	·· 02.0
04	238 37.8	01.6
05	253 37.7	01.3
06	268 37.6	N22 00.9
07	283 37.5	00.6
08	298 37.5	22 00.3
09	313 37.4	21 59.9
10	328 37.3	59.6
11	343 37.2	59.2
12	358 37.1	N21 58.9
13	13 37.1	58.5
14	28 37.0	58.2
15	43 36.9	·· 57.8
16	58 36.8	57.5
17	73 36.7	57.1
18	88 36.7	N21 56.8
19	103 36.6	56.4
20	118 36.5	56.1
21	133 36.4	·· 55.7
22	148 36.4	55.4
23	163 36.3	55.0
13 00	178 36.2	N21 54.7
01	193 36.1	54.3
02	208 36.0	54.0
03	223 36.0	·· 53.6
04	238 35.9	53.3
05	253 35.8	52.9
06	268 35.7	N21 52.5
07	283 35.7	52.2
08	298 35.6	51.8
09	313 35.5	·· 51.5
10	328 35.4	51.1
11	343 35.4	50.7
12	358 35.3	N21 50.4
13	13 35.2	50.0
14	28 35.1	49.7
15	43 35.1	·· 49.3
16	58 35.0	48.9
17	73 34.9	48.6
18	88 34.8	N21 48.2
19	103 34.8	47.8
20	118 34.7	47.5
21	133 34.6	·· 47.1
22	148 34.6	46.7
23	163 34.5	46.4
14 00	178 34.4	N21 46.0
01	193 34.3	45.6
02	208 34.3	45.3
03	223 34.2	·· 44.9
04	238 34.1	44.5
05	253 34.1	44.1
06	268 34.0	N21 43.8
07	283 33.9	43.4
08	298 33.8	43.0
09	313 33.8	·· 42.6
10	328 33.7	42.3
11	343 33.6	41.9
12	358 33.6	N21 41.5
13	13 33.5	41.1
14	28 33.4	40.8
15	43 33.4	·· 40.4
16	58 33.3	40.0
17	73 33.2	39.6
18	88 33.1	N21 39.2
19	103 33.1	38.9
20	118 33.0	38.5
21	133 32.9	·· 38.1
22	148 32.9	37.7
23	163 32.8	37.3

Day 15

d h	G.H.A.	Dec.
15 00	178 32.7	N21 36.9
01	193 32.7	36.6
02	208 32.6	36.2
03	223 32.5	·· 35.8
04	238 32.5	35.4
05	253 32.4	35.0
06	268 32.3	N21 34.6
07	283 32.3	34.2
08	298 32.2	33.8
09	313 32.2	·· 33.5
10	328 32.1	33.1
11	343 32.0	32.7
12	358 32.0	N21 32.3
13	13 31.9	31.9
14	28 31.8	31.5
15	43 31.8	·· 31.1
16	58 31.7	30.7
17	73 31.6	30.3
18	88 31.6	N21 29.9
19	103 31.5	29.5
20	118 31.5	29.1
21	133 31.4	·· 28.7
22	148 31.3	28.3
23	163 31.3	27.9
16 00	178 31.2	N21 27.5
01	193 31.1	27.1
02	208 31.1	26.7
03	223 31.0	·· 26.3
04	238 31.0	25.9
05	253 30.9	25.5
06	268 30.8	N21 25.1
07	283 30.8	24.7
08	298 30.7	24.3
09	313 30.7	·· 23.9
10	328 30.6	23.5
11	343 30.5	23.1
12	358 30.5	N21 22.7
13	13 30.4	22.3
14	28 30.4	21.9
15	43 30.3	·· 21.5
16	58 30.3	21.0
17	73 30.2	20.6
18	88 30.1	N21 20.2
19	103 30.1	19.8
20	118 30.0	19.4
21	133 30.0	·· 19.0
22	148 29.9	18.6
23	163 29.9	18.2
17 00	178 29.8	N21 17.7
01	193 29.7	17.3
02	208 29.7	16.9
03	223 29.6	·· 16.5
04	238 29.6	16.1
05	253 29.5	15.7
06	268 29.5	N21 15.2
07	283 29.4	14.8
08	298 29.4	14.4
09	313 29.3	·· 14.0
10	328 29.3	13.6
11	343 29.2	13.1
12	358 29.1	N21 12.7
13	13 29.1	12.3
14	28 29.0	11.9
15	43 29.0	·· 11.4
16	58 28.9	11.0
17	73 28.9	10.6
18	88 28.8	N21 10.2
19	103 28.8	09.7
20	118 28.7	09.3
21	133 28.7	·· 08.9
22	148 28.6	08.4
23	163 28.6	08.0

		G.H.A.	Dec.			G.H.A.	Dec.			G.H.A.	Dec.
d h		° ′	° ′	d h		° ′	° ′	d h		° ′	° ′
18	00	178 28.5	N21 07.6	21	00	178 25.5	N20 35.0	24	00	178 23.7	N19 59.3
	01	193 28.5	07.2		01	193 25.5	34.5		01	193 23.7	58.8
	02	208 28.4	06.7		02	208 25.4	34.1		02	208 23.7	58.3
	03	223 28.4 ··	06.3		03	223 25.4 ··	33.6		03	223 23.7 ··	57.8
	04	238 28.3	05.9		04	238 25.4	33.1		04	238 23.7	57.2
	05	253 28.3	05.4		05	253 25.4	32.6		05	253 23.7	56.7
	06	268 28.2	N21 05.0		06	268 25.3	N20 32.2		06	268 23.7	N19 56.2
	07	283 28.2	04.6		07	283 25.3	31.7		07	283 23.6	55.7
	08	298 28.1	04.1		08	298 25.3	31.2		08	298 23.6	55.2
	09	313 28.1 ··	03.7		09	313 25.2 ··	30.7		09	313 23.6 ··	54.6
	10	328 28.0	03.3		10	328 25.2	30.2		10	328 23.6	54.1
	11	343 28.0	02.8		11	343 25.2	29.8		11	343 23.6	53.6
	12	358 27.9	N21 02.4		12	358 25.1	N20 29.3		12	358 23.6	N19 53.1
	13	13 27.9	01.9		13	13 25.1	28.8		13	13 23.6	52.5
	14	28 27.8	01.5		14	28 25.1	28.3		14	28 23.6	52.0
	15	43 27.8 ··	01.1		15	43 25.0 ··	27.8		15	43 23.5 ··	51.5
	16	58 27.7	00.6		16	58 25.0	27.3		16	58 23.5	51.0
	17	73 27.7	21 00.2		17	73 25.0	26.9		17	73 23.5	50.4
	18	88 27.7	N20 59.7		18	88 25.0	N20 26.4		18	88 23.5	N19 49.9
	19	103 27.6	59.3		19	103 24.9	25.9		19	103 23.5	49.4
	20	118 27.6	58.9		20	118 24.9	25.4		20	118 23.5	48.9
	21	133 27.5 ··	58.4		21	133 24.9 ··	24.9		21	133 23.5 ··	48.3
	22	148 27.5	58.0		22	148 24.8	24.4		22	148 23.5	47.8
	23	163 27.4	57.5		23	163 24.8	23.9		23	163 23.5	47.3
19	00	178 27.4	N20 57.1	22	00	178 24.8	N20 23.5	25	00	178 23.4	N19 46.7
	01	193 27.3	56.6		01	193 24.8	23.0		01	193 23.4	46.2
	02	208 27.3	56.2		02	208 24.7	22.5		02	208 23.4	45.7
	03	223 27.3 ··	55.7		03	223 24.7 ··	22.0		03	223 23.4 ··	45.1
	04	238 27.2	55.3		04	238 24.7	21.5		04	238 23.4	44.6
	05	253 27.2	54.8		05	253 24.7	21.0		05	253 23.4	44.1
	06	268 27.1	N20 54.4		06	268 24.6	N20 20.5		06	268 23.4	N19 43.5
	07	283 27.1	54.0		07	283 24.6	20.0		07	283 23.4	43.0
	08	298 27.0	53.5		08	298 24.6	19.5		08	298 23.4	42.5
	09	313 27.0 ··	53.1		09	313 24.5 ··	19.0		09	313 23.4 ··	41.9
	10	328 26.9	52.6		10	328 24.5	18.5		10	328 23.4	41.4
	11	343 26.9	52.1		11	343 24.5	18.0		11	343 23.4	40.9
	12	358 26.9	N20 51.7		12	358 24.5	N20 17.5		12	358 23.3	N19 40.3
	13	13 26.8	51.2		13	13 24.4	17.0		13	13 23.3	39.8
	14	28 26.8	50.8		14	28 24.4	16.5		14	28 23.3	39.3
	15	43 26.7 ··	50.3		15	43 24.4 ··	16.1		15	43 23.3 ··	38.7
	16	58 26.7	49.9		16	58 24.4	15.6		16	58 23.3	38.2
	17	73 26.7	49.4		17	73 24.4	15.1		17	73 23.3	37.6
	18	88 26.6	N20 49.0		18	88 24.3	N20 14.6		18	88 23.3	N19 37.1
	19	103 26.6	48.5		19	103 24.3	14.1		19	103 23.3	36.6
	20	118 26.5	48.1		20	118 24.3	13.6		20	118 23.3	36.0
	21	133 26.5 ··	47.6		21	133 24.3 ··	13.1		21	133 23.3 ··	35.5
	22	148 26.5	47.1		22	148 24.2	12.6		22	148 23.3	34.9
	23	163 26.4	46.7		23	163 24.2	12.1		23	163 23.3	34.4
20	00	178 26.4	N20 46.2	23	00	178 24.2	N20 11.5	26	00	178 23.3	N19 33.8
	01	193 26.3	45.8		01	193 24.2	11.0		01	193 23.3	33.3
	02	208 26.3	45.3		02	208 24.2	10.5		02	208 23.3	32.7
	03	223 26.3 ··	44.8		03	223 24.1 ··	10.0		03	223 23.3 ··	32.2
	04	238 26.2	44.4		04	238 24.1	09.5		04	238 23.3	31.7
	05	253 26.2	43.9		05	253 24.1	09.0		05	253 23.3	31.1
	06	268 26.2	N20 43.5		06	268 24.1	N20 08.5		06	268 23.3	N19 30.6
	07	283 26.1	43.0		07	283 24.1	08.0		07	283 23.3	30.0
	08	298 26.1	42.5		08	298 24.0	07.5		08	298 23.3	29.5
	09	313 26.0 ··	42.1		09	313 24.0 ··	07.0		09	313 23.3 ··	28.9
	10	328 26.0	41.6		10	328 24.0	06.5		10	328 23.3	28.4
	11	343 26.0	41.1		11	343 24.0	06.0		11	343 23.3	27.8
	12	358 25.9	N20 40.7		12	358 24.0	N20 05.5		12	358 23.3	N19 27.3
	13	13 25.9	40.2		13	13 23.9	05.0		13	13 23.3	26.7
	14	28 25.9	39.7		14	28 23.9	04.5		14	28 23.3	26.2
	15	43 25.8 ··	39.3		15	43 23.9 ··	03.9		15	43 23.3 ··	25.6
	16	58 25.8	38.8		16	58 23.9	03.4		16	58 23.3	25.1
	17	73 25.8	38.3		17	73 23.9	02.9		17	73 23.3	24.5
	18	88 25.7	N20 37.8		18	88 23.8	N20 02.4		18	88 23.3	N19 23.9
	19	103 25.7	37.4		19	103 23.8	01.9		19	103 23.3	23.4
	20	118 25.6	36.9		20	118 23.8	01.4		20	118 23.3	22.8
	21	133 25.6 ··	36.4		21	133 23.8 ··	00.9		21	133 23.3 ··	22.3
	22	148 25.6	36.0		22	148 23.8	20 00.3		22	148 23.3	21.7
	23	163 25.5	35.5		23	163 23.8	19 59.8		23	163 23.3	21.2

27 — G.H.A. / Dec.

d/h	G.H.A.	Dec.
00	178 23.3	N19 20.6
01	193 23.3	20.1
02	208 23.3	19.5
03	223 23.3	·· 18.9
04	238 23.3	18.4
05	253 23.3	17.8
06	268 23.3	N19 17.3
07	283 23.3	16.7
08	298 23.3	16.1
09	313 23.3	·· 15.6
10	328 23.3	15.0
11	343 23.3	14.4
12	358 23.3	N19 13.9
13	13 23.3	13.3
14	28 23.3	12.7
15	43 23.3	·· 12.2
16	58 23.3	11.6
17	73 23.3	11.0
18	88 23.3	N19 10.5
19	103 23.3	09.9
20	118 23.4	09.3
21	133 23.4	·· 08.8
22	148 23.4	08.2
23	163 23.4	07.6

28

d/h	G.H.A.	Dec.
00	178 23.4	N19 07.1
01	193 23.4	06.5
02	208 23.4	05.9
03	223 23.4	·· 05.3
04	238 23.4	04.8
05	253 23.4	04.2
06	268 23.4	N19 03.6
07	283 23.4	03.1
08	298 23.5	02.5
09	313 23.5	·· 01.9
10	328 23.5	01.3
11	343 23.5	00.8
12	358 23.5	N19 00.2
13	13 23.5	18 59.6
14	28 23.5	59.0
15	43 23.5	·· 58.4
16	58 23.6	57.9
17	73 23.6	57.3
18	88 23.6	N18 56.7
19	103 23.6	56.1
20	118 23.6	55.5
21	133 23.6	·· 55.0
22	148 23.6	54.4
23	163 23.6	53.8

29

d/h	G.H.A.	Dec.
00	178 23.7	N18 53.2
01	193 23.7	52.6
02	208 23.7	52.0
03	223 23.7	·· 51.4
04	238 23.7	50.9
05	253 23.7	50.3
06	268 23.7	N18 49.7
07	283 23.8	49.1
08	298 23.8	48.5
09	313 23.8	·· 47.9
10	328 23.8	47.3
11	343 23.8	46.7
12	358 23.8	N18 46.2
13	13 23.9	45.6
14	28 23.9	45.0
15	43 23.9	·· 44.4
16	58 23.9	43.8
17	73 23.9	43.2
18	88 24.0	N18 42.6
19	103 24.0	42.0
20	118 24.0	41.4
21	133 24.0	·· 40.8
22	148 24.0	40.2
23	163 24.1	39.6

30

d/h	G.H.A.	Dec.
00	178 24.1	N18 39.0
01	193 24.1	38.4
02	208 24.1	37.8
03	223 24.1	·· 37.2
04	238 24.2	36.6
05	253 24.2	36.0
06	268 24.2	N18 35.4
07	283 24.2	34.8
08	298 24.2	34.2
09	313 24.3	·· 33.6
10	328 24.3	33.0
11	343 24.3	32.4
12	358 24.3	N18 31.8
13	13 24.4	31.2
14	28 24.4	30.6
15	43 24.4	·· 30.0
16	58 24.4	29.4
17	73 24.5	28.8
18	88 24.5	N18 28.2
19	103 24.5	27.6
20	118 24.5	27.0
21	133 24.6	·· 26.4
22	148 24.6	25.8
23	163 24.6	25.2

31

d/h	G.H.A.	Dec.
00	178 24.6	N18 24.6
01	193 24.7	23.9
02	208 24.7	23.3
03	223 24.7	·· 22.7
04	238 24.8	22.1
05	253 24.8	21.5
06	268 24.8	N18 20.9
07	283 24.8	20.3
08	298 24.9	19.7
09	313 24.9	·· 19.0
10	328 24.9	18.4
11	343 25.0	17.8
12	358 25.0	N18 17.2
13	13 25.0	16.6
14	28 25.0	16.0
15	43 25.1	·· 15.4
16	58 25.1	14.7
17	73 25.1	14.1
18	88 25.2	N18 13.5
19	103 25.2	12.9
20	118 25.2	12.3
21	133 25.3	·· 11.6
22	148 25.3	11.0
23	163 25.3	10.4

AUGUST 1

d/h	G.H.A.	Dec.
00	178 25.4	N18 09.8
01	193 25.4	09.2
02	208 25.4	08.5
03	223 25.5	·· 07.9
04	238 25.5	07.3
05	253 25.5	06.7
06	268 25.6	N18 06.0
07	283 25.6	05.4
08	298 25.6	04.8
09	313 25.7	·· 04.2
10	328 25.7	03.5
11	343 25.7	02.9
12	358 25.8	N18 02.3
13	13 25.8	01.6
14	28 25.8	01.0
15	43 25.9	18 00.4
16	58 25.9	17 59.8
17	73 26.0	59.1
18	88 26.0	N17 58.5
19	103 26.0	57.9
20	118 26.1	57.2
21	133 26.1	·· 56.6
22	148 26.1	56.0
23	163 26.2	55.3

2

d/h	G.H.A.	Dec.
00	178 26.2	N17 54.7
01	193 26.3	54.1
02	208 26.3	53.4
03	223 26.3	·· 52.8
04	238 26.4	52.2
05	253 26.4	51.5
06	268 26.5	N17 50.9
07	283 26.5	50.2
08	298 26.5	49.6
09	313 26.6	·· 49.0
10	328 26.6	48.3
11	343 26.7	47.7
12	358 26.7	N17 47.1
13	13 26.8	46.4
14	28 26.8	45.8
15	43 26.8	·· 45.1
16	58 26.9	44.5
17	73 26.9	43.8
18	88 27.0	N17 43.2
19	103 27.0	42.6
20	118 27.1	41.9
21	133 27.1	·· 41.3
22	148 27.2	40.6
23	163 27.2	40.0

3

d/h	G.H.A.	Dec.
00	178 27.2	N17 39.3
01	193 27.3	38.7
02	208 27.3	38.0
03	223 27.4	·· 37.4
04	238 27.4	36.7
05	253 27.5	36.1
06	268 27.5	N17 35.4
07	283 27.6	34.8
08	298 27.6	34.1
09	313 27.7	·· 33.5
10	328 27.7	32.8
11	343 27.8	32.2
12	358 27.8	N17 31.5
13	13 27.9	30.9
14	28 27.9	30.2
15	43 28.0	·· 29.6
16	58 28.0	28.9
17	73 28.1	28.3
18	88 28.1	N17 27.6
19	103 28.2	27.0
20	118 28.2	26.3
21	133 28.3	·· 25.6
22	148 28.3	25.0
23	163 28.4	24.3

4

d/h	G.H.A.	Dec.
00	178 28.4	N17 23.7
01	193 28.5	23.0
02	208 28.5	22.4
03	223 28.6	·· 21.7
04	238 28.6	21.0
05	253 28.7	20.4
06	268 28.7	N17 19.7
07	283 28.8	19.1
08	298 28.8	18.4
09	313 28.9	·· 17.7
10	328 28.9	17.1
11	343 29.0	16.4
12	358 29.1	N17 15.7
13	13 29.1	15.1
14	28 29.2	14.4
15	43 29.2	·· 13.7
16	58 29.3	13.1
17	73 29.3	12.4
18	88 29.4	N17 11.7
19	103 29.4	11.1
20	118 29.5	10.4
21	133 29.6	·· 09.7
22	148 29.6	09.1
23	163 29.7	08.4

d	h	G.H.A. ° ′	Dec. ° ′
5	00	178 29.7	N17 07.7
	01	193 29.8	07.1
	02	208 29.8	06.4
	03	223 29.9 ··	05.7
	04	238 30.0	05.1
	05	253 30.0	04.4
	06	268 30.1	N17 03.7
	07	283 30.1	03.0
	08	298 30.2	02.4
	09	313 30.3 ··	01.7
	10	328 30.3	01.0
	11	343 30.4	17 00.3
	12	358 30.4	N16 59.7
	13	13 30.5	59.0
	14	28 30.6	58.3
	15	43 30.6 ··	57.6
	16	58 30.7	57.0
	17	73 30.8	56.3
	18	88 30.8	N16 55.6
	19	103 30.9	54.9
	20	118 30.9	54.2
	21	133 31.0 ··	53.6
	22	148 31.1	52.9
	23	163 31.1	52.2
6	00	178 31.2	N16 51.5
	01	193 31.3	50.8
	02	208 31.3	50.2
	03	223 31.4 ··	49.5
	04	238 31.5	48.8
	05	253 31.5	48.1
	06	268 31.6	N16 47.4
	07	283 31.7	46.7
	08	298 31.7	46.1
	09	313 31.8 ··	45.4
	10	328 31.9	44.7
	11	343 31.9	44.0
	12	358 32.0	N16 43.3
	13	13 32.1	42.6
	14	28 32.1	41.9
	15	43 32.2 ··	41.3
	16	58 32.3	40.6
	17	73 32.3	39.9
	18	88 32.4	N16 39.2
	19	103 32.5	38.5
	20	118 32.5	37.8
	21	133 32.6 ··	37.1
	22	148 32.7	36.4
	23	163 32.8	35.7
7	00	178 32.8	N16 35.0
	01	193 32.9	34.3
	02	208 33.0	33.7
	03	223 33.0 ··	33.0
	04	238 33.1	32.3
	05	253 33.2	31.6
	06	268 33.3	N16 30.9
	07	283 33.3	30.2
	08	298 33.4	29.5
	09	313 33.5 ··	28.8
	10	328 33.5	28.1
	11	343 33.6	27.4
	12	358 33.7	N16 26.7
	13	13 33.8	26.0
	14	28 33.8	25.3
	15	43 33.9 ··	24.6
	16	58 34.0	23.9
	17	73 34.1	23.2
	18	88 34.1	N16 22.5
	19	103 34.2	21.8
	20	118 34.3	21.1
	21	133 34.4 ··	20.4
	22	148 34.4	19.7
	23	163 34.5	19.0

d	h	G.H.A. ° ′	Dec. ° ′
8	00	178 34.6	N16 18.3
	01	193 34.7	17.6
	02	208 34.8	16.9
	03	223 34.8 ··	16.2
	04	238 34.9	15.5
	05	253 35.0	14.8
	06	268 35.1	N16 14.1
	07	283 35.1	13.4
	08	298 35.2	12.6
	09	313 35.3 ··	11.9
	10	328 35.4	11.2
	11	343 35.5	10.5
	12	358 35.5	N16 09.8
	13	13 35.6	09.1
	14	28 35.7	08.4
	15	43 35.8 ··	07.7
	16	58 35.9	07.0
	17	73 35.9	06.3
	18	88 36.0	N16 05.6
	19	103 36.1	04.8
	20	118 36.2	04.1
	21	133 36.3 ··	03.4
	22	148 36.4	02.7
	23	163 36.4	02.0
9	00	178 36.5	N16 01.3
	01	193 36.6	16 00.6
	02	208 36.7	15 59.8
	03	223 36.8 ··	59.1
	04	238 36.9	58.4
	05	253 36.9	57.7
	06	268 37.0	N15 57.0
	07	283 37.1	56.3
	08	298 37.2	55.5
	09	313 37.3 ··	54.8
	10	328 37.4	54.1
	11	343 37.5	53.4
	12	358 37.5	N15 52.7
	13	13 37.6	51.9
	14	28 37.7	51.2
	15	43 37.8 ··	50.5
	16	58 37.9	49.8
	17	73 38.0	49.1
	18	88 38.1	N15 48.3
	19	103 38.1	47.6
	20	118 38.2	46.9
	21	133 38.3 ··	46.2
	22	148 38.4	45.4
	23	163 38.5	44.7
10	00	178 38.6	N15 44.0
	01	193 38.7	43.3
	02	208 38.8	42.5
	03	223 38.9 ··	41.8
	04	238 39.0	41.1
	05	253 39.0	40.4
	06	268 39.1	N15 39.6
	07	283 39.2	38.9
	08	298 39.3	38.2
	09	313 39.4 ··	37.5
	10	328 39.5	36.7
	11	343 39.6	36.0
	12	358 39.7	N15 35.3
	13	13 39.8	34.5
	14	28 39.9	33.8
	15	43 40.0 ··	33.1
	16	58 40.1	32.3
	17	73 40.2	31.6
	18	88 40.2	N15 30.9
	19	103 40.3	30.1
	20	118 40.4	29.4
	21	133 40.5 ··	28.7
	22	148 40.6	27.9
	23	163 40.7	27.2

d	h	G.H.A. ° ′	Dec. ° ′
11	00	178 40.8	N15 26.5
	01	193 40.9	25.7
	02	208 41.0	25.0
	03	223 41.1 ··	24.3
	04	238 41.2	23.5
	05	253 41.3	22.8
	06	268 41.4	N15 22.1
	07	283 41.5	21.3
	08	298 41.6	20.6
	09	313 41.7 ··	19.8
	10	328 41.8	19.1
	11	343 41.9	18.4
	12	358 42.0	N15 17.6
	13	13 42.1	16.9
	14	28 42.2	16.1
	15	43 42.3 ··	15.4
	16	58 42.4	14.7
	17	73 42.5	13.9
	18	88 42.6	N15 13.2
	19	103 42.7	12.4
	20	118 42.8	11.7
	21	133 42.9 ··	10.9
	22	148 43.0	10.2
	23	163 43.1	09.5
12	00	178 43.2	N15 08.7
	01	193 43.3	08.0
	02	208 43.4	07.2
	03	223 43.5 ··	06.5
	04	238 43.6	05.7
	05	253 43.7	05.0
	06	268 43.8	N15 04.2
	07	283 43.9	03.5
	08	298 44.0	02.7
	09	313 44.1 ··	02.0
	10	328 44.2	01.2
	11	343 44.3	15 00.5
	12	358 44.4	N14 59.7
	13	13 44.5	59.0
	14	28 44.6	58.2
	15	43 44.7 ··	57.5
	16	58 44.8	56.7
	17	73 44.9	56.0
	18	88 45.1	N14 55.2
	19	103 45.2	54.5
	20	118 45.3	53.7
	21	133 45.4 ··	53.0
	22	148 45.5	52.2
	23	163 45.6	51.4
13	00	178 45.7	N14 50.7
	01	193 45.8	49.9
	02	208 45.9	49.2
	03	223 46.0 ··	48.4
	04	238 46.1	47.7
	05	253 46.2	46.9
	06	268 46.3	N14 46.2
	07	283 46.5	45.4
	08	298 46.6	44.6
	09	313 46.7 ··	43.9
	10	328 46.8	43.1
	11	343 46.9	42.4
	12	358 47.0	N14 41.6
	13	13 47.1	40.8
	14	28 47.2	40.1
	15	43 47.3 ··	39.3
	16	58 47.4	38.6
	17	73 47.6	37.8
	18	88 47.7	N14 37.0
	19	103 47.8	36.3
	20	118 47.9	35.5
	21	133 48.0 ··	34.7
	22	148 48.1	34.0
	23	163 48.2	33.2

d h	G.H.A.	Dec.	d h	G.H.A.	Dec.	d h	G.H.A.	Dec.
14 00	178 48.3	N14 32.4	17 00	178 57.1	N13 36.3	20 00	179 07.1	N12 38.3
01	193 48.5	31.7	01	193 57.3	35.5	01	194 07.2	37.4
02	208 48.6	30.9	02	208 57.4	34.7	02	209 07.4	36.6
03	223 48.7 ··	30.1	03	223 57.5 ··	33.9	03	224 07.5 ··	35.8
04	238 48.8	29.4	04	238 57.7	33.1	04	239 07.7	35.0
05	253 48.9	28.6	05	253 57.8	32.4	05	254 07.8	34.2
06	268 49.0	N14 27.8	06	268 57.9	N13 31.6	06	269 08.0	N12 33.3
07	283 49.1	27.1	07	283 58.1	30.8	07	284 08.1	32.5
08	298 49.3	26.3	08	298 58.2	30.0	08	299 08.3	31.7
09	313 49.4 ··	25.5	09	313 58.3 ··	29.2	09	314 08.4 ··	30.9
10	328 49.5	24.8	10	328 58.4	28.4	10	329 08.6	30.1
11	343 49.6	24.0	11	343 58.6	27.6	11	344 08.7	29.2
12	358 49.7	N14 23.2	12	358 58.7	N13 26.8	12	359 08.9	N12 28.4
13	13 49.8	22.5	13	13 58.8	26.0	13	14 09.0	27.6
14	28 50.0	21.7	14	28 59.0	25.2	14	29 09.2	26.8
15	43 50.1 ··	20.9	15	43 59.1 ··	24.4	15	44 09.3 ··	25.9
16	58 50.2	20.1	16	58 59.2	23.6	16	59 09.5	25.1
17	73 50.3	19.4	17	73 59.4	22.8	17	74 09.6	24.3
18	88 50.4	N14 18.6	18	88 59.5	N13 22.0	18	89 09.8	N12 23.5
19	103 50.5	17.8	19	103 59.7	21.2	19	104 09.9	22.6
20	118 50.7	17.1	20	118 59.8	20.4	20	119 10.1	21.8
21	133 50.8 ··	16 3	21	133 59.9 ··	19.6	21	134 10.2 ··	21.0
22	148 50.9	15.5	22	149 00.1	18.8	22	149 10.4	20.2
23	163 51.0	14.7	23	164 00.2	18.0	23	164 10.5	19.3
15 00	178 51.1	N14 14.0	18 00	179 00.3	N13 17.2	21 00	179 10.7	N12 18.5
01	193 51.3	13.2	01	194 00.5	16.4	01	194 10.8	17.7
02	208 51.4	12.4	02	209 00.6	15.6	02	209 11.0	16.8
03	223 51.5 ··	11.6	03	224 00.7 ··	14.8	03	224 11.1 ··	16.0
04	238 51.6	10.9	04	239 00.9	14.0	04	239 11.3	15.2
05	253 51.7	10.1	05	254 01.0	13.2	05	254 11.4	14.4
06	268 51.9	N14 09.3	06	269 01.1	N13 12.4	06	269 11.6	N12 13 5
07	283 52.0	08 5	07	284 01.3	11.6	07	284 11.7	12.7
08	298 52 1	07.8	08	299 01.4	10.8	08	299 11.9	11.9
09	313 52.2 ··	07.0	09	314 01.6 ··	09.9	09	314 12.0 ··	11.0
10	328 52.3	06.2	10	329 01.7	09.1	10	329 12.2	10.2
11	343 52.5	05.4	11	344 01.8	08.3	11	344 12.3	09.4
12	358 52.6	N14 04.6	12	359 02.0	N13 07.5	12	359 12.5	N12 08.5
13	13 52.7	03.9	13	14 02.1	06.7	13	14 12.6	07.7
14	28 52.8	03.1	14	29 02.3	05.9	14	29 12.8	06.9
15	43 53.0 ··	02.3	15	44 02.4 ··	05.1	15	44 12.9 ··	06.1
16	58 53.1	01.5	16	59 02.5	04.3	16	59 13.1	05.2
17	73 53.2	00.7	17	74 02.7	03.5	17	74 13.3	04.4
18	88 53.3	N14 00.0	18	89 02.8	N13 02 7	18	89 13.4	N12 03.6
19	103 53 4	13 59.2	19	104 02.9	01.9	19	104 13.6	02.7
20	118 53.6	58.4	20	119 03.1	01.1	20	119 13.7	01.9
21	133 53.7 ··	57.6	21	134 03.2	13 00.3	21	134 13.9 ··	01.0
22	148 53.8	56.8	22	149 03.4	12 59.4	22	149 14.0	12 00.2
23	163 53.9	56.0	23	164 03.5	58.6	23	164 14.2	11 59.4
16 00	178 54.1	N13 55.3	19 00	179 03.6	N12 57.8	22 00	179 14.3	N11 58.5
01	193 54.2	54.5	01	194 03.8	57.0	01	194 14.5	57.7
02	208 54.3	53.7	02	209 03.9	56.2	02	209 14.7	56.9
03	223 54.4 ··	52.9	03	224 04.1 ··	55.4	03	224 14.8 ··	56.0
04	238 54.6	52.1	04	239 04.2	54.6	04	239 15.0	55.2
05	253 54.7	51.3	05	254 04.4	53.8	05	254 15.1	54.4
06	268 54.8	N13 50.5	06	269 04.5	N12 53.0	06	269 15.3	N11 53.5
07	283 54.9	49.8	07	284 04.6	52.1	07	284 15.4	52.7
08	298 55.1	49.0	08	299 04.8	51.3	08	299 15.6	51.8
09	313 55.2 ··	48.2	09	314 04.9 ··	50.5	09	314 15.8 ··	51.0
10	328 55.3	47.4	10	329 05.1	49.7	10	329 15.9	50.2
11	343 55.5	46.6	11	344 05.2	48.9	11	344 16.1	49.3
12	358 55.6	N13 45.8	12	359 05.4	N12 48.1	12	359 16.2	N11 48.5
13	13 55.7	45.0	13	14 05.5	47.3	13	14 16.4	47.7
14	28 55.8	44.2	14	29 05.6	46.4	14	29 16.5	46.8
15	43 56.0 ··	43.4	15	44 05.8 ··	45.6	15	44 16.7 ··	46.0
16	58 56.1	42.7	16	59 05.9	44.8	16	59 16.9	45.1
17	73 56.2	41.9	17	74 06.1	44.0	17	74 17.0	44.3
18	88 56.4	N13 41.1	18	89 06.2	N12 43.2	18	89 17.2	N11 43.5
19	103 56.5	40.3	19	104 06.4	42.4	19	104 17.3	42.6
20	118 56.6	39.5	20	119 06.5	41.5	20	119 17.5	41.8
21	133 56.7 ··	38.7	21	134 06.7 ··	40.7	21	134 17.7 ··	40.9
22	148 56.9	37.9	22	149 06.8	39.9	22	149 17.8	40.1
23	163 57.0	37.1	23	164 06.9	39.1	23	164 18.0	39.2

d h	G.H.A. ° '	Dec. ° '
23 00	179 18.1	N11 38.4
01	194 18.3	37.6
02	209 18.5	36.7
03	224 18.6	·· 35.9
04	239 18.8	35.0
05	254 18.9	34.2
06	269 19.1	N11 33.3
07	284 19.3	32.5
08	299 19.4	31.6
09	314 19.6	·· 30.8
10	329 19.8	29.9
11	344 19.9	29.1
12	359 20.1	N11 28.3
13	14 20.2	27.4
14	29 20.4	26.6
15	44 20.6	·· 25.7
16	59 20.7	24.9
17	74 20.9	24.0
18	89 21.1	N11 23.2
19	104 21.2	22.3
20	119 21.4	21.5
21	134 21.5	·· 20.6
22	149 21.7	19.8
23	164 21.9	18.9
24 00	179 22.0	N11 18.1
01	194 22.2	17.2
02	209 22.4	16.4
03	224 22.5	·· 15.5
04	239 22.7	14.7
05	254 22.9	13.8
06	269 23.0	N11 13.0
07	284 23.2	12.1
08	299 23.4	11.2
09	314 23.5	·· 10.4
10	329 23.7	09.5
11	344 23.9	08.7
12	359 24.0	N11 07.8
13	14 24.2	07.0
14	29 24.4	06.1
15	44 24.5	·· 05.3
16	59 24.7	04.4
17	74 24.9	03.6
18	89 25.0	N11 02.7
19	104 25.2	01.8
20	119 25.4	01.0
21	134 25.5	11 00.1
22	149 25.7	10 59.3
23	164 25.9	58.4
25 00	179 26.0	N10 57.6
01	194 26.2	56.7
02	209 26.4	55.8
03	224 26.6	·· 55.0
04	239 26.7	54.1
05	254 26.9	53.3
06	269 27.1	N10 52.4
07	284 27.2	51.5
08	299 27.4	50.7
09	314 27.6	·· 49.8
10	329 27.7	49.0
11	344 27.9	48.1
12	359 28.1	N10 47.2
13	14 28.3	46.4
14	29 28.4	45.5
15	44 28.6	·· 44.6
16	59 28.8	43.8
17	74 28.9	42.9
18	89 29.1	N10 42.1
19	104 29.3	41.2
20	119 29.5	40.3
21	134 29.6	·· 39.5
22	149 29.8	38.6
23	164 30.0	37.7

d h	G.H.A. ° '	Dec. ° '
26 00	179 30.2	N10 36.9
01	194 30.3	36.0
02	209 30.5	35.1
03	224 30.7	·· 34.3
04	239 30.9	33.4
05	254 31.0	32.5
06	269 31.2	N10 31.7
07	284 31.4	30.8
08	299 31.5	29.9
09	314 31.7	·· 29.1
10	329 31.9	28.2
11	344 32.1	27.3
12	359 32.2	N10 26.5
13	14 32.4	25.6
14	29 32.6	24.7
15	44 32.8	·· 23.9
16	59 33.0	23.0
17	74 33.1	22.1
18	89 33.3	N10 21.2
19	104 33.5	20.4
20	119 33.7	19.5
21	134 33.8	·· 18.6
22	149 34.0	17.8
23	164 34.2	16.9
27 00	179 34.4	N10 16.0
01	194 34.5	15.1
02	209 34.7	14.3
03	224 34.9	·· 13.4
04	239 35.1	12.5
05	254 35.3	11.7
06	269 35.4	N10 10.8
07	284 35.6	09.9
08	299 35.8	09.0
09	314 36.0	·· 08.2
10	329 36.1	07.3
11	344 36.3	06.4
12	359 36.5	N10 05.5
13	14 36.7	04.7
14	29 36.9	03.8
15	44 37.0	·· 02.9
16	59 37.2	02.0
17	74 37.4	01.1
18	89 37.6	N10 00.3
19	104 37.8	9 59.4
20	119 37.9	58.5
21	134 38.1	·· 57.6
22	149 38.3	56.8
23	164 38.5	55.9
28 00	179 38.7	N 9 55.0
01	194 38.8	54.1
02	209 39.0	53.2
03	224 39.2	·· 52.4
04	239 39.4	51.5
05	254 39.6	50.6
06	269 39.8	N 9 49.7
07	284 39.9	48.8
08	299 40.1	48.0
09	314 40.3	·· 47.1
10	329 40.5	46.2
11	344 40.7	45.3
12	359 40.8	N 9 44.4
13	14 41.0	43.6
14	29 41.2	42.7
15	44 41.4	·· 41.8
16	59 41.6	40.9
17	74 41.8	40.0
18	89 41.9	N 9 39.1
19	104 42.1	38.3
20	119 42.3	37.4
21	134 42.5	·· 36.5
22	149 42.7	35.6
23	164 42.9	34.7

d h	G.H.A. ° '	Dec. ° '
29 00	179 43.1	N 9 33.8
01	194 43.2	32.9
02	209 43.4	32.1
03	224 43.6	·· 31.2
04	239 43.8	30.3
05	254 44.0	29.4
06	269 44.2	N 9 28.5
07	284 44.3	27.6
08	299 44.5	26.7
09	314 44.7	·· 25.9
10	329 44.9	25.0
11	344 45.1	24.1
12	359 45.3	N 9 23.2
13	14 45.5	22.3
14	29 45.7	21.4
15	44 45.8	·· 20.5
16	59 46.0	19.6
17	74 46.2	18.7
18	89 46.4	N 9 17.9
19	104 46.6	17.0
20	119 46.8	16.1
21	134 47.0	·· 15.2
22	149 47.2	14.3
23	164 47.3	13.4
30 00	179 47.5	N 9 12.5
01	194 47.7	11.6
02	209 47.9	10.7
03	224 48.1	·· 09.8
04	239 48.3	08.9
05	254 48.5	08.0
06	269 48.7	N 9 07.2
07	284 48.8	06.3
08	299 49.0	05.4
09	314 49.2	·· 04.5
10	329 49.4	03.6
11	344 49.6	02.7
12	359 49.8	N 9 01.8
13	14 50.0	00.9
14	29 50.2	9 00.0
15	44 50.4	8 59.1
16	59 50.6	58.2
17	74 50.7	57.3
18	89 50.9	N 8 56.4
19	104 51.1	55.5
20	119 51.3	54.6
21	134 51.5	·· 53.7
22	149 51.7	52.8
23	164 51.9	51.9
31 00	179 52.1	N 8 51.0
01	194 52.3	50.1
02	209 52.5	49.2
03	224 52.7	·· 48.3
04	239 52.8	47.4
05	254 53.0	46.5
06	269 53.2	N 8 45.6
07	284 53.4	44.7
08	299 53.6	43.8
09	314 53.8	·· 42.9
10	329 54.0	42.0
11	344 54.2	41.1
12	359 54.4	N 8 40.2
13	14 54.6	39.3
14	29 54.8	38.4
15	44 55.0	·· 37.5
16	59 55.2	36.6
17	74 55.4	35.7
18	89 55.6	N 8 34.8
19	104 55.7	33.9
20	119 55.9	33.0
21	134 56.1	·· 32.1
22	149 56.3	31.2
23	164 56.5	30.3

SEPTEMBER

d h	G.H.A.	Dec.
1 00	179 56.7	N 8 29.4
01	194 56.9	28.5
02	209 57.1	27.6
03	224 57.3 ··	26.7
04	239 57.5	25.8
05	254 57.7	24.9
06	269 57.9	N 8 24.0
07	284 58.1	23.1
08	299 58.3	22.2
09	314 58.5 ··	21.3
10	329 58.7	20.4
11	344 58.9	19.5
12	359 59.1	N 8 18.6
13	14 59.3	17.7
14	29 59.5	16.7
15	44 59.7 ··	15.8
16	59 59.8	14.9
17	75 00.0	14.0
18	90 00.2	N 8 13.1
19	105 00.4	12.2
20	120 00.6	11.3
21	135 00.8 ··	10.4
22	150 01.0	09.5
23	165 01.2	08.6
2 00	180 01.4	N 8 07.7
01	195 01.6	06.8
02	210 01.8	05.8
03	225 02.0 ··	04.9
04	240 02.2	04.0
05	255 02.4	03.1
06	270 02.6	N 8 02.2
07	285 02.8	01.3
08	300 03.0	8 00.4
09	315 03.2	7 59.5
10	330 03.4	58.6
11	345 03.6	57.7
12	0 03.8	N 7 56.7
13	15 04.0	55.8
14	30 04.2	54.9
15	45 04.4 ··	54.0
16	60 04.6	53.1
17	75 04.8	52.2
18	90 05.0	N 7 51.3
19	105 05.2	50.4
20	120 05.4	49.4
21	135 05.6 ··	48.5
22	150 05.8	47.6
23	165 06.0	46.7
3 00	180 06.2	N 7 45.8
01	195 06.4	44.9
02	210 06.6	44.0
03	225 06.8 ··	43.0
04	240 07.0	42.1
05	255 07.2	41.2
06	270 07.4	N 7 40.3
07	285 07.6	39.4
08	300 07.8	38.5
09	315 08.0 ··	37.5
10	330 08.2	36.6
11	345 08.4	35.7
12	0 08.6	N 7 34.8
13	15 08.8	33.9
14	30 09.0	33.0
15	45 09.2 ··	32.0
16	60 09.4	31.1
17	75 09.6	30.2
18	90 09.8	N 7 29.3
19	105 10.0	28.4
20	120 10.2	27.5
21	135 10.4 ··	26.5
22	150 10.7	25.6
23	165 10.9	24.7

d h	G.H.A.	Dec.
4 00	180 11.1	N 7 23.8
01	195 11.3	22.9
02	210 11.5	21.9
03	225 11.7 ··	21.0
04	240 11.9	20.1
05	255 12.1	19.2
06	270 12.3	N 7 18.3
07	285 12.5	17.3
08	300 12.7	16.4
09	315 12.9 ··	15.5
10	330 13.1	14.6
11	345 13.3	13.7
12	0 13.5	N 7 12.7
13	15 13.7	11.8
14	30 13.9	10.9
15	45 14.1 ··	10.0
16	60 14.3	09.0
17	75 14.5	08.1
18	90 14.7	N 7 07.2
19	105 14.9	06.3
20	120 15.1	05.4
21	135 15.4 ··	04.4
22	150 15.6	03.5
23	165 15.8	02.6
5 00	180 16.0	N 7 01.7
01	195 16.2	7 00.7
02	210 16.4	6 59.8
03	225 16.6 ··	58.9
04	240 16.8	58.0
05	255 17.0	57.0
06	270 17.2	N 6 56.1
07	285 17.4	55.2
08	300 17.6	54.3
09	315 17.8 ··	53.3
10	330 18.0	52.4
11	345 18.2	51.5
12	0 18.4	N 6 50.6
13	15 18.7	49.6
14	30 18.9	48.7
15	45 19.1 ··	47.8
16	60 19.3	46.8
17	75 19.5	45.9
18	90 19.7	N 6 45.0
19	105 19.9	44.1
20	120 20.1	43.1
21	135 20.3 ··	42.2
22	150 20.5	41.3
23	165 20.7	40.3
6 00	180 20.9	N 6 39.4
01	195 21.2	38.5
02	210 21.4	37.6
03	225 21.6 ··	36.6
04	240 21.8	35.7
05	255 22.0	34.8
06	270 22.2	N 6 33.8
07	285 22.4	32.9
08	300 22.6	32.0
09	315 22.8 ··	31.0
10	330 23.0	30.1
11	345 23.2	29.2
12	0 23.5	N 6 28.3
13	15 23.7	27.3
14	30 23.9	26.4
15	45 24.1 ··	25.5
16	60 24.3	24.5
17	75 24.5	23.6
18	90 24.7	N 6 22.7
19	105 24.9	21.7
20	120 25.1	20.8
21	135 25.3 ··	19.9
22	150 25.6	18.9
23	165 25.8	18.0

d h	G.H.A.	Dec.
7 00	180 26.0	N 6 17.1
01	195 26.2	16.1
02	210 26.4	15.2
03	225 26.6 ··	14.3
04	240 26.8	13.3
05	255 27.0	12.4
06	270 27.2	N 6 11.5
07	285 27.5	10.5
08	300 27.7	09.6
09	315 27.9 ··	08.7
10	330 28.1	07.7
11	345 28.3	06.8
12	0 28.5	N 6 05.9
13	15 28.7	04.9
14	30 28.9	04.0
15	45 29.1 ··	03.0
16	60 29.4	02.1
17	75 29.6	01.2
18	90 29.8	N 6 00.2
19	105 30.0	5 59.3
20	120 30.2	58.4
21	135 30.4 ··	57.4
22	150 30.6	56.5
23	165 30.8	55.6
8 00	180 31.1	N 5 54.6
01	195 31.3	53.7
02	210 31.5	52.7
03	225 31.7 ··	51.8
04	240 31.9	50.9
05	255 32.1	49.9
06	270 32.3	N 5 49.0
07	285 32.5	48.0
08	300 32.8	47.1
09	315 33.0 ··	46.2
10	330 33.2	45.2
11	345 33.4	44.3
12	0 33.6	N 5 43.4
13	15 33.8	42.4
14	30 34.0	41.5
15	45 34.3 ··	40.5
16	60 34.5	39.6
17	75 34.7	38.7
18	90 34.9	N 5 37.7
19	105 35.1	36.8
20	120 35.3	35.8
21	135 35.5 ··	34.9
22	150 35.8	33.9
23	165 36.0	33.0
9 00	180 36.2	N 5 32.1
01	195 36.4	31.1
02	210 36.6	30.2
03	225 36.8 ··	29.2
04	240 37.1	28.3
05	255 37.3	27.4
06	270 37.5	N 5 26.4
07	285 37.7	25.5
08	300 37.9	24.5
09	315 38.1 ··	23.6
10	330 38.3	22.6
11	345 38.6	21.7
12	0 38.8	N 5 20.8
13	15 39.0	19.8
14	30 39.2	18.9
15	45 39.4 ··	17.9
16	60 39.6	17.0
17	75 39.9	16.0
18	90 40.1	N 5 15.1
19	105 40.3	14.1
20	120 40.5	13.2
21	135 40.7 ··	12.3
22	150 40.9	11.3
23	165 41.2	10.4

d	h	G.H.A.	Dec.
10	00	180 41.4	N 5 09.4
	01	195 41.6	08.5
	02	210 41.8	07.5
	03	225 42.0	· · 06.6
	04	240 42.2	05.6
	05	255 42.5	04.7
	06	270 42.7	N 5 03.8
	07	285 42.9	02.8
	08	300 43.1	01.9
	09	315 43.3	· · 00.9
	10	330 43.5	5 00.0
	11	345 43.8	4 59.0
	12	0 44.0	N 4 58.1
	13	15 44.2	57.1
	14	30 44.4	56.2
	15	45 44.6	· · 55.2
	16	60 44.8	54.3
	17	75 45.1	53.3
	18	90 45.3	N 4 52.4
	19	105 45.5	51.4
	20	120 45.7	50.5
	21	135 45.9	· · 49.5
	22	150 46.2	48.6
	23	165 46.4	47.6
11	00	180 46.6	N 4 46.7
	01	195 46.8	45.7
	02	210 47.0	44.8
	03	225 47.2	· · 43.8
	04	240 47.5	42.9
	05	255 47.7	42.0
	06	270 47.9	N 4 41.0
	07	285 48.1	40.1
	08	300 48.3	39.1
	09	315 48.6	· · 38.2
	10	330 48.8	37.2
	11	345 49.0	36.3
	12	0 49.2	N 4 35.3
	13	15 49.4	34.3
	14	30 49.6	33.4
	15	45 49.9	· · 32.4
	16	60 50.1	31.5
	17	75 50.3	30.5
	18	90 50.5	N 4 29.6
	19	105 50.7	28.6
	20	120 51.0	27.7
	21	135 51.2	· · 26.7
	22	150 51.4	25.8
	23	165 51.6	24.8
12	00	180 51.8	N 4 23.9
	01	195 52.1	22.9
	02	210 52.3	22.0
	03	225 52.5	· · 21.0
	04	240 52.7	20.1
	05	255 52.9	19.1
	06	270 53.2	N 4 18.2
	07	285 53.4	17.2
	08	300 53.6	16.3
	09	315 53.8	· · 15.3
	10	330 54.0	14.4
	11	345 54.3	13.4
	12	0 54.5	N 4 12.5
	13	15 54.7	11.5
	14	30 54.9	10.5
	15	45 55.1	· · 09.6
	16	60 55.4	08.6
	17	75 55.6	07.7
	18	90 55.8	N 4 06.7
	19	105 56.0	05.8
	20	120 56.2	04.8
	21	135 56.5	· · 03.9
	22	150 56.7	02.9
	23	165 56.9	02.0

d	h	G.H.A.	Dec.
13	00	180 57.1	N 4 01.0
	01	195 57.3	4 00.0
	02	210 57.6	3 59.1
	03	225 57.8	· · 58.1
	04	240 58.0	57.2
	05	255 58.2	56.2
	06	270 58.5	N 3 55.3
	07	285 58.7	54.3
	08	300 58.9	53.4
	09	315 59.1	· · 52.4
	10	330 59.3	51.4
	11	345 59.6	50.5
	12	0 59.8	N 3 49.5
	13	16 00.0	48.6
	14	31 00.2	47.6
	15	46 00.4	· · 46.7
	16	61 00.7	45.7
	17	76 00.9	44.7
	18	91 01.1	N 3 43.8
	19	106 01.3	42.8
	20	121 01.6	41.9
	21	136 01.8	· · 40.9
	22	151 02.0	40.0
	23	166 02.2	39.0
14	00	181 02.4	N 3 38.0
	01	196 02.7	37.1
	02	211 02.9	36.1
	03	226 03.1	· · 35.2
	04	241 03.3	34.2
	05	256 03.6	33.3
	06	271 03.8	N 3 32.3
	07	286 04.0	31.3
	08	301 04.2	30.4
	09	316 04.4	· · 29.4
	10	331 04.7	28.5
	11	346 04.9	27.5
	12	1 05.1	N 3 26.5
	13	16 05.3	25.6
	14	31 05.5	24.6
	15	46 05.8	· · 23.7
	16	61 06.0	22.7
	17	76 06.2	21.7
	18	91 06.4	N 3 20.8
	19	106 06.7	19.8
	20	121 06.9	18.9
	21	136 07.1	· · 17.9
	22	151 07.3	16.9
	23	166 07.6	16.0
15	00	181 07.8	N 3 15.0
	01	196 08.0	14.1
	02	211 08.2	13.1
	03	226 08.4	· · 12.1
	04	241 08.7	11.2
	05	256 08.9	10.2
	06	271 09.1	N 3 09.3
	07	286 09.3	08.3
	08	301 09.6	07.3
	09	316 09.8	· · 06.4
	10	331 10.0	05.4
	11	346 10.2	04.4
	12	1 10.4	N 3 03.5
	13	16 10.7	02.5
	14	31 10.9	01.6
	15	46 11.1	3 00.6
	16	61 11.3	2 59.6
	17	76 11.6	58.7
	18	91 11.8	N 2 57.7
	19	106 12.0	56.8
	20	121 12.2	55.8
	21	136 12.5	· · 54.8
	22	151 12.7	53.9
	23	166 12.9	52.9

d	h	G.H.A.	Dec.
16	00	181 13.1	N 2 51.9
	01	196 13.3	51.0
	02	211 13.6	50.0
	03	226 13.8	· · 49.0
	04	241 14.0	48.1
	05	256 14.2	47.1
	06	271 14.5	N 2 46.2
	07	286 14.7	45.2
	08	301 14.9	44.2
	09	316 15.1	· · 43.3
	10	331 15.4	42.3
	11	346 15.6	41.3
	12	1 15.8	N 2 40.4
	13	16 16.0	39.4
	14	31 16.3	38.4
	15	46 16.5	· · 37.5
	16	61 16.7	36.5
	17	76 16.9	35.6
	18	91 17.1	N 2 34.6
	19	106 17.4	33.6
	20	121 17.6	32.7
	21	136 17.8	· · 31.7
	22	151 18.0	30.7
	23	166 18.3	29.8
17	00	181 18.5	N 2 28.8
	01	196 18.7	27.8
	02	211 18.9	26.9
	03	226 19.2	· · 25.9
	04	241 19.4	24.9
	05	256 19.6	24.0
	06	271 19.8	N 2 23.0
	07	286 20.0	22.0
	08	301 20.3	21.1
	09	316 20.5	· · 20.1
	10	331 20.7	19.1
	11	346 20.9	18.2
	12	1 21.2	N 2 17.2
	13	16 21.4	16.2
	14	31 21.6	15.3
	15	46 21.8	· · 14.3
	16	61 22.1	13.3
	17	76 22.3	12.4
	18	91 22.5	N 2 11.4
	19	106 22.7	10.5
	20	121 23.0	09.5
	21	136 23.2	· · 08.5
	22	151 23.4	07.6
	23	166 23.6	06.6
18	00	181 23.8	N 2 05.6
	01	196 24.1	04.6
	02	211 24.3	03.7
	03	226 24.5	· · 02.7
	04	241 24.7	01.7
	05	256 25.0	2 00.8
	06	271 25.2	N 1 59.8
	07	286 25.4	58.8
	08	301 25.6	57.9
	09	316 25.9	· · 56.9
	10	331 26.1	55.9
	11	346 26.3	55.0
	12	1 26.5	N 1 54.0
	13	16 26.8	53.0
	14	31 27.0	52.1
	15	46 27.2	· · 51.1
	16	61 27.4	50.1
	17	76 27.6	49.2
	18	91 27.9	N 1 48.2
	19	106 28.1	47.2
	20	121 28.3	46.3
	21	136 28.5	· · 45.3
	22	151 28.8	44.3
	23	166 29.0	43.4

d h	G.H.A.	Dec.
19 00	181 29.2	N 1 42.4
01	196 29.4	41.4
02	211 29.7	40.5
03	226 29.9	·· 39.5
04	241 30.1	38.5
05	256 30.3	37.5
06	271 30.5	N 1 36.6
07	286 30.8	35.6
08	301 31.0	34.6
09	316 31.2	·· 33.7
10	331 31.4	32.7
11	346 31.7	31.7
12	1 31.9	N 1 30.8
13	16 32.1	29.8
14	31 32.3	28.8
15	46 32.6	·· 27.9
16	61 32.8	26.9
17	76 33.0	25.9
18	91 33.2	N 1 24.9
19	106 33.4	24.0
20	121 33.7	23.0
21	136 33.9	·· 22.0
22	151 34.1	21.1
23	166 34.3	20.1
20 00	181 34.6	N 1 19.1
01	196 34.8	18.2
02	211 35.0	17.2
03	226 35.2	·· 16.2
04	241 35.5	15.2
05	256 35.7	14.3
06	271 35.9	N 1 13.3
07	286 36.1	12.3
08	301 36.3	11.4
09	316 36.6	·· 10.4
10	331 36.8	09.4
11	346 37.0	08.4
12	1 37.2	N 1 07.5
13	16 37.5	06.5
14	31 37.7	05.5
15	46 37.9	·· 04.6
16	61 38.1	03.6
17	76 38.3	02.6
18	91 38.6	N 1 01.7
19	106 38.8	1 00.7
20	121 39.0	0 59.7
21	136 39.2	·· 58.7
22	151 39.5	57.8
23	166 39.7	56.8
21 00	181 39.9	N 0 55.8
01	196 40.1	54.9
02	211 40.3	53.9
03	226 40.6	·· 52.9
04	241 40.8	51.9
05	256 41.0	51.0
06	271 41.2	N 0 50.0
07	286 41.5	49.0
08	301 41.7	48.1
09	316 41.9	·· 47.1
10	331 42.1	46.1
11	346 42.3	45.1
12	1 42.6	N 0 44.2
13	16 42.8	43.2
14	31 43.0	42.2
15	46 43.2	·· 41.2
16	61 43.5	40.3
17	76 43.7	39.3
18	91 43.9	N 0 38.3
19	106 44.1	37.4
20	121 44.3	36.4
21	136 44.6	·· 35.4
22	151 44.8	34.4
23	166 45.0	33.5

d h	G.H.A.	Dec.
22 00	181 45.2	N 0 32.5
01	196 45.4	31.5
02	211 45.7	30.6
03	226 45.9	·· 29.6
04	241 46.1	28.6
05	256 46.3	27.6
06	271 46.5	N 0 26.7
07	286 46.8	25.7
08	301 47.0	24.7
09	316 47.2	·· 23.7
10	331 47.4	22.8
11	346 47.7	21.8
12	1 47.9	N 0 20.8
13	16 48.1	19.9
14	31 48.3	18.9
15	46 48.5	·· 17.9
16	61 48.8	16.9
17	76 49.0	16.0
18	91 49.2	N 0 15.0
19	106 49.4	14.0
20	121 49.6	13.0
21	136 49.9	·· 12.1
22	151 50.1	11.1
23	166 50.3	10.1
23 00	181 50.5	N 0 09.2
01	196 50.7	08.2
02	211 51.0	07.2
03	226 51.2	·· 06.2
04	241 51.4	05.3
05	256 51.6	04.3
06	271 51.8	N 0 03.3
07	286 52.0	02.3
08	301 52.3	01.4
09	316 52.5	N 0 00.4
10	331 52.7	S 0 00.6
11	346 52.9	01.6
12	1 53.1	S 0 02.5
13	16 53.4	03.5
14	31 53.6	04.5
15	46 53.8	·· 05.4
16	61 54.0	06.4
17	76 54.2	07.4
18	91 54.5	S 0 08.4
19	106 54.7	09.3
20	121 54.9	10.3
21	136 55.1	·· 11.3
22	151 55.3	12.3
23	166 55.6	13.2
24 00	181 55.8	S 0 14.2
01	196 56.0	15.2
02	211 56.2	16.2
03	226 56.4	·· 17.1
04	241 56.6	18.1
05	256 56.9	19.1
06	271 57.1	S 0 20.1
07	286 57.3	21.0
08	301 57.5	22.0
09	316 57.7	·· 23.0
10	331 57.9	23.9
11	346 58.2	24.9
12	1 58.4	S 0 25.9
13	16 58.6	26.9
14	31 58.8	27.8
15	46 59.0	·· 28.8
16	61 59.3	29.8
17	76 59.5	30.8
18	91 59.7	S 0 31.7
19	106 59.9	32.7
20	122 00.1	33.7
21	137 00.3	·· 34.7
22	152 00.6	35.6
23	167 00.8	36.6

d h	G.H.A.	Dec.
25 00	182 01.0	S 0 37.6
01	197 01.2	38.6
02	212 01.4	39.5
03	227 01.6	·· 40.5
04	242 01.9	41.5
05	257 02.1	42.5
06	272 02.3	S 0 43.4
07	287 02.5	44.4
08	302 02.7	45.4
09	317 02.9	·· 46.3
10	332 03.2	47.3
11	347 03.4	48.3
12	2 03.6	S 0 49.3
13	17 03.8	50.2
14	32 04.0	51.2
15	47 04.2	·· 52.2
16	62 04.4	53.2
17	77 04.7	54.1
18	92 04.9	S 0 55.1
19	107 05.1	56.1
20	122 05.3	57.1
21	137 05.5	·· 58.0
22	152 05.7	0 59.0
23	167 05.9	1 00.0
26 00	182 06.2	S 1 01.0
01	197 06.4	01.9
02	212 06.6	02.9
03	227 06.8	·· 03.9
04	242 07.0	04.9
05	257 07.2	05.8
06	272 07.4	S 1 06.8
07	287 07.7	07.8
08	302 07.9	08.8
09	317 08.1	·· 09.7
10	332 08.3	10.7
11	347 08.5	11.7
12	2 08.7	S 1 12.6
13	17 08.9	13.6
14	32 09.2	14.6
15	47 09.4	·· 15.6
16	62 09.6	16.5
17	77 09.8	17.5
18	92 10.0	S 1 18.5
19	107 10.2	19.5
20	122 10.4	20.4
21	137 10.6	·· 21.4
22	152 10.9	22.4
23	167 11.1	23.4
27 00	182 11.3	S 1 24.3
01	197 11.5	25.3
02	212 11.7	26.3
03	227 11.9	·· 27.3
04	242 12.1	28.2
05	257 12.3	29.2
06	272 12.6	S 1 30.2
07	287 12.8	31.2
08	302 13.0	32.1
09	317 13.2	·· 33.1
10	332 13.4	34.1
11	347 13.6	35.0
12	2 13.8	S 1 36.0
13	17 14.0	37.0
14	32 14.2	38.0
15	47 14.5	·· 38.9
16	62 14.7	39.9
17	77 14.9	40.9
18	92 15.1	S 1 41.9
19	107 15.3	42.8
20	122 15.5	43.8
21	137 15.7	·· 44.8
22	152 15.9	45.8
23	167 16.1	46.7

OCTOBER

d h	G.H.A.	Dec.
28 00	182 16.4	S 1 47.7
01	197 16.6	48.7
02	212 16.8	49.7
03	227 17.0 ··	50.6
04	242 17.2	51.6
05	257 17.4	52.6
06	272 17.6	S 1 53.5
07	287 17.8	54.5
08	302 18.0	55.5
09	317 18.2 ··	56.5
10	332 18.4	57.4
11	347 18.7	58.4
12	2 18.9	S 1 59.4
13	17 19.1	2 00.4
14	32 19.3	01.3
15	47 19.5 ··	02.3
16	62 19.7	03.3
17	77 19.9	04.3
18	92 20.1	S 2 05.2
19	107 20.3	06.2
20	122 20.5	07.2
21	137 20.7 ··	08.1
22	152 20.9	09.1
23	167 21.2	10.1
29 00	182 21.4	S 2 11.1
01	197 21.6	12.0
02	212 21.8	13.0
03	227 22.0 ··	14.0
04	242 22.2	15.0
05	257 22.4	15.9
06	272 22.6	S 2 16.9
07	287 22.8	17.9
08	302 23.0	18.8
09	317 23.2 ··	19.8
10	332 23.4	20.8
11	347 23.6	21.8
12	2 23.8	S 2 22.7
13	17 24.0	23.7
14	32 24.3	24.7
15	47 24.5 ··	25.7
16	62 24.7	26.6
17	77 24.9	27.6
18	92 25.1	S 2 28.6
19	107 25.3	29.5
20	122 25.5	30.5
21	137 25.7 ··	31.5
22	152 25.9	32.5
23	167 26.1	33.4
30 00	182 26.3	S 2 34.4
01	197 26.5	35.4
02	212 26.7	36.3
03	227 26.9 ··	37.3
04	242 27.1	38.3
05	257 27.3	39.3
06	272 27.5	S 2 40.2
07	287 27.7	41.2
08	302 27.9	42.2
09	317 28.1 ··	43.1
10	332 28.3	44.1
11	347 28.6	45.1
12	2 28.8	S 2 46.1
13	17 29.0	47.0
14	32 29.2	48.0
15	47 29.4 ··	49.0
16	62 29.6	49.9
17	77 29.8	50.9
18	92 30.0	S 2 51.9
19	107 30.2	52.9
20	122 30.4	53.8
21	137 30.6 ··	54.8
22	152 30.8	55.8
23	167 31.0	56.7

d h	G.H.A.	Dec.
1 00	182 31.2	S 2 57.7
01	197 31.4	58.7
02	212 31.6	2 59.7
03	227 31.8	3 00.6
04	242 32.0	01.6
05	257 32.2	02.6
06	272 32.4	S 3 03.5
07	287 32.6	04.5
08	302 32.8	05.5
09	317 33.0 ··	06.5
10	332 33.2	07.4
11	347 33.4	08.4
12	2 33.6	S 3 09.4
13	17 33.8	10.3
14	32 34.0	11.3
15	47 34.2 ··	12.3
16	62 34.4	13.2
17	77 34.6	14.2
18	92 34.8	S 3 15.2
19	107 35.0	16.2
20	122 35.2	17.1
21	137 35.4 ··	18.1
22	152 35.6	19.1
23	167 35.8	20.0
2 00	182 36.0	S 3 21.0
01	197 36.2	22.0
02	212 36.4	22.9
03	227 36.6 ··	23.9
04	242 36.8	24.9
05	257 37.0	25.8
06	272 37.2	S 3 26.8
07	287 37.4	27.8
08	302 37.6	28.8
09	317 37.8 ··	29.7
10	332 38.0	30.7
11	347 38.2	31.7
12	2 38.4	S 3 32.6
13	17 38.6	33.6
14	32 38.8	34.6
15	47 39.0 ··	35.5
16	62 39.2	36.5
17	77 39.4	37.5
18	92 39.6	S 3 38.4
19	107 39.8	39.4
20	122 39.9	40.4
21	137 40.1 ··	41.3
22	152 40.3	42.3
23	167 40.5	43.3
3 00	182 40.7	S 3 44.2
01	197 40.9	45.2
02	212 41.1	46.2
03	227 41.3 ··	47.2
04	242 41.5	48.1
05	257 41.7	49.1
06	272 41.9	S 3 50.1
07	287 42.1	51.0
08	302 42.3	52.0
09	317 42.5 ··	53.0
10	332 42.7	53.9
11	347 42.9	54.9
12	2 43.1	S 3 55.9
13	17 43.3	56.8
14	32 43.5	57.8
15	47 43.7 ··	58.8
16	62 43.8	3 59.7
17	77 44.0	4 00.7
18	92 44.2	S 4 01.7
19	107 44.4	02.6
20	122 44.6	03.6
21	137 44.8 ··	04.6
22	152 45.0	05.5
23	167 45.2	06.5

d h	G.H.A.	Dec.
4 00	182 45.4	S 4 07.5
01	197 45.6	08.4
02	212 45.8	09.4
03	227 46.0 ··	10.4
04	242 46.2	11.3
05	257 46.4	12.3
06	272 46.5	S 4 13.3
07	287 46.7	14.2
08	302 46.9	15.2
09	317 47.1 ··	16.1
10	332 47.3	17.1
11	347 47.5	18.1
12	2 47.7	S 4 19.0
13	17 47.9	20.0
14	32 48.1	21.0
15	47 48.3 ··	21.9
16	62 48.5	22.9
17	77 48.6	23.9
18	92 48.8	S 4 24.8
19	107 49.0	25.8
20	122 49.2	26.8
21	137 49.4 ··	27.7
22	152 49.6	28.7
23	167 49.8	29.7
5 00	182 50.0	S 4 30.6
01	197 50.2	31.6
02	212 50.4	32.5
03	227 50.5 ··	33.5
04	242 50.7	34.5
05	257 50.9	35.4
06	272 51.1	S 4 36.4
07	287 51.3	37.4
08	302 51.5	38.3
09	317 51.7 ··	39.3
10	332 51.9	40.2
11	347 52.1	41.2
12	2 52.2	S 4 42.2
13	17 52.4	43.1
14	32 52.6	44.1
15	47 52.8 ··	45.1
16	62 53.0	46.0
17	77 53.2	47.0
18	92 53.4	S 4 47.9
19	107 53.5	48.9
20	122 53.7	49.9
21	137 53.9 ··	50.8
22	152 54.1	51.8
23	167 54.3	52.8
6 00	182 54.5	S 4 53.7
01	197 54.7	54.7
02	212 54.9	55.6
03	227 55.0 ··	56.6
04	242 55.2	57.6
05	257 55.4	58.5
06	272 55.6	S 4 59.5
07	287 55.8	5 00.4
08	302 56.0	01.4
09	317 56.1 ··	02.4
10	332 56.3	03.3
11	347 56.5	04.3
12	2 56.7	S 5 05.2
13	17 56.9	06.2
14	32 57.1	07.2
15	47 57.2 ··	08.1
16	62 57.4	09.1
17	77 57.6	10.0
18	92 57.8	S 5 11.0
19	107 58.0	12.0
20	122 58.2	12.9
21	137 58.3 ··	13.9
22	152 58.5	14.8
23	167 58.7	15.8

d h	G.H.A.	Dec.
7 00	182 58.9	S 5 16.8
01	197 59.1	17.7
02	212 59.3	18.7
03	227 59.4	·· 19.6
04	242 59.6	20.6
05	257 59.8	21.6
06	273 00.0	S 5 22.5
07	288 00.2	23.5
08	303 00.3	24.4
09	318 00.5	·· 25.4
10	333 00.7	26.3
11	348 00.9	27.3
12	3 01.1	S 5 28.3
13	18 01.2	29.2
14	33 01.4	30.2
15	48 01.6	·· 31.1
16	63 01.8	32.1
17	78 02.0	33.0
18	93 02.1	S 5 34.0
19	108 02.3	35.0
20	123 02.5	35.9
21	138 02.7	·· 36.9
22	153 02.9	37.8
23	168 03.0	38.8
8 00	183 03.2	S 5 39.7
01	198 03.4	40.7
02	213 03.6	41.6
03	228 03.7	·· 42.6
04	243 03.9	43.6
05	258 04.1	44.5
06	273 04.3	S 5 45.5
07	288 04.5	46.4
08	303 04.6	47.4
09	318 04.8	·· 48.3
10	333 05.0	49.3
11	348 05.2	50.2
12	3 05.3	S 5 51.2
13	18 05.5	52.1
14	33 05.7	53.1
15	48 05.9	·· 54.1
16	63 06.0	55.0
17	78 06.2	56.0
18	93 06.4	S 5 56.9
19	108 06.6	57.9
20	123 06.7	58.8
21	138 06.9	5 59.8
22	153 07.1	6 00.7
23	168 07.3	01.7
9 00	183 07.4	S 6 02.6
01	198 07.6	03.6
02	213 07.8	04.5
03	228 08.0	·· 05.5
04	243 08.1	06.4
05	258 08.3	07.4
06	273 08.5	S 6 08.3
07	288 08.6	09.3
08	303 08.8	10.3
09	318 09.0	·· 11.2
10	333 09.2	12.2
11	348 09.3	13.1
12	3 09.5	S 6 14.1
13	18 09.7	15.0
14	33 09.8	16.0
15	48 10.0	·· 16.9
16	63 10.2	17.9
17	78 10.3	18.8
18	93 10.5	S 6 19.8
19	108 10.7	20.7
20	123 10.9	21.7
21	138 11.0	·· 22.6
22	153 11.2	23.6
23	168 11.4	24.5

d h	G.H.A.	Dec.
10 00	183 11.5	S 6 25.5
01	198 11.7	26.4
02	213 11.9	27.3
03	228 12.0	·· 28.3
04	243 12.2	29.2
05	258 12.4	30.2
06	273 12.5	S 6 31.1
07	288 12.7	32.1
08	303 12.9	33.0
09	318 13.0	·· 34.0
10	333 13.2	34.9
11	348 13.4	35.9
12	3 13.5	S 6 36.8
13	18 13.7	37.8
14	33 13.9	38.7
15	48 14.0	·· 39.7
16	63 14.2	40.6
17	78 14.4	41.6
18	93 14.5	S 6 42.5
19	108 14.7	43.5
20	123 14.9	44.4
21	138 15.0	·· 45.3
22	153 15.2	46.3
23	168 15.3	47.2
11 00	183 15.5	S 6 48.2
01	198 15.7	49.1
02	213 15.8	50.1
03	228 16.0	·· 51.0
04	243 16.2	52.0
05	258 16.3	52.9
06	273 16.5	S 6 53.9
07	288 16.7	54.8
08	303 16.8	55.7
09	318 17.0	·· 56.7
10	333 17.1	57.6
11	348 17.3	58.6
12	3 17.5	S 6 59.5
13	18 17.6	7 00.5
14	33 17.8	01.4
15	48 18.0	·· 02.3
16	63 18.1	03.3
17	78 18.3	04.2
18	93 18.4	S 7 05.2
19	108 18.6	06.1
20	123 18.8	07.1
21	138 18.9	·· 08.0
22	153 19.1	09.0
23	168 19.2	09.9
12 00	183 19.4	S 7 10.8
01	198 19.6	11.8
02	213 19.7	12.7
03	228 19.9	·· 13.6
04	243 20.0	14.6
05	258 20.2	15.5
06	273 20.4	S 7 16.5
07	288 20.5	17.4
08	303 20.7	18.3
09	318 20.8	·· 19.3
10	333 21.0	20.2
11	348 21.2	21.2
12	3 21.3	S 7 22.1
13	18 21.5	23.0
14	33 21.6	24.0
15	48 21.8	·· 24.9
16	63 21.9	25.9
17	78 22.1	26.8
18	93 22.2	S 7 27.7
19	108 22.4	28.7
20	123 22.6	29.6
21	138 22.7	·· 30.5
22	153 22.9	31.5
23	168 23.0	32.4

d h	G.H.A.	Dec.
13 00	183 23.2	S 7 33.4
01	198 23.3	34.3
02	213 23.5	35.2
03	228 23.6	·· 36.2
04	243 23.8	37.1
05	258 23.9	38.0
06	273 24.1	S 7 39.0
07	288 24.2	39.9
08	303 24.4	40.8
09	318 24.6	·· 41.8
10	333 24.7	42.7
11	348 24.9	43.7
12	3 25.0	S 7 44.6
13	18 25.2	45.5
14	33 25.3	46.5
15	48 25.5	·· 47.4
16	63 25.6	48.3
17	78 25.8	49.3
18	93 25.9	S 7 50.2
19	108 26.1	51.1
20	123 26.2	52.1
21	138 26.4	·· 53.0
22	153 26.5	53.9
23	168 26.7	54.9
14 00	183 26.8	S 7 55.8
01	198 27.0	56.7
02	213 27.1	57.7
03	228 27.3	·· 58.6
04	243 27.4	7 59.5
05	258 27.5	8 00.4
06	273 27.7	S 8 01.4
07	288 27.8	02.3
08	303 28.0	03.2
09	318 28.1	·· 04.2
10	333 28.3	05.1
11	348 28.4	06.0
12	3 28.6	S 8 07.0
13	18 28.7	07.9
14	33 28.9	08.8
15	48 29.0	·· 09.8
16	63 29.2	10.7
17	78 29.3	11.6
18	93 29.4	S 8 12.5
19	108 29.6	13.5
20	123 29.7	14.4
21	138 29.9	·· 15.3
22	153 30.0	16.3
23	168 30.2	17.2
15 00	183 30.3	S 8 18.1
01	198 30.5	19.0
02	213 30.6	20.0
03	228 30.7	·· 20.9
04	243 30.9	21.8
05	258 31.0	22.7
06	273 31.2	S 8 23.7
07	288 31.3	24.6
08	303 31.4	25.5
09	318 31.6	·· 26.5
10	333 31.7	27.4
11	348 31.9	28.3
12	3 32.0	S 8 29.2
13	18 32.1	30.2
14	33 32.3	31.1
15	48 32.4	·· 32.0
16	63 32.6	32.9
17	78 32.7	33.9
18	93 32.8	S 8 34.8
19	108 33.0	35.7
20	123 33.1	36.6
21	138 33.3	·· 37.5
22	153 33.4	38.5
23	168 33.5	39.4

d h	G.H.A.	Dec.
16 00	183 33.7	S 8 40.3
01	198 33.8	41.2
02	213 34.0	42.2
03	228 34.1	·· 43.1
04	243 34.2	44.0
05	258 34.4	44.9
06	273 34.5	S 8 45.9
07	288 34.6	46.8
08	303 34.8	47.7
09	318 34.9	·· 48.6
10	333 35.0	49.5
11	348 35.2	50.5
12	3 35.3	S 8 51.4
13	18 35.4	52.3
14	33 35.6	53.2
15	48 35.7	·· 54.1
16	63 35.8	55.1
17	78 36.0	56.0
18	93 36.1	S 8 56.9
19	108 36.2	57.8
20	123 36.4	58.7
21	138 36.5	8 59.6
22	153 36.6	.9 00.6
23	168 36.8	01.5
17 00	183 36.9	S 9 02.4
01	198 37.0	03.3
02	213 37.2	04.2
03	228 37.3	·· 05.2
04	243 37.4	06.1
05	258 37.6	07.0
06	273 37.7	S 9 07.9
07	288 37.8	08.8
08	303 38.0	09.7
09	318 38.1	·· 10.7
10	333 38.2	11.6
11	348 38.3	12.5
12	3 38.5	S 9 13.4
13	18 38.6	14.3
14	33 38.7	15.2
15	48 38.9	·· 16.1
16	63 39.0	17.1
17	78 39.1	18.0
18	93 39.2	S 9 18.9
19	108 39.4	19.8
20	123 39.5	20.7
21	138 39.6	·· 21.6
22	153 39.7	22.5
23	168 39.9	23.4
18 00	183 40.0	S 9 24.4
01	198 40.1	25.3
02	213 40.2	26.2
03	228 40.4	·· 27.1
04	243 40.5	28.0
05	258 40.6	28.9
06	273 40.7	S 9 29.8
07	288 40.9	30.7
08	303 41.0	31.6
09	318 41.1	·· 32.6
10	333 41.2	33.5
11	348 41.4	34.4
12	3 41.5	S 9 35.3
13	18 41.6	36.2
14	33 41.7	37.1
15	48 41.8	·· 38.0
16	63 42.0	38.9
17	78 42.1	39.8
18	93 42.2	S 9 40.7
19	108 42.3	41.6
20	123 42.4	42.6
21	138 42.6	·· 43.5
22	153 42.7	44.4
23	168 42.8	45.3

d h	G.H.A.	Dec.
19 00	183 42.9	S 9 46.2
01	198 43.0	47.1
02	213 43.2	48.0
03	228 43.3	·· 48.9
04	243 43.4	49.8
05	258 43.5	50.7
06	273 43.6	S 9 51.6
07	288 43.8	52.5
08	303 43.9	53.4
09	318 44.0	·· 54.3
10	333 44.1	55.2
11	348 44.2	56.1
12	3 44.3	S 9 57.0
13	18 44.5	57.9
14	33 44.6	58.8
15	48 44.7	9 59.7
16	63 44.8	10 00.7
17	78 44.9	01.6
18	93 45.0	S10 02.5
19	108 45.1	03.4
20	123 45.3	04.3
21	138 45.4	·· 05.2
22	153 45.5	06.1
23	168 45.6	07.0
20 00	183 45.7	S10 07.9
01	198 45.8	08.8
02	213 45.9	09.7
03	228 46.0	·· 10.6
04	243 46.2	11.5
05	258 46.3	12.4
06	273 46.4	S10 13.3
07	288 46.5	14.2
08	303 46.6	15.1
09	318 46.7	·· 16.0
10	333 46.8	16.9
11	348 46.9	17.7
12	3 47.0	S10 18.6
13	18 47.1	19.5
14	33 47.3	20.4
15	48 47.4	·· 21.3
16	63 47.5	22.2
17	78 47.6	23.1
18	93 47.7	S10 24.0
19	108 47.8	24.9
20	123 47.9	25.8
21	138 48.0	·· 26.7
22	153 48.1	27.6
23	168 48.2	28.5
21 00	183 48.3	S10 29.4
01	198 48.4	30.3
02	213 48.5	31.2
03	228 48.6	·· 32.1
04	243 48.7	33.0
05	258 48.9	33.9
06	273 49.0	S10 34.8
07	288 49.1	35.6
08	303 49.2	36.5
09	318 49.3	·· 37.4
10	333 49.4	38.3
11	348 49.5	39.2
12	3 49.6	S10 40.1
13	18 49.7	41.0
14	33 49.8	41.9
15	48 49.9	·· 42.8
16	63 50.0	43.7
17	78 50.1	44.6
18	93 50.2	S10 45.4
19	108 50.3	46.3
20	123 50.4	47.2
21	138 50.5	·· 48.1
22	153 50.6	49.0
23	168 50.7	49.9

d h	G.H.A.	Dec.
22 00	183 50.8	S10 50.8
01	198 50.9	51.7
02	213 51.0	52.5
03	228 51.1	·· 53.4
04	243 51.2	54.3
05	258 51.3	55.2
06	273 51.4	S10 56.1
07	288 51.5	57.0
08	303 51.6	57.9
09	318 51.7	·· 58.7
10	333 51.8	10 59.6
11	348 51.8	11 00.5
12	3 51.9	S11 01.4
13	18 52.0	02.3
14	33 52.1	03.2
15	48 52.2	·· 04.1
16	63 52.3	04.9
17	78 52.4	05.8
18	93 52.5	S11 06.7
19	108 52.6	07.6
20	123 52.7	08.5
21	138 52.8	·· 09.3
22	153 52.9	10.2
23	168 53.0	11.1
23 00	183 53.1	S11 12.0
01	198 53.2	12.9
02	213 53.2	13.8
03	228 53.3	·· 14.6
04	243 53.4	15.5
05	258 53.5	16.4
06	273 53.6	S11 17.3
07	288 53.7	18.1
08	303 53.8	19.0
09	318 53.9	·· 19.9
10	333 54.0	20.8
11	348 54.1	21.7
12	3 54.1	S11 22.5
13	18 54.2	23.4
14	33 54.3	24.3
15	48 54.4	·· 25.2
16	63 54.5	26.0
17	78 54.6	26.9
18	93 54.7	S11 27.8
19	108 54.7	28.7
20	123 54.8	29.5
21	138 54.9	·· 30.4
22	153 55.0	31.3
23	168 55.1	32.2
24 00	183 55.2	S11 33.0
01	198 55.3	33.9
02	213 55.3	34.8
03	228 55.4	·· 35.7
04	243 55.5	36.5
05	258 55.6	37.4
06	273 55.7	S11 38.3
07	288 55.7	39.2
08	303 55.8	40.0
09	318 55.9	·· 40.9
10	333 56.0	41.8
11	348 56.1	42.6
12	3 56.2	S11 43.5
13	18 56.2	44.4
14	33 56.3	45.2
15	48 56.4	·· 46.1
16	63 56.5	47.0
17	78 56.6	47.8
18	93 56.6	S11 48.7
19	108 56.7	49.6
20	123 56.8	50.5
21	138 56.9	·· 51.3
22	153 56.9	52.2
23	168 57.0	53.1

d	h	G.H.A. ° ′	Dec. ° ′
25	00	183 57.1	S11 53.9
	01	198 57.2	54.8
	02	213 57.2	55.7
	03	228 57.3	·· 56.5
	04	243 57.4	57.4
	05	258 57.5	58.2
	06	273 57.5	S11 59.1
	07	288 57.6	12 00.0
	08	303 57.7	00.8
	09	318 57.8	·· 01.7
	10	333 57.8	02.6
	11	348 57.9	03.4
	12	3 58.0	S12 04.3
	13	18 58.1	05.2
	14	33 58.1	06.0
	15	48 58.2	·· 06.9
	16	63 58.3	07.7
	17	78 58.3	08.6
	18	93 58.4	S12 09.5
	19	108 58.5	10.3
	20	123 58.6	11.2
	21	138 58.6	·· 12.0
	22	153 58.7	12.9
	23	168 58.8	13.8
26	00	183 58.8	S12 14.6
	01	198 58.9	15.5
	02	213 59.0	16.3
	03	228 59.0	·· 17.2
	04	243 59.1	18.0
	05	258 59.2	18.9
	06	273 59.2	S12 19.8
	07	288 59.3	20.6
	08	303 59.4	21.5
	09	318 59.4	·· 22.3
	10	333 59.5	23.2
	11	348 59.6	24.0
	12	3 59.6	S12 24.9
	13	18 59.7	25.7
	14	33 59.8	26.6
	15	48 59.8	·· 27.5
	16	63 59.9	28.3
	17	79 00.0	29.2
	18	94 00.0	S12 30.0
	19	109 00.1	30.9
	20	124 00.1	31.7
	21	139 00.2	·· 32.6
	22	154 00.3	33.4
	23	169 00.3	34.3
27	00	184 00.4	S12 35.1
	01	199 00.4	36.0
	02	214 00.5	36.8
	03	229 00.6	·· 37.7
	04	244 00.6	38.5
	05	259 00.7	39.4
	06	274 00.7	S12 40.2
	07	289 00.8	41.1
	08	304 00.9	41.9
	09	319 00.9	·· 42.8
	10	334 01.0	43.6
	11	349 01.0	44.5
	12	4 01.1	S12 45.3
	13	19 01.2	46.2
	14	34 01.2	47.0
	15	49 01.3	·· 47.8
	16	64 01.3	48.7
	17	79 01.4	49.5
	18	94 01.4	S12 50.4
	19	109 01.5	51.2
	20	124 01.5	52.1
	21	139 01.6	·· 52.9
	22	154 01.6	53.8
	23	169 01.7	54.6

d	h	G.H.A. ° ′	Dec. ° ′
28	00	184 01.8	S12 55.4
	01	199 01.8	56.3
	02	214 01.9	57.1
	03	229 01.9	·· 58.0
	04	244 02.0	58.8
	05	259 02.0	12 59.6
	06	274 02.1	S13 00.5
	07	289 02.1	01.3
	08	304 02.2	02.2
	09	319 02.2	·· 03.0
	10	334 02.3	03.8
	11	349 02.3	04.7
	12	4 02.4	S13 05.5
	13	19 02.4	06.4
	14	34 02.5	07.2
	15	49 02.5	·· 08.0
	16	64 02.6	08.9
	17	79 02.6	09.7
	18	94 02.7	S13 10.5
	19	109 02.7	11.4
	20	124 02.7	12.2
	21	139 02.8	·· 13.1
	22	154 02.8	13.9
	23	169 02.9	14.7
29	00	184 02.9	S13 15.6
	01	199 03.0	16.4
	02	214 03.0	17.2
	03	229 03.1	·· 18.1
	04	244 03.1	18.9
	05	259 03.1	19.7
	06	274 03.2	S13 20.5
	07	289 03.2	21.4
	08	304 03.3	22.2
	09	319 03.3	·· 23.0
	10	334 03.4	23.9
	11	349 03.4	24.7
	12	4 03.4	S13 25.5
	13	19 03.5	26.4
	14	34 03.5	27.2
	15	49 03.6	·· 28.0
	16	64 03.6	28.8
	17	79 03.6	29.7
	18	94 03.7	S13 30.5
	19	109 03.7	31.3
	20	124 03.8	32.2
	21	139 03.8	·· 33.0
	22	154 03.8	33.8
	23	169 03.9	34.6
30	00	184 03.9	S13 35.5
	01	199 03.9	36.3
	02	214 04.0	37.1
	03	229 04.0	·· 37.9
	04	244 04.1	38.8
	05	259 04.1	39.6
	06	274 04.1	S13 40.4
	07	289 04.2	41.2
	08	304 04.2	42.0
	09	319 04.2	·· 42.9
	10	334 04.3	43.7
	11	349 04.3	44.5
	12	4 04.3	S13 45.3
	13	19 04.4	46.2
	14	34 04.4	47.0
	15	49 04.4	·· 47.8
	16	64 04.5	48.6
	17	79 04.5	49.4
	18	94 04.5	S13 50.2
	19	109 04.5	51.1
	20	124 04.6	51.9
	21	139 04.6	·· 52.7
	22	154 04.6	53.5
	23	169 04.7	54.3

NOVEMBER

d	h	G.H.A. ° ′	Dec. ° ′
31	00	184 04.7	S13 55.2
	01	199 04.7	56.0
	02	214 04.8	56.8
	03	229 04.8	·· 57.6
	04	244 04.8	58.4
	05	259 04.8	13 59.2
	06	274 04.9	S14 00.0
	07	289 04.9	00.9
	08	304 04.9	01.7
	09	319 04.9	·· 02.5
	10	334 05.0	03.3
	11	349 05.0	04.1
	12	4 05.0	S14 04.9
	13	19 05.0	05.7
	14	34 05.1	06.5
	15	49 05.1	·· 07.3
	16	64 05.1	08.2
	17	79 05.1	09.0
	18	94 05.2	S14 09.8
	19	109 05.2	10.6
	20	124 05.2	11.4
	21	139 05.2	·· 12.2
	22	154 05.2	13.0
	23	169 05.3	13.8
1	00	184 05.3	S14 14.6
	01	199 05.3	15.4
	02	214 05.3	16.2
	03	229 05.3	·· 17.0
	04	244 05.4	17.8
	05	259 05.4	18.6
	06	274 05.4	S14 19.5
	07	289 05.4	20.3
	08	304 05.4	21.1
	09	319 05.5	·· 21.9
	10	334 05.5	22.7
	11	349 05.5	23.5
	12	4 05.5	S14 24.3
	13	19 05.5	25.1
	14	34 05.5	25.9
	15	49 05.6	·· 26.7
	16	64 05.6	27.5
	17	79 05.6	28.3
	18	94 05.6	S14 29.1
	19	109 05.6	29.9
	20	124 05.6	30.7
	21	139 05.6	·· 31.5
	22	154 05.7	32.3
	23	169 05.7	33.1
2	00	184 05.7	S14 33.9
	01	199 05.7	34.7
	02	214 05.7	35.5
	03	229 05.7	·· 36.3
	04	244 05.7	37.0
	05	259 05.7	37.8
	06	274 05.8	S14 38.6
	07	289 05.8	39.4
	08	304 05.8	40.2
	09	319 05.8	·· 41.0
	10	334 05.8	41.8
	11	349 05.8	42.6
	12	4 05.8	S14 43.4
	13	19 05.8	44.2
	14	34 05.8	45.0
	15	49 05.8	·· 45.8
	16	64 05.8	46.6
	17	79 05.8	47.3
	18	94 05.9	S14 48.1
	19	109 05.9	48.9
	20	124 05.9	49.7
	21	139 05.9	·· 50.5
	22	154 05.9	51.3
	23	169 05.9	52.1

d h	G.H.A. ° '	Dec. ° '
3 00	184 05.9	S14 52.9
01	199 05.9	53.7
02	214 05.9	54.4
03	229 05.9 ··	55.2
04	244 05.9	56.0
05	259 05.9	56.8
06	274 05.9	S14 57.6
07	289 05.9	58.4
08	304 05.9	59.1
09	319 05.9	14 59.9
10	334 05.9	15 00.7
11	349 05.9	01.5
12	4 05.9	S15 02.3
13	19 05.9	03.1
14	34 05.9	03.8
15	49 05.9 ··	04.6
16	64 05.9	05.4
17	79 05.9	06.2
18	94 05.9	S15 07.0
19	109 05.9	07.7
20	124 05.9	08.5
21	139 05.9 ··	09.3
22	154 05.9	10.1
23	169 05.9	10.9
4 00	184 05.9	S15 11.6
01	199 05.9	12.4
02	214 05.9	13.2
03	229 05.9 ··	14.0
04	244 05.9	14.7
05	259 05.9	15.5
06	274 05.9	S15 16.3
07	289 05.8	17.1
08	304 05.8	17.8
09	319 05.8 ··	18.6
10	334 05.8	19.4
11	349 05.8	20.1
12	4 05.8	S15 20.9
13	19 05.8	21.7
14	34 05.8	22.5
15	49 05.8 ··	23.2
16	64 05.8	24.0
17	79 05.8	24.8
18	94 05.8	S15 25.5
19	109 05.7	26.3
20	124 05.7	27.1
21	139 05.7 ··	27.8
22	154 05.7	28.6
23	169 05.7	29.4
5 00	184 05.7	S15 30.1
01	199 05.7	30.9
02	214 05.7	31.7
03	229 05.6 ··	32.4
04	244 05.6	33.2
05	259 05.6	34.0
06	274 05.6	S15 34.7
07	289 05.6	35.5
08	304 05.6	36.3
09	319 05.6 ··	37.0
10	334 05.5	37.8
11	349 05.5	38.5
12	4 05.5	S15 39.3
13	19 05.5	40.1
14	34 05.5	40.8
15	49 05.5 ··	41.6
16	64 05.4	42.3
17	79 05.4	43.1
18	94 05.4	S15 43.9
19	109 05.4	44.6
20	124 05.4	45.4
21	139 05.3 ··	46.1
22	154 05.3	46.9
23	169 05.3	47.7

d h	G.H.A. ° '	Dec. ° '
6 00	184 05.3	S15 48.4
01	199 05.3	49.2
02	214 05.2	49.9
03	229 05.2 ··	50.7
04	244 05.2	51.4
05	259 05.2	52.2
06	274 05.1	S15 52.9
07	289 05.1	53.7
08	304 05.1	54.4
09	319 05.1 ··	55.2
10	334 05.1	55.9
11	349 05.0	56.7
12	4 05.0	S15 57.4
13	19 05.0	58.2
14	34 05.0	58.9
15	49 04.9	15 59.7
16	64 04.9	16 00.4
17	79 04.9	01.2
18	94 04.8	S16 01.9
19	109 04.8	02.7
20	124 04.8	03.4
21	139 04.8 ··	04.2
22	154 04.7	04.9
23	169 04.7	05.7
7 00	184 04.7	S16 06.4
01	199 04.6	07.1
02	214 04.6	07.9
03	229 04.6 ··	08.6
04	244 04.6	09.4
05	259 04.5	10.1
06	274 04.5	S16 10.9
07	289 04.5	11.6
08	304 04.4	12.3
09	319 04.4 ··	13.1
10	334 04.4	13.8
11	349 04.3	14.6
12	4 04.3	S16 15.3
13	19 04.3	16.0
14	34 04.2	16.8
15	49 04.2 ··	17.5
16	64 04.2	18.3
17	79 04.1	19.0
18	94 04.1	S16 19.7
19	109 04.1	20.5
20	124 04.0	21.2
21	139 04.0 ··	21.9
22	154 03.9	22.7
23	169 03.9	23.4
8 00	184 03.9	S16 24.1
01	199 03.8	24.9
02	214 03.8	25.6
03	229 03.8 ··	26.3
04	244 03.7	27.1
05	259 03.7	27.8
06	274 03.6	S16 28.5
07	289 03.6	29.3
08	304 03.6	30.0
09	319 03.5 ··	30.7
10	334 03.5	31.4
11	349 03.4	32.2
12	4 03.4	S16 32.9
13	19 03.3	33.6
14	34 03.3	34.3
15	49 03.3 ··	35.1
16	64 03.2	35.8
17	79 03.2	36.5
18	94 03.1	S16 37.2
19	109 03.1	38.0
20	124 03.0	38.7
21	139 03.0 ··	39.4
22	154 02.9	40.1
23	169 02.9	40.9

d h	G.H.A. ° '	Dec. ° '
9 00	184 02.9	S16 41.6
01	199 02.8	42.3
02	214 02.8	43.0
03	229 02.7 ··	43.7
04	244 02.7	44.5
05	259 02.6	45.2
06	274 02.6	S16 45.9
07	289 02.5	46.6
08	304 02.5	47.3
09	319 02.4 ··	48.1
10	334 02.4	48.8
11	349 02.3	49.5
12	4 02.3	S16 50.2
13	19 02.2	50.9
14	34 02.2	51.6
15	49 02.1 ··	52.4
16	64 02.1	53.1
17	79 02.0	53.8
18	94 02.0	S16 54.5
19	109 01.9	55.2
20	124 01.9	55.9
21	139 01.8 ··	56.6
22	154 01.7	57.3
23	169 01.7	58.0
10 00	184 01.6	S16 58.8
01	199 01.6	16 59.5
02	214 01.5	17 00.2
03	229 01.5 ··	00.9
04	244 01.4	0I.6
05	259 01.4	02.3
06	274 01.3	S17 03.0
07	289 01.2	03.7
08	304 01.2	04.4
09	319 01.1 ··	05.1
10	334 01.1	05.8
11	349 01.0	06.5
12	4 00.9	S17 07.2
13	19 00.9	07.9
14	34 00.8	08.6
15	49 00.8 ··	09.3
16	64 00.7	10.0
17	79 00.6	10.7
18	94 00.6	S17 11.4
19	109 00.5	12.1
20	124 00.5	12.8
21	139 00.4 ··	13.5
22	154 00.3	14.2
23	169 00.3	14.9
11 00	184 00.2	S17 15.6
01	199 00.1	16.3
02	214 00.1	17.0
03	229 00.0 ··	17.7
04	243 59.9	18.4
05	258 59.9	19.1
06	273 59.8	S17 19.8
07	288 59.8	20.5
08	303 59.7	21.2
09	318 59.6 ··	21.9
10	333 59.6	22.6
11	348 59.5	23.3
12	3 59.4	S17 24.0
13	18 59.3	24.7
14	33 59.3	25.3
15	48 59.2 ··	26.0
16	63 59.1	26.7
17	78 59.1	27.4
18	93 59.0	S17 28.1
19	108 58.9	28.8
20	123 58.9	29.5
21	138 58.8 ··	30.2
22	153 58.7	30.9
23	168 58.6	31.5

	G.H.A.	Dec.			G.H.A.	Dec.			G.H.A.	Dec.
d h	° ′	° ′		**d h**	° ′	° ′		**d h**	° ′	° ′
12 00	183 58.6	S17 32.2		**15** 00	183 52.4	S18 20.1		**18** 00	183 44.3	S19 05.1
01	198 58.5	32.9		01	198 52.3	20.8		01	198 44.2	05.8
02	213 58.4	33.6		02	213 52.2	21.4		02	213 44.1	06.4
03	228 58.3 ··	34.3		03	228 52.1 ··	22.1		03	228 43.9 ··	07.0
04	243 58.3	35.0		04	243 52.0	22.7		04	243 43.8	07.6
05	258 58.2	35.6		05	258 51.9	23.4		05	258 43.7	08.2
06	273 58.1	S17 36.3		06	273 51.8	S18 24.0		06	273 43.6	S19 08.8
07	288 58.1	37.0		07	288 51.7	24.6		07	288 43.4	09.4
08	303 58.0	37.7		08	303 51.6	25.3		08	303 43.3	10.0
09	318 57.9 ··	38.4		09	318 51.5 ··	25.9		09	318 43.2 ··	10.6
10	333 57.8	39.0		10	333 51.4	26.6		10	333 43.0	11.2
11	348 57.7	39.7		11	348 51.3	27.2		11	348 42.9	11.8
12	3 57.7	S17 40.4		12	3 51.2	S18 27.8		12	3 42.8	S19 12.4
13	18 57.6	41.1		13	18 51.1	28.5		13	18 42.6	13.0
14	33 57.5	41.8		14	33 51.0	29.1		14	33 42.5	13.5
15	48 57.4 ··	42.4		15	48 50.9 ··	29.8		15	48 42.4 ··	14.1
16	63 57.4	43.1		16	63 50.8	30.4		16	63 42.3	14.7
17	78 57.3	43.8		17	78 50.7	31.0		17	78 42.1	15.3
18	93 57.2	S17 44.5		18	93 50.6	S18 31.7		18	93 42.0	S19 15.9
19	108 57.1	45.1		19	108 50.4	32.3		19	108 41.9	16.5
20	123 57.0	45.8		20	123 50.3	32.9		20	123 41.7	17.1
21	138 57.0 ··	46.5		21	138 50.2 ··	33.6		21	138 41.6 ··	17.7
22	153 56.9	47.2		22	153 50.1	34.2		22	153 41.5	18.3
23	168 56.8	47.8		23	168 50.0	34.8		23	168 41.3	18.9
13 00	183 56.7	S17 48.5		**16** 00	183 49.9	S18 35.5		**19** 00	183 41.2	S19 19.5
01	198 56.6	49.2		01	198 49.8	36.1		01	198 41.1	20.1
02	213 56.6	49.8		02	213 49.7	36.7		02	213 40.9	20.7
03	228 56.5 ··	50.5		03	228 49.6 ··	37.4		03	228 40.8 ··	21.2
04	243 56.4	51.2		04	243 49.5	38.0		04	243 40.7	21.8
05	258 56.3	51.9		05	258 49.4	38.6		05	258 40.5	22.4
06	273 56.2	S17 52.5		06	273 49.3	S18 39.3		06	273 40.4	S19 23.0
07	288 56.1	53.2		07	288 49.1	39.9		07	288 40.2	23.6
08	303 56.1	53.9		08	303 49.0	40.5		08	303 40.1	24.2
09	318 56.0 ··	54.5		09	318 48.9 ··	41.1		09	318 40.0 ··	24.8
10	333 55.9	55.2		10	333 48.8	41.8		10	333 39.8	25.3
11	348 55.8	55.9		11	348 48.7	42.4		11	348 39.7	25.9
12	3 55.7	S17 56.5		12	3 48.6	S18 43.0		12	3 39.6	S19 26.5
13	18 55.6	57.2		13	18 48.5	43.6		13	18 39.4	27.1
14	33 55.5	57.9		14	33 48.4	44.3		14	33 39.3	27.7
15	48 55.5 ··	58.5		15	48 48.3 ··	44.9		15	48 39.1 ··	28.3
16	63 55.4	59.2		16	63 48.1	45.5		16	63 39.0	28.8
17	78 55.3	17 59.9		17	78 48.0	46.1		17	78 38.9	29.4
18	93 55.2	S18 00.5		18	93 47.9	S18 46.8		18	93 38.7	S19 30.0
19	108 55.1	01.2		19	108 47.8	47.4		19	108 38.6	30.6
20	123 55.0	01.8		20	123 47.7	48.0		20	123 38.4	31.2
21	138 54.9 ··	02.5		21	138 47.6 ··	48.6		21	138 38.3 ··	31.7
22	153 54.8	03.2		22	153 47.4	49.2		22	153 38.2	32.3
23	168 54.8	03.8		23	168 47.3	49.9		23	168 38.0	32.9
14 00	183 54.7	S18 04.5		**17** 00	183 47.2	S18 50.5		**20** 00	183 37.9	S19 33.5
01	198 54.6	05.1		01	198 47.1	51.1		01	198 37.7	34.0
02	213 54.5	05.8		02	213 47.0	51.7		02	213 37.6	34.6
03	228 54.4 ··	06.5		03	228 46.9 ··	52.3		03	228 37.4 ··	35.2
04	243 54.3	07.1		04	243 46.7	52.9		04	243 37.3	35.8
05	258 54.2	07.8		05	258 46.6	53.6		05	258 37.1	36.3
06	273 54.1	S18 08.4		06	273 46.5	S18 54.2		06	273 37.0	S19 36.9
07	288 54.0	09.1		07	288 46.4	54.8		07	288 36.9	37.5
08	303 53.9	09.7		08	303 46.3	55.4		08	303 36.7	38.0
09	318 53.8 ··	10.4		09	318 46.2 ··	56.0		09	318 36.6 ··	38.6
10	333 53.7	11.0		10	333 46.0	56.6		10	333 36.4	39.2
11	348 53.7	11.7		11	348 45.9	57.2		11	348 36.3	39.8
12	3 53.6	S18 12.3		12	3 45.8	S18 57.9		12	3 36.1	S19 40.3
13	18 53.5	13.0		13	18 45.7	58.5		13	18 36.0	40.9
14	33 53.4	13.6		14	33 45.5	59.1		14	33 35.8	41.5
15	48 53.3 ··	14.3		15	48 45.4	18 59.7		15	48 35.7 ··	42.0
16	63 53.2	15.0		16	63 45.3	19 00.3		16	63 35.5	42.6
17	78 53.1	15.6		17	78 45.2	00.9		17	78 35.4	43.2
18	93 53.0	S18 16.3		18	93 45.1	S19 01.5		18	93 35.2	S19 43.7
19	108 52.9	16.9		19	108 44.9	02.1		19	108 35.1	44.3
20	123 52.8	17.5		20	123 44.8	02.7		20	123 34.9	44.8
21	138 52.7 ··	18.2		21	138 44.7 ··	03.3		21	138 34.8 ··	45.4
22	153 52.6	18.8		22	153 44.6	03.9		22	153 34.6	46.0
23	168 52.5	19.5		23	168 44.4	04.5		23	168 34.5	46.5

21

d h	G.H.A.	Dec.
21 00	183 34.3	S19 47.1
01	198 34.2	47.7
02	213 34.0	48.2
03	228 33.9 ··	48.8
04	243 33.7	49.3
05	258 33.6	49.9
06	273 33.4	S19 50.4
07	288 33.3	51.0
08	303 33.1	51.6
09	318 33.0 ··	52.1
10	333 32.8	52.7
11	348 32.6	53.2
12	3 32.5	S19 53.8
13	18 32.3	54.3
14	33 32.2	54.9
15	48 32.0 ··	55.4
16	63 31.9	56.0
17	78 31.7	56.5
18	93 31.5	S19 57.1
19	108 31.4	57.6
20	123 31.2	58.2
21	138 31.1 ··	58.7
22	153 30.9	.59.3
23	168 30.7	19 59.8
22 00	183 30.6	S20 00.4
01	198 30.4	00.9
02	213 30.3	01.4
03	228 30.1 ··	02.0
04	243 29.9	02.5
05	258 29.8	03.1
06	273 29.6	S20 03.6
07	288 29.5	04.2
08	303 29.3	04.7
09	318 29.1 ··	05.2
10	333 29.0	05.8
11	348 28.8	06.3
12	3 28.6	S20 06.9
13	18 28.5	07.4
14	33 28.3	07.9
15	48 28.1 ··	08.5
16	63 28.0	09.0
17	78 27.8	09.5
18	93 27.6	S20 10.1
19	108 27.5	10.6
20	123 27.3	11.1
21	138 27.1 ··	11.7
22	153 27.0	12.2
23	168 26.8	12.7
23 00	183 26.6	S20 13.3
01	198 26.5	13.8
02	213 26.3	14.3
03	228 26.1 ··	14.9
04	243 26.0	15.4
05	258 25.8	15.9
06	273 25.6	S20 16.4
07	288 25.5	17.0
08	303 25.3	17.5
09	318 25.1 ··	18.0
10	333 24.9	18.5
11	348 24.8	19.1
12	3 24.6	S20 19.6
13	18 24.4	20.1
14	33 24.2	20.6
15	48 24.1 ··	21.1
16	63 23.9	21.7
17	78 23.7	22.2
18	93 23.5	S20 22.7
19	108 23.4	23.2
20	123 23.2	23.7
21	138 23.0 ··	24.3
22	153 22.8	24.8
23	168 22.7	25.3

d h	G.H.A.	Dec.
24 00	183 22.5	S20 25.8
01	198 22.3	26.3
02	213 22.1	26.8
03	228 22.0 ··	27.3
04	243 21.8	27.8
05	258 21.6	28.4
06	273 21.4	S20 28.9
07	288 21.2	29.4
08	303 21.1	29.9
09	318 20.9 ··	30.4
10	333 20.7	30.9
11	348 20.5	31.4
12	3 20.3	S20 31.9
13	18 20.2	32.4
14	33 20.0	32.9
15	48 19.8 ··	33.4
16	63 19.6	33.9
17	78 19.4	34.4
18	93 19.3	S20 34.9
19	108 19.1	35.5
20	123 18.9	36.0
21	138 18.7 ··	36.5
22	153 18.5	37.0
23	168 18.3	37.5
25 00	183 18.1	S20 38.0
01	198 18.0	38.4
02	213 17.8	38.9
03	228 17.6 ··	39.4
04	243 17.4	39.9
05	258 17.2	40.4
06	273 17.0	S20 40.9
07	288 16.8	41.4
08	303 16.7	41.9
09	318 16.5 ··	42.4
10	333 16.3	42.9
11	348 16.1	43.4
12	3 15.9	S20 43.9
13	18 15.7	44.4
14	33 15.5	44.9
15	48 15.3 ··	45.4
16	63 15.1	45.8
17	78 15.0	46.3
18	93 14.8	S20 46.8
19	108 14.6	47.3
20	123 14.4	47.8
21	138 14.2 ··	48.3
22	153 14.0	48.8
23	168 13.8	49.2
26 00	183 13.6	S20 49.7
01	198 13.4	50.2
02	213 13.2	50.7
03	228 13.0 ··	51.2
04	243 12.8	51.6
05	258 12.6	52.1
06	273 12.4	S20 52.6
07	288 12.2	53.1
08	303 12.1	53.6
09	318 11.9 ··	54.0
10	333 11.7	54.5
11	348 11.5	55.0
12	3 11.3	S20 55.5
13	18 11.1	55.9
14	33 10.9	56.4
15	48 10.7 ··	56.9
16	63 10.5	57.4
17	78 10.3	57.8
18	93 10.1	S20 58.3
19	108 09.9	58.8
20	123 09.7	59.2
21	138 09.5	20 59.7
22	153 09.3	21 00.2
23	168 09.1	00.6

d h	G.H.A.	Dec.
27 00	183 08.9	S21 01.1
01	198 08.7	01.6
02	213 08.5	02.0
03	228 08.3 ··	02.5
04	243 08.1	03.0
05	258 07.9	03.4
06	273 07.7	S21 03.9
07	288 07.5	04.4
08	303 07.3	04.8
09	318 07.1 ··	05.3
10	333 06.9	05.7
11	348 06.6	06.2
12	3 06.4	S21 06.7
13	18 06.2	07.1
14	33 06.0	07.6
15	48 05.8 ··	08.0
16	63 05.6	08.5
17	78 05.4	08.9
18	93 05.2	S21 09.4
19	108 05.0	09.8
20	123 04.8	10.3
21	138 04.6 ··	10.7
22	153 04.4	11.2
23	168 04.2	11.6
28 00	183 04.0	S21 12.1
01	198 03.8	12.5
02	213 03.5	13.0
03	228 03.3 ··	13.4
04	243 03.1	13.9
05	258 02.9	14.3
06	273 02.7	S21 14.8
07	288 02.5	15.2
08	303 02.3	15.7
09	318 02.1 ··	16.1
10	333 01.9	16.6
11	348 01.6	17.0
12	3 01.4	S21 17.4
13	18 01.2	17.9
14	33 01.0	18.3
15	48 00.8 ··	18.8
16	63 00.6	19.2
17	78 00.4	19.6
18	93 00.2	S21 20.1
19	107 59.9	20.5
20	122 59.7	21.0
21	137 59.5 ··	21.4
22	152 59.3	21.8
23	167 59.1	22.3
29 00	182 58.9	S21 22.7
01	197 58.7	23.1
02	212 58.4	23.6
03	227 58.2 ··	24.0
04	242 58.0	24.4
05	257 57.8	24.8
06	272 57.6	S21 25.3
07	287 57.3	25.7
08	302 57.1	26.1
09	317 56.9 ··	26.6
10	332 56.7	27.0
11	347 56.5	27.4
12	2 56.3	S21 27.8
13	17 56.0	28.3
14	32 55.8	28.7
15	47 55.6 ··	29.1
16	62 55.4	29.5
17	77 55.2	29.9
18	92 54.9	S21 30.4
19	107 54.7	30.8
20	122 54.5	31.2
21	137 54.3 ··	31.6
22	152 54.0	32.0
23	167 53.8	32.5

d h	G.H.A.	Dec.
30 00	182 53.6	S21 32.9
01	197 53.4	33.3
02	212 53.1	33.7
03	227 52.9 ··	34.1
04	242 52.7	34.5
05	257 52.5	34.9
06	272 52.3	S21 35.4
07	287 52.0	35.8
08	302 51.8	36.2
09	317 51.6 ··	36.6
10	332 51.3	37.0
11	347 51.1	37.4
12	2 50.9	S21 37.8
13	17 50.7	38.2
14	32 50.4	38.6
15	47 50.2 ··	39.0
16	62 50.0	39.4
17	77 49.8	39.8
18	92 49.5	S21 40.2
19	107 49.3	40.6
20	122 49.1	41.1
21	137 48.8 ··	41.5
22	152 48.6	41.9
23	167 48.4	42.3
1 00	182 48.2	S21 42.7
01	197 47.9	43.1
02	212 47.7	43.4
03	227 47.5 ··	43.8
04	242 47.2	44.2
05	257 47.0	44.6
06	272 46.8	S21 45.0
07	287 46.5	45.4
08	302 46.3	45.8
09	317 46.1 ··	46.2
10	332 45.8	46.6
11	347 45.6	47.0
12	2 45.4	S21 47.4
13	17 45.1	47.8
14	32 44.9	48.2
15	47 44.7 ··	48.6
16	62 44.4	48.9
17	77 44.2	49.3
18	92 44.0	S21 49.7
19	107 43.7	50.1
20	122 43.5	50.5
21	137 43.3 ··	50.9
22	152 43.0	51.3
23	167 42.8	51.6
2 00	182 42.6	S21 52.0
01	197 42.3	52.4
02	212 42.1	52.8
03	227 41.8 ··	53.2
04	242 41.6	53.5
05	257 41.4	53.9
06	272 41.1	S21 54.3
07	287 40.9	54.7
08	302 40.7	55.0
09	317 40.4 ··	55.4
10	332 40.2	55.8
11	347 39.9	56.2
12	2 39.7	S21 56.5
13	17 39.5	56.9
14	32 39.2	57.3
15	47 39.0 ··	57.7
16	62 38.7	58.0
17	77 38.5	58.4
18	92 38.2	S21 58.8
19	107 38.0	59.1
20	122 37.8	59.5
21	137 37.5	21 59.9
22	152 37.3	22 00.2
23	167 37.0	00.6

d h	G.H.A.	Dec.
3 00	182 36.8	S22 01.0
01	197 36.5	01.3
02	212 36.3	01.7
03	227 36.1 ··	02.0
04	242 35.8	02.4
05	257 35.6	02.8
06	272 35.3	S22 03.1
07	287 35.1	03.5
08	302 34.8	03.8
09	317 34.6 ··	04.2
10	332 34.3	04.6
11	347 34.1	04.9
12	2 33.9	S22 05.3
13	17 33.6	05.6
14	32 33.4	06.0
15	47 33.1 ··	06.3
16	62 32.9	06.7
17	77 32.6	07.0
18	92 32.4	S22 07.4
19	107 32.1	07.7
20	122 31.9	08.1
21	137 31.6 ··	08.4
22	152 31.4	08.8
23	167 31.1	09.1
4 00	182 30.9	S22 09.5
01	197 30.6	09.8
02	212 30.4	10.2
03	227 30.1 ··	10.5
04	242 29.9	10.9
05	257 29.6	11.2
06	272 29.4	S22 11.5
07	287 29.1	11.9
08	302 28.9	12.2
09	317 28.6 ··	12.6
10	332 28.4	12.9
11	347 28.1	13.2
12	2 27.9	S22 13.6
13	17 27.6	13.9
14	32 27.4	14.3
15	47 27.1 ··	14.6
16	62 26.9	14.9
17	77 26.6	15.3
18	92 26.4	S22 15.6
19	107 26.1	15.9
20	122 25.8	16.3
21	137 25.6 ··	16.6
22	152 25.3	16.9
23	167 25.1	17.2
5 00	182 24.8	S22 17.6
01	197 24.6	17.9
02	212 24.3	18.2
03	227 24.1 ··	18.6
04	242 23.8	18.9
05	257 23.5	19.2
06	272 23.3	S22 19.5
07	287 23.0	19.9
08	302 22.8	20.2
09	317 22.5 ··	20.5
10	332 22.3	20.8
11	347 22.0	21.1
12	2 21.7	S22 21.5
13	17 21.5	21.8
14	32 21.2	22.1
15	47 21.0 ··	22.4
16	62 20.7	22.7
17	77 20.5	23.0
18	92 20.2	S22 23.4
19	107 19.9	23.7
20	122 19.7	24.0
21	137 19.4 ··	24.3
22	152 19.2	24.6
23	167 18.9	24.9

d h	G.H.A.	Dec.
6 00	182 18.6	S22 25.2
01	197 18.4	25.5
02	212 18.1	25.9
03	227 17.8 ··	26.2
04	242 17.6	26.5
05	257 17.3	26.8
06	272 17.1	S22 27.1
07	287 16.8	27.4
08	302 16.5	27.7
09	317 16.3 ··	28.0
10	332 16.0	28.3
11	347 15.7	28.6
12	2 15.5	S22 29.0
13	17 15.2	29.2
14	32 15.0	29.5
15	47 14.7 ··	29.8
16	62 14.4	30.1
17	77 14.2	30.4
18	92 13.9	S22 30.7
19	107 13.6	31.0
20	122 13.4	31.3
21	137 13.1 ··	31.6
22	152 12.8	31.9
23	167 12.6	32.2
7 00	182 12.3	S22 32.5
01	197 12.0	32.7
02	212 11.8	33.0
03	227 11.5 ··	33.3
04	242 11.2	33.6
05	257 11.0	33.9
06	272 10.7	S22 34.2
07	287 10.4	34.5
08	302 10.2	34.8
09	317 09.9 ··	35.1
10	332 09.6	35.3
11	347 09.4	35.6
12	2 09.1	S22 35.9
13	17 08.8	36.2
14	32 08.6	36.5
15	47 08.3 ··	36.7
16	62 08.0	37.0
17	77 07.7	37.3
18	92 07.5	S22 37.6
19	107 07.2	37.9
20	122 06.9	38.1
21	137 06.7 ··	38.4
22	152 06.4	38.7
23	167 06.1	39.0
8 00	182 05.9	S22 39.2
01	197 05.6	39.5
02	212 05.3	39.8
03	227 05.0 ··	40.1
04	242 04.8	40.3
05	257 04.5	40.6
06	272 04.2	S22 40.9
07	287 04.0	41.1
08	302 03.7	41.4
09	317 03.4 ··	41.7
10	332 03.1	41.9
11	347 02.9	42.2
12	2 02.6	S22 42.5
13	17 02.3	42.7
14	32 02.0	43.0
15	47 01.8 ··	43.3
16	62 01.5	43.5
17	77 01.2	43.8
18	92 00.9	S22 44.0
19	107 00.7	44.3
20	122 00.4	44.6
21	137 00.1 ··	44.8
22	151 59.8	45.1
23	166 59.6	45.3

d h	G.H.A.	Dec.
9 00	181 59.3	S22 45.6
01	196 59.0	45.8
02	211 58.7	46.1
03	226 58.5	·· 46.3
04	241 58.2	46.6
05	256 57.9	46.8
06	271 57.6	S22 47.1
07	286 57.4	47.3
08	301 57.1	47.6
09	316 56.8	·· 47.8
10	331 56.5	48.1
11	346 56.2	48.3
12	1 56.0	S22 48.6
13	16 55.7	48.8
14	31 55.4	49.1
15	46 55.1	·· 49.3
16	61 54.8	49.6
17	76 54.6	49.8
18	91 54.3	S22 50.0
19	106 54.0	50.3
20	121 53.7	50.5
21	136 53.5	·· 50.8
22	151 53.2	51.0
23	166 52.9	51.2
10 00	181 52.6	S22 51.5
01	196 52.3	51.7
02	211 52.0	51.9
03	226 51.8	·· 52.2
04	241 51.5	52.4
05	256 51.2	52.6
06	271 50.9	S22 52.9
07	286 50.6	53.1
08	301 50.4	53.3
09	316 50.1	·· 53.6
10	331 49.8	53.8
11	346 49.5	54.0
12	1 49.2	S22 54.3
13	16 48.9	54.5
14	31 48.7	54.7
15	46 48.4	·· 54.9
16	61 48.1	55.2
17	76 47.8	55.4
18	91 47.5	S22 55.6
19	106 47.2	55.8
20	121 47.0	56.0
21	136 46.7	·· 56.3
22	151 46.4	56.5
23	166 46.1	56.7
11 00	181 45.8	S22 56.9
01	196 45.5	57.1
02	211 45.3	57.3
03	226 45.0	·· 57.6
04	241 44.7	57.8
05	256 44.4	58.0
06	271 44.1	S22 58.2
07	286 43.8	58.4
08	301 43.5	58.6
09	316 43.3	·· 58.8
10	331 43.0	59.1
11	346 42.7	59.3
12	1 42.4	S22 59.5
13	16 42.1	59.7
14	31 41.8	22 59.9
15	46 41.5	23 00.1
16	61 41.3	00.3
17	76 41.0	00.5
18	91 40.7	S23 00.7
19	106 40.4	00.9
20	121 40.1	01.1
21	136 39.8	·· 01.3
22	151 39.5	01.5
23	166 39.2	01.7

d h	G.H.A.	Dec.
12 00	181 39.0	S23 01.9
01	196 38.7	02.1
02	211 38.4	02.3
03	226 38.1	·· 02.5
04	241 37.8	02.7
05	256 37.5	02.9
06	271 37.2	S23 03.1
07	286 36.9	03.3
08	301 36.6	03.5
09	316 36.3	·· 03.7
10	331 36.1	03.8
11	346 35.8	04.0
12	1 35.5	S23 04.2
13	16 35.2	04.4
14	31 34.9	04.6
15	46 34.6	·· 04.8
16	61 34.3	05.0
17	76 34.0	05.2
18	91 33.7	S23 05.3
19	106 33.4	05.5
20	121 33.2	05.7
21	136 32.9	·· 05.9
22	151 32.6	06.1
23	166 32.3	06.3
13 00	181 32.0	S23 06.4
01	196 31.7	06.6
02	211 31.4	06.8
03	226 31.1	·· 07.0
04	241 30.8	07.1
05	256 30.5	07.3
06	271 30.2	S23 07.5
07	286 29.9	07.7
08	301 29.6	07.8
09	316 29.3	·· 08.0
10	331 29.1	08.2
11	346 28.8	08.4
12	1 28.5	S23 08.5
13	16 28.2	08.7
14	31 27.9	08.9
15	46 27.6	·· 09.0
16	61 27.3	09.2
17	76 27.0	09.4
18	91 26.7	S23 09.5
19	106 26.4	09.7
20	121 26.1	09.9
21	136 25.8	·· 10.0
22	151 25.5	10.2
23	166 25.2	10.4
14 00	181 24.9	S23 10.5
01	196 24.6	10.7
02	211 24.3	10.8
03	226 24.0	·· 11.0
04	241 23.7	11.1
05	256 23.5	11.3
06	271 23.2	S23 11.5
07	286 22.9	11.6
08	301 22.6	11.8
09	316 22.3	·· 11.9
10	331 22.0	12.1
11	346 21.7	12.2
12	1 21.4	S23 12.4
13	16 21.1	12.5
14	31 20.8	12.7
15	46 20.5	·· 12.8
16	61 20.2	13.0
17	76 19.9	13.1
18	91 19.6	S23 13.3
19	106 19.3	13.4
20	121 19.0	13.6
21	136 18.7	·· 13.7
22	151 18.4	13.8
23	166 18.1	14.0

d h	G.H.A.	Dec.
15 00	181 17.8	S23 14.1
01	196 17.5	14.3
02	211 17.2	14.4
03	226 16.9	·· 14.5
04	241 16.6	14.7
05	256 16.3	14.8
06	271 16.0	S23 15.0
07	286 15.7	15.1
08	301 15.4	15.2
09	316 15.1	·· 15.4
10	331 14.8	15.5
11	346 14.5	15.6
12	1 14.2	S23 15.8
13	16 13.9	15.9
14	31 13.6	16.0
15	46 13.3	·· 16.1
16	61 13.0	16.3
17	76 12.7	16.4
18	91 12.4	S23 16.5
19	106 12.1	16.7
20	121 11.8	16.8
21	136 11.5	·· 16.9
22	151 11.2	17.0
23	166 10.9	17.2
16 00	181 10.6	S23 17.3
01	196 10.3	17.4
02	211 10.0	17.5
03	226 09.7	·· 17.6
04	241 09.4	17.8
05	256 09.1	17.9
06	271 08.8	S23 18.0
07	286 08.5	18.1
08	301 08.2	18.2
09	316 07.9	·· 18.3
10	331 07.6	18.4
11	346 07.3	18.6
12	1 07.0	S23 18.7
13	16 06.7	18.8
14	31 06.4	18.9
15	46 06.1	·· 19.0
16	61 05.8	19.1
17	76 05.5	19.2
18	91 05.2	S23 19.3
19	106 04.9	19.4
20	121 04.6	19.5
21	136 04.3	·· 19.6
22	151 04.0	19.8
23	166 03.7	19.9
17 00	181 03.4	S23 20.0
01	196 03.1	20.1
02	211 02.7	20.2
03	226 02.4	·· 20.3
04	241 02.1	20.4
05	256 01.8	20.5
06	271 01.5	S23 20.6
07	286 01.2	20.7
08	301 00.9	20.8
09	316 00.6	·· 20.8
10	331 00.3	20.9
11	346 00.0	21.0
12	0 59.7	S23 21.1
13	15 59.4	21.2
14	30 59.1	21.3
15	45 58.8	·· 21.4
16	60 58.5	21.5
17	75 58.2	21.6
18	90 57.9	S23 21.7
19	105 57.6	21.8
20	120 57.3	21.8
21	135 57.0	·· 21.9
22	150 56.7	22.0
23	165 56.4	22.1

d h	G.H.A.	Dec.
18 00	180 56.0	S23 22.2
01	195 55.7	22.3
02	210 55.4	22.3
03	225 55.1 ··	22.4
04	240 54.8	22.5
05	255 54.5	22.6
06	270 54.2	S23 22.7
07	285 53.9	22.7
08	300 53.6	22.8
09	315 53.3 ··	22.9
10	330 53.0	23.0
11	345 52.7	23.0
12	0 52.4	S23 23.1
13	15 52.1	23.2
14	30 51.8	23.3
15	45 51.5 ··	23.3
16	60 51.1	23.4
17	75 50.8	23.5
18	90 50.5	S23 23.5
19	105 50.2	23.6
20	120 49.9	23.7
21	135 49.6 ··	23.7
22	150 49.3	23.8
23	165 49.0	23.9
19 00	180 48.7	S23 23.9
01	195 48.4	24.0
02	210 48.1	24.1
03	225 47.8 ··	24.1
04	240 47.5	24.2
05	255 47.2	24.2
06	270 46.8	S23 24.3
07	285 46.5	24.3
08	300 46.2	24.4
09	315 45.9 ··	24.5
10	330 45.6	24.5
11	345 45.3	24.6
12	0 45.0	S23 24.6
13	15 44.7	24.7
14	30 44.4	24.7
15	45 44.1 ··	24.8
16	60 43.8	24.8
17	75 43.5	24.9
18	90 43.1	S23 24.9
19	105 42.8	25.0
20	120 42.5	25.0
21	135 42.2 ··	25.1
22	150 41.9	25.1
23	165 41.6	25.2
20 00	180 41.3	S23 25.2
01	195 41.0	25.2
02	210 40.7	25.3
03	225 40.4 ··	25.3
04	240 40.1	25.4
05	255 39.7	25.4
06	270 39.4	S23 25.5
07	285 39.1	25.5
08	300 38.8	25.5
09	315 38.5 ··	25.6
10	330 38.2	25.6
11	345 37.9	25.6
12	0 37.6	S23 25.7
13	15 37.3	25.7
14	30 37.0	25.7
15	45 36.7 ··	25.8
16	60 36.3	25.8
17	75 36.0	25.8
18	90 35.7	S23 25.9
19	105 35.4	25.9
20	120 35.1	25.9
21	135 34.8 ··	25.9
22	150 34.5	26.0
23	165 34.2	26.0

d h	G.H.A.	Dec.
21 00	180 33.9	S23 26.0
01	195 33.6	26.0
02	210 33.2	26.1
03	225 32.9 ··	26.1
04	240 32.6	26.1
05	255 32.3	26.1
06	270 32.0	S23 26.1
07	285 31.7	26.2
08	300 31.4	26.2
09	315 31.1 ··	26.2
10	330 30.8	26.2
11	345 30.5	26.2
12	0 30.1	S23 26.2
13	15 29.8	26.3
14	30 29.5	26.3
15	45 29.2 ··	26.3
16	60 28.9	26.3
17	75 28.6	26.3
18	90 28.3	S23 26.3
19	105 28.0	26.3
20	120 27.7	26.3
21	135 27.3 ··	26.3
22	150 27.0	26.3
23	165 26.7	26.3
22 00	180 26.4	S23 26.4
01	195 26.1	26.4
02	210 25.8	26.4
03	225 25.5 ··	26.4
04	240 25.2	26.4
05	255 24.9	26.4
06	270 24.5	S23 26.4
07	285 24.2	26.4
08	300 23.9	26.4
09	315 23.6 ··	26.4
10	330 23.3	26.4
11	345 23.0	26.4
12	0 22.7	S23 26.3
13	15 22.4	26.3
14	30 22.1	26.3
15	45 21.8 ··	26.3
16	60 21.4	26.3
17	75 21.1	26.3
18	90 20.8	S23 26.3
19	105 20.5	26.3
20	120 20.2	26.3
21	135 19.9 ··	26.3
22	150 19.6	26.2
23	165 19.3	26.2
23 00	180 18.9	S23 26.2
01	195 18.6	26.2
02	210 18.3	26.2
03	225 18.0 ··	26.2
04	240 17.7	26.2
05	255 17.4	26.2
06	270 17.1	S23 26.1
07	285 16.8	26.1
08	300 16.5	26.1
09	315 16.1 ··	26.0
10	330 15.8	26.0
11	345 15.5	26.0
12	0 15.2	S23 26.0
13	15 14.9	26.0
14	30 14.6	25.9
15	45 14.3 ··	25.9
16	60 14.0	25.9
17	75 13.7	25.8
18	90 13.3	S23 25.8
19	105 13.0	25.8
20	120 12.7	25.7
21	135 12.4 ··	25.7
22	150 12.1	25.7
23	165 11.8	25.7

d h	G.H.A.	Dec.
24 00	180 11.5	S23 25.6
01	195 11.2	25.6
02	210 10.9	25.5
03	225 10.5 ··	25.5
04	240 10.2	25.5
05	255 09.9	25.4
06	270 09.6	S23 25.4
07	285 09.3	25.4
08	300 09.0	25.3
09	315 08.7 ··	25.3
10	330 08.4	25.2
11	345 08.1	25.2
12	0 07.7	S23 25.1
13	15 07.4	25.1
14	30 07.1	25.0
15	45 06.8 ··	25.0
16	60 06.5	25.0
17	75 06.2	24.9
18	90 05.9	S23 24.9
19	105 05.6	24.8
20	120 05.2	24.8
21	135 04.9 ··	24.7
22	150 04.6	24.6
23	165 04.3	24.6
25 00	180 04.0	S23 24.5
01	195 03.7	24.5
02	210 03.4	24.4
03	225 03.1 ··	24.4
04	240 02.8	24.3
05	255 02.4	24.3
06	270 02.1	S23 24.2
07	285 01.8	24.1
08	300 01.5	24.1
09	315 01.2 ··	24.0
10	330 00.9	24.0
11	345 00.6	23.9
12	0 00.3	S23 23.8
13	15 00.0	23.8
14	29 59.7	23.7
15	44 59.3 ··	23.6
16	59 59.0	23.6
17	74 58.7	23.5
18	89 58.4	S23 23.4
19	104 58.1	23.4
20	119 57.8	23.3
21	134 57.5 ··	23.2
22	149 57.2	23.1
23	164 56.9	23.1
26 00	179 56.5	S23 23.0
01	194 56.2	22.9
02	209 55.9	22.8
03	224 55.6 ··	22.8
04	239 55.3	22.7
05	254 55.0	22.6
06	269 54.7	S23 22.5
07	284 54.4	22.5
08	299 54.1	22.4
09	314 53.7 ··	22.3
10	329 53.4	22.2
11	344 53.1	22.1
12	359 52.8	S23 22.0
13	14 52.5	22.0
14	29 52.2	21.9
15	44 51.9 ··	21.8
16	59 51.6	21.7
17	74 51.3	21.6
18	89 51.0	S23 21.5
19	104 50.6	21.4
20	119 50.3	21.3
21	134 50.0 ··	21.3
22	149 49.7	21.2
23	164 49.4	21.1

d h	G.H.A. ° '	Dec. ° '
27 00	179 49.1	S23 21.0
01	194 48.8	20.9
02	209 48.5	20.8
03	224 48.2 ··	20.7
04	239 47.9	20.6
05	254 47.6	20.5
06	269 47.2	S23 20.4
07	284 46.9	20.3
08	299 46.6	20.2
09	314 46.3 ··	20.1
10	329 46.0	20.0
11	344 45.7	19.9
12	359 45.4	S23 19.8
13	14 45.1	19.7
14	29 44.8	19.6
15	44 44.5 ··	19.5
16	59 44.2	19.4
17	74 43.8	19.3
18	89 43.5	S23 19.2
19	104 43.2	19.1
20	119 42.9	18.9
21	134 42.6 ··	18.8
22	149 42.3	18.7
23	164 42.0	18.6
28 00	179 41.7	S23 18.5
01	194 41.4	18.4
02	209 41.1	18.3
03	224 40.8 ··	18.2
04	239 40.5	18.0
05	254 40.1	17.9
06	269 39.8	S23 17.8
07	284 39.5	17.7
08	299 39.2	17.6
09	314 38.9 ··	17.4
10	329 38.6	17.3
11	344 38.3	17.2
12	359 38.0	S23 17.1
13	14 37.7	17.0
14	29 37.4	16.8
15	44 37.1 ··	16.7
16	59 36.8	16.6
17	74 36.5	16.5
18	89 36.2	S23 16.3
19	104 35.8	16.2
20	119 35.5	16.1
21	134 35.2 ··	15.9
22	149 34.9	15.8
23	164 34.6	15.7
29 00	179 34.3	S23 15.5
01	194 34.0	15.4
02	209 33.7	15.3
03	224 33.4 ··	15.1
04	239 33.1	15.0
05	254 32.8	14.9
06	269 32.5	S23 14.7
07	284 32.2	14.6
08	299 31.9	14.5
09	314 31.6 ··	14.3
10	329 31.3	14.2
11	344 30.9	14.0
12	359 30.6	S23 13.9
13	14 30.3	13.7
14	29 30.0	13.6
15	44 29.7 ··	13.5
16	59 29.4	13.3
17	74 29.1	13.2
18	89 28.8	S23 13.0
19	104 28.5	12.9
20	119 28.2	12.7
21	134 27.9 ··	12.6
22	149 27.6	12.4
23	164 27.3	12.3

d h	G.H.A. ° '	Dec. ° '
30 00	179 27.0	S23 12.1
01	194 26.7	12.0
02	209 26.4	11.8
03	224 26.1 ··	11.7
04	239 25.8	11.5
05	254 25.5	11.4
06	269 25.2	S23 11.2
07	284 24.9	11.0
08	299 24.6	10.9
09	314 24.3 ··	10.7
10	329 23.9	10.6
11	344 23.6	10.4
12	359 23.3	S23 10.2
13	14 23.0	10.1
14	29 22.7	09.9
15	44 22.4 ··	09.7
16	59 22.1	09.6
17	74 21.8	09.4
18	89 21.5	S23 09.3
19	104 21.2	09.1
20	119 20.9	08.9
21	134 20.6 ··	08.7
22	149 20.3	08.6
23	164 20.0	08.4
31 00	179 19.7	S23 08.2
01	194 19.4	08.1
02	209 19.1	07.9
03	224 18.8 ··	07.7
04	239 18.5	07.5
05	254 18.2	07.4
06	269 17.9	S23 07.2
07	284 17.6	07.0
08	299 17.3	06.8
09	314 17.0 ··	06.7
10	329 16.7	06.5
11	344 16.4	06.3
12	359 16.1	S23 06.1
13	14 15.8	05.9
14	29 15.5	05.8
15	44 15.2 ··	05.6
16	59 14.9	05.4
17	74 14.6	05.2
18	89 14.3	S23 05.0
19	104 14.0	04.8
20	119 13.7	04.6
21	134 13.4 ··	04.5
22	149 13.1	04.3
23	164 12.8	04.1

Table 8

DAILY LONG-TERM ALMANAC OF THE SUN FOR THE YEARS 1979, 1983, 1987, 1991, 1995, 1999, 2003, ETC.*

This table is a compilation of the columns for Greenwich Hour Angle and declination of the sun from the 1979 *Nautical Almanac*. For use beyond 1979, a correction may be necessary, depending upon the degree of precision desired. The amount of change in *4 years* in the GHA and declination (DEC) is given for each 3-day interval in Table 9.

Since the correction factors are changing slowly with the passage of time, this table is *not* a perpetual almanac, but an almanac designed for use in practical navigation for a period of about 30 years.

*See note at end of table on use of 4-year correction factors.

d h	G.H.A.	Dec.
1 00	179 12.5	S23 03.9
01	194 12.2	03.7
02	209 11.9	03.5
03	224 11.6	·· 03.3
04	239 11.3	03.1
05	254 11.0	02.9
06	269 10.7	S23 02.7
07	284 10.4	02.5
08	299 10.1	02.3
09	314 09.8	·· 02.1
10	329 09.5	01.9
11	344 09.2	01.8
12	359 08.9	S23 01.6
13	14 08.6	01.4
14	29 08.3	01.1
15	44 08.0	·· 00.9
16	59 07.7	00.7
17	74 07.4	00.5
18	89 07.2	S23 00.3
19	104 06.9	23 00.1
20	119 06.6	22 59.9
21	134 06.3	·· 59.7
22	149 06.0	59.5
23	164 05.7	59.3
2 00	179 05.4	S22 59.1
01	194 05.1	58.9
02	209 04.8	58.7
03	224 04.5	·· 58.5
04	239 04.2	58.2
05	254 03.9	58.0
06	269 03.6	S22 57.8
07	284 03.3	57.6
08	299 03.0	57.4
09	314 02.7	·· 57.2
10	329 02.4	57.0
11	344 02.1	56.7
12	359 01.8	S22 56.5
13	14 01.6	56.3
14	29 01.3	56.1
15	44 01.0	·· 55.9
16	59 00.7	55.6
17	74 00.4	55.4
18	89 00.1	S22 55.2
19	103 59.8	55.0
20	118 59.5	54.7
21	133 59.2	·· 54.5
22	148 58.9	54.3
23	163 58.6	54.1
3 00	178 58.3	S22 53.8
01	193 58.0	53.6
02	208 57.8	53.4
03	223 57.5	·· 53.1
04	238 57.2	52.9
05	253 56.9	52.7
06	268 56.6	S22 52.4
07	283 56.3	52.2
08	298 56.0	52.0
09	313 55.7	·· 51.7
10	328 55.4	51.5
11	343 55.1	51.3
12	358 54.8	S22 51.0
13	13 54.6	50.8
14	28 54.3	50.6
15	43 54.0	·· 50.3
16	58 53.7	50.1
17	73 53.4	49.8
18	88 53.1	S22 49.6
19	103 52.8	49.3
20	118 52.5	49.1
21	133 52.2	·· 48.9
22	148 52.0	48.6
23	163 51.7	48.4

d h	G.H.A.	Dec.
4 00	178 51.4	S22 48.1
01	193 51.1	47.9
02	208 50.8	47.6
03	223 50.5	·· 47.4
04	238 50.2	47.1
05	253 49.9	46.9
06	268 49.7	S22 46.6
07	283 49.4	46.4
08	298 49.1	46.1
09	313 48.8	·· 45.9
10	328 48.5	45.6
11	343 48.2	45.3
12	358 47.9	S22 45.1
13	13 47.7	44.8
14	28 47.4	44.6
15	43 47.1	·· 44.3
16	58 46.8	44.1
17	73 46.5	43.8
18	88 46.2	S22 43.5
19	103 46.0	43.3
20	118 45.7	43.0
21	133 45.4	·· 42.7
22	148 45.1	42.5
23	163 44.8	42.2
5 00	178 44.5	S22 41.9
01	193 44.3	41.7
02	208 44.0	41.4
03	223 43.7	·· 41.1
04	238 43.4	40.9
05	253 43.1	40.6
06	268 42.8	S22 40.3
07	283 42.6	40.1
08	298 42.3	39.8
09	313 42.0	·· 39.5
10	328 41.7	39.2
11	343 41.4	39.0
12	358 41.2	S22 38.7
13	13 40.9	38.4
14	28 40.6	38.1
15	43 40.3	·· 37.9
16	58 40.0	37.6
17	73 39.8	37.3
18	88 39.5	S22 37.0
19	103 39.2	36.7
20	118 38.9	36.5
21	133 38.6	·· 36.2
22	148 38.4	35.9
23	163 38.1	35.6
6 00	178 37.8	S22 35.3
01	193 37.5	35.0
02	208 37.2	34.8
03	223 37.0	·· 34.5
04	238 36.7	34.2
05	253 36.4	33.9
06	268 36.1	S22 33.6
07	283 35.9	33.3
08	298 35.6	33.0
09	313 35.3	·· 32.7
10	328 35.0	32.4
11	343 34.7	32.1
12	358 34.5	S22 31.8
13	13 34.2	31.6
14	28 33.9	31.3
15	43 33.6	·· 31.0
16	58 33.4	30.7
17	73 33.1	30.4
18	88 32.8	S22 30.1
19	103 32.5	29.8
20	118 32.3	29.5
21	133 32.0	·· 29.2
22	148 31.7	28.9
23	163 31.5	28.6

d h	G.H.A.	Dec.
7 00	178 31.2	S22 28.3
01	193 30.9	28.0
02	208 30.6	27.6
03	223 30.4	·· 27.3
04	238 30.1	27.0
05	253 29.8	26.7
06	268 29.5	S22 26.4
07	283 29.3	26.1
08	298 29.0	25.8
09	313 28.7	·· 25.5
10	328 28.5	25.2
11	343 28.2	24.9
12	358 27.9	S22 24.6
13	13 27.6	24.2
14	28 27.4	23.9
15	43 27.1	·· 23.6
16	58 26.8	23.3
17	73 26.6	23.0
18	88 26.3	S22 22.7
19	103 26.0	22.3
20	118 25.8	22.0
21	133 25.5	·· 21.7
22	148 25.2	21.4
23	163 24.9	21.1
8 00	178 24.7	S22 20.7
01	193 24.4	20.4
02	208 24.1	20.1
03	223 23.9	·· 19.8
04	238 23.6	19.5
05	253 23.3	19.1
06	268 23.1	S22 18.8
07	283 22.8	18.5
08	298 22.5	18.1
09	313 22.3	·· 17.8
10	328 22.0	17.5
11	343 21.7	17.2
12	358 21.5	S22 16.8
13	13 21.2	16.5
14	28 21.0	16.2
15	43 20.7	·· 15.8
16	58 20.4	15.5
17	73 20.2	15.2
18	88 19.9	S22 14.8
19	103 19.6	14.5
20	118 19.4	14.2
21	133 19.1	·· 13.8
22	148 18.8	13.5
23	163 18.6	13.1
9 00	178 18.3	S22 12.8
01	193 18.1	12.5
02	208 17.8	12.1
03	223 17.5	·· 11.8
04	238 17.3	11.4
05	253 17.0	11.1
06	268 16.7	S22 10.7
07	283 16.5	10.4
08	298 16.2	10.1
09	313 16.0	·· 09.7
10	328 15.7	09.4
11	343 15.4	09.0
12	358 15.2	S22 08.7
13	13 14.9	08.3
14	28 14.7	08.0
15	43 14.4	·· 07.6
16	58 14.1	07.3
17	73 13.9	06.9
18	88 13.6	S22 06.5
19	103 13.4	06.2
20	118 13.1	05.8
21	133 12.9	·· 05.5
22	148 12.6	05.1
23	163 12.3	04.8

SUN tables (G.H.A. and Dec.)

d h	G.H.A.	Dec.
10 00	178 12.1	S22 04.4
01	193 11.8	04.1
02	208 11.6	03.7
03	223 11.3	·· 03.3
04	238 11.1	03.0
05	253 10.8	02.6
06	268 10.5	S22 02.2
07	283 10.3	01.9
08	298 10.0	01.5
09	313 09.8	·· 01.2
10	328 09.5	00.8
11	343 09.3	00.4
12	358 09.0	S22 00.1
13	13 08.8	21 59.7
14	28 08.5	59.3
15	43 08.3	·· 59.0
16	58 08.0	58.6
17	73 07.8	58.2
18	88 07.5	S21 57.8
19	103 07.2	57.5
20	118 07.0	57.1
21	133 06.7	·· 56.7
22	148 06.5	56.3
23	163 06.2	56.0
11 00	178 06.0	S21 55.6
01	193 05.7	55.2
02	208 05.5	54.8
03	223 05.2	·· 54.5
04	238 05.0	54.1
05	253 04.7	53.7
06	268 04.5	S21 53.3
07	283 04.2	52.9
08	298 04.0	52.6
09	313 03.7	·· 52.2
10	328 03.5	51.8
11	343 03.2	51.4
12	358 03.0	S21 51.0
13	13 02.7	50.6
14	28 02.5	50.3
15	43 02.3	·· 49.9
16	58 02.0	49.5
17	73 01.8	49.1
18	88 01.5	S21 48.7
19	103 01.3	48.3
20	118 01.0	47.9
21	133 00.8	·· 47.5
22	148 00.5	47.1
23	163 00.3	46.7
12 00	178 00.0	S21 46.3
01	192 59.8	46.0
02	207 59.6	45.6
03	222 59.3	·· 45.2
04	237 59.1	44.8
05	252 58.8	44.4
06	267 58.6	S21 44.0
07	282 58.3	43.6
08	297 58.1	43.2
09	312 57.9	·· 42.8
10	327 57.6	42.4
11	342 57.4	42.0
12	357 57.1	S21 41.6
13	12 56.9	41.2
14	27 56.6	40.8
15	42 56.4	·· 40.4
16	57 56.2	39.9
17	72 55.9	39.5
18	87 55.7	S21 39.1
19	102 55.4	38.7
20	117 55.2	38.3
21	132 55.0	·· 37.9
22	147 54.7	37.5
23	162 54.5	37.1

d h	G.H.A.	Dec.
13 00	177 54.2	S21 36.7
01	192 54.0	36.3
02	207 53.8	35.9
03	222 53.5	·· 35.4
04	237 53.3	35.0
05	252 53.1	34.6
06	267 52.8	S21 34.2
07	282 52.6	33.8
08	297 52.3	33.4
09	312 52.1	·· 32.9
10	327 51.9	32.5
11	342 51.6	32.1
12	357 51.4	S21 31.7
13	12 51.2	31.3
14	27 50.9	30.8
15	42 50.7	·· 30.4
16	57 50.5	30.0
17	72 50.2	29.6
18	87 50.0	S21 29.2
19	102 49.8	28.7
20	117 49.5	28.3
21	132 49.3	·· 27.9
22	147 49.1	27.4
23	162 48.8	27.0
14 00	177 48.6	S21 26.6
01	192 48.4	26.2
02	207 48.1	25.7
03	222 47.9	·· 25.3
04	237 47.7	24.9
05	252 47.4	24.4
06	267 47.2	S21 24.0
07	282 47.0	23.6
08	297 46.8	23.1
09	312 46.5	·· 22.7
10	327 46.3	22.3
11	342 46.1	21.8
12	357 45.8	S21 21.4
13	12 45.6	21.0
14	27 45.4	20.5
15	42 45.2	·· 20.1
16	57 44.9	19.6
17	72 44.7	19.2
18	87 44.5	S21 18.8
19	102 44.2	18.3
20	117 44.0	17.9
21	132 43.8	·· 17.4
22	147 43.6	17.0
23	162 43.3	16.5
15 00	177 43.1	S21 16.1
01	192 42.9	15.6
02	207 42.7	15.2
03	222 42.4	·· 14.7
04	237 42.2	14.3
05	252 42.0	13.8
06	267 41.8	S21 13.4
07	282 41.5	12.9
08	297 41.3	12.5
09	312 41.1	·· 12.0
10	327 40.9	11.6
11	342 40.7	11.1
12	357 40.4	S21 10.7
13	12 40.2	10.2
14	27 40.0	09.8
15	42 39.8	·· 09.3
16	57 39.6	08.9
17	72 39.3	08.4
18	87 39.1	S21 07.9
19	102 38.9	07.5
20	117 38.7	07.0
21	132 38.5	·· 06.6
22	147 38.2	06.1
23	162 38.0	05.6

d h	G.H.A.	Dec.
16 00	177 37.8	S21 05.2
01	192 37.6	04.7
02	207 37.4	04.2
03	222 37.1	·· 03.8
04	237 36.9	03.3
05	252 36.7	02.9
06	267 36.5	S21 02.4
07	282 36.3	01.9
08	297 36.1	01.5
09	312 35.8	·· 01.0
10	327 35.6	00.5
11	342 35.4	21 00.0
12	357 35.2	S20 59.6
13	12 35.0	59.1
14	27 34.8	58.6
15	42 34.6	·· 58.2
16	57 34.3	57.7
17	72 34.1	57.2
18	87 33.9	S20 56.7
19	102 33.7	56.3
20	117 33.5	55.8
21	132 33.3	·· 55.3
22	147 33.1	54.8
23	162 32.8	54.3
17 00	177 32.6	S20 53.9
01	192 32.4	53.4
02	207 32.2	52.9
03	222 32.0	·· 52.4
04	237 31.8	51.9
05	252 31.6	51.5
06	267 31.4	S20 51.0
07	282 31.2	50.5
08	297 31.0	50.0
09	312 30.7	·· 49.5
10	327 30.5	49.0
11	342 30.3	48.5
12	357 30.1	S20 48.1
13	12 29.9	47.6
14	27 29.7	47.1
15	42 29.5	·· 46.6
16	57 29.3	46.1
17	72 29.1	45.6
18	87 28.9	S20 45.1
19	102 28.7	44.6
20	117 28.5	44.1
21	132 28.3	·· 43.6
22	147 28.1	43.1
23	162 27.9	42.6
18 00	177 27.6	S20 42.1
01	192 27.4	41.7
02	207 27.2	41.2
03	222 27.0	·· 40.7
04	237 26.8	40.2
05	252 26.6	39.7
06	267 26.4	S20 39.2
07	282 26.2	38.7
08	297 26.0	38.2
09	312 25.8	·· 37.7
10	327 25.6	37.2
11	342 25.4	36.6
12	357 25.2	S20 36.1
13	12 25.0	35.6
14	27 24.8	35.1
15	42 24.6	·· 34.6
16	57 24.4	34.1
17	72 24.2	33.6
18	87 24.0	S20 33.1
19	102 23.8	32.6
20	117 23.6	32.1
21	132 23.4	·· 31.6
22	147 23.2	31.1
23	162 23.0	30.6

d h	G.H.A.	Dec.
19 00	177 22.8	S20 30.0
01	192 22.6	29.5
02	207 22.4	29.0
03	222 22.2 ··	28.5
04	237 22.0	28.0
05	252 21.9	27.5
06	267 21.7	S20 27.0
07	282 21.5	26.4
08	297 21.3	25.9
09	312 21.1 ··	25.4
10	327 20.9	24.9
11	342 20.7	24.4
12	357 20.5	S20 23.8
13	12 20.3	23.3
14	27 20.1	22.8
15	42 19.9 ··	22.3
16	57 19.7	21.8
17	72 19.5	21.2
18	87 19.3	S20 20.7
19	102 19.1	20.2
20	117 19.0	19.7
21	132 18.8 ··	19.1
22	147 18.6	18.6
23	162 18.4	18.1
20 00	177 18.2	S20 17.5
01	192 18.0	17.0
02	207 17.8	16.5
03	222 17.6 ··	16.0
04	237 17.4	15.4
05	252 17.3	14.9
06	267 17.1	S20 14.4
07	282 16.9	13.8
08	297 16.7	13.3
09	312 16.5 ··	12.8
10	327 16.3	12.2
11	342 16.1	11.7
12	357 15.9	S20 11.2
13	12 15.8	10.6
14	27 15.6	10.1
15	42 15.4 ··	09.5
16	57 15.2	09.0
17	72 15.0	08.5
18	87 14.8	S20 07.9
19	102 14.6	07.4
20	117 14.5	06.8
21	132 14.3 ··	06.3
22	147 14.1	05.8
23	162 13.9	05.2
21 00	177 13.7	S20 04.7
01	192 13.6	04.1
02	207 13.4	03.6
03	222 13.2 ··	03.0
04	237 13.0	02.5
05	252 12.8	01.9
06	267 12.6	S20 01.4
07	282 12.5	00.8
08	297 12.3	20 00.3
09	312 12.1	19 59.7
10	327 11.9	59.2
11	342 11.8	58.6
12	357 11.6	S19 58.1
13	12 11.4	57.5
14	27 11.2	57.0
15	42 11.0 ··	56.4
16	57 10.9	55.9
17	72 10.7	55.3
18	87 10.5	S19 54.8
19	102 10.3	54.2
20	117 10.2	53.7
21	132 10.0 ··	53.1
22	147 09.8	52.5
23	162 09.6	52.0

d h	G.H.A.	Dec.
22 00	177 09.5	S19 51.4
01	192 09.3	50.9
02	207 09.1	50.3
03	222 08.9 ··	49.7
04	237 08.8	49.2
05	252 08.6	48.6
06	267 08.4	S19 48.0
07	282 08.2	47.5
08	297 08.1	46.9
09	312 07.9 ··	46.4
10	327 07.7	45.8
11	342 07.6	45.2
12	357 07.4	S19 44.7
13	12 07.2	44.1
14	27 07.0	43.5
15	42 06.9 ··	42.9
16	57 06.7	42.4
17	72 06.5	41.8
18	87 06.4	S19 41.2
19	102 06.2	40.7
20	117 06.0	40.1
21	132 05.9 ··	39.5
22	147 05.7	38.9
23	162 05.5	38.4
23 00	177 05.4	S19 37.8
01	192 05.2	37.2
02	207 05.0	36.6
03	222 04.9 ··	36.1
04	237 04.7	35.5
05	252 04.5	34.9
06	267 04.4	S19 34.3
07	282 04.2	33.8
08	297 04.0	33.2
09	312 03.9 ··	32.6
10	327 03.7	32.0
11	342 03.6	31.4
12	357 03.4	S19 30.8
13	12 03.2	30.3
14	27 03.1	29.7
15	42 02.9 ··	29.1
16	57 02.7	28.5
17	72 02.6	27.9
18	87 02.4	S19 27.3
19	102 02.3	26.8
20	117 02.1	26.2
21	132 01.9 ··	25.6
22	147 01.8	25.0
23	162 01.6	24.4
24 00	177 01.5	S19 23.8
01	192 01.3	23.2
02	207 01.1	22.6
03	222 01.0 ··	22.0
04	237 00.8	21.4
05	252 00.7	20.9
06	267 00.5	S19 20.3
07	282 00.4	19.7
08	297 00.2	19.1
09	312 00.0 ··	18.5
10	326 59.9	17.9
11	341 59.7	17.3
12	356 59.6	S19 16.7
13	11 59.4	16.1
14	26 59.3	15.5
15	41 59.1 ··	14.9
16	56 59.0	14.3
17	71 58.8	13.7
18	86 58.7	S19 13.1
19	101 58.5	12.5
20	116 58.4	11.9
21	131 58.2 ··	11.3
22	146 58.1	10.7
23	161 57.9	10.1

d h	G.H.A.	Dec.
25 00	176 57.8	S19 09.5
01	191 57.6	08.9
02	206 57.5	08.3
03	221 57.3 ··	07.6
04	236 57.2	07.0
05	251 57.0	06.4
06	266 56.9	S19 05.8
07	281 56.7	05.2
08	296 56.6	04.6
09	311 56.4 ··	04.0
10	326 56.3	03.4
11	341 56.1	02.8
12	356 56.0	S19 02.2
13	11 55.8	01.5
14	26 55.7	00.9
15	41 55.5	19 00.3
16	56 55.4	18 59.7
17	71 55.2	59.1
18	86 55.1	S18 58.5
19	101 55.0	57.9
20	116 54.8	57.2
21	131 54.7 ··	56.6
22	146 54.5	56.0
23	161 54.4	55.4
26 00	176 54.2	S18 54.8
01	191 54.1	54.1
02	206 54.0	53.5
03	221 53.8 ··	52.9
04	236 53.7	52.3
05	251 53.5	51.7
06	266 53.4	S18 51.0
07	281 53.3	50.4
08	296 53.1	49.8
09	311 53.0 ··	49.2
10	326 52.8	48.5
11	341 52.7	47.9
12	356 52.6	S18 47.2
13	11 52.4	46.7
14	26 52.3	46.0
15	41 52.1 ··	45.4
16	56 52.0	44.8
17	71 51.9	44.1
18	86 51.7	S18 43.5
19	101 51.6	42.9
20	116 51.5	42.3
21	131 51.3 ··	41.6
22	146 51.2	41.0
23	161 51.1	40.4
27 00	176 50.9	S18 39.7
01	191 50.8	39.1
02	206 50.7	38.5
03	221 50.5 ··	37.8
04	236 50.4	37.2
05	251 50.3	36.6
06	266 50.1	S18 35.9
07	281 50.0	35.3
08	296 49.9	34.6
09	311 49.7 ··	34.0
10	326 49.6	33.4
11	341 49.5	32.7
12	356 49.3	S18 32.1
13	11 49.2	31.4
14	26 49.1	30.8
15	41 49.0 ··	30.2
16	56 48.8	29.5
17	71 48.7	28.9
18	86 48.6	S18 28.2
19	101 48.4	27.6
20	116 48.3	26.9
21	131 48.2 ··	26.3
22	146 48.1	25.6
23	161 47.9	25.0

	G.H.A.	Dec.		G.H.A.	Dec.		G.H.A.	Dec.
28 00	176 47.8	S18 24.3	**31** 00	176 39.7	S17 36.2	**3** 00	176 33.4	S16 45.3
01	191 47.7	23.7	01	191 39.6	35.6	01	191 33.3	44.6
02	206 47.6	23.1	02	206 39.5	34.9	02	206 33.3	43.9
03	221 47.4	·· 22.4	03	221 39.4	·· 34.2	03	221 33.2	·· 43.1
04	236 47.3	21.8	04	236 39.3	33.5	04	236 33.1	42.4
05	251 47.2	21.1	05	251 39.2	32.8	05	251 33.0	41.7
06	266 47.1	S18 20.5	06	266 39.1	S17 32.1	06	266 33.0	S16 41.0
07	281 46.9	19.8	07	281 39.0	31.4	07	281 32.9	40.2
08	296 46.8	19.1	08	296 38.9	30.7	08	296 32.8	39.5
09	311 46.7	·· 18.5	09	311 38.8	·· 30.0	09	311 32.8	·· 38.8
10	326 46.6	17.8	10	326 38.7	29.3	10	326 32.7	38.0
11	341 46.4	17.2	11	341 38.6	28.6	11	341 32.6	37.3
12	356 46.3	S18 16.5	12	356 38.5	S17 27.9	12	356 32.5	S16 36.6
13	11 46.2	15.9	13	11 38.4	27.2	13	11 32.5	35.8
14	26 46.1	15.2	14	26 38.3	26.6	14	26 32.4	35.1
15	41 46.0	·· 14.6	15	41 38.2	·· 25.9	15	41 32.3	·· 34.4
16	56 45.8	13.9	16	56 38.1	25.2	16	56 32.3	33.6
17	71 45.7	13.3	17	71 38.0	24.5	17	71 32.2	32.9
18	86 45.6	S18 12.6	18	86 37.9	S17 23.8	18	86 32.1	S16 32.2
19	101 45.5	11.9	19	101 37.8	23.1	19	101 32.1	31.4
20	116 45.4	11.3	20	116 37.7	22.4	20	116 32.0	30.7
21	131 45.2	·· 10.6	21	131 37.7	·· 21.7	21	131 31.9	·· 30.0
22	146 45.1	10.0	22	146 37.6	21.0	22	146 31.9	29.2
23	161 45.0	09.3	23	161 37.5	20.3	23	161 31.8	28.5
29 00	176 44.9	S18 08.6	**1** 00	176 37.4	S17 19.6	**4** 00	176 31.7	S16 27.7
01	191 44.8	08.0	01	191 37.3	18.9	01	191 31.7	27.0
02	206 44.7	07.3	02	206 37.2	18.2	02	206 31.6	26.3
03	221 44.5	·· 06.6	03	221 37.1	·· 17.5	03	221 31.5	·· 25.5
04	236 44.4	06.0	04	236 37.0	16.8	04	236 31.5	24.8
05	251 44.3	05.3	05	251 36.9	16.1	05	251 31.4	24.1
06	266 44.2	S18 04.7	06	266 36.8	S17 15.4	06	266 31.4	S16 23.3
07	281 44.1	04.0	07	281 36.7	14.6	07	281 31.3	22.6
08	296 44.0	03.3	08	296 36.7	13.9	08	296 31.2	21.8
09	311 43.9	·· 02.7	09	311 36.6	·· 13.2	09	311 31.2	·· 21.1
10	326 43.7	02.0	10	326 36.5	12.5	10	326 31.1	20.3
11	341 43.6	01.3	11	341 36.4	11.8	11	341 31.0	19.6
12	356 43.5	S18 00.7	12	356 36.3	S17 11.1	12	356 31.0	S16 18.9
13	11 43.4	18 00.0	13	11 36.2	10.4	13	11 30.9	18.1
14	26 43.3	17 59.3	14	26 36.1	09.7	14	26 30.9	17.4
15	41 43.2	·· 58.6	15	41 36.0	·· 09.0	15	41 30.8	·· 16.6
16	56 43.1	58.0	16	56 36.0	08.3	16	56 30.7	15.9
17	71 43.0	57.3	17	71 35.9	07.6	17	71 30.7	15.1
18	86 42.8	S17 56.6	18	86 35.8	S17 06.9	18	86 30.6	S16 14.4
19	101 42.7	56.0	19	101 35.7	06.2	19	101 30.6	13.6
20	116 42.6	55.3	20	116 35.6	05.4	20	116 30.5	12.9
21	131 42.5	·· 54.6	21	131 35.5	·· 04.7	21	131 30.4	·· 12.1
22	146 42.4	53.9	22	146 35.5	04.0	22	146 30.4	11.4
23	161 42.3	53.3	23	161 35.4	03.3	23	161 30.3	10.6
30 00	176 42.2	S17 52.6	**2** 00	176 35.3	S17 02.6	**5** 00	176 30.3	S16 09.9
01	191 42.1	51.9	01	191 35.2	01.9	01	191 30.2	09.1
02	206 42.0	51.2	02	206 35.1	01.2	02	206 30.2	08.4
03	221 41.9	·· 50.6	03	221 35.0	17 00.4	03	221 30.1	·· 07.6
04	236 41.7	49.9	04	236 35.0	16 59.7	04	236 30.1	06.9
05	251 41.6	49.2	05	251 34.9	59.0	05	251 30.0	06.1
06	266 41.5	S17 48.5	06	266 34.8	S16 58.3	06	266 29.9	S16 05.4
07	281 41.4	47.9	07	281 34.7	57.6	07	281 29.9	04.6
08	296 41.3	47.2	08	296 34.6	56.9	08	296 29.8	03.9
09	311 41.2	·· 46.5	09	311 34.6	·· 56.1	09	311 29.8	·· 03.1
10	326 41.1	45.8	10	326 34.5	55.4	10	326 29.7	02.4
11	341 41.0	45.1	11	341 34.4	54.7	11	341 29.7	01.6
12	356 40.9	S17 44.5	12	356 34.3	S16 54.0	12	356 29.6	S16 00.9
13	11 40.8	43.8	13	11 34.2	53.3	13	11 29.6	16 00.1
14	26 40.7	43.1	14	26 34.2	52.6	14	26 29.5	15 59.4
15	41 40.6	·· 42.4	15	41 34.1	·· 51.8	15	41 29.5	·· 58.6
16	56 40.5	41.7	16	56 34.0	51.1	16	56 29.4	57.8
17	71 40.4	41.0	17	71 33.9	50.4	17	71 29.4	57.1
18	86 40.3	S17 40.4	18	86 33.9	S16 49.7	18	86 29.3	S15 56.3
19	101 40.2	39.7	19	101 33.8	48.9	19	101 29.3	55.6
20	116 40.1	39.0	20	116 33.7	48.2	20	116 29.2	54.8
21	131 40.0	·· 38.3	21	131 33.6	·· 47.5	21	131 29.2	·· 54.0
22	146 39.9	37.6	22	146 33.6	46.8	22	146 29.1	53.3
23	161 39.8	36.9	23	161 33.5	46.0	23	161 29.1	52.5

FEBRUARY

d h	G.H.A. ° ′	Dec. ° ′
6 00	176 29.0	S15 51.8
01	191 29.0	51.0
02	206 28.9	50.2
03	221 28.9	·· 49.5
04	236 28.8	48.7
05	251 28.8	48.0
06	266 28.7	S15 47.2
07	281 28.7	46.4
08	296 28.6	45.7
09	311 28.6	·· 44.9
10	326 28.6	44.1
11	341 28.5	43.4
12	356 28.5	S15 42.6
13	11 28.4	41.8
14	26 28.4	41.1
15	41 28.3	·· 40.3
16	56 28.3	39.5
17	71 28.3	38.8
18	86 28.2	S15 38.0
19	101 28.2	37.2
20	116 28.1	36.4
21	131 28.1	·· 35.7
22	146 28.0	34.9
23	161 28.0	34.1
7 00	176 28.0	S15 33.4
01	191 27.9	32.6
02	206 27.9	31.8
03	221 27.9	·· 31.0
04	236 27.8	30.3
05	251 27.8	29.5
06	266 27.7	S15 28.7
07	281 27.7	27.9
08	296 27.7	27.2
09	311 27.6	·· 26.4
10	326 27.6	25.6
11	341 27.6	24.8
12	356 27.5	S15 24.1
13	11 27.5	23.3
14	26 27.5	22.5
15	41 27.4	·· 21.7
16	56 27.4	20.9
17	71 27.3	20.2
18	86 27.3	S15 19.4
19	101 27.3	18.6
20	116 27.2	17.8
21	131 27.2	·· 17.0
22	146 27.2	16.3
23	161 27.2	15.5
8 00	176 27.1	S15 14.7
01	191 27.1	13.9
02	206 27.1	13.1
03	221 27.0	·· 12.3
04	236 27.0	11.6
05	251 27.0	10.8
06	266 26.9	S15 10.0
07	281 26.9	09.2
08	296 26.9	08.4
09	311 26.9	·· 07.6
10	326 26.8	06.8
11	341 26.8	06.0
12	356 26.8	S15 05.3
13	11 26.7	04.5
14	26 26.7	03.7
15	41 26.7	·· 02.9
16	56 26.7	02.1
17	71 26.6	01.3
18	86 26.6	S15 00.5
19	101 26.6	14 59.7
20	116 26.6	58.9
21	131 26.5	·· 58.1
22	146 26.5	57.4
23	161 26.5	56.6

d h	G.H.A. ° ′	Dec. ° ′
9 00	176 26.5	S14 55.8
01	191 26.5	55.0
02	206 26.4	54.2
03	221 26.4	·· 53.4
04	236 26.4	52.6
05	251 26.4	51.8
06	266 26.3	S14 51.0
07	281 26.3	50.2
08	296 26.3	49.4
09	311 26.3	·· 48.6
10	326 26.3	47.8
11	341 26.3	47.0
12	356 26.2	S14 46.2
13	11 26.2	45.4
14	26 26.2	44.6
15	41 26.2	·· 43.8
16	56 26.2	43.0
17	71 26.1	42.2
18	86 26.1	S14 41.4
19	101 26.1	40.6
20	116 26.1	39.8
21	131 26.1	·· 39.0
22	146 26.1	38.2
23	161 26.1	37.4
10 00	176 26.0	S14 36.6
01	191 26.0	35.8
02	206 26.0	35.0
03	221 26.0	·· 34.2
04	236 26.0	33.4
05	251 26.0	32.6
06	266 26.0	S14 31.8
07	281 25.9	30.9
08	296 25.9	30.1
09	311 25.9	·· 29.3
10	326 25.9	28.5
11	341 25.9	27.7
12	356 25.9	S14 26.9
13	11 25.9	26.1
14	26 25.9	25.3
15	41 25.9	·· 24.5
16	56 25.9	23.7
17	71 25.8	22.9
18	86 25.8	S14 22.0
19	101 25.8	21.2
20	116 25.8	20.4
21	131 25.8	·· 19.6
22	146 25.8	18.8
23	161 25.8	18.0
11 00	176 25.8	S14 17.2
01	191 25.8	16.3
02	206 25.8	15.5
03	221 25.8	·· 14.7
04	236 25.8	13.9
05	251 25.8	13.1
06	266 25.8	S14 12.3
07	281 25.8	11.5
08	296 25.8	10.6
09	311 25.7	·· 09.8
10	326 25.7	09.0
11	341 25.7	08.2
12	356 25.7	S14 07.4
13	11 25.7	06.5
14	26 25.7	05.7
15	41 25.7	·· 04.9
16	56 25.7	04.1
17	71 25.7	03.3
18	86 25.7	S14 02.4
19	101 25.7	01.6
20	116 25.7	00.8
21	131 25.7	14 00.0
22	146 25.7	13 59.2
23	161 25.7	58.3

d h	G.H.A. ° ′	Dec. ° ′
12 00	176 25.7	S13 57.5
01	191 25.7	56.7
02	206 25.7	55.9
03	221 25.7	·· 55.0
04	236 25.7	54.2
05	251 25.8	53.4
06	266 25.8	S13 52.6
07	281 25.8	51.7
08	296 25.8	50.9
09	311 25.8	·· 50.1
10	326 25.8	49.2
11	341 25.8	48.4
12	356 25.8	S13 47.6
13	11 25.8	46.8
14	26 25.8	45.9
15	41 25.8	·· 45.1
16	56 25.8	44.3
17	71 25.8	43.4
18	86 25.8	S13 42.6
19	101 25.8	41.8
20	116 25.8	40.9
21	131 25.9	·· 40.1
22	146 25.9	39.3
23	161 25.9	38.4
13 00	176 25.9	S13 37.6
01	191 25.9	36.8
02	206 25.9	35.9
03	221 25.9	·· 35.1
04	236 25.9	34.3
05	251 25.9	33.4
06	266 25.9	S13 32.6
07	281 26.0	31.8
08	296 26.0	30.9
09	311 26.0	·· 30.1
10	326 26.0	29.3
11	341 26.0	28.4
12	356 26.0	S13 27.6
13	11 26.0	26.7
14	26 26.0	25.9
15	41 26.1	·· 25.1
16	56 26.1	24.2
17	71 26.1	23.4
18	86 26.1	S13 22.5
19	101 26.1	21.7
20	116 26.1	20.9
21	131 26.2	·· 20.0
22	146 26.2	19.2
23	161 26.2	18.3
14 00	176 26.2	S13 17.5
01	191 26.2	16.7
02	206 26.2	15.8
03	221 26.3	·· 15.0
04	236 26.3	14.1
05	251 26.3	13.3
06	266 26.3	S13 12.4
07	281 26.3	11.6
08	296 26.4	10.7
09	311 26.4	·· 09.9
10	326 26.4	09.1
11	341 26.4	08.2
12	356 26.4	S13 07.4
13	11 26.5	06.5
14	26 26.5	05.7
15	41 26.5	·· 04.8
16	56 26.5	04.0
17	71 26.6	03.1
18	86 26.6	S13 02.3
19	101 26.6	01.6
20	116 26.6	13 00.6
21	131 26.6	12 59.7
22	146 26.7	58.9
23	161 26.7	58.0

15 (d h)

h	G.H.A.	Dec.
00	176 26.7	S12 57.2
01	191 26.7	56.3
02	206 26.8	55.5
03	221 26.8 ··	54.6
04	236 26.8	53.8
05	251 26.9	52.9
06	266 26.9	S12 52.0
07	281 26.9	51.2
08	296 26.9	50.3
09	311 27.0 ··	49.5
10	326 27.0	48.6
11	341 27.0	47.8
12	356 27.0	S12 46.9
13	11 27.1	46.1
14	26 27.1	45.2
15	41 27.1 ··	44.4
16	56 27.2	43.5
17	71 27.2	42.6
18	86 27.2	S12 41.8
19	101 27.3	40.9
20	116 27.3	40.1
21	131 27.3 ··	39.2
22	146 27.4	38.3
23	161 27.4	37.5

16 (d h)

h	G.H.A.	Dec.
00	176 27.4	S12 36.6
01	191 27.4	35.8
02	206 27.5	34.9
03	221 27.5 ··	34.0
04	236 27.5	33.2
05	251 27.6	32.3
06	266 27.6	S12 31.5
07	281 27.7	30.6
08	296 27.7	29.7
09	311 27.7 ··	28.9
10	326 27.8	28.0
11	341 27.8	27.1
12	356 27.8	S12 26.3
13	11 27.9	25.4
14	26 27.9	24.5
15	41 27.9 ··	23.7
16	56 28.0	22.8
17	71 28.0	21.9
18	86 28.1	S12 21.1
19	101 28.1	20.2
20	116 28.1 ··	19.3
21	131 28.2 ··	18.5
22	146 28.2	17.6
23	161 28.2	16.7

17 (d h)

h	G.H.A.	Dec.
00	176 28.3	S12 15.9
01	191 28.3	15.0
02	206 28.4	14.1
03	221 28.4 ··	13.3
04	236 28.4	12.4
05	251 28.5	11.5
06	266 28.5	S12 10.7
07	281 28.6	09.8
08	296 28.6	08.9
09	311 28.7 ··	08.0
10	326 28.7	07.2
11	341 28.7	06.3
12	356 28.8	S12 05.4
13	11 28.8	04.6
14	26 28.9	03.7
15	41 28.9 ··	02.8
16	56 29.0	01.9
17	71 29.0	01.1
18	86 29.1	S12 00.2
19	101 29.1	11 59.3
20	116 29.1	58.4
21	131 29.2 ··	57.6
22	146 29.2	56.7
23	161 29.3	55.8

18 (d h)

h	G.H.A.	Dec.
00	176 29.3	S11 54.9
01	191 29.4	54.1
02	206 29.4	53.2
03	221 29.5 ··	52.3
04	236 29.5	51.4
05	251 29.6	50.6
06	266 29.6	S11 49.7
07	281 29.7	48.8
08	296 29.7	47.9
09	311 29.8 ··	47.0
10	326 29.8	46.2
11	341 29.9	45.3
12	356 29.9	S11 44.4
13	11 30.0	43.5
14	26 30.0	42.6
15	41 30.1 ··	41.8
16	56 30.1	40.9
17	71 30.2	40.0
18	86 30.2	S11 39.1
19	101 30.3	38.2
20	116 30.3	37.3
21	131 30.4 ··	36.5
22	146 30.4	35.6
23	161 30.5	34.7

19 (d h)

h	G.H.A.	Dec.
00	176 30.5	S11 33.8
01	191 30.6	32.9
02	206 30.7	32.0
03	221 30.7 ··	31.2
04	236 30.8	30.3
05	251 30.8	29.4
06	266 30.9	S11 28.5
07	281 30.9	27.6
08	296 31.0	26.7
09	311 31.0 ··	25.8
10	326 31.1	24.9
11	341 31.2	24.1
12	356 31.2	S11 23.2
13	11 31.3	22.3
14	26 31.3	21.4
15	41 31.4 ··	20.5
16	56 31.4	19.6
17	71 31.5	18.7
18	86 31.6	S11 17.8
19	101 31.6	16.9
20	116 31.7 ··	16.1
21	131 31.7 ··	15.2
22	146 31.8	14.3
23	161 31.9	13.4

20 (d h)

h	G.H.A.	Dec.
00	176 31.9	S11 12.5
01	191 32.0	11.6
02	206 32.0	10.7
03	221 32.1 ··	09.8
04	236 32.2	08.9
05	251 32.2	08.0
06	266 32.3	S11 07.1
07	281 32.4	06.2
08	296 32.4	05.3
09	311 32.5 ··	04.5
10	326 32.5	03.6
11	341 32.6	02.7
12	356 32.7	S11 01.8
13	11 32.7	00.9
14	26 32.8	11 00.0
15	41 32.9	10 59.1
16	56 32.9	58.2
17	71 33.0	57.3
18	86 33.1	S10 56.4
19	101 33.1	55.5
20	116 33.2	54.6
21	131 33.3 ··	53.7
22	146 33.3	52.8
23	161 33.4	51.9

21 (d h)

h	G.H.A.	Dec.
00	176 33.5	S10 51.0
01	191 33.5	50.1
02	206 33.6	49.2
03	221 33.7 ··	48.3
04	236 33.7	47.4
05	251 33.8	46.5
06	266 33.9	S10 45.6
07	281 33.9	44.7
08	296 34.0	43.8
09	311 34.1 ··	42.9
10	326 34.2	42.0
11	341 34.2	41.1
12	356 34.3	S10 40.2
13	11 34.4	39.3
14	26 34.4	38.4
15	41 34.5 ··	37.5
16	56 34.6	36.6
17	71 34.7	35.7
18	86 34.7	S10 34.8
19	101 34.8	33.9
20	116 34.9	33.0
21	131 34.9 ··	32.1
22	146 35.0	31.2
23	161 35.1	30.3

22 (d h)

h	G.H.A.	Dec.
00	176 35.2	S10 29.3
01	191 35.2	28.4
02	206 35.3	27.5
03	221 35.4 ··	26.6
04	236 35.5	25.7
05	251 35.5	24.8
06	266 35.6	S10 23.9
07	281 35.7	23.0
08	296 35.8	22.1
09	311 35.8 ··	21.2
10	326 35.9	20.3
11	341 36.0	19.4
12	356 36.1	S10 18.5
13	11 36.2	17.5
14	26 36.2	16.6
15	41 36.3 ··	15.7
16	56 36.4	14.8
17	71 36.5	13.9
18	86 36.5	S10 13.0
19	101 36.6	12.1
20	116 36.7	11.2
21	131 36.8 ··	10.3
22	146 36.9	09.4
23	161 36.9	08.4

23 (d h)

h	G.H.A.	Dec.
00	176 37.0	S10 07.5
01	191 37.1	06.6
02	206 37.2	05.7
03	221 37.3 ··	04.8
04	236 37.4	03.9
05	251 37.4	03.0
06	266 37.5	S10 02.0
07	281 37.6	01.1
08	296 37.7	10 00.2
09	311 37.8	9 59.3
10	326 37.8	58.4
11	341 37.9	57.5
12	356 38.0	S 9 56.6
13	11 38.1	55.6
14	26 38.2	54.7
15	41 38.3 ··	53.8
16	56 38.4	52.9
17	71 38.4	52.0
18	86 38.5	S 9 51.1
19	101 38.6	50.1
20	116 38.7	49.2
21	131 38.8 ··	48.3
22	146 38.9	47.4
23	161 39.0	46.5

		G.H.A.	Dec.
d 24	h 00	176 39.0	S 9 45.6
	01	191 39.1	44.6
	02	206 39.2	43.7
	03	221 39.3	·· 42.8
	04	236 39.4	41.9
	05	251 39.5	41.0
	06	266 39.6	S 9 40.0
	07	281 39.7	39.1
	08	296 39.7	38.2
	09	311 39.8	·· 37.3
	10	326 39.9	36.4
	11	341 40.0	35.4
	12	356 40.1	S 9 34.5
	13	11 40.2	33.6
	14	26 40.3	32.7
	15	41 40.4	·· 31.7
	16	56 40.5	30.8
	17	71 40.6	29.9
	18	86 40.6	S 9 29.0
	19	101 40.7	28.0
	20	116 40.8	27.1
	21	131 40.9	·· 26.2
	22	146 41.0	25.3
	23	161 41.1	24.4
25	00	176 41.2	S 9 23.4
	01	191 41.3	22.5
	02	206 41.4	21.6
	03	221 41.5	·· 20.7
	04	236 41.6	19.7
	05	251 41.7	18.8
	06	266 41.8	S 9 17.9
	07	281 41.9	16.9
	08	296 41.9	16.0
	09	311 42.0	·· 15.1
	10	326 42.1	14.2
	11	341 42.2	13.2
	12	356 42.3	S 9 12.3
	13	11 42.4	11.4
	14	26 42.5	10.5
	15	41 42.6	·· 09.5
	16	56 42.7	08.6
	17	71 42.8	07.7
	18	86 42.9	S 9 06.7
	19	101 43.0	05.8
	20	116 43.1	04.9
	21	131 43.2	·· 04.0
	22	146 43.3	03.0
	23	161 43.4	02.1
26	00	176 43.5	S 9 01.2
	01	191 43.6	9 00.2
	02	206 43.7	8 59.3
	03	221 43.8	·· 58.4
	04	236 43.9	57.4
	05	251 44.0	56.5
	06	266 44.1	S 8 55.6
	07	281 44.2	54.6
	08	296 44.3	53.7
	09	311 44.4	·· 52.8
	10	326 44.5	51.8
	11	341 44.6	50.9
	12	356 44.7	S 8 50.0
	13	11 44.8	49.0
	14	26 44.9	48.1
	15	41 45.0	·· 47.2
	16	56 45.1	46.2
	17	71 45.2	45.3
	18	86 45.3	S 8 44.4
	19	101 45.4	43.4
	20	116 45.5	42.5
	21	131 45.6	·· 41.6
	22	146 45.7	40.6
	23	161 45.8	39.7

		G.H.A.	Dec.
d 27	h 00	176 45.9	S 8 38.8
	01	191 46.1	37.8
	02	206 46.2	36.9
	03	221 46.3	·· 36.0
	04	236 46.4	35.0
	05	251 46.5	34.1
	06	266 46.6	S 8 33.1
	07	281 46.7	32.2
	08	296 46.8	31.3
	09	311 46.9	·· 30.3
	10	326 47.0	29.4
	11	341 47.1	28.5
	12	356 47.2	S 8 27.5
	13	11 47.3	26.6
	14	26 47.4	25.6
	15	41 47.5	·· 24.7
	16	56 47.7	23.8
	17	71 47.8	22.8
	18	86 47.9	S 8 21.9
	19	101 48.0	20.9
	20	116 48.1	20.0
	21	131 48.2	·· 19.1
	22	146 48.3	18.1
	23	161 48.4	17.2
28	00	176 48.5	S 8 16.2
	01	191 48.6	15.3
	02	206 48.8	14.4
	03	221 48.9	·· 13.4
	04	236 49.0	12.5
	05	251 49.1	11.5
	06	266 49.2	S 8 10.6
	07	281 49.3	09.6
	08	296 49.4	08.7
	09	311 49.5	·· 07.8
	10	326 49.7	06.8
	11	341 49.8	05.9
	12	356 49.9	S 8 04.9
	13	11 50.0	04.0
	14	26 50.1	03.0
	15	41 50.2	·· 02.1
	16	56 50.3	01.1
	17	71 50.4	8 00.2
	18	86 50.6	S 7 59.3
	19	101 50.7	58.3
	20	116 50.8	57.4
	21	131 50.9	·· 56.4
	22	146 51.0	55.5
	23	161 51.1	54.5
MARCH 1	00	176 51.3	S 7 53.6
	01	191 51.4	52.6
	02	206 51.5	51.7
	03	221 51.6	·· 50.7
	04	236 51.7	49.8
	05	251 51.8	48.9
	06	266 52.0	S 7 47.9
	07	281 52.1	47.0
	08	296 52.2	46.0
	09	311 52.3	·· 45.1
	10	326 52.4	44.1
	11	341 52.6	43.2
	12	356 52.7	S 7 42.2
	13	11 52.8	41.3
	14	26 52.9	40.3
	15	41 53.0	·· 39.4
	16	56 53.1	38.4
	17	71 53.3	37.5
	18	86 53.4	S 7 36.5
	19	101 53.5	35.6
	20	116 53.6	34.6
	21	131 53.7	·· 33.7
	22	146 53.9	32.7
	23	161 54.0	31.8

		G.H.A.	Dec.
d 2	h 00	176 54.1	S 7 30.8
	01	191 54.2	29.9
	02	206 54.4	28.9
	03	221 54.5	·· 28.0
	04	236 54.6	27.0
	05	251 54.7	26.1
	06	266 54.8	S 7 25.1
	07	281 55.0	24.2
	08	296 55.1	23.2
	09	311 55.2	·· 22.3
	10	326 55.3	21.3
	11	341 55.5	20.4
	12	356 55.6	S 7 19.4
	13	11 55.7	18.4
	14	26 55.8	17.5
	15	41 56.0	·· 16.5
	16	56 56.1	15.6
	17	71 56.2	14.6
	18	86 56.3	S 7 13.7
	19	101 56.5	12.7
	20	116 56.6	11.8
	21	131 56.7	·· 10.8
	22	146 56.8	09.9
	23	161 57.0	08.9
3	00	176 57.1	S 7 08.0
	01	191 57.2	07.0
	02	206 57.4	06.0
	03	221 57.5	·· 05.1
	04	236 57.6	04.1
	05	251 57.7	03.2
	06	266 57.9	S 7 02.2
	07	281 58.0	01.3
	08	296 58.1	7 00.3
	09	311 58.2	6 59.3
	10	326 58.4	58.4
	11	341 58.5	57.4
	12	356 58.6	S 6 56.5
	13	11 58.8	55.5
	14	26 58.9	54.6
	15	41 59.0	·· 53.6
	16	56 59.2	52.6
	17	71 59.3	51.7
	18	86 59.4	S 6 50.7
	19	101 59.5	49.8
	20	116 59.7	48.8
	21	131 59.8	·· 47.9
	22	146 59.9	46.9
	23	162 00.1	45.9
4	00	177 00.2	S 6 45.0
	01	192 00.3	44.0
	02	207 00.5	43.1
	03	222 00.6	·· 42.1
	04	237 00.7	41.1
	05	252 00.9	40.2
	06	267 01.0	S 6 39.2
	07	282 01.1	38.3
	08	297 01.3	37.3
	09	312 01.4	·· 36.3
	10	327 01.5	35.4
	11	342 01.7	34.4
	12	357 01.8	S 6 33.5
	13	12 01.9	32.5
	14	27 02.1	31.5
	15	42 02.2	·· 30.6
	16	57 02.3	29.6
	17	72 02.5	28.7
	18	87 02.6	S 6 27.7
	19	102 02.8	26.7
	20	117 02.9	25.8
	21	132 03.0	·· 24.8
	22	147 03.2	23.8
	23	162 03.3	22.9

d h	G.H.A.	Dec.		d h	G.H.A.	Dec.		d h	G.H.A.	Dec.
5 00	177 03.4	S 6 21.9		**8** 00	177 13.8	S 5 12.2		**11** 00	177 25.1	S 4 01.9
01	192 03.6	20.9		01	192 14.0	11.2		01	192 25.3	4 00.9
02	207 03.7	20.0		02	207 14.1	10.3		02	207 25.4	3 59.9
03	222 03.8	·· 19.0		03	222 14.3	·· 09.3		03	222 25.6	·· 59.0
04	237 04.0	18.1		04	237 14.4	08.3		04	237 25.8	58.0
05	252 04.1	17.1		05	252 14.6	07.3		05	252 25.9	57.0
06	267 04.3	S 6 16.1		06	267 14.7	S 5 06.4		06	267 26.1	S 3 56.0
07	282 04.4	15.2		07	282 14.9	05.4		07	282 26.3	55.0
08	297 04.5	14.2		08	297 15.0	04.4		08	297 26.4	54.1
09	312 04.7	·· 13.2		09	312 15.2	·· 03.5		09	312 26.6	·· 53.1
10	327 04.8	12.3		10	327 15.3	02.5		10	327 26.7	52.1
11	342 05.0	11.3		11	342 15.5	01.5		11	342 26.9	51.1
12	357 05.1	S 6 10.3		12	357 15.6	S 5 00.5		12	357 27.1	S 3 50.1
13	12 05.2	09.4		13	12 15.8	4 59.6		13	12 27.2	49.2
14	27 05.4	08.4		14	27 15.9	58.6		14	27 27.4	48.2
15	42 05.5	·· 07.5		15	42 16.1	·· 57.6		15	42 27.6	·· 47.2
16	57 05.7	06.5		16	57 16.3	56.6		16	57 27.7	46.2
17	72 05.8	05.5		17	72 16.4	55.7		17	72 27.9	45.2
18	87 05.9	S 6 04.6		18	87 16.6	S 4 54.7		18	87 28.1	S 3 44.2
19	102 06.1	03.6		19	102 16.7	53.7		19	102 28.2	43.3
20	117 06.2	02.6		20	117 16.9	52.7		20	117 28.4	42.3
21	132 06.4	·· 01.7		21	132 17.0	·· 51.8		21	132 28.6	·· 41.3
22	147 06.5	6 00.7		22	147 17.2	50.8		22	147 28.7	40.3
23	162 06.6	5 59.7		23	162 17.3	49.8		23	162 28.9	39.3
6 00	177 06.8	S 5 58.8		**9** 00	177 17.5	S 4 48.8		**12** 00	177 29.1	S 3 38.4
01	192 06.9	57.8		01	192 17.6	47.9		01	192 29.2	37.4
02	207 07.1	56.8		02	207 17.8	46.9		02	207 29.4	36.4
03	222 07.2	·· 55.9		03	222 18.0	·· 45.9		03	222 29.6	·· 35.4
04	237 07.4	54.9		04	237 18.1	44.9		04	237 29.7	34.4
05	252 07.5	53.9		05	252 18.3	44.0		05	252 29.9	33.4
06	267 07.6	S 5 53.0		06	267 18.4	S 4 43.0		06	267 30.1	S 3 32.5
07	282 07.8	52.0		07	282 18.6	42.0		07	282 30.2	31.5
08	297 07.9	51.0		08	297 18.7	41.0		08	297 30.4	30.5
09	312 08.1	·· 50.1		09	312 18.9	·· 40.1		09	312 30.6	·· 29.5
10	327 08.2	49.1		10	327 19.0	39.1		10	327 30.7	28.5
11	342 08.4	48.1		11	342 19.2	38.1		11	342 30.9	27.6
12	357 08.5	S 5 47.2		12	357 19.4	S 4 37.1		12	357 31.1	S 3 26.6
13	12 08.6	46.2		13	12 19.5	36.1		13	12 31.2	25.6
14	27 08.8	45.2		14	27 19.7	35.2		14	27 31.4	24.6
15	42 08.9	·· 44.2		15	42 19.8	·· 34.2		15	42 31.6	·· 23.6
16	57 09.1	43.3		16	57 20.0	33.2		16	57 31.7	22.6
17	72 09.2	42.3		17	72 20.1	32.2		17	72 31.9	21.7
18	87 09.4	S 5 41.3		18	87 20.3	S 4 31.3		18	87 32.1	S 3 20.7
19	102 09.5	40.4		19	102 20.5	30.3		19	102 32.2	19.7
20	117 09.7	39.4		20	117 20.6	29.3		20	117 32.4	18.7
21	132 09.8	·· 38.4		21	132 20.8	·· 28.3		21	132 32.6	·· 17.7
22	147 10.0	37.5		22	147 20.9	27.4		22	147 32.7	16.7
23	162 10.1	36.5		23	162 21.1	26.4		23	162 32.9	15.8
7 00	177 10.2	S 5 35.5		**10** 00	177 21.3	S 4 25.4		**13** 00	177 33.1	S 3 14.8
01	192 10.4	34.6		01	192 21.4	24.4		01	192 33.2	13.8
02	207 10.5	33.6		02	207 21.6	23.4		02	207 33.4	12.8
03	222 10.7	·· 32.6		03	222 21.7	·· 22.5		03	222 33.6	·· 11.8
04	237 10.8	31.6		04	237 21.9	21.5		04	237 33.8	10.8
05	252 11.0	30.7		05	252 22.0	20.5		05	252 33.9	09.9
06	267 11.1	S 5 29.7		06	267 22.2	S 4 19.5		06	267 34.1	S 3 08.9
07	282 11.3	28.7		07	282 22.4	18.5		07	282 34.3	07.9
08	297 11.4	27.8		08	297 22.5	17.6		08	297 34.4	06.9
09	312 11.6	·· 26.8		09	312 22.7	·· 16.6		09	312 34.6	·· 05.9
10	327 11.7	25.8		10	327 22.8	15.6		10	327 34.8	04.9
11	342 11.9	24.8		11	342 23.0	14.6		11	342 34.9	04.0
12	357 12.0	S 5 23.9		12	357 23.2	S 4 13.7		12	357 35.1	S 3 03.0
13	12 12.2	22.9		13	12 23.3	12.7		13	12 35.3	02.0
14	27 12.3	21.9		14	27 23.5	11.7		14	27 35.5	01.0
15	42 12.5	·· 21.0		15	42 23.7	·· 10.7		15	42 35.6	3 00.0
16	57 12.6	20.0		16	57 23.8	09.7		16	57 35.8	2 59.0
17	72 12.8	19.0		17	72 24.0	08.8		17	72 36.0	58.0
18	87 12.9	S 5 18.0		18	87 24.1	S 4 07.8		18	87 36.1	S 2 57.1
19	102 13.1	17.1		19	102 24.3	06.8		19	102 36.3	56.1
20	117 13.2	16.1		20	117 24.5	05.8		20	117 36.5	55.1
21	132 13.4	·· 15.1		21	132 24.6	·· 04.8		21	132 36.7	·· 54.1
22	147 13.5	14.2		22	147 24.8	03.9		22	147 36.8	53.1
23	162 13.7	13.2		23	162 24.9	02.9		23	162 37.0	52.1

	G.H.A.	Dec.
14 00	177 37.2	S 2 51.2
01	192 37.3	50.2
02	207 37.5	49.2
03	222 37.7 ··	48.2
04	237 37.9	47.2
05	252 38.0	46.2
06	267 38.2	S 2 45.2
07	282 38.4	44.3
08	297 38.6	43.3
09	312 38.7 ··	42.3
10	327 38.9	41.3
11	342 39.1	40.3
12	357 39.2	S 2 39.3
13	12 39.4	38.3
14	27 39.6	37.4
15	42 39.8 ··	36.4
16	57 39.9	35.4
17	72 40.1	34.4
18	87 40.3	S 2 33.4
19	102 40.5	32.4
20	117 40.6	31.4
21	132 40.8 ··	30.5
22	147 41.0	29.5
23	162 41.2	28.5
15 00	177 41.3	S 2 27.5
01	192 41.5	26.5
02	207 41.7	25.5
03	222 41.9 ··	24.5
04	237 42.0	23.6
05	252 42.2	22.6
06	267 42.4	S 2 21.6
07	282 42.6	20.6
08	297 42.7	19.6
09	312 42.9 ··	18.6
10	327 43.1	17.6
11	342 43.3	16.7
12	357 43.4	S 2 15.7
13	12 43.6	14.7
14	27 43.8	13.7
15	42 44.0 ··	12.7
16	57 44.1	11.7
17	72 44.3	10.7
18	87 44.5	S 2 09.7
19	102 44.7	08.8
20	117 44.8	07.8
21	132 45.0 ··	06.8
22	147 45.2	05.8
23	162 45.4	04.8
16 00	177 45.6	S 2 03.8
01	192 45.7	02.8
02	207 45.9	01.9
03	222 46.1	2 00.9
04	237 46.3	1 59.9
05	252 46.4	58.9
06	267 46.6	S 1 57.9
07	282 46.8	56.9
08	297 47.0	55.9
09	312 47.2 ··	54.9
10	327 47.3	54.0
11	342 47.5	53.0
12	357 47.7	S 1 52.0
13	12 47.9	51.0
14	27 48.0	50.0
15	42 48.2 ··	49.0
16	57 48.4	48.0
17	72 48.6	47.0
18	87 48.8	S 1 46.1
19	102 48.9	45.1
20	117 49.1	44.1
21	132 49.3 ··	43.1
22	147 49.5	42.1
23	162 49.7	41.1

	G.H.A.	Dec.
17 00	177 49.8	S 1 40.1
01	192 50.0	39.1
02	207 50.2	38.2
03	222 50.4 ··	37.2
04	237 50.5	36.2
05	252 50.7	35.2
06	267 50.9	S 1 34.2
07	282 51.1	33.2
08	297 51.3	32.2
09	312 51.4 ··	31.2
10	327 51.6	30.3
11	342 51.8	29.3
12	357 52.0	S 1 28.3
13	12 52.2	27.3
14	27 52.3	26.3
15	42 52.5 ··	25.3
16	57 52.7	24.3
17	72 52.9	23.3
18	87 53.1	S 1 22.4
19	102 53.3	21.4
20	117 53.4	20.4
21	132 53.6 ··	19.4
22	147 53.8	18.4
23	162 54.0	17.4
18 00	177 54.2	S 1 16.4
01	192 54.3	15.4
02	207 54.5	14.5
03	222 54.7 ··	13.5
04	237 54.9	12.5
05	252 55.1	11.5
06	267 55.2	S 1 10.5
07	282 55.4	09.5
08	297 55.6	08.5
09	312 55.8 ··	07.5
10	327 56.0	06.5
11	342 56.2	05.6
12	357 56.3	S 1 04.6
13	12 56.5	03.6
14	27 56.7	02.6
15	42 56.9 ··	01.6
16	57 57.1	1 00.6
17	72 57.2	0 59.6
18	87 57.4	S 0 58.6
19	102 57.6	57.7
20	117 57.8	56.7
21	132 58.0 ··	55.7
22	147 58.2	54.7
23	162 58.3	53.7
19 00	177 58.5	S 0 52.7
01	192 58.7	51.7
02	207 58.9	50.7
03	222 59.1 ··	49.8
04	237 59.3	48.8
05	252 59.4	47.8
06	267 59.6	S 0 46.8
07	282 59.8	45.8
08	298 00.0	44.8
09	313 00.2 ··	43.8
10	328 00.4	42.8
11	343 00.5	41.8
12	358 00.7	S 0 40.9
13	13 00.9	39.9
14	28 01.1	38.9
15	43 01.3 ··	37.9
16	58 01.5	36.9
17	73 01.6	35.9
18	88 01.8	S 0 34.9
19	103 02.0	33.9
20	118 02.2	33.0
21	133 02.4 ··	32.0
22	148 02.6	31.0
23	163 02.8	30.0

	G.H.A.	Dec.
20 00	178 02.9	S 0 29.0
01	193 03.1	28.0
02	208 03.3	27.0
03	223 03.5 ··	26.0
04	238 03.7	25.1
05	253 03.9	24.1
06	268 04.0	S 0 23.1
07	283 04.2	22.1
08	298 04.4	21.1
09	313 04.6 ··	20.1
10	328 04.8	19.1
11	343 05.0	18.1
12	358 05.2	S 0 17.1
13	13 05.3	16.2
14	28 05.5	15.2
15	43 05.7 ··	14.2
16	58 05.9	13.2
17	73 06.1	12.2
18	88 06.3	S 0 11.2
19	103 06.4	10.2
20	118 06.6	09.2
21	133 06.8 ··	08.3
22	148 07.0	07.3
23	163 07.2	06.3
21 00	178 07.4	S 0 05.3
01	193 07.6	04.3
02	208 07.7	03.3
03	223 07.9 ··	02.3
04	238 08.1	01.3
05	253 08.3	S 0 00.4
06	268 08.5	N 0 00.6
07	283 08.7	01.6
08	298 08.9	02.6
09	313 09.0 ··	03.6
10	328 09.2	04.6
11	343 09.4	05.6
12	358 09.6	N 0 06.6
13	13 09.8	07.5
14	28 10.0	08.5
15	43 10.2 ··	09.5
16	58 10.4	10.5
17	73 10.5	11.5
18	88 10.7	N 0 12.5
19	103 10.9	13.5
20	118 11.1	14.5
21	133 11.3 ··	15.4
22	148 11.5	16.4
23	163 11.7	17.4
22 00	178 11.8	N 0 18.4
01	193 12.0	19.4
02	208 12.2	20.4
03	223 12.4 ··	21.4
04	238 12.6	22.4
05	253 12.8	23.3
06	268 13.0	N 0 24.3
07	283 13.2	25.3
08	298 13.3	26.3
09	313 13.5 ··	27.3
10	328 13.7	28.3
11	343 13.9	29.3
12	358 14.1	N 0 30.2
13	13 14.3	31.2
14	28 14.5	32.2
15	43 14.7 ··	33.2
16	58 14.8	34.2
17	73 15.0	35.2
18	88 15.2	N 0 36.2
19	103 15.4	37.2
20	118 15.6	38.1
21	133 15.8 ··	39.1
22	148 16.0	40.1
23	163 16.2	41.1

d h	G.H.A.	Dec.		d h	G.H.A.	Dec.		d h	G.H.A.	Dec.
	° ′	° ′			° ′	° ′			° ′	° ′
23 00	178 16.3	N 0 42.1		26 00	178 29.9	N 1 53.0		29 00	178 43.5	N 3 03.5
01	193 16.5	43.1		01	193 30.1	54.0		01	193 43.7	04.5
02	208 16.7	44.1		02	208 30.3	55.0		02	208 43.9	05.5
03	223 16.9	·· 45.0		03	223 30.5	·· 55.9		03	223 44.1	·· 06.5
04	238 17.1	46.0		04	238 30.6	56.9		04	238 44.2	07.4
05	253 17.3	47.0		05	253 30.8	57.9		05	253 44.4	08.4
06	268 17.5	N 0 48.0		06	268 31.0	N 1 58.9		06	268 44.6	N 3 09.4
07	283 17.7	49.0		07	283 31.2	1 59.9		07	283 44.8	10.4
08	298 17.8	50.0		08	298 31.4	2 00.9		08	298 45.0	11.3
09	313 18.0	·· 51.0		09	313 31.6	·· 01.8		09	313 45.2	·· 12.3
10	328 18.2	52.0		10	328 31.8	02.8		10	328 45.4	13.3
11	343 18.4	52.9		11	343 32.0	03.8		11	343 45.6	14.3
12	358 18.6	N 0 53.9		12	358 32.2	N 2 04.8		12	358 45.8	N 3 15.2
13	13 18.8	54.9		13	13 32.3	05.8		13	13 45.9	16.2
14	28 19.0	55.9		14	28 32.5	06.7		14	28 46.1	17.2
15	43 19.2	·· 56.9		15	43 32.7	·· 07.7		15	43 46.3	·· 18.2
16	58 19.3	57.9		16	58 32.9	08.7		16	58 46.5	19.1
17	73 19.5	58.9		17	73 33.1	09.7		17	73 46.7	20.1
18	88 19.7	N 0 59.8		18	88 33.3	N 2 10.7		18	88 46.9	N 3 21.1
19	103 19.9	1 00.8		19	103 33.5	11.7		19	103 47.1	22.1
20	118 20.1	01.8		20	110 33.7	12.6		20	118 47.3	23.0
21	133 20.3	·· 02.8		21	133 33.9	·· 13.6		21	133 47.4	·· 24.0
22	148 20.5	03.8		22	148 34.0	14.6		22	148 47.6	25.0
23	163 20.7	04.8		23	163 34.2	15.6		23	163 47.8	26.0
24 00	178 20.8	N 1 05.8		27 00	178 34.4	N 2 16.6		30 00	178 48.0	N 3 26.9
01	193 21.0	06.7		01	193 34.6	17.5		01	193 48.2	27.9
02	208 21.2	07.7		02	208 34.8	18.5		02	208 48.4	28.9
03	223 21.4	·· 08.7		03	223 35.0	·· 19.5		03	223 48.6	·· 29.8
04	238 21.6	09.7		04	238 35.2	20.5		04	238 48.8	30.8
05	253 21.8	10.7		05	253 35.4	21.5		05	253 49.0	31.8
06	268 22.0	N 1 11.7		06	268 35.6	N 2 22.4		06	268 49.1	N 3 32.8
07	283 22.2	12.7		07	283 35.7	23.4		07	283 49.3	33.7
08	298 22.3	13.6		08	298 35.9	24.4		08	298 49.5	34.7
09	313 22.5	·· 14.6		09	313 36.1	·· 25.4		09	313 49.7	·· 35.7
10	328 22.7	15.6		10	328 36.3	26.4		10	328 49.9	36.7
11	343 22.9	16.6		11	343 36.5	27.3		11	343 50.1	37.6
12	358 23.1	N 1 17.6		12	358 36.7	N 2 28.3		12	358 50.3	N 3 38.6
13	13 23.3	18.6		13	13 36.9	29.3		13	13 50.5	39.6
14	28 23.5	19.6		14	28 37.1	30.3		14	28 50.6	40.5
15	43 23.7	·· 20.5		15	43 37.3	·· 31.3		15	43 50.8	·· 41.5
16	58 23.9	21.5		16	58 37.4	32.2		16	58 51.0	42.5
17	73 24.0	22.5		17	73 37.6	33.2		17	73 51.2	43.5
18	88 24.2	N 1 23.5		18	88 37.8	N 2 34.2		18	88 51.4	N 3 44.4
19	103 24.4	24.5		19	103 38.0	35.2		19	103 51.6	45.4
20	118 24.6	25.5		20	118 38.2	36.2		20	118 51.8	46.4
21	133 24.8	·· 26.4		21	133 38.4	·· 37.1		21	133 52.0	·· 47.3
22	148 25.0	27.4		22	148 38.6	38.1		22	148 52.1	48.3
23	163 25.2	28.4		23	163 38.8	39.1		23	163 52.3	49.3
25 00	178 25.4	N 1 29.4		28 00	178 39.0	N 2 40.1		31 00	178 52.5	N 3 50.3
01	193 25.5	30.4		01	193 39.1	41.1		01	193 52.7	51.2
02	208 25.7	31.4		02	208 39.3	42.0		02	208 52.9	52.2
03	223 25.9	·· 32.3		03	223 39.5	·· 43.0		03	223 53.1	·· 53.2
04	238 26.1	33.3		04	238 39.7	44.0		04	238 53.3	54.1
05	253 26.3	34.3		05	253 39.9	45.0		05	253 53.5	55.1
06	268 26.5	N 1 35.3		06	268 40.1	N 2 45.9		06	268 53.6	N 3 56.1
07	283 26.7	36.3		07	283 40.3	46.9		07	283 53.8	57.0
08	298 26.9	37.3		08	298 40.5	47.9		08	298 54.0	58.0
09	313 27.1	·· 38.2		09	313 40.7	·· 48.9		09	313 54.2	3 59.0
10	328 27.2	39.2		10	328 40.8	49.9		10	328 54.4	4 00.0
11	343 27.4	40.2		11	343 41.0	50.8		11	343 54.6	00.9
12	358 27.6	N 1 41.2		12	358 41.2	N 2 51.8		12	358 54.8	N 4 01.9
13	13 27.8	42.2		13	13 41.4	52.8		13	13 55.0	02.9
14	28 28.0	43.2		14	28 41.6	53.8		14	28 55.1	03.8
15	43 28.2	·· 44.1		15	43 41.8	·· 54.7		15	43 55.3	·· 04.8
16	58 28.4	45.1		16	58 42.0	55.7		16	58 55.5	05.8
17	73 28.6	46.1		17	73 42.2	56.7		17	73 55.7	06.7
18	88 28.8	N 1 47.1		18	88 42.4	N 2 57.7		18	88 55.9	N 4 07.7
19	103 28.9	48.1		19	103 42.5	58.7		19	103 56.1	08.7
20	118 29.1	49.1		20	118 42.7	2 59.6		20	118 56.3	09.6
21	133 29.3	·· 50.0		21	133 42.9	3 00.6		21	133 56.5	·· 10.6
22	148 29.5	51.0		22	148 43.1	01.6		22	148 56.6	11.6
23	163 29.7	52.0		23	163 43.3	02.6		23	163 56.8	12.5

APRIL

d	h	G.H.A. (° ′)	Dec. (° ′)
1	00	178 57.0	N 4 13.5
	01	193 57.2	14.5
	02	208 57.4	15.4
	03	223 57.6	·· 16.4
	04	238 57.8	17.4
	05	253 57.9	18.3
	06	268 58.1	N 4 19.3
	07	283 58.3	20.3
	08	298 58.5	21.2
	09	313 58.7	·· 22.2
	10	328 58.9	23.2
	11	343 59.1	24.1
	12	358 59.3	N 4 25.1
	13	13 59.4	26.1
	14	28 59.6	27.0
	15	43 59.8	·· 28.0
	16	59 00.0	29.0
	17	74 00.2	29.9
	18	89 00.4	N 4 30.9
	19	104 00.6	31.9
	20	119 00.7	32.8
	21	134 00.9	·· 33.8
	22	149 01.1	34.8
	23	164 01.3	35.7
2	00	179 01.5	N 4 36.7
	01	194 01.7	37.7
	02	209 01.9	38.6
	03	224 02.0	·· 39.6
	04	239 02.2	40.5
	05	254 02.4	41.5
	06	269 02.6	N 4 42.5
	07	284 02.8	43.4
	08	299 03.0	44.4
	09	314 03.2	·· 45.4
	10	329 03.3	46.3
	11	344 03.5	47.3
	12	359 03.7	N 4 48.2
	13	14 03.9	49.2
	14	29 04.1	50.2
	15	44 04.3	·· 51.1
	16	59 04.5	52.1
	17	74 04.6	53.1
	18	89 04.8	N 4 54.0
	19	104 05.0	55.0
	20	119 05.2	55.9
	21	134 05.4	·· 56.9
	22	149 05.6	57.9
	23	164 05.7	58.8
3	00	179 05.9	N 4 59.8
	01	194 06.1	5 00.7
	02	209 06.3	01.7
	03	224 06.5	·· 02.7
	04	239 06.7	03.6
	05	254 06.9	04.6
	06	269 07.0	N 5 05.5
	07	284 07.2	06.5
	08	299 07.4	07.5
	09	314 07.6	·· 08.4
	10	329 07.8	09.4
	11	344 08.0	10.3
	12	359 08.1	N 5 11.3
	13	14 08.3	12.2
	14	29 08.5	13.2
	15	44 08.7	·· 14.2
	16	59 08.9	15.1
	17	74 09.1	16.1
	18	89 09.3	N 5 17.0
	19	104 09.4	18.0
	20	119 09.6	19.0
	21	134 09.8	·· 19.9
	22	149 10.0	20.9
	23	164 10.2	21.8

d	h	G.H.A. (° ′)	Dec. (° ′)
4	00	179 10.4	N 5 22.8
	01	194 10.5	23.7
	02	209 10.7	24.7
	03	224 10.9	·· 25.6
	04	239 11.1	26.6
	05	254 11.3	27.6
	06	269 11.5	N 5 28.5
	07	284 11.6	29.5
	08	299 11.8	30.4
	09	314 12.0	·· 31.4
	10	329 12.2	32.3
	11	344 12.4	33.3
	12	359 12.6	N 5 34.2
	13	14 12.7	35.2
	14	29 12.9	36.2
	15	44 13.1	·· 37.1
	16	59 13.3	38.1
	17	74 13.5	39.0
	18	89 13.6	N 5 40.0
	19	104 13.8	40.9
	20	119 14.0	41.9
	21	134 14.2	·· 42.8
	22	149 14.4	43.8
	23	164 14.6	44.7
5	00	179 14.7	N 5 45.7
	01	194 14.9	46.6
	02	209 15.1	47.6
	03	224 15.3	·· 48.5
	04	239 15.5	49.5
	05	254 15.7	50.4
	06	269 15.8	N 5 51.4
	07	284 16.0	52.4
	08	299 16.2	53.3
	09	314 16.4	·· 54.3
	10	329 16.6	55.2
	11	344 16.7	56.2
	12	359 16.9	N 5 57.1
	13	14 17.1	58.1
	14	29 17.3	5 59.0
	15	44 17.5	6 00.0
	16	59 17.6	00.9
	17	74 17.8	01.9
	18	89 18.0	N 6 02.8
	19	104 18.2	03.8
	20	119 18.4	04.7
	21	134 18.6	·· 05.6
	22	149 18.7	06.6
	23	164 18.9	07.5
6	00	179 19.1	N 6 08.5
	01	194 19.3	09.4
	02	209 19.5	10.4
	03	224 19.6	·· 11.3
	04	239 19.8	12.3
	05	254 20.0	13.2
	06	269 20.2	N 6 14.2
	07	284 20.4	15.1
	08	299 20.5	16.1
	09	314 20.7	·· 17.0
	10	329 20.9	18.0
	11	344 21.1	18.9
	12	359 21.3	N 6 19.9
	13	14 21.4	20.8
	14	29 21.6	21.7
	15	44 21.8	·· 22.7
	16	59 22.0	23.6
	17	74 22.2	24.6
	18	89 22.3	N 6 25.5
	19	104 22.5	26.5
	20	119 22.7	27.4
	21	134 22.9	·· 28.4
	22	149 23.0	29.3
	23	164 23.2	30.3

d	h	G.H.A. (° ′)	Dec. (° ′)
7	00	179 23.4	N 6 31.2
	01	194 23.6	32.1
	02	209 23.8	33.1
	03	224 23.9	·· 34.0
	04	239 24.1	35.0
	05	254 24.3	35.9
	06	269 24.5	N 6 36.9
	07	284 24.7	37.8
	08	299 24.8	38.7
	09	314 25.0	·· 39.7
	10	329 25.2	40.6
	11	344 25.4	41.6
	12	359 25.5	N 6 42.5
	13	14 25.7	43.4
	14	29 25.9	44.4
	15	44 26.1	·· 45.3
	16	59 26.3	46.3
	17	74 26.4	47.2
	18	89 26.6	N 6 48.1
	19	104 26.8	49.1
	20	119 27.0	50.0
	21	134 27.1	·· 51.0
	22	149 27.3	51.9
	23	164 27.5	52.8
8	00	179 27.7	N 6 53.8
	01	194 27.8	54.7
	02	209 28.0	55.7
	03	224 28.2	·· 56.6
	04	239 28.4	57.5
	05	254 28.6	58.5
	06	269 28.7	N 6 59.4
	07	284 28.9	7 00.3
	08	299 29.1	01.3
	09	314 29.3	·· 02.2
	10	329 29.4	03.2
	11	344 29.6	04.1
	12	359 29.8	N 7 05.0
	13	14 30.0	06.0
	14	29 30.1	06.9
	15	44 30.3	·· 07.8
	16	59 30.5	08.8
	17	74 30.7	09.7
	18	89 30.8	N 7 10.6
	19	104 31.0	11.6
	20	119 31.2	12.5
	21	134 31.4	·· 13.5
	22	149 31.5	14.4
	23	164 31.7	15.3
9	00	179 31.9	N 7 16.3
	01	194 32.1	17.2
	02	209 32.2	18.1
	03	224 32.4	·· 19.1
	04	239 32.6	20.0
	05	254 32.8	20.9
	06	269 32.9	N 7 21.9
	07	284 33.1	22.8
	08	299 33.3	23.7
	09	314 33.4	·· 24.6
	10	329 33.6	25.6
	11	344 33.8	26.5
	12	359 34.0	N 7 27.4
	13	14 34.1	28.4
	14	29 34.3	29.3
	15	44 34.5	·· 30.2
	16	59 34.7	31.2
	17	74 34.8	32.1
	18	89 35.0	N 7 33.0
	19	104 35.2	34.0
	20	119 35.4	34.9
	21	134 35.5	·· 35.8
	22	149 35.7	36.7
	23	164 35.9	37.7

d h	G.H.A.	Dec.
10 00	179 36.0	N 7 38.6
01	194 36.2	39.5
02	209 36.4	40.5
03	224 36.6 ··	41.4
04	239 36.7	42.3
05	254 36.9	43.2
06	269 37.1	N 7 44.2
07	284 37.2	45.1
08	299 37.4	46.0
09	314 37.6 ··	46.9
10	329 37.8	47.9
11	344 37.9	48.8
12	359 38.1	N 7 49.7
13	14 38.3	50.7
14	29 38.4	51.6
15	44 38.6 ··	52.5
16	59 38.8	53.4
17	74 38.9	54.4
18	89 39.1	N 7 55.3
19	104 39.3	56.2
20	119 39.5	57.1
21	134 39.6 ··	58.1
22	149 39.8	59.0
23	164 40.0	7 59.9
11 00	179 40.1	N 8 00.8
01	194 40.3	01.7
02	209 40.5	02.7
03	224 40.6 ··	03.6
04	239 40.8	04.5
05	254 41.0	05.4
06	269 41.1	N 8 06.4
07	284 41.3	07.3
08	299 41.5	08.2
09	314 41.6 ··	09.1
10	329 41.8	10.0
11	344 42.0	11.0
12	359 42.2	N 8 11.9
13	14 42.3	12.8
14	29 42.5	13.7
15	44 42.7 ··	14.6
16	59 42.8	15.6
17	74 43.0	16.5
18	89 43.2	N 8 17.4
19	104 43.3	18.3
20	119 43.5	19.2
21	134 43.7 ··	20.2
22	149 43.8	21.1
23	164 44.0	22.0
12 00	179 44.2	N 8 22.9
01	194 44.3	23.8
02	209 44.5	24.7
03	224 44.7 ··	25.7
04	239 44.8	26.6
05	254 45.0	27.5
06	269 45.2	N 8 28.4
07	284 45.3	29.3
08	299 45.5	30.2
09	314 45.6 ··	31.2
10	329 45.8	32.1
11	344 46.0	33.0
12	359 46.1	N 8 33.9
13	14 46.3	34.8
14	29 46.5	35.7
15	44 46.6 ··	36.6
16	59 46.8	37.6
17	74 47.0	38.5
18	89 47.1	N 8 39.4
19	104 47.3	40.3
20	119 47.5	41.2
21	134 47.6 ··	42.1
22	149 47.8	43.0
23	164 47.9	43.9

d h	G.H.A.	Dec.
13 00	179 48.1	N 8 44.8
01	194 48.3	45.8
02	209 48.4	46.7
03	224 48.6 ··	47.6
04	239 48.8	48.5
05	254 48.9	49.4
06	269 49.1	N 8 50.3
07	284 49.2	51.2
08	299 49.4	52.1
09	314 49.6 ··	53.0
10	329 49.7	54.0
11	344 49.9	54.9
12	359 50.1	N 8 55.8
13	14 50.2	56.7
14	29 50.4	57.6
15	44 50.5 ··	58.5
16	59 50.7	8 59.4
17	74 50.9	9 00.3
18	89 51.0	N 9 01.2
19	104 51.2	02.1
20	119 51.3	03.0
21	134 51.5 ··	03.9
22	149 51.7	04.8
23	164 51.8	05.7
14 00	179 52.0	N 9 06.6
01	194 52.1	07.6
02	209 52.3	08.5
03	224 52.5 ··	09.4
04	239 52.6	10.3
05	254 52.8	11.2
06	269 52.9	N 9 12.1
07	284 53.1	13.0
08	299 53.3	13.9
09	314 53.4 ··	14.8
10	329 53.6	15.7
11	344 53.7	16.6
12	359 53.9	N 9 17.5
13	14 54.0	18.4
14	29 54.2	19.3
15	44 54.4 ··	20.2
16	59 54.5	21.1
17	74 54.7	22.0
18	89 54.8	N 9 22.9
19	104 55.0	23.8
20	119 55.1	24.7
21	134 55.3 ··	25.6
22	149 55.5	26.5
23	164 55.6	27.4
15 00	179 55.8	N 9 28.3
01	194 55.9	29.2
02	209 56.1	30.1
03	224 56.2 ··	31.0
04	239 56.4	31.9
05	254 56.5	32.8
06	269 56.7	N 9 33.7
07	284 56.9	34.6
08	299 57.0	35.5
09	314 57.2 ··	36.4
10	329 57.3	37.3
11	344 57.5	38.2
12	359 57.6	N 9 39.1
13	14 57.8	40.0
14	29 57.9	40.9
15	44 58.1 ··	41.7
16	59 58.2	42.6
17	74 58.4	43.5
18	89 58.5	N 9 44.4
19	104 58.7	45.3
20	119 58.9	46.2
21	134 59.0 ··	47.1
22	149 59.2	48.0
23	164 59.3	48.9

d h	G.H.A.	Dec.
16 00	179 59.5	N 9 49.8
01	194 59.6	50.7
02	209 59.8	51.6
03	224 59.9 ··	52.5
04	240 00.1	53.4
05	255 00.2	54.2
06	270 00.4	N 9 55.1
07	285 00.5	56.0
08	300 00.7	56.9
09	315 00.8 ··	57.8
10	330 01.0	58.7
11	345 01.1	9 59.6
12	0 01.3	N10 00.5
13	15 01.4	01.4
14	30 01.6	02.3
15	45 01.7 ··	03.1
16	60 01.9	04.0
17	75 02.0	04.9
18	90 02.2	N10 05.8
19	105 02.3	06.7
20	120 02.5	07.6
21	135 02.6 ··	08.5
22	150 02.8	09.3
23	165 02.9	10.2
17 00	180 03.1	N10 11.1
01	195 03.2	12.0
02	210 03.4	12.9
03	225 03.5 ··	13.8
04	240 03.7	14.7
05	255 03.8	15.5
06	270 04.0	N10 16.4
07	285 04.1	17.3
08	300 04.4	18.2
09	315 04.4 ··	19.1
10	330 04.5	20.0
11	345 04.7	20.8
12	0 04.8	N10 21.7
13	15 05.0	22.6
14	30 05.1	23.5
15	45 05.3 ··	24.4
16	60 05.4	25.2
17	75 05.6	26.1
18	90 05.7	N10 27.0
19	105 05.9	27.9
20	120 06.0	28.8
21	135 06.1 ··	29.6
22	150 06.3	30.5
23	165 06.4	31.4
18 00	180 06.6	N10 32.3
01	195 06.7	33.2
02	210 06.9	34.0
03	225 07.0 ··	34.9
04	240 07.2	35.8
05	255 07.3	36.7
06	270 07.4	N10 37.5
07	285 07.6	38.4
08	300 07.7	39.3
09	315 07.9 ··	40.2
10	330 08.0	41.0
11	345 08.2	41.9
12	0 08.3	N10 42.8
13	15 08.4	43.7
14	30 08.6	44.5
15	45 08.7 ··	45.4
16	60 08.9	46.3
17	75 09.0	47.2
18	90 09.1	N10 48.0
19	105 09.3	48.9
20	120 09.4	49.8
21	135 09.6 ··	50.7
22	150 09.7	51.5
23	165 09.9	52.4

d	h	G.H.A. °	G.H.A. ′	Dec. °	Dec. ′
19	00	180	10.0	N10	53.3
	01	195	10.1		54.1
	02	210	10.3		55.0
	03	225	10.4	· ·	55.9
	04	240	10.5		56.7
	05	255	10.7		57.6
	06	270	10.8	N10	58.5
	07	285	11.0	10	59.4
	08	300	11.1	11	00.2
	09	315	11.2	· ·	01.1
	10	330	11.4		02.0
	11	345	11.5		02.8
	12	0	11.7	N11	03.7
	13	15	11.8		04.6
	14	30	11.9		05.4
	15	45	12.1	· ·	06.3
	16	60	12.2		07.2
	17	75	12.3		08.0
	18	90	12.5	N11	08.9
	19	105	12.6		09.8
	20	120	12.8		10.6
	21	135	12.9	· ·	11.5
	22	150	13.0		12.3
	23	165	13.2		13.2
20	00	180	13.3	N11	14.1
	01	195	13.4		14.9
	02	210	13.6		15.8
	03	225	13.7	· ·	16.7
	04	240	13.8		17.5
	05	255	14.0		18.4
	06	270	14.1	N11	19.3
	07	285	14.2		20.1
	08	300	14.4		21.0
	09	315	14.5	· ·	21.8
	10	330	14.6		22.7
	11	345	14.8		23.6
	12	0	14.9	N11	24.4
	13	15	15.0		25.3
	14	30	15.2		26.1
	15	45	15.3	· ·	27.0
	16	60	15.4		27.8
	17	75	15.6		28.7
	18	90	15.7	N11	29.6
	19	105	15.8		30.4
	20	120	16.0		31.3
	21	135	16.1	· ·	32.1
	22	150	16.2		33.0
	23	165	16.4		33.8
21	00	180	16.5	N11	34.7
	01	195	16.6		35.6
	02	210	16.8		36.4
	03	225	16.9	· ·	37.3
	04	240	17.0		38.1
	05	255	17.1		39.0
	06	270	17.3	N11	39.8
	07	285	17.4		40.7
	08	300	17.5		41.5
	09	315	17.7	· ·	42.4
	10	330	17.8		43.2
	11	345	17.9		44.1
	12	0	18.1	N11	44.9
	13	15	18.2		45.8
	14	30	18.3		46.6
	15	45	18.4	· ·	47.5
	16	60	18.6		48.3
	17	75	18.7		49.2
	18	90	18.8	N11	50.0
	19	105	18.9		50.9
	20	120	19.1		51.7
	21	135	19.2	· ·	52.6
	22	150	19.3		53.4
	23	165	19.5		54.3
22	00	180	19.6	N11	55.1
	01	195	19.7		56.0
	02	210	19.8		56.8
	03	225	20.0	· ·	57.7
	04	240	20.1		58.5
	05	255	20.2	11	59.4
	06	270	20.3	N12	00.2
	07	285	20.5		01.1
	08	300	20.6		01.9
	09	315	20.7	· ·	02.7
	10	330	20.8		03.6
	11	345	21.0		04.4
	12	0	21.1	N12	05.3
	13	15	21.2		06.1
	14	30	21.3		07.0
	15	45	21.5	· ·	07.8
	16	60	21.6		08.6
	17	75	21.7		09.5
	18	90	21.8	N12	10.3
	19	105	21.9		11.2
	20	120	22.1		12.0
	21	135	22.2	· ·	12.9
	22	150	22.3		13.7
	23	165	22.4		14.5
23	00	180	22.6	N12	15.4
	01	195	22.7		16.2
	02	210	22.8		17.1
	03	225	22.9	· ·	17.9
	04	240	23.0		18.7
	05	255	23.2		19.6
	06	270	23.3	N12	20.4
	07	285	23.4		21.2
	08	300	23.5		22.1
	09	315	23.6	· ·	22.9
	10	330	23.8		23.7
	11	345	23.9		24.6
	12	0	24.0	N12	25.4
	13	15	24.1		26.3
	14	30	24.2		27.1
	15	45	24.4	· ·	27.9
	16	60	24.5		28.8
	17	75	24.6		29.6
	18	90	24.7	N12	30.4
	19	105	24.8		31.3
	20	120	24.9		32.1
	21	135	25.1	· ·	32.9
	22	150	25.2		33.7
	23	165	25.3		34.6
24	00	180	25.4	N12	35.4
	01	195	25.5		36.2
	02	210	25.6		37.1
	03	225	25.8	· ·	37.9
	04	240	25.9		38.7
	05	255	26.0		39.6
	06	270	26.1	N12	40.4
	07	285	26.2		41.2
	08	300	26.3		42.0
	09	315	26.5	· ·	42.9
	10	330	26.6		43.7
	11	345	26.7		44.5
	12	0	26.8	N12	45.4
	13	15	26.9		46.2
	14	30	27.0		47.0
	15	45	27.1	· ·	47.8
	16	60	27.3		48.7
	17	75	27.4		49.5
	18	90	27.5	N12	50.3
	19	105	27.6		51.1
	20	120	27.7		52.0
	21	135	27.8	· ·	52.8
	22	150	27.9		53.6
	23	165	28.0		54.4
25	00	180	28.2	N12	55.2
	01	195	28.3		56.1
	02	210	28.4		56.9
	03	225	28.5	· ·	57.7
	04	240	28.6		58.5
	05	255	28.7	12	59.3
	06	270	28.8	N13	00.2
	07	285	28.9		01.0
	08	300	29.0		01.8
	09	315	29.1	· ·	02.6
	10	330	29.3		03.4
	11	345	29.4		04.3
	12	0	29.5	N13	05.1
	13	15	29.6		05.9
	14	30	29.7		06.7
	15	45	29.8	· ·	07.5
	16	60	29.9		08.3
	17	75	30.0		09.2
	18	90	30.1	N13	10.0
	19	105	30.2		10.8
	20	120	30.3		11.6
	21	135	30.5	· ·	12.4
	22	150	30.6		13.2
	23	165	30.7		14.0
26	00	180	30.8	N13	14.9
	01	195	30.9		15.7
	02	210	31.0		16.5
	03	225	31.1	· ·	17.3
	04	240	31.2		18.1
	05	255	31.3		18.9
	06	270	31.4	N13	19.7
	07	285	31.5		20.5
	08	300	31.6		21.4
	09	315	31.7	· ·	22.2
	10	330	31.8		23.0
	11	345	31.9		23.8
	12	0	32.0	N13	24.6
	13	15	32.1		25.4
	14	30	32.2		26.2
	15	45	32.3	· ·	27.0
	16	60	32.5		27.8
	17	75	32.6		28.6
	18	90	32.7	N13	29.4
	19	105	32.8		30.2
	20	120	32.9		31.0
	21	135	33.0	· ·	31.8
	22	150	33.1		32.7
	23	165	33.2		33.5
27	00	180	33.3	N13	34.3
	01	195	33.4		35.1
	02	210	33.5		35.9
	03	225	33.6	· ·	36.7
	04	240	33.7		37.5
	05	255	33.8		38.3
	06	270	33.9	N13	39.1
	07	285	34.0		39.9
	08	300	34.1		40.7
	09	315	34.2	· ·	41.5
	10	330	34.3		42.3
	11	345	34.4		43.1
	12	0	34.5	N13	43.9
	13	15	34.6		44.7
	14	30	34.7		45.5
	15	45	34.8	· ·	46.3
	16	60	34.9		47.1
	17	75	35.0		47.9
	18	90	35.1	N13	48.7
	19	105	35.2		49.5
	20	120	35.3		50.3
	21	135	35.4	· ·	51.1
	22	150	35.5		51.8
	23	165	35.6		52.6

d h	G.H.A.	Dec.
28 00	180 35.6	N13 53.4
01	195 35.7	54.2
02	210 35.8	55.0
03	225 35.9 ··	55.8
04	240 36.0	56.6
05	255 36.1	57.4
06	270 36.2	N13 58.2
07	285 36.3	59.0
08	300 36.4	13 59.8
09	315 36.5	14 00.6
10	330 36.6	01.4
11	345 36.7	02.2
12	0 36.8	N14 02.9
13	15 36.9	03.7
14	30 37.0	04.5
15	45 37.1 ··	05.3
16	60 37.2	06.1
17	75 37.3	06.9
18	90 37.3	N14 07.7
19	105 37.4	08.5
20	120 37.5	09.2
21	135 37.6 ··	10.0
22	150 37.7	10.8
23	165 37.8	11.6
29 00	180 37.9	N14 12.4
01	195 38.0	13.2
02	210 38.1	14.0
03	225 38.2 ··	14.7
04	240 38.3	15.5
05	255 38.4	16.3
06	270 38.4	N14 17.1
07	285 38.5	17.9
08	300 38.6	18.7
09	315 38.7 ··	19.4
10	330 38.8	20.2
11	345 38.9	21.0
12	0 39.0	N14 21.8
13	15 39.1	22.6
14	30 39.2	23.3
15	45 39.2 ··	24.1
16	60 39.3	24.9
17	75 39.4	25.7
18	90 39.5	N14 26.4
19	105 39.6	27.2
20	120 39.7	28.0
21	135 39.8 ··	28.8
22	150 39.9	29.5
23	165 39.9	30.3
30 00	180 40.0	N14 31.1
01	195 40.1	31.9
02	210 40.2	32.6
03	225 40.3 ··	33.4
04	240 40.4	34.2
05	255 40.5	35.0
06	270 40.6	N14 35.7
07	285 40.6	36.5
08	300 40.7	37.3
09	315 40.8 ··	38.1
10	330 40.9	38.8
11	345 41.0	39.6
12	0 41.0	N14 40.4
13	15 41.1	41.1
14	30 41.2	41.9
15	45 41.3 ··	42.7
16	60 41.4	43.4
17	75 41.5	44.2
18	90 41.5	N14 45.0
19	105 41.6	45.7
20	120 41.7	46.5
21	135 41.8 ··	47.3
22	150 41.9	48.0
23	165 41.9	48.8

MAY

d h	G.H.A.	Dec.
1 00	180 42.0	N14 49.6
01	195 42.1	50.3
02	210 42.2	51.1
03	225 42.3 ··	51.9
04	240 42.3	52.6
05	255 42.4	53.4
06	270 42.5	N14 54.2
07	285 42.6	54.9
08	300 42.7	55.7
09	315 42.7 ··	56.4
10	330 42.8	57.2
11	345 42.9	58.0
12	0 43.0	N14 58.7
13	15 43.1	14 59.5
14	30 43.1	15 00.2
15	45 43.2 ··	01.0
16	60 43.3	01.8
17	75 43.4	02.5
18	90 43.4	N15 03.3
19	105 43.5	04.0
20	120 43.6	04.0
21	135 43.7 ··	05.5
22	150 43.7	06.3
23	165 43.8	07.1
2 00	180 43.9	N15 07.8
01	195 44.0	08.6
02	210 44.0	09.3
03	225 44.1 ··	10.1
04	240 44.2	10.8
05	255 44.3	11.6
06	270 44.3	N15 12.3
07	285 44.4	13.1
08	300 44.5	13.8
09	315 44.6 ··	14.6
10	330 44.6	15.3
11	345 44.7	16.1
12	0 44.8	N15 16.8
13	15 44.8	17.6
14	30 44.9	18.3
15	45 45.0 ··	19.1
16	60 45.1	19.8
17	75 45.1	20.6
18	90 45.2	N15 21.3
19	105 45.3	22.1
20	120 45.3	22.8
21	135 45.4 ··	23.6
22	150 45.5	24.3
23	165 45.6	25.0
3 00	180 45.6	N15 25.8
01	195 45.7	26.5
02	210 45.8	27.3
03	225 45.8 ··	28.0
04	240 45.9	28.8
05	255 46.0	29.5
06	270 46.0	N15 30.2
07	285 46.1	31.0
08	300 46.2	31.7
09	315 46.2 ··	32.5
10	330 46.3	33.2
11	345 46.4	33.9
12	0 46.4	N15 34.7
13	15 46.5	35.4
14	30 46.6	36.2
15	45 46.6 ··	36.9
16	60 46.7	37.6
17	75 46.8	38.4
18	90 46.8	N15 39.1
19	105 46.9	39.8
20	120 47.0	40.6
21	135 47.0 ··	41.3
22	150 47.1	42.0
23	165 47.2	42.8

d h	G.H.A.	Dec.
4 00	180 47.2	N15 43.5
01	195 47.3	44.2
02	210 47.4	45.0
03	225 47.4 ··	45.7
04	240 47.5	46.4
05	255 47.5	47.2
06	270 47.6	N15 47.9
07	285 47.7	48.6
08	300 47.7	49.4
09	315 47.8 ··	50.1
10	330 47.9	50.8
11	345 47.9	51.5
12	0 48.0	N15 52.3
13	15 48.0	53.0
14	30 48.1	53.7
15	45 48.2 ··	54.5
16	60 48.2	55.2
17	75 48.3	55.9
18	90 48.3	N15 56.6
19	105 48.4	57.4
20	120 48.5	58.1
21	135 48.5 ··	58.8
22	150 48.6	15 59.5
23	165 48.6	16 00.3
5 00	180 48.7	N16 01.0
01	195 48.8	01.7
02	210 48.8	02.4
03	225 48.9 ··	03.1
04	240 48.9	03.9
05	255 49.0	04.6
06	270 49.0	N16 05.3
07	285 49.1	06.0
08	300 49.2	06.7
09	315 49.2 ··	07.5
10	330 49.3	08.2
11	345 49.3	08.9
12	0 49.4	N16 09.6
13	15 49.4	10.3
14	30 49.5	11.0
15	45 49.6 ··	11.8
16	60 49.6	12.5
17	75 49.7	13.2
18	90 49.7	N16 13.9
19	105 49.8	14.6
20	120 49.8	15.3
21	135 49.9 ··	16.0
22	150 49.9	16.8
23	165 50.0	17.5
6 00	180 50.0	N16 18.2
01	195 50.1	18.9
02	210 50.1	19.6
03	225 50.2 ··	20.3
04	240 50.2	21.0
05	255 50.3	21.7
06	270 50.4	N16 22.4
07	285 50.4	23.1
08	300 50.5	23.8
09	315 50.5 ··	24.6
10	330 50.6	25.3
11	345 50.6	26.0
12	0 50.7	N16 26.7
13	15 50.7	27.4
14	30 50.8	28.1
15	45 50.8 ··	28.8
16	60 50.9	29.5
17	75 50.9	30.2
18	90 51.0	N16 30.9
19	105 51.0	31.6
20	120 51.0	32.3
21	135 51.1 ··	33.0
22	150 51.1	33.7
23	165 51.2	34.4

	G.H.A.	Dec.
d h	° ′	° ′
7 00	180 51.2	N16 35.1
01	195 51.3	35.8
02	210 51.3	36.5
03	225 51.4 ··	37.2
04	240 51.4	37.9
05	255 51.5	38.6
06	270 51.5	N16 39.3
07	285 51.6	40.0
08	300 51.6	40.7
09	315 51.7 ··	41.4
10	330 51.7	42.1
11	345 51.7	42.8
12	0 51.8	N16 43.5
13	15 51.8	44.2
14	30 51.9	44.8
15	45 51.9 ··	45.5
16	60 52.0	46.2
17	75 52.0	46.9
18	90 52.0	N16 47.6
19	105 52.1	48.3
20	120 52.1	49.0
21	135 52.2 ··	49.7
22	150 52.2	50.4
23	165 52.3	51.1
8 00	180 52.3	N16 51.8
01	195 52.3	52.4
02	210 52.4	53.1
03	225 52.4 ··	53.8
04	240 52.5	54.5
05	255 52.5	55.2
06	270 52.5	N16 55.9
07	285 52.6	56.6
08	300 52.6	57.2
09	315 52.7 ··	57.9
10	330 52.7	58.6
11	345 52.7	16 59.3
12	0 52.8	N17 00.0
13	15 52.8	00.7
14	30 52.9	01.3
15	45 52.9 ··	02.0
16	60 52.9	02.7
17	75 53.0	03.4
18	90 53.0	N17 04.1
19	105 53.0	04.7
20	120 53.1	05.4
21	135 53.1 ··	06.1
22	150 53.1	06.8
23	165 53.2	07.4
9 00	180 53.2	N17 08.1
01	195 53.3	08.8
02	210 53.3	09.5
03	225 53.3 ··	10.1
04	240 53.4	10.8
05	255 53.4	11.5
06	270 53.4	N17 12.2
07	285 53.5	12.8
08	300 53.5	13.5
09	315 53.5 ··	14.2
10	330 53.6	14.9
11	345 53.6	15.5
12	0 53.6	N17 16.2
13	15 53.7	16.9
14	30 53.7	17.5
15	45 53.7 ··	18.2
16	60 53.8	18.9
17	75 53.8	19.5
18	90 53.8	N17 20.2
19	105 53.8	20.9
20	120 53.9	21.5
21	135 53.9 ··	22.2
22	150 53.9	22.9
23	165 54.0	23.5

	G.H.A.	Dec.
d h	° ′	° ′
10 00	180 54.0	N17 24.2
01	195 54.0	24.9
02	210 54.1	25.5
03	225 54.1 ··	26.2
04	240 54.1	26.9
05	255 54.1	27.5
06	270 54.2	N17 28.2
07	285 54.2	28.8
08	300 54.2	29.5
09	315 54.3 ··	30.2
10	330 54.3	30.8
11	345 54.3	31.5
12	0 54.3	N17 32.1
13	15 54.4	32.8
14	30 54.4	33.5
15	45 54.4 ··	34.1
16	60 54.4	34.8
17	75 54.5	35.4
18	90 54.5	N17 36.1
19	105 54.5	36.7
20	120 54.5	37.4
21	135 54.6 ··	38.0
22	150 54.6	38.7
23	165 54.6	39.3
11 00	180 54.6	N17 40.0
01	195 54.7	40.7
02	210 54.7	41.3
03	225 54.7 ··	42.0
04	240 54.7	42.6
05	255 54.8	43.3
06	270 54.8	N17 43.9
07	285 54.8	44.6
08	300 54.8	45.2
09	315 54.8 ··	45.8
10	330 54.9	46.5
11	345 54.9	47.1
12	0 54.9	N17 47.8
13	15 54.9	48.4
14	30 54.9	49.1
15	45 55.0 ··	49.7
16	60 55.0	50.4
17	75 55.0	51.0
18	90 55.0	N17 51.7
19	105 55.0	52.3
20	120 55.1	52.9
21	135 55.1 ··	53.6
22	150 55.1	54.2
23	165 55.1	54.9
12 00	180 55.1	N17 55.5
01	195 55.2	56.1
02	210 55.2	56.8
03	225 55.2 ··	57.4
04	240 55.2	58.1
05	255 55.2	58.7
06	270 55.2	N17 59.3
07	285 55.3	18 00.0
08	300 55.3	00.6
09	315 55.3 ··	01.2
10	330 55.3	01.9
11	345 55.3	02.5
12	0 55.3	N18 03.1
13	15 55.3	03.8
14	30 55.4	04.4
15	45 55.4 ··	05.0
16	60 55.4	05.7
17	75 55.4	06.3
18	90 55.4	N18 06.9
19	105 55.4	07.6
20	120 55.4	08.2
21	135 55.5 ··	08.8
22	150 55.5	09.4
23	165 55.5	10.1

	G.H.A.	Dec.
d h	° ′	° ′
13 00	180 55.5	N18 10.7
01	195 55.5	11.3
02	210 55.5	11.9
03	225 55.5 ··	12.6
04	240 55.5	13.2
05	255 55.5	13.8
06	270 55.6	N18 14.4
07	285 55.6	15.1
08	300 55.6	15.7
09	315 55.6 ··	16.3
10	330 55.6	16.9
11	345 55.6	17.6
12	0 55.6	N18 18.2
13	15 55.6	18.8
14	30 55.6	19.4
15	45 55.6 ··	20.0
16	60 55.6	20.7
17	75 55.7	21.3
18	90 55.7	N18 21.9
19	105 55.7	22.5
20	120 55.7	23.1
21	135 55.7 ··	23.7
22	150 55.7	24.4
23	165 55.7	25.0
14 00	180 55.7	N18 25.6
01	195 55.7	26.2
02	210 55.7	26.8
03	225 55.7 ··	27.4
04	240 55.7	28.0
05	255 55.7	28.6
06	270 55.7	N18 29.3
07	285 55.7	29.9
08	300 55.7	30.5
09	315 55.7 ··	31.1
10	330 55.7	31.7
11	345 55.7	32.3
12	0 55.7	N18 32.9
13	15 55.7	33.5
14	30 55.8	34.1
15	45 55.8 ··	34.7
16	60 55.8	35.3
17	75 55.8	35.9
18	90 55.8	N18 36.5
19	105 55.8	37.2
20	120 55.8	37.8
21	135 55.8 ··	38.4
22	150 55.8	39.0
23	165 55.8	39.6
15 00	180 55.8	N18 40.2
01	195 55.8	40.8
02	210 55.8	41.4
03	225 55.8 ··	42.0
04	240 55.8	42.6
05	255 55.8	43.2
06	270 55.8	N18 43.8
07	285 55.8	44.4
08	300 55.7	45.0
09	315 55.7 ··	45.6
10	330 55.7	46.1
11	345 55.7	46.7
12	0 55.7	N18 47.3
13	15 55.7	47.9
14	30 55.7	48.5
15	45 55.7 ··	49.1
16	60 55.7	49.7
17	75 55.7	50.3
18	90 55.7	N18 50.9
19	105 55.7	51.5
20	120 55.7	52.1
21	135 55.7 ··	52.7
22	150 55.7	53.3
23	165 55.7	53.8

d	h	G.H.A. °	′	Dec. °	′
16	00	180	55.7	N18	54.4
	01	195	55.7		55.0
	02	210	55.7		55.6
	03	225	55.7	··	56.2
	04	240	55.7		56.8
	05	255	55.6		57.4
	06	270	55.6	N18	57.9
	07	285	55.6		58.5
	08	300	55.6		59.1
	09	315	55.6	18	59.7
	10	330	55.6	19	00.3
	11	345	55.6		00.9
	12	0	55.6	N19	01.4
	13	15	55.6		02.0
	14	30	55.6		02.6
	15	45	55.6	··	03.2
	16	60	55.5		03.8
	17	75	55.5		04.3
	18	90	55.5	N19	04.9
	19	105	55.5		05.5
	20	120	55.5		06.1
	21	135	55.5	··	06.7
	22	150	55.5		07.2
	23	165	55.5		07.8
17	00	180	55.5	N19	08.4
	01	195	55.4		09.0
	02	210	55.4		09.5
	03	225	55.4	··	10.1
	04	240	55.4		10.7
	05	255	55.4		11.2
	06	270	55.4	N19	11.8
	07	285	55.4		12.4
	08	300	55.3		13.0
	09	315	55.3	··	13.5
	10	330	55.3		14.1
	11	345	55.3		14.7
	12	0	55.3	N19	15.2
	13	15	55.3		15.8
	14	30	55.3		16.4
	15	45	55.2	··	16.9
	16	60	55.2		17.5
	17	75	55.2		18.1
	18	90	55.2	N19	18.6
	19	105	55.2		19.2
	20	120	55.2		19.8
	21	135	55.1	··	20.3
	22	150	55.1		20.9
	23	165	55.1		21.4
18	00	180	55.1	N19	22.0
	01	195	55.1		22.6
	02	210	55.0		23.1
	03	225	55.0	··	23.7
	04	240	55.0		24.2
	05	255	55.0		24.8
	06	270	55.0	N19	25.4
	07	285	55.0		25.9
	08	300	54.9		26.5
	09	315	54.9	··	27.0
	10	330	54.9		27.6
	11	345	54.9		28.1
	12	0	54.8	N19	28.7
	13	15	54.8		29.2
	14	30	54.8		29.8
	15	45	54.8	··	30.3
	16	60	54.8		30.9
	17	75	54.7		31.4
	18	90	54.7	N19	32.0
	19	105	54.7		32.5
	20	120	54.7		33.1
	21	135	54.6	··	33.6
	22	150	54.6		34.2
	23	165	54.6		34.7

d	h	G.H.A. °	′	Dec. °	′
19	00	180	54.6	N19	35.3
	01	195	54.5		35.8
	02	210	54.5		36.4
	03	225	54.5	··	36.9
	04	240	54.5		37.5
	05	255	54.4		38.0
	06	270	54.4	N19	38.6
	07	285	54.4		39.1
	08	300	54.4		39.6
	09	315	54.3	··	40.2
	10	330	54.3		40.7
	11	345	54.3		41.3
	12	0	54.3	N19	41.8
	13	15	54.2		42.4
	14	30	54.2		42.9
	15	45	54.2	··	43.4
	16	60	54.2		44.0
	17	75	54.1		44.5
	18	90	54.1	N19	45.0
	19	105	54.1		45.6
	20	120	54.0		46.1
	21	135	54.0		46.6
	22	150	54.0		47.2
	23	165	53.9		47.7
20	00	180	53.9	N19	48.3
	01	195	53.9		48.8
	02	210	53.9		49.3
	03	225	53.8	··	49.8
	04	240	53.8		50.4
	05	255	53.8		50.9
	06	270	53.7	N19	51.4
	07	285	53.7		52.0
	08	300	53.7		52.5
	09	315	53.6	··	53.0
	10	330	53.6		53.6
	11	345	53.6		54.1
	12	0	53.5	N19	54.6
	13	15	53.5		55.1
	14	30	53.5		55.7
	15	45	53.4	··	56.2
	16	60	53.4		56.7
	17	75	53.4		57.2
	18	90	53.3	N19	57.7
	19	105	53.3		58.3
	20	120	53.3		58.8
	21	135	53.2	··	59.3
	22	150	53.2	19	59.8
	23	165	53.2	20	00.4
21	00	180	53.1	N20	00.9
	01	195	53.1		01.4
	02	210	53.1		01.9
	03	225	53.0	··	02.4
	04	240	53.0		02.9
	05	255	52.9		03.5
	06	270	52.9	N20	04.0
	07	285	52.9		04.5
	08	300	52.8		05.0
	09	315	52.8	··	05.5
	10	330	52.8		06.0
	11	345	52.7		06.5
	12	0	52.7	N20	07.1
	13	15	52.6		07.6
	14	30	52.6		08.1
	15	45	52.6	··	08.6
	16	60	52.5		09.1
	17	75	52.5		09.6
	18	90	52.4	N20	10.1
	19	105	52.4		10.6
	20	120	52.4		11.1
	21	135	52.3	··	11.6
	22	150	52.3		12.1
	23	165	52.2		12.7

d	h	G.H.A. °	′	Dec. °	′
22	00	180	52.2	N20	13.2
	01	195	52.1		13.7
	02	210	52.1		14.2
	03	225	52.1	··	14.7
	04	240	52.0		15.2
	05	255	52.0		15.7
	06	270	51.9	N20	16.2
	07	285	51.9		16.7
	08	300	51.8		17.2
	09	315	51.8	··	17.7
	10	330	51.8		18.2
	11	345	51.7		18.7
	12	0	51.7	N20	19.2
	13	15	51.6		19.7
	14	30	51.6		20.2
	15	45	51.5	··	20.7
	16	60	51.5		21.2
	17	75	51.4		21.6
	18	90	51.4	N20	22.1
	19	105	51.4		22.6
	20	120	51.3		23.1
	21	135	51.3	··	23.6
	22	150	51.2		24.1
	23	165	51.2		24.6
23	00	180	51.1	N20	25.1
	01	195	51.1		25.6
	02	210	51.0		26.1
	03	225	51.0	··	26.6
	04	240	50.9		27.0
	05	255	50.9		27.5
	06	270	50.8	N20	28.0
	07	285	50.8		28.5
	08	300	50.7		29.0
	09	315	50.7	··	29.5
	10	330	50.6		30.0
	11	345	50.6		30.4
	12	0	50.5	N20	30.9
	13	15	50.5		31.4
	14	30	50.4		31.9
	15	45	50.4	··	32.4
	16	60	50.3		32.9
	17	75	50.3		33.3
	18	90	50.2	N20	33.8
	19	105	50.2		34.3
	20	120	50.1		34.8
	21	135	50.1	··	35.2
	22	150	50.0		35.7
	23	165	50.0		36.2
24	00	180	49.9	N20	36.7
	01	195	49.9		37.2
	02	210	49.8		37.6
	03	225	49.8	··	38.1
	04	240	49.7		38.6
	05	255	49.6		39.0
	06	270	49.6	N20	39.5
	07	285	49.5		40.0
	08	300	49.5		40.5
	09	315	49.4	··	40.9
	10	330	49.4		41.4
	11	345	49.3		41.9
	12	0	49.3	N20	42.3
	13	15	49.2		42.8
	14	30	49.2		43.3
	15	45	49.1	··	43.7
	16	60	49.0		44.2
	17	75	49.0		44.7
	18	90	48.9	N20	45.1
	19	105	48.9		45.6
	20	120	48.8		46.1
	21	135	48.8	··	46.5
	22	150	48.7		47.0
	23	165	48.6		47.4

d h	G.H.A.	Dec.
25 00	180 48.6	N20 47.9
01	195 48.5	48.4
02	210 48.5	48.8
03	225 48.4	·· 49.3
04	240 48.3	49.7
05	255 48.3	50.2
06	270 48.2	N20 50.7
07	285 48.2	51.1
08	300 48.1	51.6
09	315 48.0	·· 52.0
10	330 48.0	52.5
11	345 47.9	52.9
12	0 47.9	N20 53.4
13	15 47.8	53.8
14	30 47.7	54.3
15	45 47.7	·· 54.7
16	60 47.6	55.2
17	75 47.6	55.6
18	90 47.5	N20 56.1
19	105 47.4	56.5
20	120 47.4	57.0
21	135 47.3	·· 57.4
22	150 47.2	57.9
23	165 47.2	58.3
26 00	180 47.1	N20 58.8
01	195 47.1	59.2
02	210 47.0	20 59.7
03	225 46.9	21 00.1
04	240 46.9	00.6
05	255 46.8	01.0
06	270 46.7	N21 01.4
07	285 46.7	01.9
08	300 46.6	02.3
09	315 46.5	·· 02.8
10	330 46.5	03.2
11	345 46.4	03.6
12	0 46.3	N21 04.1
13	15 46.3	04.5
14	30 46.2	05.0
15	45 46.1	·· 05.4
16	60 46.1	05.8
17	75 46.0	06.3
18	90 45.9	N21 06.7
19	105 45.9	07.1
20	120 45.8	07.6
21	135 45.7	·· 08.0
22	150 45.7	08.4
23	165 45.6	08.9
27 00	180 45.5	N21 09.3
01	195 45.5	09.7
02	210 45.4	10.2
03	225 45.3	·· 10.6
04	240 45.3	11.0
05	255 45.2	11.4
06	270 45.1	N21 11.9
07	285 45.0	12.3
08	300 45.0	12.7
09	315 44.9	·· 13.1
10	330 44.8	13.6
11	345 44.8	14.0
12	0 44.7	N21 14.4
13	15 44.6	14.8
14	30 44.5	15.3
15	45 44.5	·· 15.7
16	60 44.4	16.1
17	75 44.3	16.5
18	90 44.3	N21 16.9
19	105 44.2	17.4
20	120 44.1	17.8
21	135 44.0	·· 18.2
22	150 44.0	18.6
23	165 43.9	19.0

d h	G.H.A.	Dec.
28 00	180 43.8	N21 19.4
01	195 43.7	19.9
02	210 43.7	20.3
03	225 43.6	·· 20.7
04	240 43.5	21.1
05	255 43.4	21.5
06	270 43.4	N21 21.9
07	285 43.3	22.3
08	300 43.2	22.7
09	315 43.1	·· 23.2
10	330 43.1	23.6
11	345 43.0	24.0
12	0 42.9	N21 24.4
13	15 42.8	24.8
14	30 42.8	25.2
15	45 42.7	·· 25.6
16	60 42.6	26.0
17	75 42.5	26.4
18	90 42.5	N21 26.8
19	105 42.4	27.2
20	120 42.3	27.6
21	135 42.2	·· 28.0
22	150 42.1	28.4
23	165 42.1	28.8
29 00	180 42.0	N21 29.2
01	195 41.9	29.6
02	210 41.8	30.0
03	225 41.8	·· 30.4
04	240 41.7	30.8
05	255 41.6	31.2
06	270 41.5	N21 31.6
07	285 41.4	32.0
08	300 41.4	32.4
09	315 41.3	·· 32.8
10	330 41.2	33.2
11	345 41.1	33.6
12	0 41.0	N21 34.0
13	15 41.0	34.4
14	30 40.9	34.8
15	45 40.8	·· 35.2
16	60 40.7	35.5
17	75 40.6	35.9
18	90 40.5	N21 36.3
19	105 40.5	36.7
20	120 40.4	37.1
21	135 40.3	·· 37.5
22	150 40.2	37.9
23	165 40.1	38.3
30 00	180 40.1	N21 38.6
01	195 40.0	39.0
02	210 39.9	39.4
03	225 39.8	·· 39.8
04	240 39.7	40.2
05	255 39.6	40.5
06	270 39.6	N21 40.9
07	285 39.5	41.3
08	300 39.4	41.7
09	315 39.3	·· 42.1
10	330 39.2	42.4
11	345 39.1	42.8
12	0 39.0	N21 43.2
13	15 39.0	43.6
14	30 38.9	44.0
15	45 38.8	·· 44.3
16	60 38.7	44.7
17	75 38.6	45.1
18	90 38.5	N21 45.4
19	105 38.4	45.8
20	120 38.4	46.2
21	135 38.3	·· 46.6
22	150 38.2	46.9
23	165 38.1	47.3

JUNE

d h	G.H.A.	Dec.
31 00	180 38.0	N21 47.7
01	195 37.9	48.0
02	210 37.8	48.4
03	225 37.7	·· 48.8
04	240 37.7	49.1
05	255 37.6	49.5
06	270 37.5	N21 49.9
07	285 37.4	50.2
08	300 37.3	50.6
09	315 37.2	·· 51.0
10	330 37.1	51.3
11	345 37.0	51.7
12	0 36.9	N21 52.0
13	15 36.9	52.4
14	30 36.8	52.8
15	45 36.7	·· 53.1
16	60 36.6	53.5
17	75 36.5	53.8
18	90 36.4	N21 54.2
19	105 36.3	54.6
20	120 36.2	54.9
21	135 36.1	·· 55.3
22	150 36.0	55.6
23	165 35.9	56.0
1 00	180 35.9	N21 56.3
01	195 35.8	56.7
02	210 35.7	57.0
03	225 35.6	·· 57.4
04	240 35.5	57.7
05	255 35.4	58.1
06	270 35.3	N21 58.4
07	285 35.2	58.8
08	300 35.1	59.1
09	315 35.0	·· 59.5
10	330 34.9	21 59.8
11	345 34.8	22 00.2
12	0 34.7	N22 00.5
13	15 34.7	00.9
14	30 34.6	01.2
15	45 34.5	· 01.5
16	60 34.4	01.9
17	75 34.3	02.2
18	90 34.2	N22 02.6
19	105 34.1	02.9
20	120 34.0	03.2
21	135 33.9	·· 03.6
22	150 33.8	03.9
23	165 33.7	04.3
2 00	180 33.6	N22 04.6
01	195 33.5	04.9
02	210 33.4	05.3
03	225 33.3	·· 05.6
04	240 33.2	05.9
05	255 33.1	06.3
06	270 33.0	N22 06.6
07	285 32.9	06.9
08	300 32.8	07.3
09	315 32.7	·· 07.6
10	330 32.6	07.9
11	345 32.5	08.3
12	0 32.4	N22 08.6
13	15 32.4	08.9
14	30 32.3	09.2
15	45 32.2	·· 09.6
16	60 32.1	09.9
17	75 32.0	10.2
18	90 31.9	N22 10.6
19	105 31.8	10.9
20	120 31.7	11.2
21	135 31.6	·· 11.5
22	150 31.5	11.8
23	165 31.4	12.2

d h	G.H.A. ° '	Dec. ° '
3 00	180 31.3	N22 12.5
01	195 31.2	12.8
02	210 31.1	13.1
03	225 31.0 ··	13.4
04	240 30.9	13.8
05	255 30.8	14.1
06	270 30.7	N22 14.4
07	285 30.6	14.7
08	300 30.5	15.0
09	315 30.4 ··	15.3
10	330 30.3	15.7
11	345 30.2	16.0
12	0 30.1	N22 16.3
13	15 30.0	16.6
14	30 29.9	16.9
15	45 29.8 ··	17.2
16	60 29.7	17.5
17	75 29.6	17.8
18	90 29.4	N22 18.2
19	105 29.3	18.5
20	120 29.2	18.8
21	135 29.1 ··	19.1
22	150 29.0	19.4
23	165 28.9	19.7
4 00	180 28.8	N22 20.0
01	195 28.7	20.3
02	210 28.6	20.6
03	225 28.5 ··	20.9
04	240 28.4	21.2
05	255 28.3	21.5
06	270 28.2	N22 21.8
07	285 28.1	22.1
08	300 28.0	22.4
09	315 27.9 ··	22.7
10	330 27.8	23.0
11	345 27.7	23.3
12	0 27.6	N22 23.6
13	15 27.5	23.9
14	30 27.4	24.2
15	45 27.3 ··	24.5
16	60 27.2	24.8
17	75 27.1	25.1
18	90 27.0	N22 25.4
19	105 26.8	25.7
20	120 26.7	25.9
21	135 26.6 ··	26.2
22	150 26.5	26.5
23	165 26.4	26.8
5 00	180 26.3	N22 27.1
01	195 26.2	27.4
02	210 26.1	27.7
03	225 26.0 ··	28.0
04	240 25.9	28.2
05	255 25.8	28.5
06	270 25.7	N22 28.8
07	285 25.6	29.1
08	300 25.5	29.4
09	315 25.3 ··	29.7
10	330 25.2	29.9
11	345 25.1	30.2
12	0 25.0	N22 30.5
13	15 24.9	30.8
14	30 24.8	31.1
15	45 24.7 ··	31.3
16	60 24.6	31.6
17	75 24.5	31.9
18	90 24.4	N22 32.2
19	105 24.3	32.5
20	120 24.2	32.7
21	135 24.0 ··	33.0
22	150 23.9	33.3
23	165 23.8	33.5

d h	G.H.A. ° '	Dec. ° '
6 00	180 23.7	N22 33.8
01	195 23.6	34.1
02	210 23.5	34.4
03	225 23.4 ··	34.6
04	240 23.3	34.9
05	255 23.2	35.2
06	270 23.1	N22 35.4
07	285 22.9	35.7
08	300 22.8	36.0
09	315 22.7 ··	36.2
10	330 22.6	36.5
11	345 22.5	36.8
12	0 22.4	N22 37.0
13	15 22.3	37.3
14	30 22.2	37.6
15	45 22.1 ··	37.8
16	60 21.9	38.1
17	75 21.8	38.3
18	90 21.7	N22 38.6
19	105 21.6	38.9
20	120 21.5	39.1
21	135 21.4 ··	39.4
22	150 21.3	39.6
23	165 21.2	39.9
7 00	180 21.0	N22 40.1
01	195 20.9	40.4
02	210 20.8	40.7
03	225 20.7 ··	40.9
04	240 20.6	41.2
05	255 20.5	41.4
06	270 20.4	N22 41.7
07	285 20.2	41.9
08	300 20.1	42.2
09	315 20.0 ··	42.4
10	330 19.9	42.7
11	345 19.8	42.9
12	0 19.7	N22 43.2
13	15 19.6	43.4
14	30 19.4	43.7
15	45 19.3 ··	43.9
16	60 19.2	44.1
17	75 19.1	44.4
18	90 19.0	N22 44.6
19	105 18.9	44.9
20	120 18.8	45.1
21	135 18.6 ··	45.4
22	150 18.5	45.6
23	165 18.4	45.8
8 00	180 18.3	N22 46.1
01	195 18.2	46.3
02	210 18.1	46.5
03	225 17.9 ··	46.8
04	240 17.8	47.0
05	255 17.7	47.3
06	270 17.6	N22 47.5
07	285 17.5	47.7
08	300 17.4	48.0
09	315 17.2 ··	48.2
10	330 17.1	48.4
11	345 17.0	48.7
12	0 16.9	N22 48.9
13	15 16.8	49.1
14	30 16.7	49.3
15	45 16.5 ··	49.6
16	60 16.4	49.8
17	75 16.3	50.0
18	90 16.2	N22 50.3
19	105 16.1	50.5
20	120 16.0	50.7
21	135 15.8 ··	50.9
22	150 15.7	51.2
23	165 15.6	51.4

d h	G.H.A. ° '	Dec. ° '
9 00	180 15.5	N22 51.6
01	195 15.4	51.8
02	210 15.2	52.0
03	225 15.1 ··	52.3
04	240 15.0	52.5
05	255 14.9	52.7
06	270 14.8	N22 52.9
07	285 14.6	53.1
08	300 14.5	53.4
09	315 14.4 ··	53.6
10	330 14.3	53.8
11	345 14.2	54.0
12	0 14.0	N22 54.2
13	15 13.9	54.4
14	30 13.8	54.6
15	45 13.7 ··	54.8
16	60 13.6	55.1
17	75 13.4	55.3
18	90 13.3	N22 55.5
19	105 13.2	55.7
20	120 13.1	55.9
21	135 13.0 ··	56.1
22	150 12.8	56.3
23	165 12.7	56.5
10 00	180 12.6	N22 56.7
01	195 12.5	56.9
02	210 12.4	57.1
03	225 12.2 ··	57.3
04	240 12.1	57.5
05	255 12.0	57.7
06	270 11.9	N22 57.9
07	285 11.8	58.1
08	300 11.6	58.3
09	315 11.5 ··	58.5
10	330 11.4	58.7
11	345 11.3	58.9
12	0 11.1	N22 59.1
13	15 11.0	59.3
14	30 10.9	59.5
15	45 10.8 ··	59.7
16	60 10.6	22 59.9
17	75 10.5	23 00.1
18	90 10.4	N23 00.3
19	105 10.3	00.5
20	120 10.2	00.7
21	135 10.0 ··	00.9
22	150 09.9	01.1
23	165 09.8	01.3
11 00	180 09.7	N23 01.4
01	195 09.5	01.6
02	210 09.4	01.8
03	225 09.3 ··	02.0
04	240 09.2	02.2
05	255 09.0	02.4
06	270 08.9	N23 02.6
07	285 08.8	02.7
08	300 08.7	02.9
09	315 08.5 ··	03.1
10	330 08.4	03.3
11	345 08.3	03.5
12	0 08.2	N23 03.7
13	15 08.1	03.8
14	30 07.9	04.0
15	45 07.8 ··	04.2
16	60 07.7	04.4
17	75 07.6	04.5
18	90 07.4	N23 04.7
19	105 07.3	04.9
20	120 07.2	05.1
21	135 07.1 ··	05.2
22	150 06.9	05.4
23	165 06.8	05.6

	G.H.A.	Dec.		G.H.A.	Dec.		G.H.A.	Dec.
12 00	180 06.7	N23 05.8	**15** 00	179 57.4	N23 16.3	**18** 00	179 47.8	N23 23.1
01	195 06.5	05.9	01	194 57.3	16.4	01	194 47.7	23.1
02	210 06.4	06.1	02	209 57.1	16.5	02	209 47.5	23.2
03	225 06.3 ··	06.3	03	224 57.0 ··	16.6	03	224 47.4 ··	23.3
04	240 06.2	06.4	04	239 56.9	16.7	04	239 47.3	23.3
05	255 06.0	06.6	05	254 56.8	16.9	05	254 47.1	23.4
06	270 05.9	N23 06.8	06	269 56.6	N23 17.0	06	269 47.0	N23 23.5
07	285 05.8	06.9	07	284 56.5	17.1	07	284 46.9	23.5
08	300 05.7	07.1	08	299 56.4	17.2	08	299 46.7	23.6
09	315 05.5 ··	07.3	09	314 56.2 ··	17.3	09	314 46.6 ··	23.7
10	330 05.4	07.4	10	329 56.1	17.4	10	329 46.4	23.7
11	345 05.3	07.6	11	344 56.0	17.5	11	344 46.3	23.8
12	0 05.2	N23 07.8	12	359 55.8	N23 17.7	12	359 46.2	N23 23.8
13	15 05.0	07.9	13	14 55.7	17.8	13	14 46.0	23.9
14	30 04.9	08.1	14	29 55.6	17.9	14	29 45.9	24.0
15	45 04.8 ··	08.3	15	44 55.4 ··	18.0	15	14 45.8 ··	24.0
16	60 04.7	08.4	16	59 55.3	18.1	16	59 45.6	24.1
17	75 04.5	08.6	17	74 55.2	18.2	17	74 45.5	24.1
18	90 04.4	N23 08.7	18	89 55.0	N23 18.3	18	89 45.4	N23 24.2
19	105 04.3	08.9	19	104 54.9	18.4	19	104 45.2	24.2
20	120 04.1	09.0	20	119 54.8	18.5	20	119 45.1	24.3
21	135 04.0 ··	09.2	21	134 54.6 ··	18.6	21	134 45.0 ··	24.4
22	150 03.9	09.4	22	149 54.5	18.7	22	149 44.8	24.4
23	165 03.8	09.5	23	164 54.4	18.8	23	164 44.7	24.5
13 00	180 03.6	N23 09.7	**16** 00	179 54.2	N23 18.9	**19** 00	179 44.5	N23 24.5
01	195 03.5	09.8	01	194 54.1	19.0	01	194 44.4	24.6
02	210 03.4	10.0	02	209 54.0	19.1	02	209 44.3	24.6
03	225 03.2 ··	10.1	03	224 53.8 ··	19.2	03	224 44.1 ··	24.7
04	240 03.1	10.3	04	239 53.7	19.3	04	239 44.0	24.7
05	255 03.0	10.4	05	254 53.6	19.4	05	254 43.9	24.8
06	270 02.9	N23 10.6	06	269 53.4	N23 19.5	06	269 43.7	N23 24.8
07	285 02.7	10.7	07	284 53.3	19.6	07	284 43.6	24.9
08	300 02.6	10.9	08	299 53.2	19.7	08	299 43.5	24.9
09	315 02.5 ··	11.0	09	314 53.0 ··	19.8	09	314 43.3 ··	24.9
10	330 02.3	11.2	10	329 52.9	19.9	10	329 43.2	25.0
11	345 02.2	11.3	11	344 52.8	20.0	11	344 43.1	25.0
12	0 02.1	N23 11.5	12	359 52.6	N23 20.1	12	359 42.9	N23 25.1
13	15 02.0	11.6	13	14 52.5	20.2	13	14 42.8	25.1
14	30 01.8	11.8	14	29 52.4	20.3	14	29 42.6	25.2
15	45 01.7 ··	11.9	15	44 52.2 ··	20.4	15	44 42.5 ··	25.2
16	60 01.6	12.0	16	59 52.1	20.5	16	59 42.4	25.2
17	75 01.4	12.2	17	74 52.0	20.6	17	74 42.2	25.3
18	90 01.3	N23 12.3	18	89 51.8	N23 20.7	18	89 42.1	N23 25.3
19	105 01.2	12.5	19	104 51.7	20.8	19	104 42.0	25.4
20	120 01.1	12.6	20	119 51.6	20.9	20	119 41.8	25.4
21	135 00.9 ··	12.8	21	134 51.4 ··	20.9	21	134 41.7 ··	25.5
22	150 00.8	12.9	22	149 51.3	21.0	22	149 41.5	25.5
23	165 00.7	13.0	23	164 51.2	21.1	23	164 41.4	25.5
14 00	180 00.5	N23 13.2	**17** 00	179 51.0	N23 21.2	**20** 00	179 41.3	N23 25.5
01	195 00.4	13.3	01	194 50.9	21.3	01	194 41.1	25.6
02	210 00.3	13.4	02	209 50.8	21.4	02	209 41.0	25.6
03	225 00.2 ··	13.6	03	224 50.6 ··	21.5	03	224 40.9 ··	25.6
04	240 00.0	13.7	04	239 50.5	21.5	04	239 40.7	25.7
05	254 59.9	13.8	05	254 50.4	21.6	05	254 40.6	25.7
06	269 59.8	N23 14.0	06	269 50.2	N23 21.7	06	269 40.5	N23 25.7
07	284 59.6	14.1	07	284 50.1	21.8	07	284 40.3	25.8
08	299 59.5	14.2	08	299 50.0	21.9	08	299 40.2	25.8
09	314 59.4 ··	14.4	09	314 49.8 ··	22.0	09	314 40.0 ··	25.8
10	329 59.2	14.5	10	329 49.7	22.0	10	329 39.9	25.8
11	344 59.1	14.6	11	344 49.6	22.1	11	344 39.8	25.9
12	359 59.0	N23 14.8	12	359 49.4	N23 22.2	12	359 39.6	N23 25.9
13	14 58.8	14.9	13	14 49.3	22.3	13	14 39.5	25.9
14	29 58.7	15.0	14	29 49.2	22.3	14	29 39.4	25.9
15	44 58.6 ··	15.2	15	44 49.0 ··	22.4	15	44 39.2 ··	26.0
16	59 58.5	15.3	16	59 48.9	22.5	16	59 39.1	26.0
17	74 58.3	15.4	17	74 48.7	22.6	17	74 39.0	26.0
18	89 58.2	N23 15.5	18	89 48.6	N23 22.6	18	89 38.8	N23 26.0
19	104 58.1	15.7	19	104 48.5	22.7	19	104 38.7	26.1
20	119 57.9	15.8	20	119 48.3	22.8	20	119 38.5	26.1
21	134 57.8 ··	15.9	21	134 48.2 ··	22.9	21	134 38.4 ··	26.1
22	149 57.7	16.0	22	149 48.1	22.9	22	149 38.3	26.1
23	164 57.5	16.1	23	164 47.9	23.0	23	164 38.1	26.1

21

d h	G.H.A.	Dec.
00	179 38.0	N23 26.2
01	194 37.9	26.2
02	209 37.7	26.2
03	224 37.6 ··	26.2
04	239 37.4	26.2
05	254 37.3	26.2
06	269 37.2	N23 26.2
07	284 37.0	26.3
08	299 36.9	26.3
09	314 36.8 ··	26.3
10	329 36.6	26.3
11	344 36.5	26.3
12	359 36.3	N23 26.3
13	14 36.2	26.3
14	29 36.1	26.3
15	44 35.9 ··	26.3
16	59 35.8	26.3
17	74 35.7	26.3
18	89 35.5	N23 26.3
19	104 35.4	26.3
20	119 35.3	26.4
21	134 35.1 ··	26.4
22	149 35.0	26.4
23	164 34.8	26.4

22

d h	G.H.A.	Dec.
00	179 34.7	N23 26.4
01	194 34.6	26.4
02	209 34.4	26.4
03	224 34.3 ··	26.4
04	239 34.2	26.3
05	254 34.0	26.3
06	269 33.9	N23 26.3
07	284 33.7	26.3
08	299 33.6	26.3
09	314 33.5 ··	26.3
10	329 33.3	26.3
11	344 33.2	26.3
12	359 33.1	N23 26.3
13	14 32.9	26.3
14	29 32.8	26.3
15	44 32.7 ··	26.3
16	59 32.5	26.3
17	74 32.4	26.2
18	89 32.2	N23 26.2
19	104 32.1	26.2
20	119 32.0	26.2
21	134 31.8 ··	26.2
22	149 31.7	26.2
23	164 31.6	26.2

23

d h	G.H.A.	Dec.
00	179 31.4	N23 26.1
01	194 31.3	26.1
02	209 31.1	26.1
03	224 31.0 ··	26.1
04	239 30.9	26.1
05	254 30.7	26.0
06	269 30.6	N23 26.0
07	284 30.5	26.0
08	299 30.3	26.0
09	314 30.2 ··	26.0
10	329 30.1	25.9
11	344 29.9	25.9
12	359 29.8	N23 25.9
13	14 29.6	25.9
14	29 29.5	25.8
15	44 29.4 ··	25.8
16	59 29.2	25.8
17	74 29.1	25.7
18	89 29.0	N23 25.7
19	104 28.8	25.7
20	119 28.7	25.7
21	134 28.6 ··	25.6
22	149 28.4	25.6
23	164 28.3	25.6

24

d h	G.H.A.	Dec.
00	179 28.2	N23 25.5
01	194 28.0	25.5
02	209 27.9	25.4
03	224 27.7 ··	25.4
04	239 27.6	25.4
05	254 27.5	25.3
06	269 27.3	N23 25.3
07	284 27.2	25.3
08	299 27.1	25.2
09	314 26.9 ··	25.2
10	329 26.8	25.1
11	344 26.7	25.1
12	359 26.5	N23 25.1
13	14 26.4	25.0
14	29 26.3	25.0
15	44 26.1 ··	24.9
16	59 26.0	24.9
17	74 25.8	24.8
18	89 25.7	N23 24.8
19	104 25.6	24.7
20	119 25.4	24.7
21	134 25.3 ··	24.6
22	149 25.2	24.6
23	164 25.0	24.5

25

d h	G.H.A.	Dec.
00	179 24.9	N23 24.5
01	194 24.8	24.4
02	209 24.6	24.4
03	224 24.5 ··	24.3
04	239 24.4	24.3
05	254 24.2	24.2
06	269 24.1	N23 24.2
07	284 24.0	24.1
08	299 23.8	24.0
09	314 23.7 ··	24.0
10	329 23.6	23.9
11	344 23.4	23.9
12	359 23.3	N23 23.8
13	14 23.1	23.7
14	29 23.0	23.7
15	44 22.9 ··	23.6
16	59 22.7	23.6
17	74 22.6	23.5
18	89 22.5	N23 23.4
19	104 22.3	23.4
20	119 22.2	23.3
21	134 22.1 ··	23.2
22	149 21.9	23.2
23	164 21.8	23.1

26

d h	G.H.A.	Dec.
00	179 21.7	N23 23.0
01	194 21.5	23.0
02	209 21.4	22.9
03	224 21.3 ··	22.8
04	239 21.1	22.8
05	254 21.0	22.7
06	269 20.9	N23 22.6
07	284 20.7	22.5
08	299 20.6	22.5
09	314 20.5 ··	22.4
10	329 20.3	22.3
11	344 20.2	22.2
12	359 20.1	N23 22.2
13	14 19.9	22.1
14	29 19.8	22.0
15	44 19.7 ··	21.9
16	59 19.5	21.8
17	74 19.4	21.8
18	89 19.3	N23 21.7
19	104 19.2	21.6
20	119 19.0	21.5
21	134 18.9 ··	21.4
22	149 18.8	21.3
23	164 18.6	21.3

27

d h	G.H.A.	Dec.
00	179 18.5	N23 21.2
01	194 18.4	21.1
02	209 18.2	21.0
03	224 18.1 ··	20.9
04	239 18.0	20.8
05	254 17.8	20.7
06	269 17.7	N23 20.6
07	284 17.6	20.6
08	299 17.4	20.5
09	314 17.3 ··	20.4
10	329 17.2	20.3
11	344 17.0	20.2
12	359 16.9	N23 20.1
13	14 16.8	20.0
14	29 16.7	19.9
15	44 16.5 ··	19.8
16	59 16.4	19.7
17	74 16.3	19.6
18	89 16.1	N23 19.5
19	104 16.0	19.4
20	119 15.9	19.3
21	134 15.7 ··	19.2
22	149 15.6	19.1
23	164 15.5	19.0

28

d h	G.H.A.	Dec.
00	179 15.3	N23 18.9
01	194 15.2	18.8
02	209 15.1	18.7
03	224 15.0 ··	18.6
04	239 14.8	18.5
05	254 14.7	18.4
06	269 14.6	N23 18.3
07	284 14.4	18.2
08	299 14.3	18.1
09	314 14.2 ··	18.0
10	329 14.1	17.8
11	344 13.9	17.7
12	359 13.8	N23 17.6
13	14 13.7	17.5
14	29 13.5	17.4
15	44 13.4 ··	17.3
16	59 13.3	17.2
17	74 13.2	17.1
18	89 13.0	N23 16.9
19	104 12.9	16.8
20	119 12.8	16.7
21	134 12.6 ··	16.6
22	149 12.5	16.5
23	164 12.4	16.3

29

d h	G.H.A.	Dec.
00	179 12.3	N23 16.2
01	194 12.1	16.1
02	209 12.0	16.0
03	224 11.9 ··	15.9
04	239 11.7	15.7
05	254 11.6	15.6
06	269 11.5	N23 15.5
07	284 11.4	15.4
08	299 11.2	15.2
09	314 11.1 ··	15.1
10	329 11.0	15.0
11	344 10.9	14.9
12	359 10.7	N23 14.7
13	14 10.6	14.6
14	29 10.5	14.5
15	44 10.3 ··	14.3
16	59 10.2	14.2
17	74 10.1	14.1
18	89 10.0	N23 14.0
19	104 09.8	13.8
20	119 09.7	13.7
21	134 09.6 ··	13.6
22	149 09.5	13.4
23	164 09.3	13.3

JULY

SUN

d h	G.H.A.	Dec.
30 00	179 09.2	N23 13.1
01	194 09.1	13.0
02	209 09.0	12.9
03	224 08.8	·· 12.7
04	239 08.7	12.6
05	254 08.6	12.5
06	269 08.5	N23 12.3
07	284 08.3	12.2
08	299 08.2	12.0
09	314 08.1	·· 11.9
10	329 08.0	11.7
11	344 07.8	11.6
12	359 07.7	N23 11.5
13	14 07.6	11.3
14	29 07.5	11.2
15	44 07.3	·· 11.0
16	59 07.2	10.9
17	74 07.1	10.7
18	89 07.0	N23 10.6
19	104 06.9	10.4
20	119 06.7	10.3
21	134 06.6	·· 10.1
22	149 06.5	10.0
23	164 06.4	09.8
1 00	179 06.2	N23 09.7
01	194 06.1	09.5
02	209 06.0	09.4
03	224 05.9	·· 09.2
04	239 05.7	09.0
05	254 05.6	08.9
06	269 05.5	N23 08.7
07	284 05.4	08.6
08	299 05.3	08.4
09	314 05.1	·· 08.2
10	329 05.0	08.1
11	344 04.9	07.9
12	359 04.8	N23 07.8
13	14 04.6	07.6
14	29 04.5	07.4
15	44 04.4	·· 07.3
16	59 04.3	07.1
17	74 04.2	06.9
18	89 04.0	N23 06.8
19	104 03.9	06.6
20	119 03.8	06.4
21	134 03.7	·· 06.3
22	149 03.5	06.1
23	164 03.4	05.9
2 00	179 03.3	N23 05.8
01	194 03.2	05.6
02	209 03.1	05.4
03	224 02.9	·· 05.3
04	239 02.8	05.1
05	254 02.7	04.9
06	269 02.6	N23 04.7
07	284 02.5	04.6
08	299 02.3	04.4
09	314 02.2	·· 04.2
10	329 02.1	04.0
11	344 02.0	03.9
12	359 01.9	N23 03.7
13	14 01.7	03.5
14	29 01.6	03.3
15	44 01.5	·· 03.1
16	59 01.4	03.0
17	74 01.3	02.8
18	89 01.1	N23 02.6
19	104 01.0	02.4
20	119 00.9	02.2
21	134 00.8	·· 02.0
22	149 00.7	01.9
23	164 00.5	01.7

d h	G.H.A.	Dec.
3 00	179 00.4	N23 01.5
01	194 00.3	01.3
02	209 00.2	01.1
03	224 00.1	·· 00.9
04	239 00.0	00.7
05	253 59.9	00.5
06	268 59.7	N23 00.3
07	283 59.6	23 00.1
08	298 59.5	22 59.9
09	313 59.4	·· 59.7
10	328 59.3	59.6
11	343 59.2	59.4
12	358 59.0	N22 59.2
13	13 58.9	59.0
14	28 58.8	58.8
15	43 58.7	·· 58.6
16	58 58.6	58.4
17	73 58.5	58.2
18	88 58.4	N22 58.0
19	103 58.2	57.8
20	118 58.1	57.6
21	133 58.0	·· 57.4
22	148 57.9	57.2
23	163 57.8	57.0
4 00	178 57.7	N22 56.8
01	193 57.6	56.6
02	208 57.4	56.4
03	223 57.3	·· 56.1
04	238 57.2	55.9
05	253 57.1	55.7
06	268 57.0	N22 55.5
07	283 56.9	55.3
08	298 56.8	55.1
09	313 56.6	·· 54.9
10	328 56.5	54.7
11	343 56.4	54.5
12	358 56.3	N22 54.3
13	13 56.2	54.1
14	28 56.1	53.8
15	43 56.0	·· 53.6
16	58 55.9	53.4
17	73 55.7	53.2
18	88 55.6	N22 53.0
19	103 55.5	52.8
20	118 55.4	52.5
21	133 55.3	·· 52.3
22	148 55.2	52.1
23	163 55.1	51.9
5 00	178 55.0	N22 51.7
01	193 54.9	51.4
02	208 54.7	51.2
03	223 54.6	·· 51.0
04	238 54.5	50.8
05	253 54.4	50.6
06	268 54.3	N22 50.3
07	283 54.2	50.1
08	298 54.1	49.9
09	313 54.0	·· 49.7
10	328 53.9	49.4
11	343 53.8	49.2
12	358 53.6	N22 49.0
13	13 53.5	48.7
14	28 53.4	48.5
15	43 53.3	·· 48.3
16	58 53.2	48.0
17	73 53.1	47.8
18	88 53.0	N22 47.6
19	103 52.9	47.3
20	118 52.8	47.1
21	133 52.7	·· 46.9
22	148 52.6	46.6
23	163 52.5	46.4

d h	G.H.A.	Dec.
6 00	178 52.3	N22 46.2
01	193 52.2	45.9
02	208 52.1	45.7
03	223 52.0	·· 45.5
04	238 51.9	45.2
05	253 51.8	45.0
06	268 51.7	N22 44.7
07	283 51.6	44.5
08	298 51.5	44.2
09	313 51.4	·· 44.0
10	328 51.3	43.8
11	343 51.2	43.5
12	358 51.1	N22 43.3
13	13 51.0	43.0
14	28 50.9	42.8
15	43 50.8	·· 42.5
16	58 50.6	42.3
17	73 50.5	42.0
18	88 50.4	N22 41.8
19	103 50.3	41.5
20	118 50.2	41.3
21	133 50.1	·· 41.0
22	148 50.0	40.8
23	163 49.9	40.5
7 00	178 49.8	N22 40.3
01	193 49.7	40.0
02	208 49.6	39.8
03	223 49.5	·· 39.5
04	238 49.4	39.3
05	253 49.3	39.0
06	268 49.2	N22 38.7
07	283 49.1	38.5
08	298 49.0	38.2
09	313 48.9	·· 38.0
10	328 48.8	37.7
11	343 48.7	37.4
12	358 48.6	N22 37.2
13	13 48.5	36.9
14	28 48.4	36.7
15	43 48.3	·· 36.4
16	58 48.2	36.1
17	73 48.1	35.9
18	88 48.0	N22 35.6
19	103 47.9	35.3
20	118 47.8	35.1
21	133 47.7	·· 34.8
22	148 47.6	34.5
23	163 47.5	34.3
8 00	178 47.4	N22 34.0
01	193 47.3	33.7
02	208 47.2	33.4
03	223 47.1	·· 33.2
04	238 47.0	32.9
05	253 46.9	32.6
06	268 46.8	N22 32.4
07	283 46.7	32.1
08	298 46.6	31.8
09	313 46.5	·· 31.5
10	328 46.4	31.2
11	343 46.3	31.0
12	358 46.2	N22 30.7
13	13 46.1	30.4
14	28 46.0	30.1
15	43 45.9	·· 29.9
16	58 45.8	29.6
17	73 45.7	29.3
18	88 45.6	N22 29.0
19	103 45.5	28.7
20	118 45.4	28.4
21	133 45.3	·· 28.2
22	148 45.2	27.9
23	163 45.1	27.6

d h	G.H.A.	Dec.
9 00	178 45.0	N22 27.3
01	193 44.9	27.0
02	208 44.8	26.7
03	223 44.7 ··	26.4
04	238 44.7	26.2
05	253 44.6	25.9
06	268 44.5 N22	25.6
07	283 44.4	25.3
08	298 44.3	25.0
09	313 44.2 ··	24.7
10	328 44.1	24.4
11	343 44.0	24.1
12	358 43.9 N22	23.8
13	13 43.8	23.5
14	28 43.7	23.2
15	43 43.6 ··	22.9
16	58 43.5	22.6
17	73 43.4	22.3
18	88 43.3 N22	22.0
19	103 43.3	21.7
20	118 43.2	21.4
21	133 43.1 ··	21.1
22	148 43.0	20.8
23	163 42.9	20.5
10 00	178 42.8 N22	20.2
01	193 42.7	19.9
02	208 42.6	19.6
03	223 42.5 ··	19.3
04	238 42.4	19.0
05	253 42.3	18.7
06	268 42.3 N22	18.4
07	283 42.2	18.1
08	298 42.1	17.8
09	313 42.0 ··	17.5
10	328 41.9	17.2
11	343 41.8	16.9
12	358 41.7 N22	16.6
13	13 41.6	16.2
14	28 41.5	15.9
15	43 41.5 ··	15.6
16	58 41.4	15.3
17	73 41.3	15.0
18	88 41.2 N22	14.7
19	103 41.1	14.4
20	118 41.0	14.0
21	133 40.9	13.7
22	148 40.8	13.4
23	163 40.8	13.1
11 00	178 40.7 N22	12.8
01	193 40.6	12.5
02	208 40.5	12.1
03	223 40.4 ··	11.8
04	238 40.3	11.5
05	253 40.2	11.2
06	268 40.2 N22	10.9
07	283 40.1	10.5
08	298 40.0	10.2
09	313 39.9 ··	09.9
10	328 39.8	09.6
11	343 39.7	09.2
12	358 39.6 N22	08.9
13	13 39.6	08.6
14	28 39.5	08.3
15	43 39.4 ··	07.9
16	58 39.3	07.6
17	73 39.2	07.3
18	88 39.1 N22	06.9
19	103 39.1	06.6
20	118 39.0	06.3
21	133 38.9 ··	05.9
22	148 38.8	05.6
23	163 38.7	05.3

d h	G.H.A.	Dec.
12 00	178 38.6	N22 04.9
01	193 38.6	04.6
02	208 38.5	04.3
03	223 38.4 ··	03.9
04	238 38.3	03.6
05	253 38.2	03.3
06	268 38.2 N22	02.9
07	283 38.1	02.6
08	298 38.0	02.2
09	313 37.9 ··	01.9
10	328 37.8	01.6
11	343 37.8	01.2
12	358 37.7 N22	00.9
13	13 37.6	00.5
14	28 37.5	22 00.2
15	43 37.4	21 59.9
16	58 37.4	59.5
17	73 37.3	59.2
18	88 37.2 N21	58.8
19	103 37.1	58.5
20	118 37.0	58.1
21	133 37.0 ··	57.8
22	148 36.9	57.4
23	163 36.8	57.1
13 00	178 36.7 N21	56.7
01	193 36.7	56.4
02	208 36.6	56.0
03	223 36.5 ··	55.7
04	238 36.4	55.3
05	253 36.3	55.0
06	268 36.3 N21	54.6
07	283 36.2	54.3
08	298 36.1	53.9
09	313 36.0 ··	53.5
10	328 36.0	53.2
11	343 35.9	52.8
12	358 35.8 N21	52.5
13	13 35.7	52.1
14	28 35.7	51.8
15	43 35.6 ··	51.4
16	58 35.5	51.0
17	73 35.4	50.7
18	88 35.4 N21	50.3
19	103 35.3	50.0
20	118 35.2	49.6
21	133 35.2 ··	49.2
22	148 35.1	48.9
23	163 35.0	48.5
14 00	178 34.9 N21	48.1
01	193 34.9	47.8
02	208 34.8	47.4
03	223 34.7 ··	47.0
04	238 34.6	46.7
05	253 34.6	46.3
06	268 34.5 N21	45.9
07	283 34.4	45.6
08	298 34.4	45.2
09	313 34.3 ··	44.8
10	328 34.2	44.4
11	343 34.1	44.1
12	358 34.1 N21	43.7
13	13 34.0	43.3
14	28 33.9	42.9
15	43 33.9 ··	42.6
16	58 33.8	42.2
17	73 33.7	41.8
18	88 33.7 N21	41.4
19	103 33.6	41.1
20	118 33.5	40.7
21	133 33.5 ··	40.3
22	148 33.4	39.9
23	163 33.3	39.5

d h	G.H.A.	Dec.
15 00	178 33.2	N21 39.2
01	193 33.2	38.8
02	208 33.1	38.4
03	223 33.0 ··	38.0
04	238 33.0	37.6
05	253 32.9	37.3
06	268 32.8 N21	36.9
07	283 32.8	36.5
08	298 32.7	36.1
09	313 32.6 ··	35.7
10	328 32.6	35.3
11	343 32.5	34.9
12	358 32.4 N21	34.5
13	13 32.4	34.2
14	28 32.3	33.8
15	43 32.3 ··	33.4
16	58 32.2	33.0
17	73 32.1	32.6
18	88 32.1 N21	32.2
19	103 32.0	31.8
20	118 31.9	31.4
21	133 31.9 ··	31.0
22	148 31.8	30.6
23	163 31.7	30.2
16 00	178 31.7 N21	29.8
01	193 31.6	29.4
02	208 31.5	29.0
03	223 31.5 ··	28.6
04	238 31.4	28.2
05	253 31.4	27.8
06	268 31.3 N21	27.4
07	283 31.2	27.0
08	298 31.2	26.6
09	313 31.1 ··	26.2
10	328 31.1	25.8
11	343 31.0	25.4
12	358 30.9 N21	25.0
13	13 30.9	24.6
14	28 30.8	24.2
15	43 30.8 ··	23.8
16	58 30.7	23.4
17	73 30.6	23.0
18	88 30.6 N21	22.6
19	103 30.5	22.2
20	118 30.5	21.8
21	133 30.4 ··	21.4
22	148 30.3	21.0
23	163 30.3	20.5
17 00	178 30.2 N21	20.1
01	193 30.2	19.7
02	208 30.1	19.3
03	223 30.1 ··	18.9
04	238 30.0	18.5
05	253 29.9	18.1
06	268 29.9 N21	17.6
07	283 29.8	17.2
08	298 29.8	16.8
09	313 29.7 ··	16.4
10	328 29.7	16.0
11	343 29.6	15.6
12	358 29.5 N21	15.1
13	13 29.5	14.7
14	28 29.4	14.3
15	43 29.4 ··	13.9
16	58 29.3	13.5
17	73 29.3	13.0
18	88 29.2 N21	12.6
19	103 29.2	12.2
20	118 29.1	11.8
21	133 29.1 ··	11.3
22	148 29.0	10.9
23	163 29.0	10.5

	G.H.A.	Dec.
18 00	178 28.9	N21 10.1
01	193 28.8	09.6
02	208 28.8	09.2
03	223 28.7	·· 08.8
04	238 28.7	08.4
05	253 28.6	07.9
06	268 28.6	N21 07.5
07	283 28.5	07.1
08	298 28.5	06.6.
09	313 28.4	·· 06.2
10	328 28.4	05.8
11	343 28.3	05.3
12	358 28.3	N21 04.9
13	13 28.2	04.5
14	28 28.2	04.0
15	43 28.1	·· 03.6
16	58 28.1	03.2
17	73 28.0	02.7
18	88 28.0	N21 02.3
19	103 27.9	01.8
20	118 27.9	01.4
21	133 27.8	·· 01.0
22	148 27.8	00.5
23	163 27.7	21 00.1
19 00	178 27.7	N20 59.6
01	193 27.6	59.2
02	208 27.6	58.8
03	223 27.6	·· 58.3
04	238 27.5	57.9
05	253 27.5	57.4
06	268 27.4	N20 57.0
07	283 27.4	56.5
08	298 27.3	56.1
09	313 27.3	·· 55.6
10	328 27.2	55.2
11	343 27.2	54.7
12	358 27.1	N20 54.3
13	13 27.1	53.9
14	28 27.1	53.4
15	43 27.0	·· 53.0
16	58 27.0	52.5
17	73 26.9	52.0
18	88 26.9	N20 51.6
19	103 26.8	51.1
20	118 26.8	50.7
21	133 26.8	·· 50.2
22	148 26.7	49.8
23	163 26.7	49.3
20 00	178 26.6	N20 48.9
01	193 26.6	48.4
02	208 26.5	48.0
03	223 26.5	·· 47.5
04	238 26.5	47.0
05	253 26.4	46.6
06	268 26.4	N20 46.1
07	283 26.3	45.7
08	298 26.3	45.2
09	313 26.3	·· 44.7
10	328 26.2	44.3
11	343 26.2	43.8
12	358 26.1	N20 43.3
13	13 26.1	42.9
14	28 26.1	42.4
15	43 26.0	·· 42.0
16	58 26.0	41.5
17	73 25.9	41.0
18	88 25.9	N20 40.6
19	103 25.9	40.1
20	118 25.8	39.6
21	133 25.8	·· 39.2
22	148 25.8	38.7
23	163 25.7	38.2

	G.H.A.	Dec.
21 00	178 25.7	N20 37.7
01	193 25.6	37.3
02	208 25.6	36.8
03	223 25.6	·· 36.3
04	238 25.5	35.9
05	253 25.5	35.4
06	268 25.5	N20 34.9
07	283 25.4	34.4
08	298 25.4	34.0
09	313 25.4	·· 33.5
10	328 25.3	33.0
11	343 25.3	32.5
12	358 25.3	N20 32.0
13	13 25.2	31.6
14	28 25.2	31.1
15	43 25.2	·· 30.6
16	58 25.1	30.1
17	73 25.1	29.7
18	88 25.1	N20 29.2
19	103 25.0	28.7
20	118 25.0	28.2
21	133 25.0	·· 27.7
22	148 24.9	27.2
23	163 24.9	26.8
22 00	178 24.9	N20 26.3
01	193 24.9	25.8
02	208 24.8	25.3
03	223 24.8	·· 24.8
04	238 24.8	24.3
05	253 24.7	23.8
06	268 24.7	N20 23.3
07	283 24.7	22.9
08	298 24.6	22.4
09	313 24.6	·· 21.9
10	328 24.6	21.4
11	343 24.6	20.9
12	358 24.5	N20 20.4
13	13 24.5	19.9
14	28 24.5	19.4
15	43 24.5	·· 18.9
16	58 24.4	18.4
17	73 24.4	17.9
18	88 24.4	N20 17.4
19	103 24.3	16.9
20	118 24.3	16.4
21	133 24.3	·· 15.9
22	148 24.3	15.5
23	163 24.2	15.0
23 00	178 24.2	N20 14.5
01	193 24.2	14.0
02	208 24.2	13.5
03	223 24.1	·· 13.0
04	238 24.1	12.5
05	253 24.1	11.9
06	268 24.1	N20 11.4
07	283 24.1	10.9
08	298 24.0	10.4
09	313 24.0	·· 09.9
10	328 24.0	09.4
11	343 24.0	08.9
12	358 23.9	N20 08.4
13	13 23.9	07.9
14	28 23.9	07.4
15	43 23.9	·· 06.9
16	58 23.9	06.4
17	73 23.8	05.9
18	88 23.8	N20 05.4
19	103 23.8	04.9
20	118 23.8	04.3
21	133 23.8	·· 03.8
22	148 23.7	03.3
23	163 23.7	02.8

	G.H.A.	Dec.
24 00	178 23.7	N20 02.3
01	193 23.7	01.8
02	208 23.7	01.3
03	223 23.6	·· 00.8
04	238 23.6	20 00.2
05	253 23.6	19 59.7
06	268 23.6	N19 59.2
07	283 23.6	58.7
08	298 23.6	58.2
09	313 23.5	·· 57.7
10	328 23.5	57.1
11	343 23.5	56.6
12	358 23.5	N19 56.1
13	13 23.5	55.6
14	28 23.5	55.1
15	43 23.5	·· 54.5
16	58 23.4	54.0
17	73 23.4	53.5
18	88 23.4	N19 53.0
19	103 23.4	52.4
20	118 23.4	51.9
21	133 23.4	·· 51.4
22	148 23.4	50.9
23	163 23.3	50.3
25 00	178 23.3	N19 49.8
01	193 23.3	49.3
02	208 23.3	48.8
03	223 23.3	·· 48.2
04	238 23.3	47.7
05	253 23.3	47.2
06	268 23.3	N19 46.6
07	283 23.3	46.1
08	298 23.2	45.6
09	313 23.2	·· 45.0
10	328 23.2	44.5
11	343 23.2	44.0
12	358 23.2	N19 43.4
13	13 23.2	42.9
14	28 23.2	42.4
15	43 23.2	·· 41.8
16	58 23.2	41.3
17	73 23.2	40.8
18	88 23.2	N19 40.2
19	103 23.1	39.7
20	118 23.1	39.2
21	133 23.1	·· 38.6
22	148 23.1	38.1
23	163 23.1	37.5
26 00	178 23.1	N19 37.0
01	193 23.1	36.5
02	208 23.1	35.9
03	223 23.1	·· 35.4
04	238 23.1	34.8
05	253 23.1	34.3
06	268 23.1	N19 33.7
07	283 23.1	33.2
08	298 23.1	32.6
09	313 23.1	·· 32.1
10	328 23.1	31.6
11	343 23.1	31.0
12	358 23.1	N19 30.5
13	13 23.1	29.9
14	28 23.0	29.4
15	43 23.0	·· 28.8
16	58 23.0	28.3
17	73 23.0	27.7
18	88 23.0	N19 27.2
19	103 23.0	26.6
20	118 23.0	26.1
21	133 23.0	·· 25.5
22	148 23.0	25.0
23	163 23.0	24.4

d	h	G.H.A. °	′	Dec. °	′
27	00	178	23.0	N19	23.8
	01	193	23.0		23.3
	02	208	23.0		22.7
	03	223	23.0	· ·	22.2
	04	238	23.0		21.6
	05	253	23.0		21.1
	06	268	23.0	N19	20.5
	07	283	23.0		20.0
	08	298	23.0		19.4
	09	313	23.0	· ·	18.8
	10	328	23.1		18.3
	11	343	23.1		17.7
	12	358	23.1	N19	17.2
	13	13	23.1		16.6
	14	28	23.1		16.0
	15	43	23.1	· ·	15.5
	16	58	23.1		14.9
	17	73	23.1		14.3
	18	88	23.1	N19	13.8
	19	103	23.1		13.2
	20	118	23.1		12.7
	21	133	23.1	· ·	12.1
	22	148	23.1		11.5
	23	163	23.1		11.0
28	00	178	23.1	N19	10.4
	01	193	23.1		09.8
	02	208	23.1		09.2
	03	223	23.1	· ·	08.7
	04	238	23.1		08.1
	05	253	23.2		07.5
	06	268	23.2	N19	07.0
	07	283	23.2		06.4
	08	298	23.2		05.8
	09	313	23.2	· ·	05.3
	10	328	23.2		04.7
	11	343	23.2		04.1
	12	358	23.2	N19	03.5
	13	13	23.2		03.0
	14	28	23.2		02.4
	15	43	23.2	· ·	01.8
	16	58	23.3		01.2
	17	73	23.3		00.7
	18	88	23.3	N19	00.1
	19	103	23.3	18	59.5
	20	118	23.3		58.9
	21	133	23.3	· ·	58.3
	22	148	23.3		57.8
	23	163	23.3		57.2
29	00	178	23.4	N18	56.6
	01	193	23.4		56.0
	02	208	23.4		55.4
	03	223	23.4	· ·	54.9
	04	238	23.4		54.3
	05	253	23.4		53.7
	06	268	23.4	N18	53.1
	07	283	23.5		52.5
	08	298	23.5		51.9
	09	313	23.5	· ·	51.4
	10	328	23.5		50.8
	11	343	23.5		50.2
	12	358	23.5	N18	49.6
	13	13	23.5		49.0
	14	28	23.6		48.4
	15	43	23.6	· ·	47.8
	16	58	23.6		47.2
	17	73	23.6		46.7
	18	88	23.6	N18	46.1
	19	103	23.7		45.5
	20	118	23.7		44.9
	21	133	23.7	· ·	44.3
	22	148	23.7		43.7
	23	163	23.7		43.1

d	h	G.H.A. °	′	Dec. °	′
30	00	178	23.7	N18	42.5
	01	193	23.8		41.9
	02	208	23.8		41.3
	03	223	23.8	· ·	40.7
	04	238	23.8		40.1
	05	253	23.8		39.5
	06	268	23.9	N18	38.9
	07	283	23.9		38.3
	08	298	23.9		37.7
	09	313	23.9	· ·	37.1
	10	328	24.0		36.5
	11	343	24.0		35.9
	12	358	24.0	N18	35.3
	13	13	24.0		34.7
	14	28	24.0		34.1
	15	43	24.1	· ·	33.5
	16	58	24.1		32.9
	17	73	24.1		32.3
	18	88	24.1	N18	31.7
	19	103	24.2		31.1
	20	118	24.2		30.5
	21	133	24.2	· ·	29.9
	22	148	24.2		29.3
	23	163	24.3		28.7
31	00	178	24.3	N18	28.1
	01	193	24.3		27.5
	02	208	24.3		26.9
	03	223	24.4	· ·	26.3
	04	238	24.4		25.7
	05	253	24.4		25.1
	06	268	24.5	N18	24.5
	07	283	24.5		23.9
	08	298	24.5		23.2
	09	313	24.5	· ·	22.6
	10	328	24.6		22.0
	11	343	24.6		21.4
	12	358	24.6	N18	20.8
	13	13	24.7		20.2
	14	28	24.7		19.6
	15	43	24.7	· ·	19.0
	16	58	24.7		18.3
	17	73	24.8		17.7
	18	88	24.8	N18	17.1
	19	103	24.8		16.5
	20	118	24.9		15.9
	21	133	24.9	· ·	15.3
	22	148	24.9		14.6
	23	163	25.0		14.0
AUGUST 1	00	178	25.0	N18	13.4
	01	193	25.0		12.8
	02	208	25.1		12.2
	03	223	25.1	· ·	11.5
	04	238	25.1		10.9
	05	253	25.2		10.3
	06	268	25.2	N18	09.7
	07	283	25.2		09.1
	08	298	25.3		08.4
	09	313	25.3	· ·	07.8
	10	328	25.3		07.2
	11	343	25.4		06.6
	12	358	25.4	N18	05.9
	13	13	25.4		05.3
	14	28	25.5		04.7
	15	43	25.5	· ·	04.1
	16	58	25.6		03.4
	17	73	25.6		02.8
	18	88	25.6	N18	02.2
	19	103	25.7		01.6
	20	118	25.7		00.9
	21	133	25.7	18	00.3
	22	148	25.8	17	59.7
	23	163	25.8		59.0

d	h	G.H.A. °	′	Dec. °	′
2	00	178	25.9	N17	58.4
	01	193	25.9		57.8
	02	208	25.9		57.1
	03	223	26.0	· ·	56.5
	04	238	26.0		55.9
	05	253	26.1		55.2
	06	268	26.1	N17	54.6
	07	283	26.1		54.0
	08	298	26.2		53.3
	09	313	26.2	· ·	52.7
	10	328	26.3		52.1
	11	343	26.3		51.4
	12	358	26.3	N17	50.8
	13	13	26.4		50.2
	14	28	26.4		49.5
	15	43	26.5	· ·	48.9
	16	58	26.5		48.2
	17	73	26.6		47.6
	18	88	26.6	N17	47.0
	19	103	26.6		46.3
	20	118	26.7		45.7
	21	133	26.7	· ·	45.0
	22	148	26.8		44.4
	23	163	26.8		43.8
3	00	178	26.9	N17	43.1
	01	193	26.9		42.5
	02	208	27.0		41.8
	03	223	27.0	· ·	41.2
	04	238	27.1		40.5
	05	253	27.1		39.9
	06	268	27.2	N17	39.2
	07	283	27.2		38.6
	08	298	27.2		37.9
	09	313	27.3	· ·	37.3
	10	328	27.3		36.7
	11	343	27.4		36.0
	12	358	27.4	N17	35.4
	13	13	27.5		34.7
	14	28	27.5		34.1
	15	43	27.6	· ·	33.4
	16	58	27.6		32.8
	17	73	27.7		32.1
	18	88	27.7	N17	31.4
	19	103	27.8		30.8
	20	118	27.8		30.1
	21	133	27.9	· ·	29.5
	22	148	27.9		28.8
	23	163	28.0		28.2
4	00	178	28.0	N17	27.5
	01	193	28.1		26.9
	02	208	28.1		26.2
	03	223	28.2	· ·	25.6
	04	238	28.3		24.9
	05	253	28.3		24.2
	06	268	28.4	N17	23.6
	07	283	28.4		22.9
	08	298	28.5		22.3
	09	313	28.5	· ·	21.6
	10	328	28.6		20.9
	11	343	28.6		20.3
	12	358	28.7	N17	19.6
	13	13	28.7		19.0
	14	28	28.8		18.3
	15	43	28.9	· ·	17.6
	16	58	28.9		17.0
	17	73	29.0		16.3
	18	88	29.0	N17	15.7
	19	103	29.1		15.0
	20	118	29.1		14.3
	21	133	29.2	· ·	13.7
	22	148	29.3		13.0
	23	163	29.3		12.3

d h	G.H.A.	Dec.		d h	G.H.A.	Dec.		d h	G.H.A.	Dec.
5 00	178 29.4	N17 11.7		8 00	178 34.2	N16 22.4		11 00	178 40.4	N15 30.8
01	193 29.4	11.0		01	193 34.3	21.7		01	193 40.5	30.0
02	208 29.5	10.3		02	208 34.4	21.0		02	208 40.6	29.3
03	223 29.5	·· 09.7		03	223 34.5	·· 20.3		03	223 40.7	·· 28.6
04	238 29.6	09.0		04	238 34.6	19.6		04	238 40.8	27.8
05	253 29.7	08.3		05	253 34.6	18.9		05	253 40.9	27.1
06	268 29.7	N17 07.6		06	268 34.7	N16 18.2		06	268 41.0	N15 26.4
07	283 29.8	07.0		07	283 34.8	17.5		07	283 41.1	25.6
08	298 29.8	06.3		08	298 34.9	16.8		08	298 41.2	24.9
09	313 29.9	·· 05.6		09	313 34.9	·· 16.1		09	313 41.3	·· 24.2
10	328 30.0	05.0		10	328 35.0	15.4		10	328 41.4	23.4
11	343 30.0	04.3		11	343 35.1	14.7		11	343 41.5	22.7
12	358 30.1	N17 03.6		12	358 35.2	N16 14.0		12	358 41.6	N15 22.0
13	13 30.1	02.9		13	13 35.3	13.3		13	13 41.7	21.2
14	28 30.2	02.3		14	28 35.3	12.6		14	28 41.8	20.5
15	43 30.3	·· 01.6		15	43 35.4	·· 11.8		15	43 41.9	·· 19.7
16	58 30.3	00.9		16	58 35.5	11.1		16	58 42.0	19.0
17	73 30.4	17 00.3		17	73 35.6	10.4		17	73 42.1	18.3
18	88 30.5	N16 59.6		18	88 35.7	N16 09.7		18	88 42.2	N15 17.5
19	103 30.5	58.9		19	103 35.8	09.0		19	103 42.3	16.8
20	118 30.6	58.2		20	118 35.8	08.3		20	118 42.4	16.0
21	133 30.7	·· 57.5		21	133 35.9	·· 07.6		21	133 42.5	·· 15.3
22	148 30.7	56.9		22	148 36.0	06.9		22	148 42.6	14.6
23	163 30.8	56.2		23	163 36.1	06.2		23	163 42.7	13.8
6 00	178 30.8	N16 55.5		9 00	178 36.2	N16 05.5		12 00	178 42.8	N15 13.1
01	193 30.9	54.8		01	193 36.3	04.7		01	193 42.9	12.3
02	208 31.0	54.2		02	208 36.3	04.0		02	208 43.0	11.6
03	223 31.0	·· 53.5		03	223 36.4	·· 03.3		03	223 43.1	·· 10.8
04	238 31.1	52.8		04	238 36.5	02.6		04	238 43.2	10.1
05	253 31.2	52.1		05	253 36.6	01.9		05	253 43.3	09.4
06	268 31.2	N16 51.4		06	268 36.7	N16 01.2		06	268 43.4	N15 08.6
07	283 31.3	50.8		07	283 36.8	16 00.5		07	283 43.5	07.9
08	298 31.4	50.1		08	298 36.8	15 59.8		08	298 43.6	07.1
09	313 31.4	·· 49.4		09	313 36.9	·· 59.0		09	313 43.7	·· 06.4
10	328 31.5	48.7		10	328 37.0	58.3		10	328 43.8	05.6
11	343 31.6	48.0		11	343 37.1	57.6		11	343 43.9	04.9
12	358 31.6	N16 47.3		12	358 37.2	N15 56.9		12	358 44.0	N15 04.1
13	13 31.7	46.7		13	13 37.3	56.2		13	13 44.1	03.4
14	28 31.8	46.0		14	28 37.4	55.4		14	28 44.2	02.6
15	43 31.8	·· 45.3		15	43 37.4	·· 54.7		15	43 44.3	·· 01.9
16	58 31.9	44.6		16	58 37.5	54.0		16	58 44.4	01.1
17	73 32.0	43.9		17	73 37.6	53.3		17	73 44.5	15 00.4
18	88 32.1	N16 43.2		18	88 37.7	N15 52.6		18	88 44.6	N14 59.6
19	103 32.1	42.5		19	103 37.8	51.9		19	103 44.7	58.9
20	118 32.2	41.8		20	118 37.9	51.1		20	118 44.9	58.1
21	133 32.3	·· 41.2		21	133 38.0	·· 50.4		21	133 45.0	·· 57.4
22	148 32.3	40.5		22	148 38.1	49.7		22	148 45.1	56.6
23	163 32.4	39.8		23	163 38.1	49.0		23	163 45.2	55.9
7 00	178 32.5	N16 39.1		10 00	178 38.2	N15 48.2		13 00	178 45.3	N14 55.1
01	193 32.5	38.4		01	193 38.3	47.5		01	193 45.4	54.4
02	208 32.6	37.7		02	208 38.4	46.8		02	208 45.5	53.6
03	223 32.7	·· 37.0		03	223 38.5	·· 46.1		03	223 45.6	·· 52.9
04	238 32.8	36.3		04	238 38.6	45.4		04	238 45.7	52.1
05	253 32.8	35.6		05	253 38.7	44.6		05	253 45.8	51.3
06	268 32.9	N16 34.9		06	268 38.8	N15 43.9		06	268 45.9	N14 50.6
07	283 33.0	34.3		07	283 38.9	43.2		07	283 46.0	49.8
08	298 33.0	33.6		08	298 39.0	42.5		08	298 46.1	49.1
09	313 33.1	·· 32.9		09	313 39.0	·· 41.7		09	313 46.2	·· 48.3
10	328 33.2	32.2		10	328 39.1	41.0		10	328 46.4	47.6
11	343 33.3	31.5		11	343 39.2	40.3		11	343 46.5	46.8
12	358 33.3	N16 30.8		12	358 39.3	N15 39.5		12	358 46.6	N14 46.0
13	13 33.4	30.1		13	13 39.4	38.8		13	13 46.7	45.3
14	28 33.5	29.4		14	28 39.5	38.1		14	28 46.8	44.5
15	43 33.6	·· 28.7		15	43 39.6	·· 37.4		15	43 46.9	·· 43.8
16	58 33.6	28.0		16	58 39.7	36.6		16	58 47.0	43.0
17	73 33.7	27.3		17	73 39.8	35.9		17	73 47.1	42.3
18	88 33.8	N16 26.6		18	88 39.9	N15 35.2		18	88 47.2	N14 41.5
19	103 33.9	25.9		19	103 40.0	34.4		19	103 47.3	40.7
20	118 33.9	25.2		20	118 40.1	33.7		20	118 47.5	40.0
21	133 34.0	·· 24.5		21	133 40.2	·· 33.0		21	133 47.6	·· 39.2
22	148 34.1	23.8		22	148 40.3	32.2		22	148 47.7	38.4
23	163 34.2	23.1		23	163 40.3	31.5		23	163 47.8	37.7

	G.H.A.	Dec.
14 00	178 47.9	N14 36.9
01	193 48.0	36.2
02	208 48.1	35.4
03	223 48.2 ··	34.6
04	238 48.3	33.9
05	253 48.5	33.1
06	268 48.6	N14 32.3
07	283 48.7	31.6
08	298 48.8	30.8
09	313 48.9 ··	30.0
10	328 49.0	29.3
11	343 49.1	28.5
12	358 49.3	N14 27.7
13	13 49.4	27.0
14	28 49.5	26.2
15	43 49.6 ··	25.4
16	58 49.7	24.7
17	73 49.8	23.9
18	88 50.0	N14 23.1
19	103 50.1	22.4
20	118 50.2	21.6
21	133 50.3 ··	20.8
22	148 50.4	20.0
23	163 50.5	19.3
15 00	178 50.7	N14 18.5
01	193 50.8	17.7
02	208 50.9	17.0
03	223 51.0 ··	16.2
04	238 51.1	15.4
05	253 51.2	14.6
06	268 51.4	N14 13.9
07	283 51.5	13.1
08	298 51.6	12.3
09	313 51.7 ··	11.5
10	328 51.8	10.8
11	343 52.0	10.0
12	358 52.1	N14 09.2
13	13 52.2	08.4
14	28 52.3	07.6
15	43 52.4 ··	06.9
16	58 52.6	06.1
17	73 52.7	05.3
18	88 52.8	N14 04.5
19	103 52.9	03.7
20	118 53.0	03.0
21	133 53.2 ··	02.2
22	148 53.3	01.4
23	163 53.4	14 00.6
16 00	178 53.5	N13 59.8
01	193 53.7	59.1
02	208 53.8	58.3
03	223 53.9 ··	57.5
04	238 54.0	56.7
05	253 54.2	55.9
06	268 54.3	N13 55.1
07	283 54.4	54.4
08	298 54.5	53.6
09	313 54.7 ··	52.8
10	328 54.8	52.0
11	343 54.9	51.2
12	358 55.0	N13 50.4
13	13 55.2	49.6
14	28 55.3	48.9
15	43 55.4 ··	48.1
16	58 55.5	47.3
17	73 55.7	46.5
18	88 55.8	N13 45.7
19	103 55.9	44.9
20	118 56.0	44.1
21	133 56.2 ··	43.3
22	148 56.3	42.6
23	163 56.4	41.8

	G.H.A.	Dec.
17 00	178 56.6	N13 41.0
01	193 56.7	40.2
02	208 56.8	39.4
03	223 56.9 ··	38.6
04	238 57.1	37.8
05	253 57.2	37.0
06	268 57.3	N13 36.2
07	283 57.5	35.4
08	298 57.6	34.6
09	313 57.7 ··	33.8
10	328 57.8	33.0
11	343 58.0	32.2
12	358 58.1	N13 31.4
13	13 58.2	30.7
14	28 58.4	29.9
15	43 58.5 ··	29.1
16	58 58.6	28.3
17	73 58.8	27.5
18	88 58.9	N13 26.7
19	103 59.0	25.9
20	118 59.2	25.1
21	133 59.3 ··	24.3
22	148 59.4	23.5
23	163 59.6	22.7
18 00	178 59.7	N13 21.9
01	193 59.8	21.1
02	209 00.0	20.3
03	224 00.1 ··	19.5
04	239 00.2	18.7
05	254 00.4	17.9
06	269 00.5	N13 17.1
07	284 00.6	16.3
08	299 00.8	15.5
09	314 00.9 ··	14.7
10	329 01.0	13.9
11	344 01.2	13.1
12	359 01.3	N13 12.2
13	14 01.4	11.4
14	29 01.6	10.6
15	44 01.7 ··	09.8
16	59 01.9	09.0
17	74 02.0	08.2
18	89 02.1	N13 07.4
19	104 02.3	06.6
20	119 02.4	05.8
21	134 02.5 ··	05.0
22	149 02.7	04.2
23	164 02.8	03.4
19 00	179 03.0	N13 02.6
01	194 03.1	01.8
02	209 03.2	01.0
03	224 03.4	13 00.1
04	239 03.5	12 59.3
05	254 03.7	58.5
06	269 03.8	N12 57.7
07	284 03.9	56.9
08	299 04.1	56.1
09	314 04.2 ··	55.3
10	329 04.4	54.5
11	344 04.5	53.7
12	359 04.6	N12 52.8
13	14 04.8	52.0
14	29 04.9	51.2
15	44 05.1 ··	50.4
16	59 05.2	49.6
17	74 05.3	48.8
18	89 05.5	N12 48.0
19	104 05.6	47.1
20	119 05.8	46.3
21	134 05.9 ··	45.5
22	149 06.1	44.7
23	164 06.2	43.9

	G.H.A.	Dec.
20 00	179 06.3	N12 43.1
01	194 06.5	42.2
02	209 06.6	41.4
03	224 06.8 ··	40.6
04	239 06.9	39.8
05	254 07.1	39.0
06	269 07.2	N12 38.2
07	284 07.4	37.3
08	299 07.5	36.5
09	314 07.6 ··	35.7
10	329 07.8	34.9
11	344 07.9	34.1
12	359 08.1	N12 33.2
13	14 08.2	32.4
14	29 08.4	31.6
15	44 08.5 ··	30.8
16	59 08.7	29.9
17	74 08.8	29.1
18	89 09.0	N12 28.3
19	104 09.1	27.5
20	119 09.3	26.6
21	134 09.4 ··	25.8
22	149 09.6	25.0
23	164 09.7	24.2
21 00	179 09.8	N12 23.3
01	194 10.0	22.5
02	209 10.1	21.7
03	224 10.3 ··	20.9
04	239 10.4	20.0
05	254 10.6	19.2
06	269 10.7	N12 18.4
07	284 10.9	17.6
08	299 11.0	16.7
09	314 11.2 ··	15.9
10	329 11.3	15.1
11	344 11.5	14.2
12	359 11.6	N12 13.4
13	14 11.8	12.6
14	29 11.9	11.8
15	44 12.1 ··	10.9
16	59 12.3	10.1
17	74 12.4	09.3
18	89 12.6	N12 08.4
19	104 12.7	07.6
20	119 12.9	06.8
21	134 13.0 ··	05.9
22	149 13.2	05.1
23	164 13.3	04.3
22 00	179 13.5	N12 03.4
01	194 13.6	02.6
02	209 13.8	01.8
03	224 13.9 ··	00.9
04	239 14.1	12 00.1
05	254 14.2	11 59.3
06	269 14.4	N11 58.4
07	284 14.6	57.6
08	299 14.7	56.8
09	314 14.9 ··	55.9
10	329 15.0	55.1
11	344 15.2	54.3
12	359 15.3	N11 53.4
13	14 15.5	52.6
14	29 15.6	51.7
15	44 15.8 ··	50.9
16	59 16.0	50.1
17	74 16.1	49.2
18	89 16.3	N11 48.4
19	104 16.4	47.5
20	119 16.6	46.7
21	134 16.7 ··	45.9
22	149 16.9	45.0
23	164 17.1	44.2

d h	G.H.A.	Dec.
23 00	179 17.2	N11 43.3
01	194 17.4	42.5
02	209 17.5	41.7
03	224 17.7 ··	40.8
04	239 17.8	40.0
05	254 18.0	39.1
06	269 18.2	N11 38.3
07	284 18.3	37.4
08	299 18.5	36.6
09	314 18.6 ··	35.8
10	329 18.8	34.9
11	344 19.0	34.1
12	359 19.1	N11 33.2
13	14 19.3	32.4
14	29 19.4	31.5
15	44 19.6 ··	30.7
16	59 19.8	29.8
17	74 19.9	29.0
18	89 20.1	N11 28.1
19	104 20.2	27.3
20	119 20.4	26.4
21	134 20.6 ··	25.6
22	149 20.7	24.8
23	164 20.9	23.9
24 00	179 21.1	N11 23.1
01	194 21.2	22.2
02	209 21.4	21.4
03	224 21.5 ··	20.5
04	239 21.7	19.7
05	254 21.9	18.8
06	269 22.0	N11 18.0
07	284 22.2	17.1
08	299 22.4	16.3
09	314 22.5 ··	15.4
10	329 22.7	14.6
11	344 22.9	13.7
12	359 23.0	N11 12.8
13	14 23.2	12.0
14	29 23.4	11.1
15	44 23.5 ··	10.3
16	59 23.7	09.4
17	74 23.9	08.6
18	89 24.0	N11 07.7
19	104 24.2	06.9
20	119 24.4	06.0
21	134 24.5 ··	05.2
22	149 24.7	04.3
23	164 24.9	03.4
25 00	179 25.0	N11 02.6
01	194 25.2	01.7
02	209 25.4	00.9
03	224 25.5	11 00.0
04	239 25.7	10 59.2
05	254 25.9	58.3
06	269 26.0	N10 57.4
07	284 26.2	56.6
08	299 26.4	55.7
09	314 26.5 ··	54.9
10	329 26.7	54.0
11	344 26.9	53.2
12	359 27.0	N10 52.3
13	14 27.2	51.4
14	29 27.4	50.6
15	44 27.5 ··	49.7
16	59 27.7	48.9
17	74 27.9	48.0
18	89 28.1	N10 47.1
19	104 28.2	46.3
20	119 28.4	45.4
21	134 28.6 ··	44.5
22	149 28.7	43.7
23	164 28.9	42.8

d h	G.H.A.	Dec.
26 00	179 29.1	N10 42.0
01	194 29.3	41.1
02	209 29.4	40.2
03	224 29.6 ··	39.4
04	239 29.8	38.5
05	254 29.9	37.6
06	269 30.1	N10 36.8
07	284 30.3	35.9
08	299 30.5	35.0
09	314 30.6 ··	34.2
10	329 30.8	33.3
11	344 31.0	32.4
12	359 31.2	N10 31.6
13	14 31.3	30.7
14	29 31.5	29.8
15	44 31.7 ··	29.0
16	59 31.9	28.1
17	74 32.0	27.2
18	89 32.2	N10 26.4
19	104 32.4	25.5
20	119 32.6	24.6
21	134 32.7 ··	23.8
22	149 32.9	22.9
23	164 33.1	22.0
27 00	179 33.3	N10 21.1
01	194 33.4	20.3
02	209 33.6	19.4
03	224 33.8 ··	18.5
04	239 34.0	17.7
05	254 34.1	16.8
06	269 34.3	N10 15.9
07	284 34.5	15.0
08	299 34.7	14.2
09	314 34.8 ··	13.3
10	329 35.0	12.4
11	344 35.2	11.5
12	359 35.4	N10 10.7
13	14 35.6	09.8
14	29 35.7	08.9
15	44 35.9 ··	08.1
16	59 36.1	07.2
17	74 36.3	06.3
18	89 36.4	N10 05.4
19	104 36.6	04.6
20	119 36.8	03.7
21	134 37.0 ··	02.8
22	149 37.2	01.9
23	164 37.3	01.0
28 00	179 37.5	N10 00.2
01	194 37.7	9 59.3
02	209 37.9	58.4
03	224 38.1 ··	57.5
04	239 38.2	56.7
05	254 38.4	55.8
06	269 38.6	N 9 54.9
07	284 38.8	54.0
08	299 39.0	53.1
09	314 39.2 ··	52.3
10	329 39.3	51.4
11	344 39.5	50.5
12	359 39.7	N 9 49.6
13	14 39.9	48.7
14	29 40.1	47.9
15	44 40.2 ··	47.0
16	59 40.4	46.1
17	74 40.6	45.2
18	89 40.8	N 9 44.3
19	104 41.0	43.5
20	119 41.2	42.6
21	134 41.3 ··	41.7
22	149 41.5	40.8
23	164 41.7	39.9

d h	G.H.A.	Dec.
29 00	179 41.9	N 9 39.0
01	194 42.1	38.2
02	209 42.3	37.3
03	224 42.4 ··	36.4
04	239 42.6	35.5
05	254 42.8	34.6
06	269 43.0	N 9 33.7
07	284 43.2	32.8
08	299 43.4	32.0
09	314 43.6 ··	31.1
10	329 43.7	30.2
11	344 43.9	29.3
12	359 44.1	N 9 28.4
13	14 44.3	27.5
14	29 44.5	26.6
15	44 44.7 ··	25.8
16	59 44.9	24.9
17	74 45.0	24.0
18	89 45.2	N 9 23.1
19	104 45.4	22.2
20	119 45.6	21.3
21	134 45.8 ··	20.4
22	149 46.0	19.5
23	164 46.2	18.6
30 00	179 46.4	N 9 17.8
01	194 46.5	16.9
02	209 46.7	16.0
03	224 46.9 ··	15.1
04	239 47.1	14.2
05	254 47.3	13.3
06	269 47.5	N 9 12.4
07	284 47.7	11.5
08	299 47.9	10.6
09	314 48.1 ··	09.7
10	329 48.2	08.8
11	344 48.4	07.9
12	359 48.6	N 9 07.1
13	14 48.8	06.2
14	29 49.0	05.3
15	44 49.2 ··	04.4
16	59 49.4	03.5
17	74 49.6	02.6
18	89 49.8	N 9 01.7
19	104 50.0	9 00.8
20	119 50.1	8 59.9
21	134 50.3 ··	59.0
22	149 50.5	58.1
23	164 50.7	57.2
31 00	179 50.9	N 8 56.3
01	194 51.1	55.4
02	209 51.3	54.5
03	224 51.5 ··	53.6
04	239 51.7	52.7
05	254 51.9	51.8
06	269 52.1	N 8 50.9
07	284 52.3	50.0
08	299 52.4	49.1
09	314 52.6 ··	48.2
10	329 52.8	47.3
11	344 53.0	46.4
12	359 53.2	N 8 45.5
13	14 53.4	44.6
14	29 53.6	43.8
15	44 53.8 ··	42.9
16	59 54.0	42.0
17	74 54.2	41.1
18	89 54.4	N 8 40.2
19	104 54.6	39.3
20	119 54.8	38.3
21	134 55.0	37.4
22	149 55.2	36.5
23	164 55.4	35.6

d h	G.H.A.	Dec.
1 00	179 55.5	N 8 34.7
01	194 55.7	33.8
02	209 55.9	32.9
03	224 56.1 ··	32.0
04	239 56.3	31.1
05	254 56.5	30.2
06	269 56.7	N 8 29.3
07	284 56.9	28.4
08	299 57.1	27.5
09	314 57.3 ··	26.6
10	329 57.5	25.7
11	344 57.7	24.8
12	359 57.9	N 8 23.9
13	14 58.1	23.0
14	29 58.3	22.1
15	44 58.5 ··	21.2
16	59 58.7	20.3
17	74 58.9	19.4
18	89 59.1	N 8 18.5
19	104 59.3	17.6
20	119 59.5	16.7
21	134 59.7 ··	15.8
22	149 59.9	14.8
23	165 00.1	13.9
2 00	180 00.3	N 8 13.0
01	195 00.5	12.1
02	210 00.7	11.2
03	225 00.9 ··	10.3
04	240 01.1	09.4
05	255 01.3	08.5
06	270 01.5	N 8 07.6
07	285 01.7	06.7
08	300 01.9	05.8
09	315 02.1 ··	04.9
10	330 02.3	03.9
11	345 02.5	03.0
12	0 02.7	N 8 02.1
13	15 02.9	01.2
14	30 03.1	8 00.3
15	45 03.3	7 59.4
16	60 03.5	58.5
17	75 03.7	57.6
18	90 03.9	N 7 56.7
19	105 04.1	55.7
20	120 04.3	54.8
21	135 04.5 ··	53.9
22	150 04.7	53.0
23	165 04.9	52.1
3 00	180 05.1	N 7 51.2
01	195 05.3	50.3
02	210 05.5	49.4
03	225 05.7 ··	48.4
04	240 05.9	47.5
05	255 06.1	46.6
06	270 06.3	N 7 45.7
07	285 06.5	44.8
08	300 06.7	43.9
09	315 06.9 ··	43.0
10	330 07.1	42.0
11	345 07.3	41.1
12	0 07.5	N 7 40.2
13	15 07.7	39.3
14	30 07.9	38.4
15	45 08.1 ··	37.5
16	60 08.3	36.6
17	75 08.5	35.6
18	90 08.7	N 7 34.7
19	105 08.9	33.8
20	120 09.1	32.9
21	135 09.3 ··	32.0
22	150 09.5	31.0
23	165 09.7	30.1

d h	G.H.A.	Dec.
4 00	180 09.9	N 7 29.2
01	195 10.1	28.3
02	210 10.3	27.4
03	225 10.5 ··	26.5
04	240 10.7	25.5
05	255 11.0	24.6
06	270 11.2	N 7 23.7
07	285 11.4	22.8
08	300 11.6	21.9
09	315 11.8 ··	20.9
10	330 12.0	20.0
11	345 12.2	19.1
12	0 12.4	N 7 18.2
13	15 12.6	17.3
14	30 12.8	16.3
15	45 13.0 ··	15.4
16	60 13.2	14.5
17	75 13.4	13.6
18	90 13.6	N 7 12.7
19	105 13.8	11.7
20	120 14.0	10.8
21	135 14.2 ··	09.9
22	150 14.5	09.0
23	165 14.7	08.0
5 00	180 14.9	N 7 07.1
01	195 15.1	06.2
02	210 15.3	05.3
03	225 15.5 ··	04.4
04	240 15.7	03.4
05	255 15.9	02.5
06	270 16.1	N 7 01.6
07	285 16.3	7 00.7
08	300 16.5	6 59.7
09	315 16.7 ··	58.8
10	330 16.9	57.9
11	345 17.1	57.0
12	0 17.4	N 6 56.0
13	15 17.6	55.1
14	30 17.8	54.2
15	45 18.0 ··	53.3
16	60 18.2	52.3
17	75 18.4	51.4
18	90 18.6	N 6 50.5
19	105 18.8	49.5
20	120 19.0	48.6
21	135 19.2 ··	47.7
22	150 19.4	46.8
23	165 19.7	45.8
6 00	180 19.9	N 6 44.9
01	195 20.1	44.0
02	210 20.3	43.1
03	225 20.5 ··	42.1
04	240 20.7	41.2
05	255 20.9	40.3
06	270 21.1	N 6 39.3
07	285 21.3	38.4
08	300 21.5	37.5
09	315 21.8 ··	36.6
10	330 22.0	35.6
11	345 22.2	34.7
12	0 22.4	N 6 33.8
13	15 22.6	32.8
14	30 22.8	31.9
15	45 23.0 ··	31.0
16	60 23.2	30.0
17	75 23.4	29.1
18	90 23.7	N 6 28.2
19	105 23.9	27.3
20	120 24.1	26.3
21	135 24.3 ··	25.4
22	150 24.5	24.5
23	165 24.7	23.5

d h	G.H.A.	Dec.
7 00	180 24.9	N 6 22.6
01	195 25.1	21.7
02	210 25.3	20.7
03	225 25.6 ··	19.8
04	240 25.8	18.9
05	255 26.0	17.9
06	270 26.2	N 6 17.0
07	285 26.4	16.1
08	300 26.6	15.1
09	315 26.8 ··	14.2
10	330 27.0	13.3
11	345 27.3	12.3
12	0 27.5	N 6 11.4
13	15 27.7	10.5
14	30 27.9	09.5
15	45 28.1 ··	08.6
16	60 28.3	07.7
17	75 28.5	06.7
18	90 28.7	N 6 05.8
19	105 29.0	04.8
20	120 29.2	03.9
21	135 29.4 ··	03.0
22	150 29.6	02.0
23	165 29.8	01.1
8 00	180 30.0	N 6 00.2
01	195 30.2	5 59.2
02	210 30.5	58.3
03	225 30.7 ··	57.4
04	240 30.9	56.4
05	255 31.1	55.5
06	270 31.3	N 5 54.5
07	285 31.5	53.6
08	300 31.7	52.7
09	315 32.0 ··	51.7
10	330 32.2	50.8
11	345 32.4	49.9
12	0 32.6	N 5 48.9
13	15 32.8	48.0
14	30 33.0	47.0
15	45 33.2 ··	46.1
16	60 33.5	45.2
17	75 33.7	44.2
18	90 33.9	N 5 43.3
19	105 34.1	42.3
20	120 34.3	41.4
21	135 34.5 ··	40.5
22	150 34.8	39.5
23	165 35.0	38.6
9 00	180 35.2	N 5 37.6
01	195 35.4	36.7
02	210 35.6	35.8
03	225 35.8 ··	34.8
04	240 36.0	33.9
05	255 36.3	32.9
06	270 36.5	N 5 32.0
07	285 36.7	31.1
08	300 36.9	30.1
09	315 37.1 ··	29.2
10	330 37.3	28.2
11	345 37.6	27.3
12	0 37.8	N 5 26.3
13	15 38.0	25.4
14	30 38.2	24.5
15	45 38.4 ··	23.5
16	60 38.6	22.6
17	75 38.9	21.6
18	90 39.1	N 5 20.7
19	105 39.3	19.7
20	120 39.5	18.8
21	135 39.7 ··	17.9
22	150 39.9	16.9
23	165 40.2	16.0

d h	G.H.A. ° '	Dec. ° '
10 00	180 40.4	N 5 15.0
01	195 40.6	14.1
02	210 40.8	13.1
03	225 41.0 ··	12.2
04	240 41.2	11.2
05	255 41.5	10.3
06	270 41.7	N 5 09.4
07	285 41.9	08.4
08	300 42.1	07.5
09	315 42.3 ··	06.5
10	330 42.6	05.6
11	345 42.8	04.6
12	0 43.0	N 5 03.7
13	15 43.2	02.7
14	30 43.4	01.8
15	45 43.6	5 00.8
16	60 43.9	4 59.9
17	75 44.1	58.9
18	90 44.3	N 4 58.0
19	105 44.5	57.1
20	120 44.7	56.1
21	135 45.0 ··	55.2
22	150 45.2	54.2
23	165 45.4	53.3
11 00	180 45.6	N 4 52.3
01	195 45.8	51.4
02	210 46.0	50.4
03	225 46.3 ··	49.5
04	240 46.5	48.5
05	255 46.7	47.6
06	270 46.9	N 4 46.6
07	285 47.1	45.7
08	300 47.4	44.7
09	315 47.6 ··	43.8
10	330 47.8	42.8
11	345 48.0	41.9
12	0 48.2	N 4 40.9
13	15 48.5	40.0
14	30 48.7	39.0
15	45 48.9 ··	38.1
16	60 49.1	37.1
17	75 49.3	36.2
18	90 49.5	N 4 35.2
19	105 49.8	34.3
20	120 50.0	33.3
21	135 50.2 ··	32.4
22	150 50.4	31.4
23	165 50.6	30.5
12 00	180 50.9	N 4 29.5
01	195 51.1	28.6
02	210 51.3	27.6
03	225 51.5 ··	26.7
04	240 51.7	25.7
05	255 52.0	24.8
06	270 52.2	N 4 23.8
07	285 52.4	22.9
08	300 52.6	21.9
09	315 52.8 ··	21.0
10	330 53.1	20.0
11	345 53.3	19.1
12	0 53.5	N 4 18.1
13	15 53.7	17.1
14	30 53.9	16.2
15	45 54.2 ··	15.2
16	60 54.4	14.3
17	75 54.6	13.3
18	90 54.8	N 4 12.4
19	105 55.0	11.4
20	120 55.3	10.5
21	135 55.5 ··	09.5
22	150 55.7	08.6
23	165 55.9	07.6

d h	G.H.A. ° '	Dec. ° '
13 00	180 56.2	N 4 06.7
01	195 56.4	05.7
02	210 56.6	04.7
03	225 56.8 ··	03.8
04	240 57.0	02.8
05	255 57.3	01.9
06	270 57.5	N 4 00.9
07	285 57.7	4 00.0
08	300 57.9	3 59.0
09	315 58.1 ··	58.1
10	330 58.4	57.1
11	345 58.6	56.1
12	0 58.8	N 3 55.2
13	15 59.0	54.2
14	30 59.2	53.3
15	45 59.5 ··	52.3
16	60 59.7	51.4
17	75 59.9	50.4
18	91 00.1	N 3 49.5
19	106 00.3	48.5
20	121 00.6	47.5
21	136 00.8 ··	46.6
22	151 01.0	45.6
23	166 01.2	44.7
14 00	181 01.5	N 3 43.7
01	196 01.7	42.8
02	211 01.9	41.8
03	226 02.1 ··	40.8
04	241 02.3	39.9
05	256 02.6	38.9
06	271 02.8	N 3 38.0
07	286 03.0	37.0
08	301 03.2	36.1
09	316 03.4 ··	35.1
10	331 03.7	34.1
11	346 03.9	33.2
12	1 04.1	N 3 32.2
13	16 04.3	31.3
14	31 04.6	30.3
15	46 04.8 ··	29.3
16	61 05.0	28.4
17	76 05.2	27.4
18	91 05.4	N 3 26.5
19	106 05.7	25.5
20	121 05.9	24.5
21	136 06.1 ··	23.6
22	151 06.3	22.6
23	166 06.5	21.7
15 00	181 06.8	N 3 20.7
01	196 07.0	19.7
02	211 07.2	18.8
03	226 07.4 ··	17.8
04	241 07.7	16.9
05	256 07.9	15.9
06	271 08.1	N 3 14.9
07	286 08.3	14.0
08	301 08.5	13.0
09	316 08.8 ··	12.1
10	331 09.0	11.1
11	346 09.2	10.1
12	1 09.4	N 3 09.2
13	16 09.7	08.2
14	31 09.9	07.3
15	46 10.1 ··	06.3
16	61 10.3	05.3
17	76 10.5	04.4
18	91 10.8	N 3 03.4
19	106 11.0	02.4
20	121 11.2	01.5
21	136 11.4	3 00.5
22	151 11.7	2 59.6
23	166 11.9	58.6

d h	G.H.A. ° '	Dec. ° '
16 00	181 12.1	N 2 57.6
01	196 12.3	56.7
02	211 12.5	55.7
03	226 12.8 ··	54.7
04	241 13.0	53.8
05	256 13.2	52.8
06	271 13.4	N 2 51.9
07	286 13.7	50.9
08	301 13.9	49.9
09	316 14.1 ··	49.0
10	331 14.3	48.0
11	346 14.5	47.0
12	1 14.8	N 2 46.1
13	16 15.0	45.1
14	31 15.2	44.2
15	46 15.4 ··	43.2
16	61 15.6	42.2
17	76 15.9	41.3
18	91 16.1	N 2 40.3
19	106 16.3	39.3
20	121 16.5	38.4
21	136 16.8 ··	37.4
22	151 17.0	36.4
23	166 17.2	35.5
17 00	181 17.4	N 2 34.5
01	196 17.6	33.5
02	211 17.9	32.6
03	226 18.1 ··	31.6
04	241 18.3	30.6
05	256 18.5	29.7
06	271 18.8	N 2 28.7
07	286 19.0	27.8
08	301 19.2	26.8
09	316 19.4 ··	25.8
10	331 19.6	24.9
11	346 19.9	23.9
12	1 20.1	N 2 22.9
13	16 20.3	22.0
14	31 20.5	21.0
15	46 20.8 ··	20.0
16	61 21.0	19.1
17	76 21.2	18.1
18	91 21.4	N 2 17.1
19	106 21.6	16.2
20	121 21.9	15.2
21	136 22.1 ··	14.2
22	151 22.3	13.3
23	166 22.5	12.3
18 00	181 22.8	N 2 11.3
01	196 23.0	10.4
02	211 23.2	09.4
03	226 23.4 ··	08.4
04	241 23.6	07.5
05	256 23.9	06.5
06	271 24.1	N 2 05.5
07	286 24.3	04.6
08	301 24.5	03.6
09	316 24.8 ··	02.6
10	331 25.0	01.7
11	346 25.2	2 00.7
12	1 25.4	N 1 59.7
13	16 25.6	58.8
14	31 25.9	57.8
15	46 26.1 ··	56.8
16	61 26.3	55.9
17	76 26.5	54.9
18	91 26.7	N 1 53.9
19	106 27.0	52.9
20	121 27.2	52.0
21	136 27.4 ··	51.0
22	151 27.6	50.0
23	166 27.9	49.1

		G.H.A.		Dec.
d h		° ′		° ′
19 00	181	28.1	N 1	48.1
01	196	28.3		47.1
02	211	28.5		46.2
03	226	28.7	··	45.2
04	241	29.0		44.2
05	256	29.2		43.3
06	271	29.4	N 1	42.3
07	286	29.6		41.3
08	301	29.9		40.4
09	316	30.1	··	39.4
10	331	30.3		38.4
11	346	30.5		37.5
12	1	30.7	N 1	36.5
13	16	31.0		35.5
14	31	31.2		34.5
15	46	31.4	··	33.6
16	61	31.6		32.6
17	76	31.8		31.6
18	91	32.1	N 1	30.7
19	106	32.3		29.7
20	121	32.5		28.7
21	136	32.7	··	27.8
22	151	32.9		26.8
23	166	33.2		25.8
20 00	181	33.4	N 1	24.8
01	196	33.6		23.9
02	211	33.8		22.9
03	226	34.1	··	21.9
04	241	34.3		21.0
05	256	34.5		20.0
06	271	34.7	N 1	19.0
07	286	34.9		18.1
08	301	35.2		17.1
09	316	35.4	··	16.1
10	331	35.6		15.1
11	346	35.8		14.2
12	1	36.0	N 1	13.2
13	16	36.3		12.2
14	31	36.5		11.3
15	46	36.7	··	10.3
16	61	36.9		09.3
17	76	37.1		08.3
18	91	37.4	N 1	07.4
19	106	37.6		06.4
20	121	37.8		05.4
21	136	38.0	··	04.5
22	151	38.3		03.5
23	166	38.5		02.5
21 00	181	38.7	N 1	01.6
01	196	38.9	1	00.6
02	211	39.1	0	59.6
03	226	39.4	··	58.6
04	241	39.6		57.7
05	256	39.8		56.7
06	271	40.0	N 0	55.7
07	286	40.2		54.8
08	301	40.5		53.8
09	316	40.7	··	52.8
10	331	40.9		51.8
11	346	41.1		50.9
12	1	41.3	N 0	49.9
13	16	41.6		48.9
14	31	41.8		47.9
15	46	42.0	··	47.0
16	61	42.2		46.0
17	76	42.4		45.0
18	91	42.7	N 0	44.1
19	106	42.9		43.1
20	121	43.1		42.1
21	136	43.3	··	41.1
22	151	43.5		40.2
23	166	43.8		39.2

		G.H.A.		Dec.
d h		° ′		° ′
22 00	181	44.0	N 0	38.2
01	196	44.2		37.3
02	211	44.4		36.3
03	226	44.6	··	35.3
04	241	44.9		34.3
05	256	45.1		33.4
06	271	45.3	N 0	32.4
07	286	45.5		31.4
08	301	45.7		30.4
09	316	45.9	··	29.5
10	331	46.2		28.5
11	346	46.4		27.5
12	1	46.6	N 0	26.6
13	16	46.8		25.6
14	31	47.0		24.6
15	46	47.3	··	23.6
16	61	47.5		22.7
17	76	47.7		21.7
18	91	47.9	N 0	20.7
19	106	48.1		19.7
20	121	48.4		18.8
21	136	48.6	··	17.8
22	151	48.8		16.8
23	166	49.0		15.9
23 00	181	49.2	N 0	14.9
01	196	49.5		13.9
02	211	49.7		12.9
03	226	49.9	··	12.0
04	241	50.1		11.0
05	256	50.3		10.0
06	271	50.5	N 0	09.0
07	286	50.8		08.1
08	301	51.0		07.1
09	316	51.2	··	06.1
10	331	51.4		05.1
11	346	51.6		04.2
12	1	51.9	N 0	03.2
13	16	52.1		02.2
14	31	52.3		01.3
15	46	52.5	N 0	00.3
16	61	52.7	S 0	00.7
17	76	52.9		01.7
18	91	53.2	S 0	02.6
19	106	53.4		03.6
20	121	53.6		04.6
21	136	53.8	··	05.6
22	151	54.0		06.5
23	166	54.3		07.5
24 00	181	54.5	S 0	08.5
01	196	54.7		09.5
02	211	54.9		10.4
03	226	55.1	··	11.4
04	241	55.3		12.4
05	256	55.6		13.4
06	271	55.8	S 0	14.3
07	286	56.0		15.3
08	301	56.2		16.3
09	316	56.4	··	17.2
10	331	56.6		18.2
11	346	56.9		19.2
12	1	57.1	S 0	20.2
13	16	57.3		21.1
14	31	57.5		22.1
15	46	57.7	··	23.1
16	61	57.9		24.1
17	76	58.2		25.0
18	91	58.4	S 0	26.0
19	106	58.6		27.0
20	121	58.8		28.0
21	136	59.0	··	28.9
22	151	59.2		29.9
23	166	59.5		30.9

		G.H.A.		Dec.
d h		° ′		° ′
25 00	181	59.7	S 0	31.9
01	196	59.9		32.8
02	212	00.1		33.8
03	227	00.3	··	34.8
04	242	00.5		35.8
05	257	00.8		36.7
06	272	01.0	S 0	37.7
07	287	01.2		38.7
08	302	01.4		39.6
09	317	01.6	··	40.6
10	332	01.8		41.6
11	347	02.0		42.6
12	2	02.3	S 0	43.5
13	17	02.5		44.5
14	32	02.7		45.5
15	47	02.9	··	46.5
16	62	03.1		47.4
17	77	03.3		48.4
18	92	03.6	S 0	49.4
19	107	03.8		50.4
20	122	04.0		51.3
21	137	04.2	··	52.3
22	152	04.4		53.3
23	167	04.6		54.3
26 00	182	04.8	S 0	55.2
01	197	05.1		56.2
02	212	05.3		57.2
03	227	05.5	··	58.2
04	242	05.7	0	59.1
05	257	05.9	1	00.1
06	272	06.1	S 1	01.1
07	287	06.3		02.1
08	302	06.6		03.0
09	317	06.8	··	04.0
10	332	07.0		05.0
11	347	07.2		06.0
12	2	07.4	S 1	06.9
13	17	07.6		07.9
14	32	07.8		08.9
15	47	08.1	··	09.8
16	62	08.3		10.8
17	77	08.5		11.8
18	92	08.7	S 1	12.8
19	107	08.9		13.7
20	122	09.1		14.7
21	137	09.3	··	15.7
22	152	09.5		16.7
23	167	09.8		17.6
27 00	182	10.0	S 1	18.6
01	197	10.2		19.6
02	212	10.4		20.6
03	227	10.6	··	21.5
04	242	10.8		22.5
05	257	11.0		23.5
06	272	11.3	S 1	24.5
07	287	11.5		25.4
08	302	11.7		26.4
09	317	11.9	··	27.4
10	332	12.1		28.4
11	347	12.3		29.3
12	2	12.5	S 1	30.3
13	17	12.7		31.3
14	32	13.0		32.3
15	47	13.2	··	33.2
16	62	13.4		34.2
17	77	13.6		35.2
18	92	13.8	S 1	36.1
19	107	14.0		37.1
20	122	14.2		38.1
21	137	14.4	··	39.1
22	152	14.6		40.0
23	167	14.9		41.0

SUN — G.H.A. and Dec.

d h	G.H.A. ° '	Dec. ° '
28 00	182 15.1	S 1 42.0
01	197 15.3	43.0
02	212 15.5	43.9
03	227 15.7	·· 44.9
04	242 15.9	45.9
05	257 16.1	46.9
06	272 16.3	S 1 47.8
07	287 16.5	48.8
08	302 16.8	49.8
09	317 17.0	·· 50.8
10	332 17.2	51.7
11	347 17.4	52.7
12	2 17.6	S 1 53.7
13	17 17.8	54.6
14	32 18.0	55.6
15	47 18.2	·· 56.6
16	62 18.4	57.6
17	77 18.6	58.5
18	92 18.9	S 1 59.5
19	107 19.1	2 00.5
20	122 19.3	01.5
21	137 19.5	·· 02.4
22	152 19.7	03.4
23	167 19.9	04.4
29 00	182 20.1	S 2 05.3
01	197 20.3	06.3
02	212 20.5	07.3
03	227 20.7	·· 08.3
04	242 20.9	09.2
05	257 21.2	10.2
06	272 21.4	S 2 11.2
07	287 21.6	12.2
08	302 21.8	13.1
09	317 22.0	·· 14.1
10	332 22.2	15.1
11	347 22.4	16.1
12	2 22.6	S 2 17.0
13	17 22.8	18.0
14	32 23.0	19.0
15	47 23.2	·· 19.9
16	62 23.4	20.9
17	77 23.7	21.9
18	92 23.9	S 2 22.9
19	107 24.1	23.8
20	122 24.3	24.8
21	137 24.5	·· 25.8
22	152 24.7	26.7
23	167 24.9	27.7
30 00	182 25.1	S 2 28.7
01	197 25.3	29.7
02	212 25.5	30.6
03	227 25.7	·· 31.6
04	242 25.9	32.6
05	257 26.1	33.6
06	272 26.3	S 2 34.5
07	287 26.5	35.5
08	302 26.8	36.5
09	317 27.0	·· 37.4
10	332 27.2	38.4
11	347 27.4	39.4
12	2 27.6	S 2 40.4
13	17 27.8	41.3
14	32 28.0	42.3
15	47 28.2	·· 43.3
16	62 28.4	44.2
17	77 28.6	45.2
18	92 28.8	S 2 46.2
19	107 29.0	47.2
20	122 29.2	48.1
21	137 29.4	·· 49.1
22	152 29.6	50.1
23	167 29.8	51.0

d h	G.H.A. ° '	Dec. ° '
1 00	182 30.0	S 2 52.0
01	197 30.2	53.0
02	212 30.4	54.0
03	227 30.7	·· 54.9
04	242 30.9	55.9
05	257 31.1	56.9
06	272 31.3	S 2 57.8
07	287 31.5	58.8
08	302 31.7	2 59.8
09	317 31.9	3 00.7
10	332 32.1	01.7
11	347 32.3	02.7
12	2 32.5	S 3 03.7
13	17 32.7	04.6
14	32 32.9	05.6
15	47 33.1	·· 06.6
16	62 33.3	07.5
17	77 33.5	08.5
18	92 33.7	S 3 09.5
19	107 33.9	10.5
20	122 34.1	11.4
21	137 34.3	·· 12.4
22	152 34.5	13.4
23	167 34.7	14.3
2 00	182 34.9	S 3 15.3
01	197 35.1	16.3
02	212 35.3	17.2
03	227 35.5	·· 18.2
04	242 35.7	19.2
05	257 35.9	20.1
06	272 36.1	S 3 21.1
07	287 36.3	22.1
08	302 36.5	23.1
09	317 36.7	·· 24.0
10	332 36.9	25.0
11	347 37.1	26.0
12	2 37.3	S 3 26.9
13	17 37.5	27.9
14	32 37.7	28.9
15	47 37.9	·· 29.8
16	62 38.1	30.8
17	77 38.3	31.8
18	92 38.5	S 3 32.7
19	107 38.7	33.7
20	122 38.9	34.7
21	137 39.1	·· 35.6
22	152 39.3	36.6
23	167 39.5	37.6
3 00	182 39.7	S 3 38.6
01	197 39.9	39.5
02	212 40.1	40.5
03	227 40.3	·· 41.5
04	242 40.5	42.4
05	257 40.7	43.4
06	272 40.9	S 3 44.4
07	287 41.1	45.3
08	302 41.3	46.3
09	317 41.5	·· 47.3
10	332 41.7	48.2
11	347 41.9	49.2
12	2 42.1	S 3 50.2
13	17 42.3	51.1
14	32 42.5	52.1
15	47 42.7	·· 53.1
16	62 42.9	54.0
17	77 43.1	55.0
18	92 43.3	S 3 56.0
19	107 43.5	56.9
20	122 43.7	57.9
21	137 43.9	·· 58.9
22	152 44.1	3 59.8
23	167 44.2	4 00.8

d h	G.H.A. ° '	Dec. ° '
4 00	182 44.4	S 4 01.8
01	197 44.6	02.7
02	212 44.8	03.7
03	227 45.0	·· 04.7
04	242 45.2	05.6
05	257 45.4	06.6
06	272 45.6	S 4 07.6
07	287 45.8	08.5
08	302 46.0	09.5
09	317 46.2	·· 10.5
10	332 46.4	11.4
11	347 46.6	12.4
12	2 46.8	S 4 13.3
13	17 47.0	14.3
14	32 47.2	15.3
15	47 47.4	·· 16.2
16	62 47.5	17.2
17	77 47.7	18.2
18	92 47.9	S 4 19.1
19	107 48.1	20.1
20	122 48.3	21.1
21	137 48.5	·· 22.0
22	152 48.7	23.0
23	167 48.9	24.0
5 00	182 49.1	S 4 24.9
01	197 49.3	25.9
02	212 49.5	26.9
03	227 49.7	·· 27.8
04	242 49.9	28.8
05	257 50.0	29.7
06	272 50.2	S 4 30.7
07	287 50.4	31.7
08	302 50.6	32.6
09	317 50.8	·· 33.6
10	332 51.0	34.6
11	347 51.2	35.5
12	2 51.4	S 4 36.5
13	17 51.6	37.4
14	32 51.8	38.4
15	47 51.9	·· 39.4
16	62 52.1	40.3
17	77 52.3	41.3
18	92 52.5	S 4 42.3
19	107 52.7	43.2
20	122 52.9	44.2
21	137 53.1	·· 45.1
22	152 53.3	46.1
23	167 53.5	47.1
6 00	182 53.6	S 4 48.0
01	197 53.8	49.0
02	212 54.0	50.0
03	227 54.2	·· 50.9
04	242 54.4	51.9
05	257 54.6	52.8
06	272 54.8	S 4 53.8
07	287 55.0	54.8
08	302 55.1	55.7
09	317 55.3	·· 56.7
10	332 55.5	57.6
11	347 55.7	58.6
12	2 55.9	S 4 59.6
13	17 56.1	5 00.5
14	32 56.3	01.5
15	47 56.4	·· 02.4
16	62 56.6	03.4
17	77 56.8	04.4
18	92 57.0	S 5 05.3
19	107 57.2	06.3
20	122 57.4	07.2
21	137 57.6	·· 08.2
22	152 57.7	09.2
23	167 57.9	10.1

d	h	G.H.A.	Dec.
d	h	° ′	° ′
7	00	182 58.1	S 5 11.1
	01	197 58.3	12.0
	02	212 58.5	13.0
	03	227 58.7	· · 14.0
	04	242 58.8	14.9
	05	257 59.0	15.9
	06	272 59.2	S 5 16.8
	07	287 59.4	17.8
	08	302 59.6	18.7
	09	317 59.8	· · 19.7
	10	332 59.9	20.7
	11	348 00.1	21.6
	12	3 00.3	S 5 22.6
	13	18 00.5	23.5
	14	33 00.7	24.5
	15	48 00.8	· · 25.4
	16	63 01.0	26.4
	17	78 01.2	27.4
	18	93 01.4	S 5 28.3
	19	108 01.6	29.3
	20	123 01.7	30.2
	21	138 01.9	· · 31.2
	22	153 02.1	32.1
	23	168 02.3	33.1
8	00	183 02.5	S 5 34.1
	01	198 02.6	35.0
	02	213 02.8	36.0
	03	228 03.0	· · 36.9
	04	243 03.2	37.9
	05	258 03.4	38.8
	06	273 03.5	S 5 39.8
	07	288 03.7	40.7
	08	303 03.9	41.7
	09	318 04.1	· · 42.7
	10	333 04.2	43.6
	11	348 04.4	44.6
	12	3 04.6	S 5 45.5
	13	18 04.8	46.5
	14	33 05.0	47.4
	15	48 05.1	· · 48.4
	16	63 05.3	49.3
	17	78 05.5	50.3
	18	93 05.7	S 5 51.2
	19	108 05.8	52.2
	20	123 06.0	53.1
	21	138 06.2	· · 54.1
	22	153 06.4	55.1
	23	168 06.5	56.0
9	00	183 06.7	S 5 57.0
	01	198 06.9	57.9
	02	213 07.1	58.9
	03	228 07.2	5 59.8
	04	243 07.4	6 00.8
	05	258 07.6	01.7
	06	273 07.8	S 6 02.7
	07	288 07.9	03.6
	08	303 08.1	04.6
	09	318 08.3	· · 05.5
	10	333 08.5	06.5
	11	348 08.6	07.4
	12	3 08.8	S 6 08.4
	13	18 09.0	09.3
	14	33 09.2	10.3
	15	48 09.3	· · 11.2
	16	63 09.5	12.2
	17	78 09.7	13.1
	18	93 09.8	S 6 14.1
	19	108 10.0	15.0
	20	123 10.2	16.0
	21	138 10.4	· · 16.9
	22	153 10.5	17.9
	23	168 10.7	18.8

d	h	G.H.A.	Dec.
d	h	° ′	° ′
10	00	183 10.9	S 6 19.8
	01	198 11.0	20.7
	02	213 11.2	21.7
	03	228 11.4	· · 22.6
	04	243 11.5	23.6
	05	258 11.7	24.5
	06	273 11.9	S 6 25.5
	07	288 12.1	26.4
	08	303 12.2	27.4
	09	318 12.4	· · 28.3
	10	333 12.6	29.3
	11	348 12.7	30.2
	12	3 12.9	S 6 31.2
	13	18 13.1	32.1
	14	33 13.2	33.1
	15	48 13.4	· · 34.0
	16	63 13.6	34.9
	17	78 13.7	35.9
	18	93 13.9	S 6 36.8
	19	108 14.1	37.8
	20	123 14.2	38.7
	21	138 14.4	· · 39.7
	22	153 14.6	40.6
	23	168 14.7	41.6
11	00	183 14.9	S 6 42.5
	01	198 15.1	43.5
	02	213 15.2	44.4
	03	228 15.4	· · 45.4
	04	243 15.6	46.3
	05	258 15.7	47.3
	06	273 15.9	S 6 48.2
	07	288 16.1	49.1
	08	303 16.2	50.1
	09	318 16.4	· · 51.0
	10	333 16.5	52.0
	11	348 16.7	52.9
	12	3 16.9	S 6 53.9
	13	18 17.0	54.8
	14	33 17.2	55.8
	15	48 17.4	· · 56.7
	16	63 17.5	57.6
	17	78 17.7	58.6
	18	93 17.8	S 6 59.5
	19	108 18.0	7 00.5
	20	123 18.2	01.4
	21	138 18.3	· · 02.4
	22	153 18.5	03.3
	23	168 18.6	04.2
12	00	183 18.8	S 7 05.2
	01	198 19.0	06.1
	02	213 19.1	07.1
	03	228 19.3	· · 08.0
	04	243 19.4	09.0
	05	258 19.6	09.9
	06	273 19.8	S 7 10.8
	07	288 19.9	11.8
	08	303 20.1	12.7
	09	318 20.2	· · 13.7
	10	333 20.4	14.6
	11	348 20.6	15.5
	12	3 20.7	S 7 16.5
	13	18 20.9	17.4
	14	33 21.0	18.4
	15	48 21.2	· · 19.3
	16	63 21.3	20.2
	17	78 21.5	21.2
	18	93 21.7	S 7 22.1
	19	108 21.8	23.1
	20	123 22.0	24.0
	21	138 22.1	· · 24.9
	22	153 22.3	25.9
	23	168 22.4	26.8

d	h	G.H.A.	Dec.
d	h	° ′	° ′
13	00	183 22.6	S 7 27.8
	01	198 22.7	28.7
	02	213 22.9	29.6
	03	228 23.0	· · 30.6
	04	243 23.2	31.5
	05	258 23.4	32.4
	06	273 23.5	S 7 33.4
	07	288 23.7	34.3
	08	303 23.8	35.3
	09	318 24.0	· · 36.2
	10	333 24.1	37.1
	11	348 24.3	38.1
	12	3 24.4	S 7 39.0
	13	18 24.6	39.9
	14	33 24.7	40.9
	15	48 24.9	· · 41.8
	16	63 25.0	42.7
	17	78 25.2	43.7
	18	93 25.3	S 7 44.6
	19	108 25.5	45.6
	20	123 25.6	46.5
	21	138 25.8	· · 47.4
	22	153 25.9	48.4
	23	168 26.1	49.3
14	00	183 26.2	S 7 50.2
	01	198 26.4	51.2
	02	213 26.5	52.1
	03	228 26.7	· · 53.0
	04	243 26.8	54.0
	05	258 27.0	54.9
	06	273 27.1	S 7 55.8
	07	288 27.3	56.8
	08	303 27.4	57.7
	09	318 27.6	· · 58.6
	10	333 27.7	7 59.6
	11	348 27.9	8 00.5
	12	3 28.0	S 8 01.4
	13	18 28.2	02.4
	14	33 28.3	03.3
	15	48 28.5	· · 04.2
	16	63 28.6	05.1
	17	78 28.7	06.1
	18	93 28.9	S 8 07.0
	19	108 29.0	07.9
	20	123 29.2	08.9
	21	138 29.3	· · 09.8
	22	153 29.5	10.7
	23	168 29.6	11.7
15	00	183 29.8	S 8 12.6
	01	198 29.9	13.5
	02	213 30.0	14.4
	03	228 30.2	· · 15.4
	04	243 30.3	16.3
	05	258 30.5	17.2
	06	273 30.6	S 8 18.2
	07	288 30.8	19.1
	08	303 30.9	20.0
	09	318 31.0	· · 20.9
	10	333 31.2	21.9
	11	348 31.3	22.8
	12	3 31.5	S 8 23.7
	13	18 31.6	24.7
	14	33 31.7	25.6
	15	48 31.9	· · 26.5
	16	63 32.0	27.4
	17	78 32.2	28.4
	18	93 32.3	S 8 29.3
	19	108 32.4	30.2
	20	123 32.6	31.1
	21	138 32.7	· · 32.1
	22	153 32.9	33.0
	23	168 33.0	33.9

16	G.H.A.		Dec.
d h	° ′		° ′
00	183 33.1	S 8	34.8
01	198 33.3		35.8
02	213 33.4		36.7
03	228 33.5	··	37.6
04	243 33.7		38.5
05	258 33.8		39.5
06	273 34.0	S 8	40.4
07	288 34.1		41.3
08	303 34.2		42.2
09	318 34.4	··	43.1
10	333 34.5		44.1
11	348 34.6		45.0
12	3 34.8	S 8	45.9
13	18 34.9		46.8
14	33 35.0		47.8
15	48 35.2	··	48.7
16	63 35.3		49.6
17	78 35.4		50.5
18	93 35.6	S 8	51.4
19	108 35.7		52.4
20	123 35.8		53.3
21	138 36.0	··	54.2
22	153 36.1		55.1
23	168 36.2		56.0
17 00	183 36.4	S 8	57.0
01	198 36.5		57.9
02	213 36.6		58.8
03	228 36.8	8	59.7
04	243 36.9	9	00.6
05	258 37.0		01.6
06	273 37.1	S 9	02.5
07	288 37.3		03.4
08	303 37.4		04.3
09	318 37.5	··	05.2
10	333 37.7		06.1
11	348 37.8		07.1
12	3 37.9	S 9	08.0
13	18 38.1		08.9
14	33 38.2		09.8
15	48 38.3	··	10.7
16	63 38.4		11.6
17	78 38.6		12.6
18	93 38.7	S 9	13.5
19	108 38.8		14.4
20	123 38.9		15.3
21	138 39.1	··	16.2
22	153 39.2		17.1
23	168 39.3		18.0
18 00	183 39.4	S 9	19.0
01	198 39.6		19.9
02	213 39.7		20.8
03	228 39.8	··	21.7
04	243 39.9		22.6
05	258 40.1		23.5
06	273 40.2	S 9	24.4
07	288 40.3		25.4
08	303 40.4		26.3
09	318 40.6	··	27.2
10	333 40.7		28.1
11	348 40.8		29.0
12	3 40.9	S 9	29.9
13	18 41.1		30.8
14	33 41.2		31.7
15	48 41.3	··	32.6
16	63 41.4		33.6
17	78 41.5		34.5
18	93 41.7	S 9	35.4
19	108 41.8		36.3
20	123 41.9		37.2
21	138 42.0	··	38.1
22	153 42.1		39.0
23	168 42.3		39.9

19	G.H.A.		Dec.
d h	° ′		° ′
00	183 42.4	S 9	40.8
01	198 42.5		41.7
02	213 42.6		42.6
03	228 42.7	··	43.5
04	243 42.9		44.5
05	258 43.0		45.4
06	273 43.1	S 9	46.3
07	288 43.2		47.2
08	303 43.3		48.1
09	318 43.4	··	49.0
10	333 43.6		49.9
11	348 43.7		50.8
12	3 43.8	S 9	51.7
13	18 43.9		52.6
14	33 44.0		53.5
15	48 44.1	··	54.4
16	63 44.2		55.3
17	78 44.4		56.2
18	93 44.5	S 9	57.1
19	108 44.6		58.0
20	123 44.7		58.9
21	138 44.8	9	59.8
22	153 44.9	10	00.7
23	168 45.0		01.6
20 00	183 45.2	S10	02.5
01	198 45.3		03.5
02	213 45.4		04.4
03	228 45.5	··	05.3
04	243 45.6		06.2
05	258 45.7		07.1
06	273 45.8	S10	08.0
07	288 45.9		08.9
08	303 46.0		09.8
09	318 46.2	··	10.7
10	333 46.3		11.6
11	348 46.4		12.5
12	3 46.5	S10	13.4
13	18 46.6		14.3
14	33 46.7		15.2
15	48 46.8	··	16.1
16	63 46.9		17.0
17	78 47.0		17.8
18	93 47.1	S10	18.7
19	108 47.2		19.6
20	123 47.3		20.5
21	138 47.5	··	21.4
22	153 47.6		22.3
23	168 47.7		23.2
21 00	183 47.8	S10	24.1
01	198 47.9		25.0
02	213 48.0		25.9
03	228 48.1	··	26.8
04	243 48.2		27.7
05	258 48.3		28.6
06	273 48.4	S10	29.5
07	288 48.5		30.4
08	303 48.6		31.3
09	318 48.7	··	32.2
10	333 48.8		33.1
11	348 48.9		34.0
12	3 49.0	S10	34.9
13	18 49.1		35.8
14	33 49.2		36.6
15	48 49.3	··	37.5
16	63 49.4		38.4
17	78 49.5		39.3
18	93 49.6	S10	40.2
19	108 49.7		41.1
20	123 49.8		42.0
21	138 49.9	··	42.9
22	153 50.0		43.8
23	168 50.1		44.7

22	G.H.A.		Dec.
d h	° ′		° ′
00	183 50.2	S10	45.6
01	198 50.3		46.4
02	213 50.4		47.3
03	228 50.5	··	48.2
04	243 50.6		49.1
05	258 50.7		50.0
06	273 50.8	S10	50.9
07	288 50.9		51.8
08	303 51.0		52.7
09	318 51.1	··	53.5
10	333 51.2		54.4
11	348 51.3		55.3
12	3 51.4	S10	56.2
13	18 51.5		57.1
14	33 51.6		58.0
15	48 51.7	··	58.9
16	63 51.8	10	59.7
17	78 51.9	11	00.6
18	93 52.0	S11	01.5
19	108 52.0		02.4
20	123 52.1		03.3
21	138 52.2	··	04.2
22	153 52.3		05.0
23	168 52.4		05.9
23 00	183 52.5	S11	06.8
01	198 52.6		07.7
02	213 52.7		08.6
03	228 52.8	··	09.5
04	243 52.9		10.3
05	258 53.0		11.2
06	273 53.1	S11	12.1
07	288 53.1		13.0
08	303 53.2		13.9
09	318 53.3	··	14.7
10	333 53.4		15.6
11	348 53.5		16.5
12	3 53.6	S11	17.4
13	18 53.7		18.3
14	33 53.8		19.1
15	48 53.9	··	20.0
16	63 53.9		20.9
17	78 54.0		21.8
18	93 54.1	S11	22.7
19	108 54.2		23.5
20	123 54.3		24.4
21	138 54.4	··	25.3
22	153 54.5		26.2
23	168 54.6		27.0
24 00	183 54.6	S11	27.9
01	198 54.7		28.8
02	213 54.8		29.7
03	228 54.9	··	30.5
04	243 55.0		31.4
05	258 55.1		32.3
06	273 55.1	S11	33.2
07	288 55.2		34.0
08	303 55.3		34.9
09	318 55.4	··	35.8
10	333 55.5		36.6
11	348 55.6		37.5
12	3 55.6	S11	38.4
13	18 55.7		39.3
14	33 55.8		40.1
15	48 55.9	··	41.0
16	63 56.0		41.9
17	78 56.0		42.7
18	93 56.1	S11	43.6
19	108 56.2		44.5
20	123 56.3		45.4
21	138 56.4	··	46.2
22	153 56.4		47.1
23	168 56.5		48.0

d	h	G.H.A.		Dec.
		° ′		° ′
25	00	183 56.6	S11	48.8
	01	198 56.7		49.7
	02	213 56.7		50.6
	03	228 56.8	··	51.4
	04	243 56.9		52.3
	05	258 57.0		53.2
	06	273 57.0	S11	54.0
	07	288 57.1		54.9
	08	303 57.2		55.8
	09	318 57.3	··	56.6
	10	333 57.4		57.5
	11	348 57.4		58.4
	12	3 57.5	S11	59.2
	13	18 57.6	12	00.1
	14	33 57.6		01.0
	15	48 57.7	··	01.8
	16	63 57.8		02.7
	17	78 57.9		03.5
	18	93 57.9	S12	04.4
	19	108 58.0		05.3
	20	123 58.1		06.1
	21	138 58.2	··	07.0
	22	153 58.2		07.9
	23	168 58.3		08.7
26	00	183 58.4	S12	09.6
	01	198 58.4		10.4
	02	213 58.5		11.3
	03	228 58.6	··	12.2
	04	243 58.6		13.0
	05	258 58.7		13.9
	06	273 58.8	S12	14.7
	07	288 58.9		15.6
	08	303 58.9		16.4
	09	318 59.0	··	17.3
	10	333 59.1		18.2
	11	348 59.1		19.0
	12	3 59.2	S12	19.9
	13	18 59.3		20.7
	14	33 59.3		21.6
	15	48 59.4	··	22.4
	16	63 59.5		23.3
	17	78 59.5		24.2
	18	93 59.6	S12	25.0
	19	108 59.7		25.9
	20	123 59.7		26.7
	21	138 59.8	··	27.6
	22	153 59.8		28.4
	23	168 59.9		29.3
27	00	184 00.0	S12	30.1
	01	199 00.0		31.0
	02	214 00.1		31.8
	03	229 00.2	··	32.7
	04	244 00.2		33.5
	05	259 00.3		34.4
	06	274 00.3	S12	35.2
	07	289 00.4		36.1
	08	304 00.5		36.9
	09	319 00.5	··	37.8
	10	334 00.6		38.6
	11	349 00.7		39.5
	12	4 00.7	S12	40.3
	13	19 00.8		41.2
	14	34 00.8		42.0
	15	49 00.9	··	42.9
	16	64 00.9		43.7
	17	79 01.0		44.6
	18	94 01.1	S12	45.4
	19	109 01.1		46.3
	20	124 01.2		47.1
	21	139 01.2	··	48.0
	22	154 01.3		48.8
	23	169 01.3		49.6

d	h	G.H.A.		Dec.
		° ′		° ′
28	00	184 01.4	S12	50.5
	01	199 01.5		51.3
	02	214 01.5		52.2
	03	229 01.6	··	53.0
	04	244 01.6		53.9
	05	259 01.7		54.7
	06	274 01.7	S12	55.6
	07	289 01.8		56.4
	08	304 01.8		57.2
	09	319 01.9	··	58.1
	10	334 01.9		58.9
	11	349 02.0	12	59.8
	12	4 02.0	S13	00.6
	13	19 02.1		01.4
	14	34 02.1		02.3
	15	49 02.2	··	03.1
	16	64 02.3		04.0
	17	79 02.3		04.8
	18	94 02.4	S13	05.6
	19	109 02.4		06.5
	20	124 02.5		07.3
	21	139 02.5	··	08.1
	22	154 02.5		09.0
	23	169 02.6		09.8
29	00	184 02.6	S13	10.7
	01	199 02.7		11.5
	02	214 02.7		12.3
	03	229 02.8	··	13.2
	04	244 02.8		14.0
	05	259 02.9		14.8
	06	274 02.9	S13	15.7
	07	289 03.0		16.5
	08	304 03.0		17.3
	09	319 03.1	··	18.2
	10	334 03.1		19.0
	11	349 03.2		19.8
	12	4 03.2	S13	20.7
	13	19 03.2		21.5
	14	34 03.3		22.3
	15	49 03.3	··	23.1
	16	64 03.4		24.0
	17	79 03.4		24.8
	18	94 03.5	S13	25.6
	19	109 03.5		26.5
	20	124 03.5		27.3
	21	139 03.6	··	28.1
	22	154 03.6		29.0
	23	169 03.7		29.8
30	00	184 03.7	S13	30.6
	01	199 03.7		31.4
	02	214 03.8		32.3
	03	229 03.8	··	33.1
	04	244 03.9		33.9
	05	259 03.9		34.7
	06	274 03.9	S13	35.6
	07	289 04.0		36.4
	08	304 04.0		37.2
	09	319 04.1	··	38.0
	10	334 04.1		38.9
	11	349 04.1		39.7
	12	4 04.2	S13	40.5
	13	19 04.2		41.3
	14	34 04.2		42.2
	15	49 04.3	··	43.0
	16	64 04.3		43.8
	17	79 04.3		44.6
	18	94 04.4	S13	45.4
	19	109 04.4		46.3
	20	124 04.4		47.1
	21	139 04.5	··	47.9
	22	154 04.5		48.7
	23	169 04.5		49.5

d	h	G.H.A.		Dec.
		° ′		° ′
31	00	184 04.6	S13	50.3
	01	199 04.6		51.2
	02	214 04.6		52.0
	03	229 04.7	··	52.8
	04	244 04.7		53.6
	05	259 04.7		54.4
	06	274 04.8	S13	55.3
	07	289 04.8		56.1
	08	304 04.8		56.9
	09	319 04.9	··	57.7
	10	334 04.9		58.5
	11	349 04.9	13	59.3
	12	4 04.9	S14	00.1
	13	19 05.0		01.0
	14	34 05.0		01.8
	15	49 05.0	··	02.6
	16	64 05.1		03.4
	17	79 05.1		04.2
	18	94 05.1	S14	05.0
	19	109 05.1		05.8
	20	124 05.2		06.6
	21	139 05.2	··	07.4
	22	154 05.2		08.3
	23	169 05.2		09.1
NOVEMBER 1	00	184 05.3	S14	09.9
	01	199 05.3		10.7
	02	214 05.3		11.5
	03	229 05.3	··	12.3
	04	244 05.4		13.1
	05	259 05.4		13.9
	06	274 05.4	S14	14.7
	07	289 05.4		15.5
	08	304 05.4		16.3
	09	319 05.5	··	17.1
	10	334 05.5		17.9
	11	349 05.5		18.7
	12	4 05.5	S14	19.5
	13	19 05.6		20.4
	14	34 05.6		21.2
	15	49 05.6	··	22.0
	16	64 05.6		22.8
	17	79 05.6		23.6
	18	94 05.6	S14	24.4
	19	109 05.7		25.2
	20	124 05.7		26.0
	21	139 05.7	··	26.8
	22	154 05.7		27.6
	23	169 05.7		28.4
2	00	184 05.8	S14	29.2
	01	199 05.8		30.0
	02	214 05.8		30.8
	03	229 05.8	··	31.6
	04	244 05.8		32.4
	05	259 05.8		33.2
	06	274 05.8	S14	34.0
	07	289 05.9		34.7
	08	304 05.9		35.5
	09	319 05.9	··	36.3
	10	334 05.9		37.1
	11	349 05.9		37.9
	12	4 05.9	S14	38.7
	13	19 05.9		39.5
	14	34 05.9		40.3
	15	49 06.0	··	41.1
	16	64 06.0		41.9
	17	79 06.0		42.7
	18	94 06.0	S14	43.5
	19	109 06.0		44.3
	20	124 06.0		45.1
	21	139 06.0	··	45.9
	22	154 06.0		46.6
	23	169 06.0		47.4

d h	G.H.A.	Dec
3 00	184 06.0	S14 48.2
01	199 06.1	49.0
02	214 06.1	49.8
03	229 06.1	·· 50.6
04	244 06.1	51.4
05	259 06.1	52.2
06	274 06.1	S14 53.0
07	289 06.1	53.7
08	304 06.1	54.5
09	319 06.1	·· 55.3
10	334 06.1	56.1
11	349 06.1	56.9
12	4 06.1	S14 57.7
13	19 06.1	58.5
14	34 06.1	14 59.2
15	49 06.1	15 00.0
16	64 06.1	00.8
17	79 06.1	01.6
18	94 06.1	S15 02.4
19	109 06.1	03.1
20	124 06.1	03.9
21	139 06.1	·· 04.7
22	154 06.1	05.5
23	169 06.1	06.3
4 00	184 06.1	S15 07.0
01	199 06.1	07.8
02	214 06.1	08.6
03	229 06.1	·· 09.4
04	244 06.1	10.2
05	259 06.1	10.9
06	274 06.1	S15 11.7
07	289 06.1	12.5
08	304 06.1	13.3
09	319 06.1	·· 14.0
10	334 06.1	14.8
11	349 06.1	15.6
12	4 06.1	S15 16.4
13	19 06.1	17.1
14	34 06.1	17.9
15	49 06.1	·· 18.7
16	64 06.1	19.5
17	79 06.1	20.2
18	94 06.1	S15 21.0
19	109 06.1	21.8
20	124 06.1	22.5
21	139 06.0	·· 23.3
22	154 06.0	24.1
23	169 06.0	24.9
5 00	184 06.0	S15 25.6
01	199 06.0	26.4
02	214 06.0	27.2
03	229 06.0	·· 27.9
04	244 06.0	28.7
05	259 06.0	29.5
06	274 06.0	S15 30.2
07	289 05.9	31.0
08	304 05.9	31.8
09	319 05.9	·· 32.5
10	334 05.9	33.3
11	349 05.9	34.1
12	4 05.9	S15 34.8
13	19 05.9	35.6
14	34 05.9	36.3
15	49 05.8	·· 37.1
16	64 05.8	37.9
17	79 05.8	38.6
18	94 05.8	S15 39.4
19	109 05.8	40.1
20	124 05.8	40.9
21	139 05.7	·· 41.7
22	154 05.7	42.4
23	169 05.7	43.2

d h	G.H.A.	Dec
6 00	184 05.7	S15 43.9
01	199 05.7	44.7
02	214 05.7	45.5
03	229 05.6	·· 46.2
04	244 05.6	47.0
05	259 05.6	47.7
06	274 05.6	S15 48.5
07	289 05.6	49.2
08	304 05.5	50.0
09	319 05.5	·· 50.7
10	334 05.5	51.5
11	349 05.5	52.3
12	4 05.5	S15 53.0
13	19 05.4	53.8
14	34 05.4	54.5
15	49 05.4	·· 55.3
16	64 05.4	56.0
17	79 05.3	56.8
18	94 05.3	S15 57.5
19	109 05.3	58.3
20	124 05.3	59.0
21	139 05.2	15 59.8
22	154 05.2	16 00.5
23	169 05.2	01.3
7 00	184 05.2	S16 02.0
01	199 05.1	02.7
02	214 05.1	03.5
03	229 05.1	·· 04.2
04	244 05.1	05.0
05	259 05.0	05.7
06	274 05.0	S16 06.5
07	289 05.0	07.2
08	304 04.9	08.0
09	319 04.9	·· 08.7
10	334 04.9	09.4
11	349 04.8	10.2
12	4 04.8	S16 10.9
13	19 04.8	11.7
14	34 04.8	12.4
15	49 04.7	·· 13.2
16	64 04.7	13.9
17	79 04.7	14.6
18	94 04.6	S16 15.4
19	109 04.6	16.1
20	124 04.6	16.8
21	139 04.5	·· 17.6
22	154 04.5	18.3
23	169 04.5	19.1
8 00	184 04.4	S16 19.8
01	199 04.4	20.5
02	214 04.3	21.3
03	229 04.3	·· 22.0
04	244 04.3	22.7
05	259 04.2	23.5
06	274 04.2	S16 24.2
07	289 04.2	24.9
08	304 04.1	25.7
09	319 04.1	·· 26.4
10	334 04.0	27.1
11	349 04.0	27.9
12	4 04.0	S16 28.6
13	19 03.9	29.3
14	34 03.9	30.0
15	49 03.8	·· 30.8
16	64 03.8	31.5
17	79 03.8	32.2
18	94 03.7	S16 33.0
19	109 03.7	33.7
20	124 03.6	34.4
21	139 03.6	·· 35.1
22	154 03.6	35.9
23	169 03.5	36.6

d h	G.H.A.	Dec
9 00	184 03.5	S16 37.3
01	199 03.4	38.0
02	214 03.4	38.8
03	229 03.3	·· 39.5
04	244 03.3	40.2
05	259 03.2	40.9
06	274 03.2	S16 41.7
07	289 03.1	42.4
08	304 03.1	43.1
09	319 03.0	·· 43.8
10	334 03.0	44.5
11	349 03.0	45.3
12	4 02.9	S16 46.0
13	19 02.9	46.7
14	34 02.8	47.4
15	49 02.8	·· 48.1
16	64 02.7	48.8
17	79 02.7	49.6
18	94 02.6	S16 50.3
19	109 02.6	51.0
20	124 02.5	51.7
21	139 02.4	·· 52.4
22	154 02.4	53.1
23	169 02.3	53.8
10 00	184 02.3	S16 54.6
01	199 02.2	55.3
02	214 02.2	56.0
03	229 02.1	·· 56.7
04	244 02.1	57.4
05	259 02.0	58.1
06	274 02.0	S16 58.8
07	289 01.9	16 59.5
08	304 01.9	17 00.2
09	319 01.8	·· 00.9
10	334 01.7	01.7
11	349 01.7	02.4
12	4 01.6	S17 03.1
13	19 01.6	03.8
14	34 01.5	04.5
15	49 01.4	·· 05.2
16	64 01.4	05.9
17	79 01.3	06.6
18	94 01.3	S17 07.3
19	109 01.2	08.0
20	124 01.1	08.7
21	139 01.1	·· 09.4
22	154 01.0	10.1
23	169 01.0	10.8
11 00	184 00.9	S17 11.5
01	199 00.8	12.2
02	214 00.8	12.9
03	229 00.7	·· 13.6
04	244 00.7	14.3
05	259 00.6	15.0
06	274 00.5	S17 15.7
07	289 00.5	16.4
08	304 00.4	17.1
09	319 00.3	·· 17.8
10	334 00.3	18.5
11	349 00.2	19.2
12	4 00.1	S17 19.9
13	19 00.1	20.6
14	34 00.0	21.3
15	48 59.9	·· 22.0
16	63 59.9	22.6
17	78 59.8	23.3
18	93 59.7	S17 24.0
19	108 59.6	24.7
20	123 59.6	25.4
21	138 59.5	·· 26.1
22	153 59.4	26.8
23	168 59.4	27.5

G.H.A. and Dec. tables

Day 12

d h	G.H.A.	Dec.
12 00	183 59.3	S17 28.2
01	198 59.2	28.9
02	213 59.2	29.5
03	228 59.1 ··	30.2
04	243 59.0	30.9
05	258 58.9	31.6
06	273 58.9	S17 32.3
07	288 58.8	33.0
08	303 58.7	33.6
09	318 58.6 ··	34.3
10	333 58.6	35.0
11	348 58.5	35.7
12	3 58.4	S17 36.4
13	18 58.3	37.1
14	33 58.3	37.7
15	48 58.2 ··	38.4
16	63 58.1	39.1
17	78 58.0	39.8
18	93 58.0	S17 40.5
19	108 57.9	41.1
20	123 57.8	41.8
21	138 57.7 ··	42.5
22	153 57.6	43.2
23	168 57.6	43.8

Day 13

d h	G.H.A.	Dec.
13 00	183 57.5	S17 44.5
01	198 57.4	45.2
02	213 57.3	45.9
03	228 57.2 ··	46.5
04	243 57.2	47.2
05	258 57.1	47.9
06	273 57.0	S17 48.6
07	288 56.9	49.2
08	303 56.8	49.9
09	318 56.7 ··	50.6
10	333 56.7	51.2
11	348 56.6	51.9
12	3 56.5	S17 52.6
13	18 56.4	53.3
14	33 56.3	53.9
15	48 56.2 ··	54.6
16	63 56.1	55.3
17	78 56.1	55.9
18	93 56.0	S17 56.6
19	108 55.9	57.3
20	123 55.8	57.9
21	138 55.7 ··	58.6
22	153 55.6	59.2
23	168 55.5	17 59.9

Day 14

d h	G.H.A.	Dec.
14 00	183 55.4	S18 00.6
01	198 55.3	01.2
02	213 55.3	01.9
03	228 55.2 ··	02.6
04	243 55.1	03.2
05	258 55.0	03.9
06	273 54.9	S18 04.5
07	288 54.8	05.2
08	303 54.7	05.9
09	318 54.6 ··	06.5
10	333 54.5	07.2
11	348 54.4	07.8
12	3 54.3	S18 08.5
13	18 54.2	09.1
14	33 54.1	09.8
15	48 54.1 ··	10.4
16	63 54.0	11.1
17	78 53.9	11.7
18	93 53.8	S18 12.4
19	108 53.7	13.1
20	123 53.6	13.7
21	138 53.5 ··	14.4
22	153 53.4	15.0
23	168 53.3	15.7

Day 15

d h	G.H.A.	Dec.
15 00	183 53.2	S18 16.3
01	198 53.1	17.0
02	213 53.0	17.6
03	228 52.9 ··	18.3
04	243 52.8	18.9
05	258 52.7	19.5
06	273 52.6	S18 20.2
07	288 52.5	20.8
08	303 52.4	21.5
09	318 52.3 ··	22.1
10	333 52.2	22.8
11	348 52.1	23.4
12	3 52.0	S18 24.1
13	18 51.9	24.7
14	33 51.8	25.3
15	48 51.7 ··	26.0
16	63 51.6	26.6
17	78 51.5	27.3
18	93 51.3	S18 27.9
19	108 51.2	28.5
20	123 51.1	29.2
21	138 51.0 ··	29.8
22	153 50.9	30.4
23	168 50.8	31.1

Day 16

d h	G.H.A.	Dec.
16 00	183 50.7	S18 31.7
01	198 50.6	32.4
02	213 50.5	33.0
03	228 50.4 ··	33.6
04	243 50.3	34.3
05	258 50.2	34.9
06	273 50.1	S18 35.5
07	288 49.9	36.2
08	303 49.8	36.8
09	318 49.7 ··	37.4
10	333 49.6	38.0
11	348 49.5	38.7
12	3 49.4	S18 39.3
13	18 49.3	39.9
14	33 49.2	40.6
15	48 49.1 ··	41.2
16	63 48.9	41.8
17	78 48.8	42.4
18	93 48.7	S18 43.1
19	108 48.6	43.7
20	123 48.5	44.3
21	138 48.4 ··	44.9
22	153 48.3	45.6
23	168 48.1	46.2

Day 17

d h	G.H.A.	Dec.
17 00	183 48.0	S18 46.8
01	198 47.9	47.4
02	213 47.8	48.1
03	228 47.7 ··	48.7
04	243 47.6	49.3
05	258 47.4	49.9
06	273 47.3	S18 50.5
07	288 47.2	51.2
08	303 47.1	51.8
09	318 47.0 ··	52.4
10	333 46.8	53.0
11	348 46.7	53.6
12	3 46.6	S18 54.2
13	18 46.5	54.8
14	33 46.4	55.5
15	48 46.2 ··	56.1
16	63 46.1	56.7
17	78 46.0	57.3
18	93 45.9	S18 57.9
19	108 45.8	58.5
20	123 45.6	59.1
21	138 45.5	18 59.7
22	153 45.4	19 00.4
23	168 45.3	01.0

Day 18

d h	G.H.A.	Dec.
18 00	183 45.1	S19 01.6
01	198 45.0	02.2
02	213 44.9	02.8
03	228 44.8 ··	03.4
04	243 44.6	04.0
05	258 44.5	04.6
06	273 44.4	S19 05.2
07	288 44.2	05.8
08	303 44.1	06.4
09	318 44.0 ··	07.0
10	333 43.9	07.6
11	348 43.7	08.2
12	3 43.6	S19 08.8
13	18 43.5	09.4
14	33 43.3	10.0
15	48 43.2 ··	10.6
16	63 43.1	11.2
17	78 43.0	11.8
18	93 42.8	S19 12.4
19	108 42.7	13.0
20	123 42.6	13.6
21	138 42.4 ··	14.2
22	153 42.3	14.8
23	168 42.2	15.4

Day 19

d h	G.H.A.	Dec.
19 00	183 42.0	S19 16.0
01	198 41.9	16.6
02	213 41.8	17.2
03	228 41.6 ··	17.8
04	243 41.5	18.4
05	258 41.4	18.9
06	273 41.2	S19 19.5
07	288 41.1	20.1
08	303 40.9	20.7
09	318 40.8 ··	21.3
10	333 40.7	21.9
11	348 40.5	22.5
12	3 40.4	S19 23.1
13	18 40.3	23.7
14	33 40.1	24.2
15	48 40.0 ··	24.8
16	63 39.8	25.4
17	78 39.7	26.0
18	93 39.6	S19 26.6
19	108 39.4	27.2
20	123 39.3	27.7
21	138 39.1 ··	28.3
22	153 39.0	28.9
23	168 38.9	29.5

Day 20

d h	G.H.A.	Dec.
20 00	183 38.7	S19 30.1
01	198 38.6	30.6
02	213 38.4	31.2
03	228 38.3 ··	31.8
04	243 38.1	32.4
05	258 38.0	32.9
06	273 37.9	S19 33.5
07	288 37.7	34.1
08	303 37.6	34.7
09	318 37.4 ··	35.2
10	333 37.2	35.8
11	348 37.1	36.4
12	3 37.0	S19 37.0
13	18 36.8	37.5
14	33 36.7	38.1
15	48 36.5 ··	38.7
16	63 36.4	39.2
17	78 36.2	39.8
18	93 36.1	S19 40.4
19	108 35.9	40.9
20	123 35.8	41.5
21	138 35.6 ··	42.1
22	153 35.5	42.6
23	168 35.3	43.2

		G.H.A.	Dec.
d	h	° ′	° ′
21	00	183 35.2	S19 43.8
	01	198 35.0	44.3
	02	213 34.9	44.9
	03	228 34.7 ··	45.5
	04	243 34.6	46.0
	05	258 34.4	46.6
	06	273 34.3	S19 47.1
	07	288 34.1	47.7
	08	303 34.0	48.3
	09	318 33.8 ··	48.8
	10	333 33.7	49.4
	11	348 33.5	49.9
	12	3 33.4	S19 50.5
	13	18 33.2	51.1
	14	33 33.1	51.6
	15	48 32.9 ··	52.2
	16	63 32.7	52.7
	17	78 32.6	53.3
	18	93 32.4	S19 53.8
	19	108 32.3	54.4
	20	123 32.1	54.9
	21	138 32.0 ··	55.5
	22	153 31.8	56.0
	23	168 31.6	56.6
22	00	183 31.5	S19 57.1
	01	198 31.3	57.7
	02	213 31.2	58.2
	03	228 31.0 ··	58.8
	04	243 30.8	59.3
	05	258 30.7	19 59.9
	06	273 30.5	S20 00.4
	07	288 30.4	01.0
	08	303 30.2	01.5
	09	318 30.0 ··	02.1
	10	333 29.9	02.6
	11	348 29.7	03.1
	12	3 29.5	S20 03.7
	13	18 29.4	04.2
	14	33 29.2	04.8
	15	48 29.1 ··	05.3
	16	63 28.9	05.8
	17	78 28.7	06.4
	18	93 28.6	S20 06.9
	19	108 28.4	07.5
	20	123 28.2	08.0
	21	138 28.1 ··	08.5
	22	153 27.9	09.1
	23	168 27.7	09.6
23	00	183 27.6	S20 10.1
	01	198 27.4	10.7
	02	213 27.2	11.2
	03	228 27.1 ··	11.7
	04	243 26.9	12.3
	05	258 26.7	12.8
	06	273 26.6	S20 13.3
	07	288 26.4	13.9
	08	303 26.2	14.4
	09	318 26.0 ··	14.9
	10	333 25.9	15.4
	11	348 25.7	16.0
	12	3 25.5	S20 16.5
	13	18 25.4	17.0
	14	33 25.2	17.5
	15	48 25.0 ··	18.1
	16	63 24.8	18.6
	17	78 24.7	19.1
	18	93 24.5	S20 19.6
	19	108 24.3	20.2
	20	123 24.2	20.7
	21	138 24.0 ··	21.2
	22	153 23.8	21.7
	23	168 23.6	22.2

		G.H.A.	Dec.
d	h	° ′	° ′
24	00	183 23.5	S20 22.8
	01	198 23.3	23.3
	02	213 23.1	23.8
	03	228 22.9 ··	24.3
	04	243 22.8	24.8
	05	258 22.6	25.3
	06	273 22.4	S20 25.9
	07	288 22.2	26.4
	08	303 22.0	26.9
	09	318 21.9 ··	27.4
	10	333 21.7	27.9
	11	348 21.5	28.4
	12	3 21.3	S20 28.9
	13	18 21.2	29.4
	14	33 21.0	30.0
	15	48 20.8 ··	30.5
	16	63 20.6	31.0
	17	78 20.4	31.5
	18	93 20.3	S20 32.0
	19	108 20.1	32.5
	20	123 19.9	33.0
	21	138 19.7 ··	33.5
	22	153 19.5	34.0
	23	168 19.3	34.5
25	00	183 19.2	S20 35.0
	01	198 19.0	35.5
	02	213 18.8	36.0
	03	228 18.6 ··	36.5
	04	243 18.4	37.0
	05	258 18.2	37.5
	06	273 18.1	S20 38.0
	07	288 17.9	38.5
	08	303 17.7	39.0
	09	318 17.5 ··	39.5
	10	333 17.3	40.0
	11	348 17.1	40.5
	12	3 16.9	S20 41.0
	13	18 16.8	41.5
	14	33 16.6	42.0
	15	48 16.4 ··	42.5
	16	63 16.2	43.0
	17	78 16.0	43.5
	18	93 15.8	S20 43.9
	19	108 15.6	44.4
	20	123 15.4	44.9
	21	138 15.2 ··	45.4
	22	153 15.1	45.9
	23	168 14.9	46.4
26	00	183 14.7	S20 46.9
	01	198 14.5	47.4
	02	213 14.3	47.9
	03	228 14.1 ··	48.3
	04	243 13.9	48.8
	05	258 13.7	49.3
	06	273 13.5	S20 49.8
	07	288 13.3	50.3
	08	303 13.1	50.8
	09	318 12.9 ··	51.2
	10	333 12.7	51.7
	11	348 12.6	52.2
	12	3 12.4	S20 52.7
	13	18 12.2	53.1
	14	33 12.0	53.6
	15	48 11.8 ··	54.1
	16	63 11.6	54.6
	17	78 11.4	55.1
	18	93 11.2	S20 55.5
	19	108 11.0	56.0
	20	123 10.8	56.5
	21	138 10.6 ··	56.9
	22	153 10.4	57.4
	23	168 10.2	57.9

		G.H.A.	Dec.
d	h	° ′	° ′
27	00	183 10.0	S20 58.4
	01	198 09.8	58.8
	02	213 09.6	59.3
	03	228 09.4	20 59.8
	04	243 09.2	21 00.2
	05	258 09.0	00.7
	06	273 08.8	S21 01.2
	07	288 08.6	01.6
	08	303 08.4	02.1
	09	318 08.2 ··	02.6
	10	333 08.0	03.0
	11	348 07.8	03.5
	12	3 07.6	S21 04.0
	13	18 07.4	04.4
	14	33 07.2	04.9
	15	48 07.0 ··	05.3
	16	63 06.8	05.8
	17	78 06.6	06.3
	18	93 06.4	S21 06.7
	19	108 06.2	07.2
	20	123 06.0	07.6
	21	138 05.8 ··	08.1
	22	153 05.6	08.5
	23	168 05.4	09.0
28	00	183 05.2	S21 09.5
	01	198 04.9	09.9
	02	213 04.7	10.4
	03	228 04.5 ··	10.8
	04	243 04.3	11.3
	05	258 04.1	11.7
	06	273 03.9	S21 12.2
	07	288 03.7	12.6
	08	303 03.5	13.1
	09	318 03.3 ··	13.5
	10	333 03.1	14.0
	11	348 02.9	14.4
	12	3 02.7	S21 14.8
	13	18 02.5	15.3
	14	33 02.2	15.7
	15	48 02.0 ··	16.2
	16	63 01.8	16.6
	17	78 01.6	17.1
	18	93 01.4	S21 17.5
	19	108 01.2	17.9
	20	123 01.0	18.4
	21	138 00.8 ··	18.8
	22	153 00.6	19.3
	23	168 00.3	19.7
29	00	183 00.1	S21 20.1
	01	197 59.9	20.6
	02	212 59.7	21.0
	03	227 59.5 ··	21.5
	04	242 59.3	21.9
	05	257 59.1	22.3
	06	272 58.8	S21 22.8
	07	287 58.6	23.2
	08	302 58.4	23.6
	09	317 58.2 ··	24.0
	10	332 58.0	24.5
	11	347 57.8	24.9
	12	2 57.6	S21 25.3
	13	17 57.3	25.8
	14	32 57.1	26.2
	15	47 56.9 ··	26.6
	16	62 56.7	27.0
	17	77 56.5	27.5
	18	92 56.2	S21 27.9
	19	107 56.0	28.3
	20	122 55.8	28.7
	21	137 55.6 ··	29.2
	22	152 55.4	29.6
	23	167 55.2	30.0

DECEMBER

d h	G.H.A.	Dec.
30 00	182 54.9	S21 30.4
01	197 54.7	30.8
02	212 54.5	31.3
03	227 54.3 ··	31.7
04	242 54.0	32.1
05	257 53.8	32.5
06	272 53.6 S21	32.9
07	287 53.4	33.4
08	302 53.2	33.8
09	317 52.9 ··	34.2
10	332 52.7	34.6
11	347 52.5	35.0
12	2 52.3 S21	35.4
13	17 52.0	35.8
14	32 51.8	36.2
15	47 51.6 ··	36.7
16	62 51.4	37.1
17	77 51.2	37.5
18	92 50.9 S21	37.9
19	107 50.7	38.3
20	122 50.5	38.7
21	137 50.2 ··	39.1
22	152 50.0	39.5
23	167 49.8	39.9
1 00	182 49.6 S21	40.3
01	197 49.3	40.7
02	212 49.1	41.1
03	227 48.9 ··	41.5
04	242 48.7	41.9
05	257 48.4	42.3
06	272 48.2 S21	42.7
07	287 48.0	43.1
08	302 47.7	43.5
09	317 47.5 ··	43.9
10	332 47.3	44.3
11	347 47.1	44.7
12	2 46.8 S21	45.1
13	17 46.6	45.5
14	32 46.4	45.9
15	47 46.1 ··	46.3
16	62 45.9	46.7
17	77 45.7	47.1
18	92 45.4 S21	47.4
19	107 45.2	47.8
20	122 45.0	48.2
21	137 44.7 ··	48.6
22	152 44.5	49.0
23	167 44.3	49.4
2 00	182 44.0 S21	49.8
01	197 43.8	50.2
02	212 43.6	50.5
03	227 43.3 ··	50.9
04	242 43.1	51.3
05	257 42.9	51.7
06	272 42.6 S21	52.1
07	287 42.4	52.5
08	302 42.2	52.8
09	317 41.9 ··	53.2
10	332 41.7	53.6
11	347 41.5	54.0
12	2 41.2 S21	54.4
13	17 41.0	54.7
14	32 40.7	55.1
15	47 40.5 ··	55.5
16	62 40.3	55.9
17	77 40.0	56.2
18	92 39.8 S21	56.6
19	107 39.6	57.0
20	122 39.3	57.3
21	137 39.1 ··	57.7
22	152 38.8	58.1
23	167 38.6	58.5

d h	G.H.A.	Dec.
3 00	182 38.4	S21 58.8
01	197 38.1	59.2
02	212 37.9	59.6
03	227 37.6 21	59.9
04	242 37.4 22	00.3
05	257 37.2	00.7
06	272 36.9 S22	01.0
07	287 36.7	01.4
08	302 36.4	01.7
09	317 36.2 ··	02.1
10	332 35.9	02.5
11	347 35.7	02.8
12	2 35.5 S22	03.2
13	17 35.2	03.5
14	32 35.0	03.9
15	47 34.7 ··	04.3
16	62 34.5	04.6
17	77 34.2	05.0
18	92 34.0 S22	05.3
19	107 33.8	05.7
20	122 33.5	06.0
21	137 33.3 ··	06.4
22	152 33.0	06.7
23	167 32.8	07.1
4 00	182 32.5 S22	07.4
01	197 32.3	07.8
02	212 32.0	08.1
03	227 31.8 ··	08.5
04	242 31.5	08.8
05	257 31.3	09.2
06	272 31.0 S22	09.5
07	287 30.8	09.9
08	302 30.5	10.2
09	317 30.3 ··	10.6
10	332 30.0	10.9
11	347 29.8	11.3
12	2 29.5 S22	11.6
13	17 29.3	11.9
14	32 29.0	12.3
15	47 28.8 ··	12.6
16	62 28.5	13.0
17	77 28.3	13.3
18	92 28.0 S22	13.6
19	107 27.8	14.0
20	122 27.5	14.3
21	137 27.3 ··	14.6
22	152 27.0	15.0
23	167 26.8	15.3
5 00	182 26.5 S22	15.6
01	197 26.3	16.0
02	212 26.0	16.3
03	227 25.8 ··	16.6
04	242 25.5	17.0
05	257 25.3	17.3
06	272 25.0 S22	17.6
07	287 24.8	18.0
08	302 24.5	18.3
09	317 24.3 ··	18.6
10	332 24.0	18.9
11	347 23.7	19.3
12	2 23.5 S22	19.6
13	17 23.2	19.9
14	32 23.0	20.2
15	47 22.7 ··	20.5
16	62 22.5	20.9
17	77 22.2	21.2
18	92 22.0 S22	21.5
19	107 21.7	21.8
20	122 21.4	22.1
21	137 21.2 ··	22.5
22	152 20.9	22.8
23	167 20.7	23.1

d h	G.H.A.	Dec.
6 00	182 20.4	S22 23.4
01	197 20.1	23.7
02	212 19.9	24.0
03	227 19.6 ··	24.4
04	242 19.4	24.7
05	257 19.1	25.0
06	272 18.9 S22	25.3
07	287 18.6	25.6
08	302 18.3	25.9
09	317 18.1 ··	26.2
10	332 17.8	26.5
11	347 17.6	26.8
12	2 17.3 S22	27.1
13	17 17.0	27.4
14	32 16.8	27.7
15	47 16.5 ··	28.0
16	62 16.2	28.3
17	77 16.0	28.6
18	92 15.7 S22	28.9
19	107 15.5	29.2
20	122 15.2	29.5
21	137 14.9 ··	29.8
22	152 14.7	30.1
23	167 14.4	30.4
7 00	182 14.1 S22	30.7
01	197 13.9	31.0
02	212 13.6	31.3
03	227 13.3 ··	31.6
04	242 13.1	31.9
05	257 12.8	32.2
06	272 12.6 S22	32.5
07	287 12.3	32.8
08	302 12.0	33.1
09	317 11.8 ··	33.4
10	332 11.5	33.7
11	347 11.2	34.0
12	2 11.0 S22	34.2
13	17 10.7	34.5
14	32 10.4	34.8
15	47 10.2 ··	35.1
16	62 09.9	35.4
17	77 09.6	35.7
18	92 09.4 S22	36.0
19	107 09.1	36.2
20	122 08.8	36.5
21	137 08.5 ··	36.8
22	152 08.3	37.1
23	167 08.0	37.4
8 00	182 07.7 S22	37.6
01	197 07.5	37.9
02	212 07.2	38.2
03	227 06.9 ··	38.5
04	242 06.7	38.7
05	257 06.4	39.0
06	272 06.1 S22	39.3
07	287 05.8	39.6
08	302 05.6	39.8
09	317 05.3 ··	40.1
10	332 05.0	40.4
11	347 04.8	40.6
12	2 04.5 S22	40.9
13	17 04.2	41.2
14	32 03.9	41.4
15	47 03.7 ··	41.7
16	62 03.4	42.0
17	77 03.1	42.2
18	92 02.9 S22	42.5
19	107 02.6	42.8
20	122 02.3	43.0
21	137 02.0 ··	43.3
22	152 01.8	43.6
23	167 01.5	43.8

Day 9

d h	G.H.A.	Dec.
9 00	182 01.2	S22 44.1
01	197 00.9	44.3
02	212 00.7	44.6
03	227 00.4 ··	44.9
04	242 00.1	45.1
05	256 59.8	45.4
06	271 59.6	S22 45.6
07	286 59.3	45.9
08	301 59.0	46.1
09	316 58.7 ··	46.4
10	331 58.5	46.6
11	346 58.2	46.9
12	1 57.9	S22 47.1
13	16 57.6	47.4
14	31 57.3	47.6
15	46 57.1 ··	47.9
16	61 56.8	48.1
17	76 56.5	48.4
18	91 56.2	S22 48.6
19	106 56.0	48.9
20	121 55.7	49.1
21	136 55.4 ··	49.4
22	151 55.1	49.6
23	166 54.8	49.8

Day 10

d h	G.H.A.	Dec.
10 00	181 54.6	S22 50.1
01	196 54.3	50.3
02	211 54.0	50.6
03	226 53.7 ··	50.8
04	241 53.4	51.0
05	256 53.2	51.3
06	271 52.9	S22 51.5
07	286 52.6	51.7
08	301 52.3	52.0
09	316 52.0 ··	52.2
10	331 51.8	52.4
11	346 51.5	52.7
12	1 51.2	S22 52.9
13	16 50.9	53.1
14	31 50.6	53.4
15	46 50.4 ··	53.6
16	61 50.1	53.8
17	76 49.8	54.1
18	91 49.5	S22 54.3
19	106 49.2	54.5
20	121 48.9	54.7
21	136 48.7 ··	55.0
22	151 48.4	55.2
23	166 48.1	55.4

Day 11

d h	G.H.A.	Dec.
11 00	181 47.8	S22 55.6
01	196 47.5	55.9
02	211 47.2	56.1
03	226 47.0 ··	56.3
04	241 46.7	56.5
05	256 46.4	56.7
06	271 46.1	S22 57.0
07	286 45.8	57.2
08	301 45.5	57.4
09	316 45.2 ··	57.6
10	331 45.0	57.8
11	346 44.7	58.0
12	1 44.4	S22 58.2
13	16 44.1	58.5
14	31 43.8	58.7
15	46 43.5 ··	58.9
16	61 43.2	59.1
17	76 43.0	59.3
18	91 42.7	S22 59.5
19	106 42.4	59.7
20	121 42.1	22 59.9
21	136 41.8	23 00.1
22	151 41.5	00.3
23	166 41.2	00.5

Day 12

d h	G.H.A.	Dec.
12 00	181 40.9	S23 00.7
01	196 40.6	00.9
02	211 40.4	01.1
03	226 40.1 ··	01.3
04	241 39.8	01.5
05	256 39.5	01.7
06	271 39.2	S23 01.9
07	286 38.9	02.1
08	301 38.6	02.3
09	316 38.3 ··	02.5
10	331 38.0	02.7
11	346 37.8	02.9
12	1 37.5	S23 03.1
13	16 37.2	03.3
14	31 36.9	03.5
15	46 36.6 ··	03.7
16	61 36.3	03.9
17	76 36.0	04.1
18	91 35.7	S23 04.3
19	106 35.4	04.4
20	121 35.1	04.6
21	136 34.8 ··	04.8
22	151 34.6	05.0
23	166 34.3	05.2

Day 13

d h	G.H.A.	Dec.
13 00	181 34.0	S23 05.4
01	196 33.7	05.6
02	211 33.4	05.7
03	226 33.1 ··	05.9
04	241 32.8	06.1
05	256 32.5	06.3
06	271 32.2	S23 06.5
07	286 31.9	06.6
08	301 31.6	06.8
09	316 31.3 ··	07.0
10	331 31.0	07.2
11	346 30.7	07.4
12	1 30.5	S23 07.5
13	16 30.2	07.7
14	31 29.9	07.9
15	46 29.6 ··	08.0
16	61 29.3	08.2
17	76 29.0	08.4
18	91 28.7	S23 08.6
19	106 28.4	08.7
20	121 28.1	08.9
21	136 27.8 ··	09.1
22	151 27.5	09.2
23	166 27.2	09.4

Day 14

d h	G.H.A.	Dec.
14 00	181 26.9	S23 09.6
01	196 26.6	09.7
02	211 26.3	09.9
03	226 26.0 ··	10.1
04	241 25.7	10.2
05	256 25.4	10.4
06	271 25.1	S23 10.5
07	286 24.8	10.7
08	301 24.5	10.9
09	316 24.2 ··	11.0
10	331 23.9	11.2
11	346 23.6	11.3
12	1 23.3	S23 11.5
13	16 23.1	11.6
14	31 22.8	11.8
15	46 22.5 ··	11.9
16	61 22.2	12.1
17	76 21.9	12.2
18	91 21.6	S23 12.4
19	106 21.3	12.5
20	121 21.0	12.7
21	136 20.7 ··	12.8
22	151 20.4	13.0
23	166 20.1	13.1

Day 15

d h	G.H.A.	Dec.
15 00	181 19.8	S23 13.3
01	196 19.5	13.4
02	211 19.2	13.6
03	226 18.9 ··	13.7
04	241 18.6	13.9
05	256 18.3	14.0
06	271 18.0	S23 14.1
07	286 17.7	14.3
08	301 17.4	14.4
09	316 17.1 ··	14.6
10	331 16.8	14.7
11	346 16.5	14.8
12	1 16.2	S23 15.0
13	16 15.9	15.1
14	31 15.6	15.2
15	46 15.3 ··	15.4
16	61 15.0	15.5
17	76 14.7	15.6
18	91 14.4	S23 15.8
19	106 14.1	15.9
20	121 13.8	16.0
21	136 13.5 ··	16.2
22	151 13.2	16.3
23	166 12.9	16.4

Day 16

d h	G.H.A.	Dec.
16 00	181 12.5	S23 16.5
01	196 12.2	16.7
02	211 11.9	16.8
03	226 11.6 ··	16.9
04	241 11.3	17.0
05	256 11.0	17.2
06	271 10.7	S23 17.3
07	286 10.4	17.4
08	301 10.1	17.5
09	316 09.8 ··	17.7
10	331 09.5	17.8
11	346 09.2	17.9
12	1 08.9	S23 18.0
13	16 08.6	18.1
14	31 08.3	18.2
15	46 08.0 ··	18.4
16	61 07.7	18.5
17	76 07.4	18.6
18	91 07.1	S23 18.7
19	106 06.8	18.8
20	121 06.5	18.9
21	136 06.2 ··	19.0
22	151 05.9	19.1
23	166 05.6	19.2

Day 17

d h	G.H.A.	Dec.
17 00	181 05.3	S23 19.3
01	196 05.0	19.5
02	211 04.7	19.6
03	226 04.3 ··	19.7
04	241 04.0	19.8
05	256 03.7	19.9
06	271 03.4	S23 20.0
07	286 03.1	20.1
08	301 02.8	20.2
09	316 02.5 ··	20.3
10	331 02.2	20.4
11	346 01.9	20.5
12	1 01.6	S23 20.6
13	16 01.3	20.7
14	31 01.0	20.8
15	46 00.7 ··	20.9
16	61 00.4	21.0
17	76 00.1	21.0
18	90 59.8	S23 21.1
19	105 59.5	21.2
20	120 59.1	21.3
21	135 58.8 ··	21.4
22	150 58.5	21.5
23	165 58.2	21.6

		G.H.A.		Dec.	

<table>
<tr><td>d h</td><td colspan=2>G.H.A.
o '</td><td colspan=2>Dec.
o '</td></tr>
</table>

d h	G.H.A. ° '	Dec. ° '	d h	G.H.A. ° '	Dec. ° '	d h	G.H.A. ° '	Dec. ° '
18 00	180 57.9	S23 21.7	21 00	180 35.6	S23 25.9	24 00	180 13.2	S23 25.8
01	195 57.6	21.8	01	195 35.3	25.9	01	195 12.9	25.8
02	210 57.3	21.9	02	210 35.0	25.9	02	210 12.6	25.7
03	225 57.0	·· 21.9	03	225 34.7	·· 25.9	03	225 12.2	·· 25.7
04	240 56.7	22.0	04	240 34.4	26.0	04	240 11.9	25.7
05	255 56.4	22.1	05	255 34.1	26.0	05	255 11.6	25.7
06	270 56.1	S23 22.2	06	270 33.8	S23 26.0	06	270 11.3	S23 25.6
07	285 55.8	22.3	07	285 33.5	26.0	07	285 11.0	25.6
08	300 55.5	22.4	08	300 33.1	26.1	08	300 10.7	25.5
09	315 55.2	·· 22.4	09	315 32.8	·· 26.1	09	315 10.4	·· 25.5
10	330 54.8	22.5	10	330 32.5	26.1	10	330 10.1	25.5
11	345 54.5	22.6	11	345 32.2	26.1	11	345 09.8	25.5
12	0 54.2	S23 22.7	12	0 31.9	S23 26.1	12	0 09.4	S23 25.4
13	15 53.9	22.7	13	15 31.6	26.2	13	15 09.1	25.4
14	30 53.6	22.8	14	30 31.3	26.2	14	30 08.8	25.3
15	45 53.3	·· 22.9	15	45 31.0	·· 26.2	15	45 08.5	·· 25.3
16	60 53.0	23.0	16	60 30.7	26.2	16	60 08.2	25.2
17	75 52.7	23.0	17	75 30.3	26.2	17	75 07.9	25.2
18	90 52.4	S23 23.1	18	90 30.0	S23 26.2	18	90 07.6	S23 25.1
19	105 52.1	23.2	19	105 29.7	26.3	19	105 07.3	25.1
20	120 51.8	23.3	20	120 29.4	26.3	20	120 07.0	25.0
21	135 51.5	·· 23.3	21	135 29.1	·· 26.3	21	135 06.6	·· 25.0
22	150 51.1	23.4	22	150 28.8	26.3	22	150 06.3	25.0
23	165 50.8	23.5	23	165 28.5	26.3	23	165 06.0	24.9
19 00	180 50.5	S23 23.5	22 00	180 28.2	S23 26.3	25 00	180 05.7	S23 24.9
01	195 50.2	23.6	01	195 27.8	26.3	01	195 05.4	24.8
02	210 49.9	23.7	02	210 27.5	26.3	02	210 05.1	24.8
03	225 49.6	·· 23.7	03	225 27.2	·· 26.3	03	225 04.8	·· 24.7
04	240 49.3	23.8	04	240 26.9	26.3	04	240 04.5	24.6
05	255 49.0	23.9	05	255 26.6	26.4	05	255 04.1	24.6
06	270 48.7	S23 23.9	06	270 26.3	S23 26.4	06	270 03.8	S23 24.5
07	285 48.4	24.0	07	285 26.0	26.4	07	285 03.5	24.5
08	300 48.1	24.1	08	300 25.7	26.4	08	300 03.2	24.4
09	315 47.7	·· 24.1	09	315 25.4	·· 26.4	09	315 02.9	·· 24.4
10	330 47.4	24.2	10	330 25.0	26.4	10	330 02.6	24.3
11	345 47.1	24.2	11	345 24.7	26.4	11	345 02.3	24.3
12	0 46.8	S23 24.3	12	0 24.4	S23 26.4	12	0 02.0	S23 24.2
13	15 46.5	24.4	13	15 24.1	26.4	13	15 01.7	24.1
14	30 46.2	24.4	14	30 23.8	26.4	14	30 01.4	24.1
15	45 45.9	·· 24.5	15	45 23.5	·· 26.4	15	45 01.0	·· 24.0
16	60 45.6	24.5	16	60 23.2	26.4	16	60 00.7	24.0
17	75 45.3	24.6	17	75 22.9	26.4	17	75 00.4	23.9
18	90 45.0	S23 24.6	18	90 22.5	S23 26.3	18	90 00.1	S23 23.8
19	105 44.6	24.7	19	105 22.2	26.3	19	104 59.8	23.8
20	120 44.3	24.7	20	120 21.9	26.3	20	119 59.5	23.7
21	135 44.0	·· 24.8	21	135 21.6	·· 26.3	21	134 59.2	·· 23.6
22	150 43.7	24.8	22	150 21.3	26.3	22	149 58.9	23.6
23	165 43.4	24.9	23	165 21.0	26.3	23	164 58.6	23.5
20 00	180 43.1	S23 24.9	23 00	180 20.7	S23 26.3	26 00	179 58.2	S23 23.4
01	195 42.8	25.0	01	195 20.4	26.3	01	194 57.9	23.4
02	210 42.5	25.0	02	210 20.0	26.3	02	209 57.6	23.3
03	225 42.2	·· 25.1	03	225 19.7	·· 26.3	03	224 57.3	·· 23.2
04	240 41.9	25.1	04	240 19.4	26.3	04	239 57.0	23.1
05	255 41.5	25.2	05	255 19.1	26.2	05	254 56.7	23.1
06	270 41.2	S23 25.2	06	270 18.8	S23 26.2	06	269 56.4	S23 23.0
07	285 40.9	25.3	07	285 18.5	26.2	07	284 56.1	22.9
08	300 40.6	25.3	08	300 18.2	26.2	08	299 55.8	22.8
09	315 40.3	·· 25.3	09	315 17.9	·· 26.2	09	314 55.5	·· 22.8
10	330 40.0	25.4	10	330 17.6	26.2	10	329 55.1	22.7
11	345 39.7	25.4	11	345 17.2	26.1	11	344 54.8	22.6
12	0 39.4	S23 25.5	12	0 16.9	S23 26.1	12	359 54.5	S23 22.5
13	15 39.1	25.5	13	15 16.6	26.1	13	14 54.2	22.5
14	30 38.8	25.5	14	30 16.3	26.1	14	29 53.9	22.4
15	45 38.4	·· 25.6	15	45 16.0	·· 26.0	15	44 53.6	·· 22.3
16	60 38.1	25.6	16	60 15.7	26.0	16	59 53.3	22.2
17	75 37.8	25.6	17	75 15.4	26.0	17	74 53.0	22.1
18	90 37.5	S23 25.7	18	90 15.1	S23 26.0	18	89 52.7	S23 22.0
19	105 37.2	25.7	19	105 14.7	26.0	19	104 52.4	22.0
20	120 36.9	25.7	20	120 14.4	25.9	20	119 52.0	21.9
21	135 36.6	·· 25.8	21	135 14.1	·· 25.9	21	134 51.7	·· 21.8
22	150 36.3	25.8	22	150 13.8	25.9	22	149 51.4	21.7
23	165 36.0	25.8	23	165 13.5	25.8	23	164 51.1	21.6

d h	G.H.A.	Dec.	d h	G.H.A.	Dec.
27 00	179 50.8	S23 21.5	30 00	179 28.8	S23 13.0
01	194 50.5	21.4	01	194 28.5	12.9
02	209 50.2	21.3	02	209 28.2	12.7
03	224 49.9	·· 21.3	03	224 27.9	·· 12.6
04	239 49.6	21.2	04	239 27.6	12.4
05	254 49.3	21.1	05	254 27.3	12.3
06	269 49.0	S23 21.0	06	269 27.0	S23 12.1
07	284 48.7	20.9	07	284 26.6	12.0
08	299 48.3	20.8	08	299 26.3	11.8
09	314 48.0	·· 20.7	09	314 26.0	·· 11.7
10	329 47.7	20.6	10	329 25.7	11.5
11	344 47.4	20.5	11	344 25.4	11.4
12	359 47.1	S23 20.4	12	359 25.1	S23 11.2
13	14 46.8	20.3	13	14 24.8	11.0
14	29 46.5	20.2	14	29 24.5	10.9
15	44 46.2	·· 20.1	15	44 24.2	·· 10.7
16	59 45.9	20.0	16	59 23.9	10.6
17	74 45.6	19.9	17	74 23.6	10.4
18	89 45.3	S23 19.8	18	89 23.3	S23 10.2
19	104 45.0	19.7	19	104 23.0	10.1
20	119 44.6	19.6	20	119 22.7	09.9
21	134 44.3	·· 19.5	21	134 22.4	·· 09.8
22	149 44.0	19.4	22	149 22.1	09.6
23	164 43.7	19.3	23	164 21.8	09.4
28 00	179 43.4	S23 19.2	31 00	179 21.5	S23 09.3
01	194 43.1	19.1	01	194 21.2	09.1
02	209 42.8	18.9	02	209 20.9	08.9
03	224 42.5	·· 18.8	03	224 20.6	·· 08.8
04	239 42.2	18.7	04	239 20.3	08.6
05	254 41.9	18.6	05	254 20.0	08.4
06	269 41.6	S23 18.5	06	269 19.7	S23 08.2
07	284 41.3	18.4	07	284 19.4	08.1
08	299 41.0	18.3	08	299 19.1	07.9
09	314 40.7	·· 18.2	09	314 18.8	·· 07.7
10	329 40.3	18.0	10	329 18.5	07.5
11	344 40.0	17.9	11	344 18.2	07.4
12	359 39.7	S23 17.8	12	359 17.9	S23 07.2
13	14 39.4	17.7	13	14 17.6	07.0
14	29 39.1	17.6	14	29 17.3	06.8
15	44 38.8	·· 17.4	15	44 17.0	·· 06.7
16	59 38.5	17.3	16	59 16.7	06.5
17	74 38.2	17.2	17	74 16.4	06.3
18	89 37.9	S23 17.1	18	89 16.1	S23 06.1
19	104 37.6	17.0	19	104 15.8	05.9
20	119 37.3	16.8	20	119 15.5	05.8
21	134 37.0	·· 16.7	21	134 15.3	·· 05.6
22	149 36.7	16.6	22	149 15.0	05.4
23	164 36.4	16.5	23	164 14.7	05.2
29 00	179 36.1	S23 16.3			
01	194 35.8	16.2			
02	209 35.5	16.1			
03	224 35.2	·· 15.9			
04	239 34.8	15.8			
05	254 34.5	15.7			
06	269 34.2	S23 15.5			
07	284 33.9	15.4			
08	299 33.6	15.3			
09	314 33.3	·· 15.1			
10	329 33.0	15.0			
11	344 32.7	14.9			
12	359 32.4	S23 14.7			
13	14 32.1	14.6			
14	29 31.8	14.5			
15	44 31.5	·· 14.3			
16	59 31.2	14.2			
17	74 30.9	14.0			
18	89 30.6	S23 13.9			
19	104 30.3	13.7			
20	119 30.0	13.6			
21	134 29.7	·· 13.5			
22	149 29.4	13.3			
23	164 29.1	13.2			

Table 9

FOUR-YEAR CHANGE IN GREENWICH HOUR ANGLE (GHA) & DECLINATION (DEC.) OF THE SUN

GHA Corr.	Date	Dec. Corr.	GHA Corr.	Date	Dec. Corr.
January				*April*	
−0′11	1	−0′32	+0′35	1	+0′52
−0.13	4	−0.35	+0.33	4	+0.51
−0.12	7	−0.39	+0.32	7	+0.49
−0.09	10	−0.42	+0.33	10	+0.47
−0.04	13	−0.44	+0.35	13	+0.44
+0.03	16	−0.46	+0.37	16	+0.41
+0.09	19	−0.48	+0.39	19	+0.39
+0.13	22	−0.49	+0.37	22	+0.37
+0.15	25	−0.52	+0.33	25	+0.36
+0.15	28	−0.54	+0.28	28	+0.34
February				*May*	
+0′13	1	−0′57	+0′24	1	+0′33
+0.14	4	−0.59	+0.20	4	+0.31
+0.15	7	−0.60	+0.19	7	+0.28
+0.19	10	−0.61	+0.19	10	+0.25
+0.24	13	−0.60	+0.21	13	+0.22
+0.29	16	−0.60	+0.23	16	+0.19
+0.34	19	−0.59	+0.23	19	+0.16
+0.36	22	−0.59	+0.20	22	+0.13
+0.36	25	−0.60	+0.16	25	+0.11
+0.34	28	−0.61	+0.11	28	+0.08
March				*June*	
+0′32	1	−0′62	+0′04	1	+0′04
+0.31	4	−0.62	+0.02	4	+0.01
+0.31	7	−0.62	+0.02	7	−0.02
+0.33	10	−0.61	+0.04	10	−0.05
+0.36	13	−0.59	+0.07	13	−0.08
+0.40	16	−0.56	+0.09	16	−0.12
+0.43	19	−0.54	+0.08	19	−0.15
+0.44	22	+0.53	+0.06	22	−0.18
+0.43	25	+0.52	+0.01	25	−0.22
+0.40	28	+0.52	−0.03	28	−0.25

GHA Corr.	Date	Dec. Corr.	GHA Corr.	Date	Dec. Corr.
	July			October	
−0′.06	1	−0′.28	0′.00	1	+0′.67
−0.07	4	−0.31	+0.02	4	+0.65
−0.05	7	−0.34	+0.05	7	+0.62
−0.01	10	−0.37	+0.07	10	+0.60
+0.03	13	−0.39	+0.08	13	+0.57
+0.05	16	−0.41	+0.07	16	+0.56
+0.06	19	−0.44	+0.04	19	+0.54
+0.04	22	−0.47	−0.01	22	+0.53
+0.01	25	−0.50	−0.05	25	+0.52
−0.02	28	−0.54	−0.08	28	+0.50
	August			November	
−0′.03	1	−0′.57	−0′.09	1	+0′.46
−0.02	4	−0.59	−0.07	4	+0.42
+0.02	7	−0.60	−0.04	7	+0.38
+0.06	10	−0.60	−0.03	10	+0.35
+0.09	13	−0.61	−0.03	13	+0.31
+0.11	16	−0.62	−0.05	16	+0.28
+0.11	19	−0.64	−0.09	19	+0.25
+0.09	22	−0.66	−0.14	22	+0.22
+0.06	25	−0.68	−0.18	25	+0.19
+0.03	28	−0.70	−0.20	28	+0.16
	September			December	
+0′.02	1	−0′.71	−0′.18	1	+0′.12
+0.04	4	−0.70	−0.15	4	+0.08
+0.08	7	−0.69	−0.11	7	+0.04
+0.11	10	−0.68	−0.09	10	0.00
+0.13	13	−0.67	−0.08	13	−0.04
+0.13	16	−0.67	−0.09	16	−0.08
+0.11	19	−0.68	−0.12	19	−0.12
+0.08	22	−0.68	−0.15	22	−0.16
+0.04	25	+0.69	−0.17	25	−0.19
+0.01	28	+0.68	−0.17	28	−0.23

Part VII

WORK FORMS

WORK FORM—LATITUDE

I Sextant:

HE	Dip		
7 ft	3'	Sextant Reading (Hs)	_____ ° _____ '
13 ft	4'	Index Correction (+ off ⁄ − on)	± _____ '
21 ft	5'	Dip (see quick table @ left)	− _____ '
		Apparent Altitude (ha)	_____ ° _____ '

ha	R		
10°	5'	Semi-Diameter (+ ☉ ⁄ − ☉)	± 16'
12°	4'	Refraction (see quick table @ left)	− _____ '
16°	3'		
21°	2'	Observed Altitude (Ho)	_____ ° _____ '
33°	1'		
63°	0'		

II Zenith Distance:

Zenith (90°)   89°   60'

− Ho   − _____ ° _____ '

Zenith Distance (z) _____ _____

III Declination:

From almanac (take value for nearest whole hour)   _____ ° _____ '

IV Latitude:

Add or subtract declination and zenith distance to find a number that squares up best with DR.

Alternatively, draw picture in any one of the three ways shown in the text.   _____ ° _____ N'S

WORK FORM—LONGITUDE

I Time:

	Hour	Minutes	Seconds
AM Altitude		—	—
PM Altitude	+	—	—
Divide	2/		
Time of Meridian Passage		—	—
Convert to GMT	±	— 00	— 00
GMT of Meridian Passage		—	—

II Greenwich Hour Angle:

Time GHA		
4m	1°	
1m	15′	
4s	1′	

GHA @ Hour _____ ° _____ ′

+ Additional minutes + _____ ° _____ ′

+ Additional seconds + _____ ° _____ ′

GHA @ Meridian Passage _____ ° _____ ′

III Effect of Vessel's Motion:

1. Effective speed N or S _____ knots (A)
2. A ÷ 6 (A divided by 6) _____ (B)
3. Table 9 Correction for 6 knots _____ ′ (C)
4. B times C = Correction for your boat _____ ′ (D)
5. Motion *toward* sun; subtract D from GHA — _____ ′

   Motion *away* from sun, add D to GHA + _____ ′
6. GHA @ Meridian Passage _____ ° _____ ′

IV Longitude:

If GHA is less than 180°, you are at a west longitude and GHA = your longitude W.

If GHA is greater than 180°, you are at an east longitude and 360° − GHA = your longitude E.

If GHA = 180°, your longitude is 180° and you are free to designate it E or W as you please.

304 **Latitude & Longitude by the Noon Sight**